Prealgebra
A Transitional Approach

Prealgebra
A Transitional Approach

Jill Parker Beer
Kent State University

Dwyn Griffin Peake
Kent State University

▲▲ **ADDISON-WESLEY**

An imprint of Addison Wesley Longman, Inc.

Reading, Massachusetts • Menlo Park, California • New York • Harlow, England
Don Mills, Ontario • Sydney • Mexico City • Madrid • Amsterdam

Sponsoring Editor: Karin E. Wagner
Developmental Editor: Maxine Effenson Chuck
Project Editor: Anne Schmidt Ryan
Design Administrator: Jess Schaal
Text and Cover Design: Lesiak/Crampton Design: Lucy Lesiak
Cover Illustration: A Second Phase Chief's Blanket, Navajo, circa 1850. 56 by 71 inches. All reproduction rights reserved by Joshua Baer & Company, Santa Fe. A Second Phase Chief's Blanket, Navajo, circa 1860. 55 by 68 inches. All reproduction rights reserved by Joshua Baer & Company, Santa Fe.
Production Administrator: Randee Wire
Compositor: Interactive Composition Corporation
Printer and Binder: R. R. Donnelley & Sons Company
Cover Printer: Phoenix Color Corporation

AMATYC Standards shown on inside front and back covers excerpted from *Crossroads in Mathematics: Standards for Introductory Mathematics Before Calculus*/American Mathematical Association of Two-Year Colleges.

Prealgebra, A Transitional Approach

Library of Congress Cataloging-in-Publication Data

Beer, Jill Parker.
   Prealgebra : a transitional approach / Jill Parker Beer,
Dwyn Griffin Peake.
        p.   cm.
   Includes index.
   ISBN 0-673-99952-1
   1. Mathematics.   I. Peake, Dwyn Griffin. II. Title.
QA39.2.B43 1996
510—DC20
                                                   96-9208
                                                      CIP

97 98 99—DOW—9 8 7 6 5 4 3 2

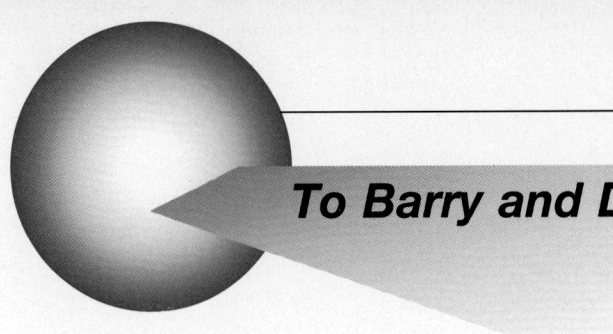

To Barry and Dick

Contents

Preface   xii
Preface to the Student   xx
Study Skills Tip: Studying Mathematics   xxii
The Mathematics Journal: Past Experiences in Mathematics   xxii

1 Sets   1

1.1 Set Theory   1

1.2 Set Relations   2

1.3 Operations with Sets   3
Venn Diagrams   4

1.4 The Numbers as Sets   7
Small Group Activity: Problem Solving with Venn Diagrams   11
Chapter 1 Review   12
Chapter 1 Practice Test   13
Study Skills Tip: Note Taking   15
The Mathematics Journal: Goals   15

2 Whole Numbers   16

2.1 Place Value and Expanded Notation   16
Expanded Notation   17

2.2 Rounding Whole Numbers   19

2.3 Addition and Subtraction of Whole Numbers   22
Addition   22
Properties of the Real Numbers under Addition   23
Subtraction   23

2.4 Multiplication and Division of Whole Numbers   25
Multiplication   25
Properties of the Real Numbers for Multiplication   26
Division   27
Division by Zero   27
The Division Algorithm   28
Prime Numbers and Prime Factorization   28

2.5 Exponents and Roots   34
Exponents   34
Roots   35

2.6 Order of Operations   37

2.7 Averages and Estimation   41
Averages   *41*
Estimation   *42*

Small Group Activity: Eratosthenes's Sieve   44
Chapter 2 Review   45
Chapter 2 Practice Test   49
Study Skills Tip: Managing Time for Study   50
Part I: Homework   50
Part II: Time Management   51
The Mathematics Journal: Time Management   52

③ Introduction to Algebra   53

3.1 Simplifying Algebraic Expressions   53
Associative Property   *54*
Distributive Property   *55*
Combining Like or Similar Terms   *56*

3.2 Exponents   60

3.3 Evaluating Algebraic Expressions   64

3.4 Translating Words into Algebraic Expressions   67

Small Group Activity: Practicing Evaluating Algebraic Expressions   71
Chapter 3 Review   72
Chapter 3 Practice Test   77
Study Skills Tip: Tests   78
Part I: Preparation   78
Part II: Top Ten Test-Taking Strategies   79
The Mathematics Journal: Study Strategies   80
Cumulative Review: Chapters 1, 2, and 3   80

④ Integers   83

4.1 The Set of Integers   84
Comparison of Integers   *84*

4.2 Absolute Value   86
Opposite or Additive Inverse   *87*

4.3 Addition of Integers   89
The Associative Property of Addition   *91*

4.4 Subtraction of Integers   96

4.5 Multiplication of Integers   100

4.6 Division of Integers   104

4.7 Order of Operations and Averages Using Integers   106
Averages   *108*

4.8 Solving Equations   112

Small Group Activity: Understanding Integer Numbers   117
Chapter 4 Review   117
Chapter 4 Practice Test   122
The Mathematics Journal: Analysis of Test Preparation   123

5 **Fractions and Mixed Numbers** 124

5.1 The Set of Rational Numbers as Fractions 124
Expressing Integers as Fractions 125
Types of Fractions 126
Positive and Negative Fractions 128

5.2 Using the Identity Property of Multiplication with Fractions 130
Reducing Fractions to Lowest Terms 133

5.3 Multiplication of Fractions and Mixed Numbers 136
Multiplying Positive and Negative Fractions 138
Multiplying Mixed Numbers 138

5.4 Division of Fractions and Mixed Numbers 143

5.5 Addition and Subtraction of Fractions and Mixed Numbers 148
Addition and Subtraction of Mixed Numbers 153

5.6 Order of Operations and Complex Fractions 159
Complex Fractions 161

5.7 Solving Equations Using the Multiplication Property of Equality 165
Using the Multiplication Property of Equality to Solve Equations 165

Small Group Activity: Using Egyptian Fractions 169
Chapter 5 Review 169
Chapter 5 Practice Test 174
The Mathematics Journal: Explaining a Procedure I 176

6 **Algebraic Expressions** 177

6.1 Exponents (Revisited) 177

6.2 Polynomials 183
Addition and Subtraction of Polynomials 184
Multiplication of Polynomials 185
Multiplication of Binomials 186
Generalized Distributive Property 186
FOIL Method for Multiplying Binomials 186
Finding Common Monomial Factors 187

6.3 Simplifying, Multiplying, and Dividing Algebraic Fractions 191
Simplifying Algebraic Functions 191
Multiplication and Division of Algebraic Fractions 192
Multiplication of Algebraic Fractions 192
Division of Algebraic Fractions 193
Combining Operations 193

6.4 Addition and Subtraction of Algebraic Fractions 195
Adding and Subtracting Algebraic Fractions with the Same Denominators 195
Adding and Subtracting Fractions with Different Denominators 196

6.5 Evaluating Algebraic Expressions 198

Small Group Activity: Pascal's Triangle 202
Chapter 6 Review 202
Chapter 6 Practice Test 207
The Mathematics Journal: Revision of Time Management Plan 209
Cumulative Review: Chapters 4, 5, and 6 210

7 **Decimals** 213

7.1 Place Value and the Comparison of Decimals 213
The Principle of Trichotomy 215
Scientific Notation 216
Rounding Decimals 217

7.2 Addition and Subtraction of Decimals 222

7.3 Multiplication of Decimals 228
Multiplying by Powers of 10 230
Dividing by Powers of 10 236

7.4 Division of Decimals 234

7.5 Conversions between Fractions and Decimals 240

Small Group Activity: Estimating Grocery Shopping 249
Chapter 7 Review 249
Chapter 7 Practice Test 254
The Mathematics Journal: Midterm Progress Report 256

8 **Equations (Revisited) and Inequalities** 257

8.1 Solving Equations Using Both the Addition and Multiplication Properties of Equality 257
Solving Equations Using the Addition Property of Equality (Revisited) 257
Solving Equations Using the Multiplication Property of Equality (Revisited) 258
Solving Equations Using Both Properties 259

8.2 Solving Equations Containing Fractions 262

8.3 Solving Equations Containing Decimals 267
Solving Equations Containing Both Decimals and Fractions 269

8.4 Solving Linear Inequalities in One Variable 271
Graphing Inequalities on the Number Line 272
Properties of Inequalities 275
Solving Inequalities Using Both Properties 277

Small Group Activity: The Coin Toss 280
Chapter 8 Review 281
Chapter 8 Practice Test 285
The Mathematics Journal: Explaining a Procedure II 287

9 **Problem Solving** 288

9.1 Engaging in Problem Solving 289
Seeing Yourself as a Problem Solver 289
The Problem-Solving Process 289

9.2 Some Strategies for Solving Logic and Mathematics Problems 291
Read to Comprehend 291
Analyze the Problem 292

9.3 Solving Problems by Using Equations and Formulas 298
Solving Problems by Using Equations 299
Solving Problems by Using Formulas 303

9.4 Using Equations in One Variable to Solve Coin and Ticket Problems (Optional) 306

Small Group Activity: Strategies for Problem Solving 309
Chapter 9 Review 309
Chapter 9 Practice Test 310
The Mathematics Journal: Thinking About Problem Solving 312
Cumulative Review: Chapters 7, 8, and 9 312

10 Ratio and Proportion 316

10.1 Ratios 316
Units of Linear Measure, Capacity, Weight, and Time 318

10.2 The Fundamental Property of Proportion 326

10.3 Applications of Proportion 331

Small Group Activity: Ratios and Fitness 338
Chapter 10 Review 338
Chapter 10 Practice Test 341
The Mathematics Journal: Present Experiences in Mathematics 342

11 Percents 343

11.1 Conversions: Fractions, Decimals, and Percents 343

11.2 Basic Percent Problems or Cases of Percent 350
The Equation Method 350
The Proportion Method 353

11.3 Percent Application Problems 358

11.4 Percent Increase and Decrease 365

11.5 Simple Interest 370

Small Group Activity: Estimating Tips 373
Chapter 11 Review 374
Chapter 11 Practice Test 377
Study Skills Tip: Preparation for the Final Exam 379
The Mathematics Journal: Mathematics in Everyday Life 380

12 Geometry and Measurement 381

12.1 Linear Measure and Applications 381
Linear Measure 382
U.S. Customary System of Linear Measure 382
Metric System of Linear Measure 383
Plane Figures 385
The Pythagorean Theorem 385
Perimeters of Polygons 387
Perimeters of Rectangles 390
Perimeters of Squares 391
Circles 393
Perimeters of Other Plane Figures 395

12.2 Square Measure and Applications 401
Area Square Measure 401
Area of a Rectangle 403
Area of a Square 405
Area of a Parallelogram 406
Area of a Triangle 407
Area of a Circle 408
Areas of Other Plane Figures 409

12.3 Cubic Measure and Applications 412
U.S. Customary and Metric Systems of Cubic Measure 412
Volume of a Right Rectangular Solid 414
Volume of a Cube 414
Volume of a Right Circular Cylinder 415
Volume of a Sphere 416

Small Group Activity: How large is a _____ ? It's about the size of a _____ .   418
Chapter 12 Review   418
Chapter 12 Practice Test   423
The Mathematics Journal: Future Experiences in Mathematics   425
Cumulative Review: Chapters 10, 11, and 12   425

13 Graphs and Charts   429

13.1 The Rectangular Coordinate Plane 429
Naming the Coordinates of Points   433
The Quadrants   434

13.2 Graphing Linear Equations   436
Horizontal and Vertical Lines   440

13.3 Interpreting Statistical Graphs and Charts   446

13.4 Constructing Statistical Graphs and Charts   452
Bar Graphs   453
Line Graphs   454
Circle Graphs or Pie Charts (Optional)   455

Small Group Activity: Graphs and Charts   462
Chapter 13 Review   462
Chapter 13 Practice Test   469
Study Skills Tip: How to Take the Final Exam   473
The Mathematics Journal: Opening-Day Advice to a Prealgebra Student   473
Practice Final Exam   474

A The Arithmetic of Whole Numbers   481

A.1 Reading and Writing Numbers 481
A.2 Addition   483
A.3 Subtraction   485
A.4 Multiplication   487
A.5 Division   489

B For Your Reference   492

B.1 Non-Linear Measures   492
B.2 Conversions between the U.S. Customary and the Metric System   492
B.3 Properties of the Real Numbers   494

Answers to Problem Sets   497
Index   515

Preface

Prealgebra: A Transitional Approach is a texbook intended for use primarily in a foundations mathematics course for incoming college students underprepared in mathematics. Designed to assist instructors as they take the first steps in making the transitions needed to incorporate the *Standards for Introductory College Mathematics* as developed by the American Mathematical Association of Two-Year Colleges (AMATYC, 1995) into their everyday teaching, the book includes topics from number theory, discrete mathematics, statistics, and geometry, as well as a review of the arithmetic of real numbers and a thorough treatment of beginning algebra. It features a metacognitive approach to problem solving, activities designed for collaboration, and detailed plans for a mathematics journal that students will keep as they use the book.

Students will find this text an effective tool as they build the foundation in mathematics they will need to successfully pursue their college and career goals. It is designed to assist them to become good problem solvers, to learn to communicate mathematically, to use modern technology appropriately, to develop good study skills in mathematics, and to improve their confidence in their abilities to "do mathematics."

The book employs a "spiral approach" to learning mathematics—a new topic is introduced early and then revisited several times to strengthen the students' depth of understanding. Chapter 1 introduces the students to set theory; Chapter 2 then applies this theory to the structure of the whole number system. Chapter 3 introduces students to the algebra of whole numbers, while Chapter 4 reviews the arithmetic of integers and extends the study of algebra to include the integers. Chapter 5 begins with a review of the arithmetic of rational numbers and then extends the study of algebra to include rational numbers; Chapter 6 continues this study with an in-depth treatment of the algebra of rational expressions. Chapter 7 completes the study of the arithmetic of the set of real numbers by reviewing decimal operations. Chapter 8 focuses on solving equations and inequalities by revisiting topics from Chapters 4, 5, 6, and 7. Chapter 9 provides students with an in-depth treatment of problem solving and its role in mathematics. Chapter 10 examines the concepts of ratio and proportion and applies them to topics from Chapters 5, 8, and 9. Chapter 11 concentrates on percents and reinforces the use of fractions (Chapter 5), decimals (Chapter 7), and equations (Chapter 8) in solving real-world applications. Chapter 12 uses algebra (Chapters 3 and 6) to evaluate geometric formulas and proportions (Chapter 10) to study measurement. The first part of Chapter 13 extends the study of equations (Chapter 8) to include those in two variables and introduces coordinate geometry by teaching students to graph these equations. And finally, the second part of Chapter 13 examines the role that graphs play in the study of statistics.

Throughout the text, the content of ***Prealgebra: A Transitional Approach*** is modeled on the 1995 AMATYC Standards for Content. In Chapters 2, 4, 5, and 7, care has been taken to ensure that students develop a good "number sense," as recommended in Standard C-1.

The study of algebra, both as a language and as a system, begins in Chapter 3 and continues for the remainder of the book. In keeping with Standard C-2, Chapters 9, 10, and 11 assist students in learning to ". . . translate problem situations into their symbolic representations and use those representations to solve problems."

By focusing on geometry and measurement, Chapter 12 ensures the desired outcomes of Standard C-3 that "students will develop a spatial and measurement sense."

The text prepares students for the study of functions (Standard C-4) by focusing on relations, teaching them to evaluate algebraic expressions in Chapters 3 and 6 and to graph linear equations in two variables in Chapter 13.

This book follows the recommendation of Standard C-5 by including the following discrete mathematics topics: set theory (Chapter 1), number theory (Chapters 2 and 5), basic counting techniques (Chapters 1 and 9), and abstract algebra, beginning with Chapter 2 where students are introduced to the group and field properties for real numbers then used throughout the remainder of the text.

In keeping with Standard C-6, Probability and Statistics, the Small Group Activity entitled "The Coin Toss" (Chapter 8) introduces students to probability, while in Chapter 13 students learn to use statistical graphs and charts to analyze real-world data.

Key Features

As illustrated in the following chart, several of the key features in *Prealgebra: A Transitional Approach* will assist instructors in implementing the AMATYC Standards.

AMATYC STANDARD

Students will develop the view that mathematics is a growing discipline, interrelated with human culture, and understand its connections to other disciplines.

KEY FEATURE

CHAPTER INTRODUCTIONS These brief essays benefit students by putting the subject matter of each chapter into an historical context and by applying the concepts to the world beyond mathematics.

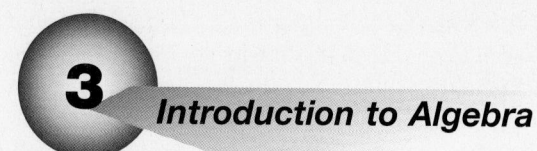

3 Introduction to Algebra

3.1 Simplifying Algebraic Expressions
3.2 Exponents
3.3 Evaluating Algebraic Expressions
3.4 Translating Words into Algebraic Expressions

Introduction to Algebra

The word *algebra* is derived from the title of a ninth century Arabic work on mathematics by Al-khwarizmi called *Al-jabr wa'l maqabalah* (the science of reuniting). Algebraic methods were used by the Babylonians about 1900±1600 B.C. Progress was slow until the seventeenth century when Descartes, a French philosopher and mathematician, introduced the modern system of notation.

Algebra is a part of mathematics. When using algebra, letters are used to hold the place of unknown numbers. In the following example $x + 6 = 10$, the letter x represents the unknown. Letters are also used to express general relationships between numbers as shown in the following example, $a + b = b + a$, where the equation holds true for all real number values of a and b. Algebra also allows us to find the values of variables in equations and inequalities, such as $x + 6 = 9$ or $7x < 96$. In this chapter, you will start to learn basic algebraic concepts.

3.1 **Simplifying Algebraic Expressions**

OBJECTIVES

- Recognize an algebraic expression.
- Name the terms of an algebraic expression.
- Recognize a variable, the coefficient of the variable, a monomial, and a constant.
- Combine like terms.
- Use the Distributive Property to simplify an expression.

An algebraic expression is a sum of terms. To simplify means to make as simple as

variable or a product of numbers and or

usually a lower case letter of the English esent an unknown number.

repeated addition: $3 + 3$ is equal to $2 \cdot 3$. e this into a general statement by using

would read $x + x = 2x$.
$+ 0 = 2(0) = 0$. $x + x = 2x$ is true for 0.
y some!

53

9 Problem Solving

9.1 Engaging in Problem Solving
9.2 Some Strategies for Solving Logic and Mathematics Problems
9.3 Solving Problems by Using Equations and Formulas
9.4 Using Equations in One Variable to Solve Coin and Ticket Problems (Optional)

Introduction to Problem Solving

Do you consider yourself a good problem solver? Before you answer that question, take an imaginary journey with us.

Suppose you are an 18-year-old college freshman, and a friend suggests that the two of you go to Florida for spring break.

Or suppose you are 30-something, married, with small children, and your spouse suggests a get-away weekend for just the two of you.

Or suppose you are a single parent, and a couple of friends suggest going to the World Series or going to New York to do some holiday shopping. Would you be interested? What if they admitted that they didn't have the trip planned yet and didn't even know whether it was possible? Would you still be interested? What if they invited you to join them in making it happen? Would you look on it as an adventure and say, "Why not?"

Improbable as any of these excursions may seem in the light of your present circumstances, we all know that thousands of college students spend spring break in Florida every year; the World Series is always sold out (a few people in attendance must be single parents); out-of-town shoppers crowd the New York stores each December; and hotels are often sold out on weekends because of their advertised get-away specials.

If you say, "OK, let's go!" what do you and your friends do? You call a few hotels, visit a travel agency, or contact AAA. In short, you begin gathering information–information you'll need to achieve your goal. Once you have the information, you'll start sorting it. You'll keep what will help you, and you'll discard the rest. Now you'll start planning the trip piece by piece. You will have to get tickets to the ball game. You will have to make living arrangements for the children. You will have to decide how you're going to finance the trip. Sometimes, your first ideas won't work. You may have to start over; you may have to devise a new plan. If that plan doesn't work, you may have to come up with another one, and another, and another. Each time, you'll refine the process; each time you'll get closer to Florida, the World Series, or New York.

You may, however, decide at any time to abort the trip. You may decide that you have other priorities, it's too expensive, you don't have the time, or you may get discouraged and quit. You may change your goals.

What have we been talking about during this exercise? Something we all do every day and several times a day. We've been talking about **problem solving.**

If we were to ask high school freshmen in an algebra class if they considered themselves good problem solvers, the majority would say, "No." If we were to ask them if they had attended a concert given by a major rock star in the past six

288

AMATYC STANDARD

Students will engage in substantial mathematical problem solving.

KEY FEATURE

Chapter 9 is devoted entirely to this topic. The metacognitive approach used is unique to this text. It is based on the premise that students are already dealing with sophisticated problems in their everyday lives. It encourages them to reflect on their own problem-solving behaviors, assists them in developing a model for problem solving, and guides them in applying appropriate strategies to mathematics.

AMATYC STANDARD

Mathematics faculty will foster interactive learning through . . . collaborative activities so that students can learn to work effectively in groups and communicate about mathematics both orally and in writing.

KEY FEATURE

SMALL GROUP ACTIVITIES One of these collaborative activities is placed in each chapter immediately preceding the chapter review. Designed for students to do in groups, these activities foster student involvement. By requiring that students work with each other to solve problems, these activities provide ways for students to learn from each other and to communicate mathematics as they talk and listen to each other while exploring a mathematical topic or problem together.

---

44   CHAPTER **2** WHOLE NUMBERS

13. $235 + 117 + 89 + 22$ (to hundreds)

14. $391 + 219 + 77 + 30$ (to hundreds)

15. $10,000 - 2500$ (to thousands)

16. $50,000 - 7500$ (to thousands)

17. $25,659 + 31,743$ (to ten thousands)

18. $72,556 + 37,921$ (to ten thousands)

19. $123 + 456 + 789$ (to thousands)

20. $879 + 564 + 321$ (to thousands)

Applications:

21. This season, Willy Joe, your football team's quarterback, completed three passes in the first game, seven in the second game, and two in the third game. Find the average number of passes he completed in these games.

22. This season, Eric Harrison, the leading rusher on the football team, ran 3 yards in the first game, 5 yards in the second game, and 100 yards in the third game. Find the average number of yards he ran in these games.

23. According to the payroll office of ADV Advertising, Jim earns $942 per month; Agnes, $1098; Mike, $1458; Kay, $1830; Patti $2376; and Gary, $3660.
A. What is the average monthly salary of this group?

B. Check the reasonableness of your answer by estimating it to hundreds.

24. Robbi, Beth, Patricia, and Geraldine are all employees of Brite State University. Robbi earns $28,500 per year; Beth, $30,258; Patricia, $31,580; and Geraldine, $22,490.
A. What is the average yearly salary of these employees?

B. Check the reasonableness of your answer by estimating their average yearly salaries to thousands.

SMALL GROUP ACTIVITY *Eratosthenes's Sieve*

Eratosthenes was a geographer and mathematician who lived during the third century B.C. He used a method he called a sieve to find the prime numbers from 2 to 100. First, he wrote down the numbers from 2 to 100. Next, he punched out the numbers which were multiples of 2—the even numbers, except 2 itself. After that, he punched out all the multiples of 3, except 3 itself. He continued in the same manner, looking for multiples of 5, then 7, etc., until he had eliminated all the composite numbers. When he finished, he noted that the results looked like a sieve and that the numbers remaining were prime.

Activity 1
Make your own sieve for the primes from 2 to 100.

Activity 2
List the prime numbers you found.

Activity 3
After completing the sieve, discuss these questions.
a. Why did you only need to search for multiples of the prime numbers? Why not multiples of 6, 10, etc.?
b. Note that 11 was the largest number you needed to find multiples of. Why?
c. If you were asked to find the prime numbers from 2 to 150, what is the largest number you would need to find multiples of? Why?

---

256   CHAPTER **7** DECIMALS

23. Simplify: $0.2 + \dfrac{4}{5} + \dfrac{7}{5}$

24. Simplify: $0.7\left(\dfrac{3}{8} + 0.125\right)$

25. Simplify: $\left(\dfrac{1}{3}\right)^2(5.4) + \left(\dfrac{1}{2}\right)^3(3.2)$

26. Express in scientific notation: 0.00000765

27. Express in standard notation: 2.48×10^6

28. Pens are sold for $23.64 a dozen. How much does each pen cost?

29. A phone call to Florida costs 22 cents for the first minute and 19 cents for each additional minute. How much will a 17-minute call cost?

30. Your math textbook cost $42.97; your English textbook, $23.45; and the three books required for your history class, $2.99, $13.54, and $10.02. How much did your textbooks cost this semester?

THE MATHEMATICS JOURNAL	
Midterm Progress Report	
Task: As you reach the midpoint in the semester, take a few minutes to reflect on how the semester is progressing for you.	
Starters: What do you want to get out of the course? Are your plans helping you accomplish your goals?	

AMATYC STANDARD

Students will acquire the ability to read, write, listen to, and speak mathematics.

KEY FEATURE

THE MATHEMATICS JOURNAL Designed to assist students in reflecting on their experiences and goals in mathematics, in monitoring their progress in the course, and in aiding them to "write mathematics," the journal entries appear throughout the book.

AMATYC STANDARD

Students will use appropriate technology to enhance their mathematical thinking and understanding and to solve mathematical problems and judge the reasonableness of their results.

KEY FEATURE

USING THE CALCULATOR Beginning in Chapter 2, this boxed feature appears in all appropriate sections of every chapter. By following the detailed hands-on instructions, students learn to use a scientific calculator to solve appropriate problems.

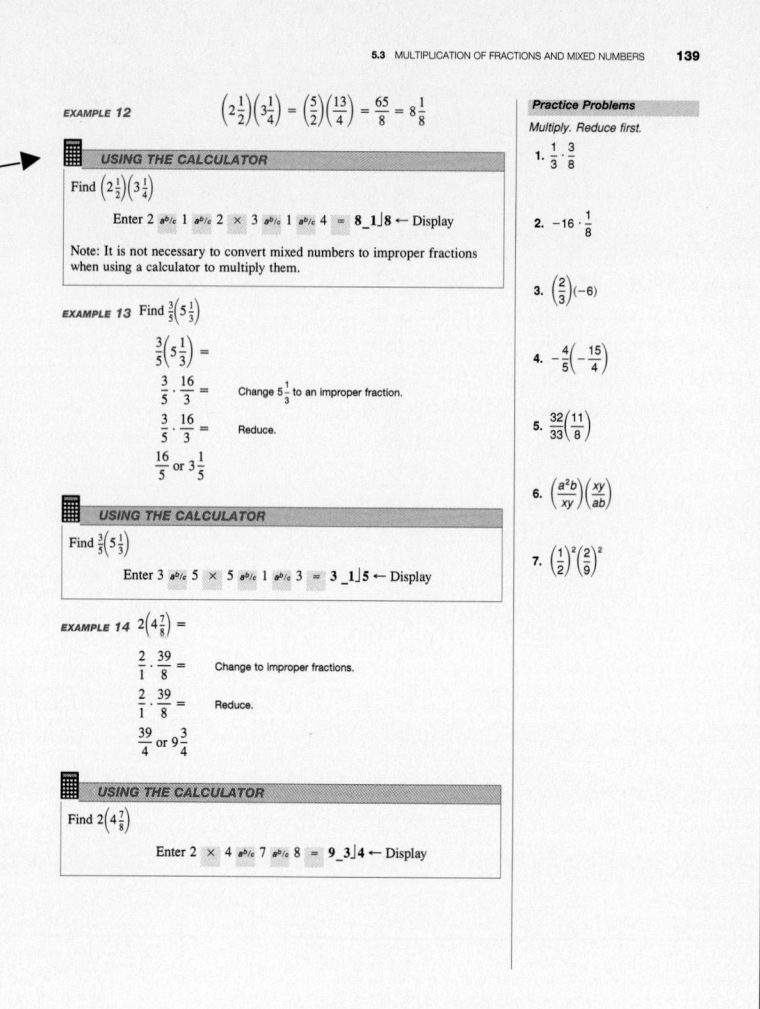

AMATYC STANDARD

Students will learn mathematics through modeling real-world situations.

KEY FEATURE

APPLICATIONS Where appropriate, application problems appear in all sections immediately following the examples. They put the content of the sections into the context of real-world problems and provide students with models for solving similar problems. Application problems can also be found in the problem sets, where they provide students the opportunity to use the models developed. To the extent possible, actual data rather than contrived data have been used in creating these problems.

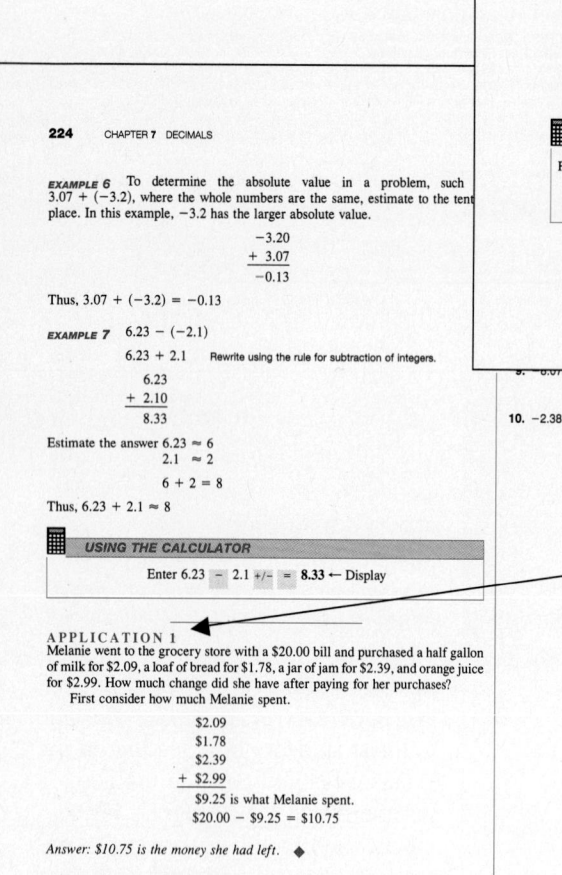

In addition to these features, *Prealgebra: A Transitional Approach* contains several other features designed to make it an effective tool for both students and instructors. Among them are:

STUDY SKILLS TIPS Addressed directly to the students, these short (150-200 word) essays suggest ways to develop or improve study skills in mathematics. Each one deals with a specific topic. This feature precedes each of the first four chapters and reappears before Chapters 11 and 12.

OBJECTIVES The objectives are listed at the beginning of each section of every chapter and again in the Chapter Reviews. They outline for students what they should be able to do with the mathematics they learn in each section.

DEFINITIONS Each new term is defined when it is introduced. These definitions provide the foundation, or base, for the mathematics the students will learn.

RULES Rules, properties, and theorems are boxed throughout the text. They provide convenient references for students.

NOTES Placed in appropriate spots throughout the text, they are used to draw the students' attention to and clarify concepts or rules that can cause confusion.

EXAMPLES These can be found in every section. They lead students step-by-step through the solutions of problems relating to the given section.

PRACTICE PROBLEMS These problems can be found in the margins of the pages of each section. They provide students with a quick check on the skills they have just mastered working through the examples. The answers to these problems appear at the end of each section for ready reference.

PROBLEM SETS After Chapter 1, Problem Sets are placed at the end of each section. They are intended to be assigned as homework so that students can get sufficient practice developing appropriate skills. Usually, they are graduated from easy to difficult. The even- and odd-numbered problems are usually paired, so that students working only the odd- or even-numbered problems will have the opportunity to test the full range and scope of the problems. Where appropriate, the problem sets include real-world problems that require that students apply the mathematics they have learned.

CHAPTER REVIEWS A chapter review that assists students in synthesizing the concepts learned and in preparing for a chapter test appears in each chapter immediately following the Small Group Activity. The first part of each review is designed to help the student recall the concepts learned in the chapter—the definitions, properties, and theorems. The second part of each review lists each objective from the chapter, followed by several appropriate problems using that objective. By working these problems and checking their answers, students learn how to structure their own personal reviews in preparation for the chapter test.

PRACTICE TESTS A practice test appears in each chapter immediately following the chapter review. Students can use the practice test to determine how well they have mastered the content of the chapter and how well prepared they are for an actual test on the chapter. Instructors may want to have students work on this practice test in pairs so that students can assist each other in preparing for the actual test.

CUMULATIVE REVIEWS   The book contains four cumulative reviews—after Chapter 3, Chapter 6, Chapter 9, and Chapter 12. Similar to the practice tests in design, they contain the concepts covered in the three chapters preceding them. They assist students in reviewing larger portions of the text in preparation for unit tests.

PRACTICE FINAL EXAM   This final exercise after Chapter 13 is a review of the entire text and is designed to assist students in preparing for the final exam.

REFERENCE MATERIALS   Appendix B, entitled "For Your Reference," contains reference materials, including U.S. customary and metric measures and the properties for real numbers. While these materials can be found throughout the text, students will find it convenient to have them also listed in one place.

ANSWERS   Answers to odd-numbered problems throughout the text are listed at the end of the book.

SUPPLEMENTS

For the Instructor

Prealgebra: A Transitional Approach is accompanied by a complete supplemental package.

ANNOTATED INSTRUCTOR'S EDITION   In this AIE version of ***Prealgebra: A Transitional Approach,*** the answers to all problems in the problem sets, chapter reviews, practice tests, cumulative reviews, and practice final exam are printed in color beside or below the problems as they occur throughout the text.

INSTRUCTOR'S MANUAL   Part I of this manual includes step-by-step solutions for all problems in problem sets, chapter reviews, practice tests, cumulative reviews, and the practice final exam. Besides a test bank consisting of chapter tests, unit tests, and a final exam, Part II includes daily planning charts indicating the recommended time to be spent on each topic, a form for a get-acquainted activity, a form for gathering background information on students, and a sample pretest and preliminary quiz that instructors can use to identify students who may have difficulty with the course. An appendix contains answers to the manual's chapter and unit tests and final exam and to the textbook's Small Group Activities.

COURSE MANAGEMENT AND TESTING SYSTEM   InterAct Math Plus combines course management and on-line testing with the features of InterAct Math Tutorial Software (see Supplements For the Student) to create an invaluable teaching resource. InterAct Math Plus is available for both Windows and Macintosh. Contact your local Addison Wesley Longman sales consultant for details.

For the Student

STUDENT'S SOLUTION MANUAL   This manual includes step-by-step solutions for all problems in the answer sets, chapter reviews, practice tests, cumulative reviews, and practice final exam.

INTERACT MATH TUTORIAL SOFTWARE   Available for both Windows and Macintosh, this tutorial software has been developed and designed by professional software engineers working closely with a team of experienced math educators.

InterAct Math Tutorial Software includes exercises that closely parallel those in the text. Each exercise has an example and an interactive guided solution designed to involve students in the solution process and help them identify precisely where they are having trouble. In addition, this software recognizes common student errors and provides students with appropriate customized feedback. It also tracks student activity and scores for each section, which can then be printed out. Finally, all questions are algorithmically based, providing students with virtually limitless potential for practice. InterAct Math Tutorial Software is free to qualifying adopters.

VIDEOTAPES   A set of videotapes covering many of the topics in the text is free to adopters. Contact your local Addison Wesley Longman sales consultant for details.

ACKNOWLEDGMENTS

Our thanks to our friends and colleagues in the Developmental Services Department of Undergraduate Studies at Kent State University for their encouragement, support, and assistance, and in particular to Gary Padak and Nancy Adams. Many thanks to our fellow instructors Jack DiAlesandro, Meg Leslie, Joe Minerovic, Mary Romans, and Diane Dulzer for their numerous contributions, valuable suggestions, continued interest, and willingness to incorporate materials from the manuscript into their teaching. Our thanks to Edwin Gibson for his comments on the chapter on problem solving and to Harold Schwartz for his extensive editorial suggestions.

Our heartfelt gratitude to Karin Wagner, our sponsoring editor at Addison Wesley Longman, for believing in us and in our project, for her support, and for her care-filled guidance. Our thanks to Jason Jordan and Maxine Effenson Chuck for their editorial assistance; to Anne Schmidt Ryan, our project editor, for the interest in and care she took of us and our manuscript; to Lynn Mooney, who edited our art manuscript; and to the production staff at Addison Wesley Longman.

Our thanks to the following reviewers for improving our manuscript by sharing their professional opinions with us:

Stanley Carter	Central Missouri State University
Denise Chellsen	Cuesta Community College
Joyce Curry-Daly	Cuesta Community College
Terry Y. Fung	Kean College
Irineu Glajar	Austin Community College
Maryann E. Justinger	Erie Community College-South
Robert Kaiden	Lorain County Community College
Judy Kasabian	El Camino College
Theodore Lai	Hudson County Community College
Sr. Margaret Mary Leslie	Kent State University
Joseph Minerovic	Kent State University
Kathy Mowers	Owensboro Community College
Mary Romans	Kent State University
Edith M. Ruben	New River Community College
Karen Schwitters	Seminole Community College
Sandra Vrem	College of the Redwoods
Paul Wright	Austin Community College
Michelle A. Wyatt	Community College of Southern Nevada

And last of all, to our families and our students, a warm, appreciative, and heartfelt "thank you!"

Jill Parker Beer
Dwyn Griffin Peake

Preface to the Student

Welcome! We've designed this text for you. We want it to serve as a support as you learn the mathematics presented in it. We urge you to make the book your own. It's yours! Make notes in it. Get as familiar as you can with it.

We'd like to highlight some of the features we've included in this text that we hope will contribute to your success in mathematics. All sections are set up similarly. Once you get familiar with the organization, you should be comfortable with it throughout the text. The content will change, but the format will remain the same.

We've listed the **Objectives** at the beginning of each section of each chapter. They tell you what mathematics you should be able to do after you have studied that section. Study the **Examples** found in every section. We've tried to include one to cover each different situation. Use them as models. After you've studied the examples, try working the **Practice Problems** that appear in almost every section. Solving these problems will tell you immediately whether you understand the concepts presented.

The **Using the Calculator** feature provides you with "hands-on" instructions. Follow them step-by-step as you punch the appropriate keys on your scientific calculator. The calculator is one of the great modern tools of mathematics. Make yours work for you.

The **Study Skills Tips** that appear throughout the text, starting with "Studying Mathematics" as part of this Preface, are intended to assist you as you develop and refine your own study skills in mathematics.

And, of course, we must mention homework. We've designed the **Problem Sets** to give you a chance to practice the mathematics you're learning. The even- and odd-numbered problems are paired by objective and level of difficulty. For the most part, the problems are graduated from easy to difficult. Try doing your homework with a friend or two; we strongly urge you to find two or three fellow students and arrange to meet regularly to study mathematics together. We encourage you to use the **Small Group Activities** that precede each chapter review. They are designed to foster cooperative learning.

Every chapter concludes with a **Chapter Review** and a **Practice Test.** The objectives are also listed again as part of each chapter review. As you study for each chapter test, see whether you can accomplish these objectives. The number in parentheses at the end of each objective in the Chapter Review refers to the section of the chapter in which that concept was covered. When reviewing, refer to the appropriate section as needed.

Use the **Cumulative Review** after each set of three chapters to prepare for unit tests. At the end of the book, you'll find a **Practice Final Exam.** Take it as part of your preparation for the upcoming final exam in the course you're taking.

Last of all, we'd like to tell you about **The Mathematics Journal,** which we suggest you keep while using this text, beginning with the journal entry at the end of this preface. Some of the entries will ask you to write about your feelings and thoughts as you progress through the book. Doing so can help you sort out your feelings about mathematics and may even provide you with new insights as to why you have them; writing down your goals and resolutions should help reinforce them. Monitoring your progress and writing about it can be excellent ways to help you see how your skills are developing. Explaining a mathematical concept in writing is a powerful way of finding out if you understand it thoroughly.

We designed each of these features to assist you with the process of learning mathematics. We wish you success. Now, let's get to work.

STUDY SKILLS TIP

Studying Mathematics

Studying mathematics is like studying another language; it has it own vocabulary that needs to be practiced. Learning mathematics is like building a wall with blocks. Each concept rests on a preceding, more basic idea. If you do not understand a basic idea, you cannot build on it, i.e., understand a more advanced concept. Never feel that you can gloss over a particular concept and think that you can just miss those problems on the test referring to it ("After all, it will only be a four-point problem."), because that same concept will be needed to understand many other concepts as the semester goes on.

Always be sure you find someone—the teacher, a tutor (if you have one), or a study buddy—who can help you understand as you go along. **Never, never** leave a question unanswered. To be sure you know the questions you need to ask, it is important to do your homework as soon as possible after each class. This way, the lecture is still fresh in your mind and you can still make sense of the notes you took. This is also a great time to revise your notes so they will be in a more helpful format when it is time to review them for a test.

Study mathematics to gain an understanding of the structure of it. Strive to learn why it works. Simply learning a series of steps or a procedure to follow in solving particular problems is not enough. You are less likely to forget how to solve problems if you understand the structure of the mathematics involved in their solutions.

THE MATHEMATICS JOURNAL

Past Experiences in Mathematics

Task: As you begin this course, reflect on your experiences in learning mathematics. Write a one-page (150-word) entry on them. Thinking and writing about what has worked for you and what hasn't will help to get you ready for this new course.

Starters: Describe your successful experiences. Describe your unsuccessful experiences. Why do you think you were more successful in certain learning situations than in others?

1

Sets

1.1 Set Theory
1.2 Set Relations
1.3 Operations with Sets
1.4 Numbers as Sets

Introduction to Set Theory

Set theory is relatively new in the history of mathematics. Georg Cantor (1845–1918) developed it and is considered the father of set theory. Cantor was of Danish ancestry, which may account for his use of the Danish letter Ø to represent the empty set. While Cantor's views were not widely accepted during his lifetime, his work on set theory today forms the cornerstone of modern mathematics. Cantor was not the only mathematician to work in this field. George Boole (1815–1867) developed the rules for operating with sets, and John Venn (1834–1923) devised a series of diagrams to represent set operations and relations.

1.1 Set Theory

OBJECTIVES

- Use the terminology of set theory correctly.
- Use set notation correctly.
- List the elements of a set.
- Determine whether a given set is finite or infinite.

We use the term **set** in everyday life. We speak of a set of dishes and we refer to a set of golf clubs. We use this term to organize things. Mathematicians apply it to mathematics. This branch of mathematics is called **set theory.**

DEFINITION **Set:** A collection of objects.

EXAMPLE 1 The set of letters that form the vowels of the alphabet. In set theory, this set is written as V = {a, e, i, o, u}.

DEFINITION **Member** or **element:** Each of the objects that belongs to the set.

EXAMPLE 2 The letter o is an element or member of the set of vowels.

DEFINITION **∈:** The Greek letter epsilon. This is used to indicate that an element belongs to a particular set.

EXAMPLE 3 o ∈ V is read "o is an element of the set V." In everyday English, we would say, "The letter o is a vowel.

DEFINITION **∉:** Means "is not an element of a particular set."

EXAMPLE 4 q ∉ V is read "q is not an element of the set V." In English, we would say, "The letter q is not a vowel."

Notes   Mathematicians have developed special notation and conventions to use with sets, namely:

* A capital letter is usually used to name a set.
* An equal sign follows the name.
* If letters are used to name the elements of the set, they are usually lowercase letters.
* The elements of the set are separated by commas.
* The entire set is enclosed in braces { }.
* The Greek letter $\in$ is used to indicate that an object belongs to a particular set.

EXAMPLE 5   Written in set notation, the set consisting of the days of the week would look like this:

D = {Sunday, Monday, Tuesday, Wednesday, Thursday, Friday, Saturday}

Notes • The choice of the letter *D* as the name of the set is arbitrary—any letter could have been used.
  • In a set, the order in which the members are listed is also arbitrary.

DEFINITION   **Equal sets:** Sets that contain exactly the same elements. The symbol = is used to indicate the equality of two sets.

EXAMPLE 6   {a, e, i, o, u} = {e, a, u, i, o}

DEFINITION   **Finite set:** A set that has a definite number of elements. (If the question "How many elements does the set have?" can be answered, the set is finite.)

EXAMPLE 7   Set D: the days of the week is a finite set because there are seven elements in it, i.e., the seven days of the week.

DEFINITION   **Infinite set:** A set which contains an unlimited number of elements.

EXAMPLE 8   The set of numbers we use for counting, {1, 2, 3, . . . }, is an infinite set.

1.2 Set Relations

OBJECTIVE
● Construct the subsets of a given set.

Sets may be studied in terms of their relationships to other sets.

DEFINITION   **Subset:** A set constructed so that all of its elements belong to another set. Set B is a subset of a set A if all the elements of B also belong to A.

EXAMPLE 1   {*o*} is a subset of the set of vowels because *o* is a vowel.

DEFINITION   **⊆:** The symbol used for the phrase "is a subset of."

EXAMPLE 2   Let S = {Monday, Tuesday, Wednesday, Thursday, Friday}

Since all the elements of the set S, of weekdays, belong to the set D of days of the week, S is a subset of D.

$$S \subseteq D.$$

EXAMPLE 3 $\{a\} \subseteq V$

DEFINITION   **$\not\subseteq$:** The symbol used for the phrase "is not a subset of the set under discussion."

EXAMPLE 4 $D \not\subseteq S$, i.e., the set of days of the week is not a subset of the set of weekdays because not all the days of the week are weekdays.

DEFINITION   **Null set** or **empty set:** The set which contains no elements. The symbols $\emptyset$ or { } are used to represent the empty set.

Note   The null set is a subset of every set.

EXAMPLE 5 $\emptyset \subseteq$ {Sunday, Monday, Tuesday}

EXAMPLE 6 $\emptyset \subseteq$ {a, e, i, o, u}

EXAMPLE 7   Let S = {a, b, c}
The subsets of S are:

$$\emptyset, \{a\}, \{b\}, \{c\}, \{a, b\}, \{a, c\}, \{b, c\}, \{a, b, c\}$$

There are 8 subsets.

Note   The original set has 3 elements. To determine the number of subsets, use 2 as a factor 3 times.

$$2 \times 2 \times 2 = 8$$

EXAMPLE 8   Let T = {w, x, y, z}
The set T has 16 subsets. ($2 \times 2 \times 2 \times 2 = 16$)

$$\emptyset, \{w\}, \{x\}, \{y\}, \{z\}, \{w, x\}, \{w, y\}, \{w, z\}, \{x, y\},$$
$$\{x, z\}, \{y, z\}, \{w, x, y\}, \{w, x, z\}, \{w, y, z\}, \{x, y, z\},$$
$$\{w, x, y, z\}$$

1.3   Operations with Sets

OBJECTIVES

- Form the intersection of two or more sets.
- Form the union of two or more sets.
- Use Venn diagrams to represent subsets.
- Use Venn diagrams to represent the union of sets.
- Use Venn diagrams for the intersection of sets.

Just as arithmetic operations, such as addition and multiplication, can be performed on numbers, so too the operations of **union** and **intersection** can be performed on sets.

DEFINITION   **Union of two sets:** A set whose elements belong to one or the other or both of the sets. The symbol ∪ is used to represent the union of two sets. (Think *U* for Union—*uniting;* i.e., joining together.)

EXAMPLE 1   Let A = {Sept., Oct., Nov.} B = {Oct., Nov., Dec.}

A ∪ B = {Sept., Oct., Nov., Dec.}

Notes   • Name each element only once.
  • The union of three or more sets is formed the same way.

DEFINITION   **Intersection of two sets:** A set whose elements belong to both the sets. The symbol ∩ is used to indicate the intersection of two sets.

Note   In everyday English, we use this concept when we refer to the place where two streets cross as the *intersection* of the two streets.

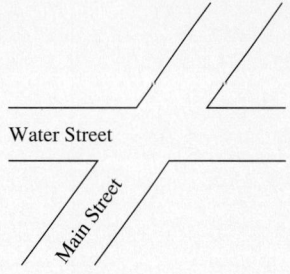

EXAMPLE 2   Using sets A and B from Example 1, A ∩ B = {Oct., Nov.}

DEFINITION   **Disjoint sets:** Sets that do not intersect. In other words, sets whose intersection is the empty set.

EXAMPLE 3   {a, e, i} ∩ {Oct., Nov.} = ∅,

so {a, e, i} and {Oct., Nov.} are disjoint sets.

Venn Diagrams

Venn diagrams provide convenient ways to visualize the operations and relations among sets. In these diagrams, a set is represented by a circle. The name of the set is written either inside or outside the circle. The circle is usually enclosed in a rectangle. The rectangle represents a **universal set.** The set represented by the circle is a subset of the universal set.

EXAMPLE 1   Use a Venn diagram to represent set A, where

A = {all students taking prealgebra}.

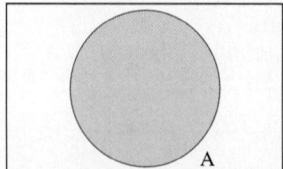

Note   The universal set in this example could be "all students."

EXAMPLE 2   Use a Venn diagram to represent set B, where

B = {all students taking English}.

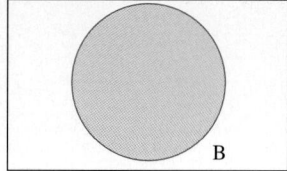

Note   Shading is often used for highlighting purposes.

EXAMPLE 3   Using sets A and B from Examples 1 and 2, draw a Venn diagram to represent A ∩ B. Shade the region.

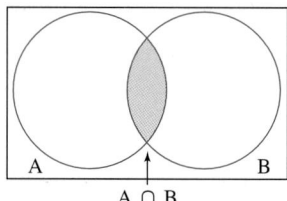

A ∩ B = {all students taking both prealgebra and English}.

Note   The circles overlap. The overlapping region represents A ∩ B.

EXAMPLE 4   Using sets A and B from Examples 1 and 2, draw a Venn diagram to represent A ∪ B. Shade the region.

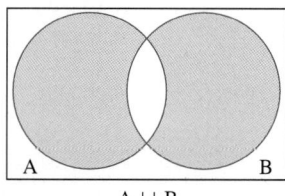

A ∪ B = {all students taking prealgebra or English}

EXAMPLE 5   Draw Venn diagrams to represent L ∩ S and L ∪ S, where

L = {1, 2, 3} and S = {3, 4, 5}.

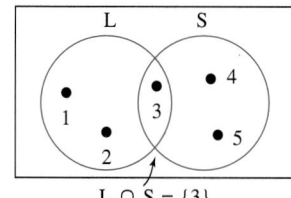
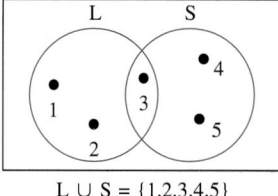

Note   When counting elements in L ∪ S, be careful not to count the number 3 twice.

EXAMPLE 6   Let F = {all women taking prealgebra}
M = {all men taking prealgebra}

Draw a Venn diagram to represent F ∩ M.

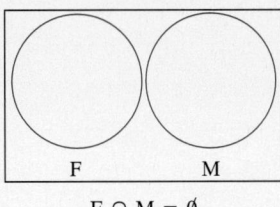

$$F \cap M = \emptyset$$

Note   These circles do not overlap. The sets are disjoint.

Now let's use Venn diagrams to represent the results of the operations of union and intersection on three sets.

EXAMPLE 7   Draw a Venn diagram to represent the intersections and unions among sets A, B, and C,

where A = {all students taking prealgebra}
B = {all students taking English}
C = {all students taking psychology}

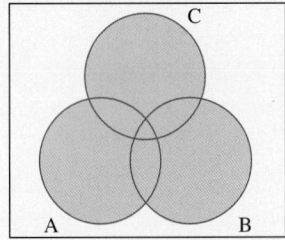

A ∪ B ∪ C = {all students taking prealgebra or English or psychology}

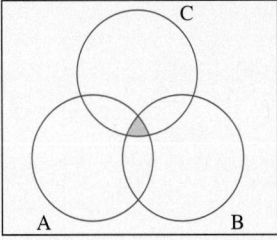

A ∩ B ∩ C = {all students taking prealgebra and English and psychology}

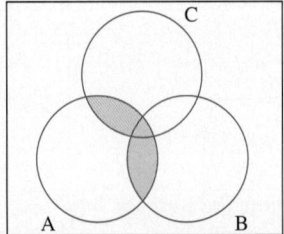

A ∩ (B ∪ C) = {all students taking English or psychology who are also taking prealgebra}

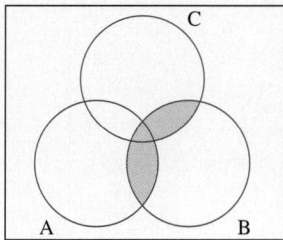

B ∩ (A ∪ C) = {all students taking prealgebra or psychology who are also taking English}

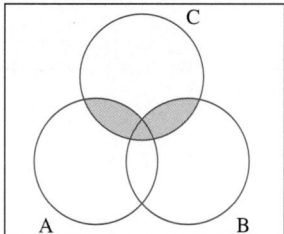

C ∩ (A ∪ B) = {all students taking prealgebra or English who are also taking psychology}

1.4 The Numbers as Sets

OBJECTIVE

● Use set theory when working with numbers.

Set theory was developed to help us organize the numbers. In this text, we will work with the set **R** of **real numbers.** These are the numbers that can be placed in **one-to-one correspondence** with the points on a line. We call it a number line. By a **one-to-one correspondence** between the real numbers and the points on a number line, we mean that for each real number we name, we can locate a unique point to pair with it, and for each point we locate, we can name a unique real number to pair with it. We use a diagram to represent the number line.

0

In Chapter 2, we will work with two subsets of the real numbers, namely, the natural numbers and the whole numbers. In Chapters 4 and 5, we will work with two more subsets of the real numbers, namely the integers and the rational numbers.

DEFINITION **Natural numbers** or **counting numbers:** The numbers we use for counting.

- In set notation: N = {1, 2, 3, . . .}
- Represented as a Venn diagram:

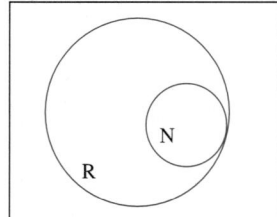

- On a number line:

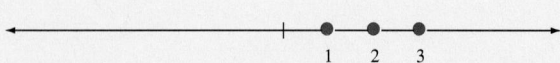

Notes
- The set of natural numbers is an infinite set because it has an unlimited number of elements.
- Ellipses (. . .) are used to denote the missing elements.
- On the number line, filled-in circles indicate the set N.

DEFINITION **Whole numbers:** The set of natural numbers and zero.

- In set notation: W = {0, 1, 2, 3, . . .}
- Represented by a Venn diagram:

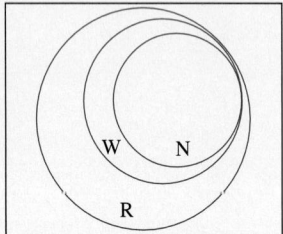

W ⊆ R and N ⊆ W

- On a number line:

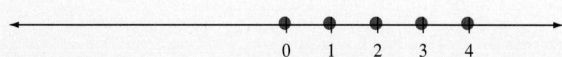

DEFINITION **Integers:** The set containing the natural numbers, zero, and the negatives of the natural numbers.

- In set notation: J = {. . . , −3, −2, −1, 0, 1, 2, 3, . . .}
- Represented by a Venn diagram:

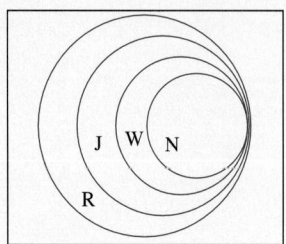

J ⊆ R, W ⊆ J, and N ⊆ J

- On a number line:

DEFINITION **Rational numbers:** The subset of real numbers whose elements can be expressed as fractions or ratios. Each member of the set can be written as one integer divided by another.

Every rational number can be written in the form $\frac{a}{b}$, where a and b are integers and b is not equal to zero.

EXAMPLE 1 The number $\frac{3}{5}$ is a rational number because 3 and 5 are integers.

EXAMPLE 2 The number 33 is a rational number because 33 could be written as $\frac{33}{1}$.

Notes
- The letter Q is often used to name the set of rational numbers. (Think of Q for quotient.)

- In set notation: $\frac{a}{b} \in Q$, if $a, b \in J$ and $b \neq 0$.

- Represented by a Venn diagram:

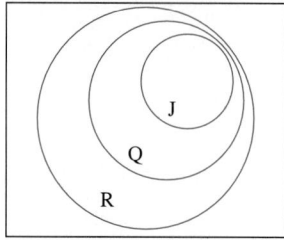

$Q \subseteq R$ and $J \subseteq Q$

DEFINITION **Irrational numbers:** Those real numbers that are not rational, i.e., cannot be expressed as the indicated quotient of two integers. There is no general agreement among mathematicians on reserving a specific letter to name the irrational numbers. In this text, therefore, we will use the letter *T*.

EXAMPLE 3 Recall the formula for finding the circumference of a circle:

$$C = d\pi$$

Circumference equals diameter times π. The number π is an irrational number. It cannot be expressed *exactly* as the quotient of two integers.

Notes
- $T \subseteq R$
- $Q \cup T = R$
- $Q \cap T = \emptyset$

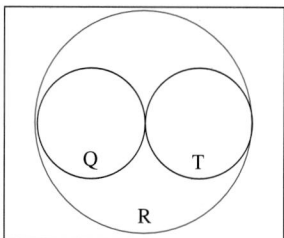

PROBLEM SET 1.1–1.4

Using sets: A = {0, 1, 2}, B = {1, 2, 3, 4}, and C = {4, 5}, follow the directions for each item.

1. List the subsets of A.

2. Form A $\cup$ B.

3. Form A $\cap$ B $\cap$ C.

4. Form A $\cup$ (B $\cap$ C).

5. List the subsets of B.   **6.** Form A ∩ B.

7. Form A ∪ B ∪ C.   **8.** Form A ∩ (B ∪ C).

Using sets E, F, and G, shade the indicated regions on the Venn diagrams.

9. E ∪ F

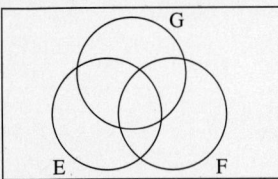

10. E ∩ F ∩ G

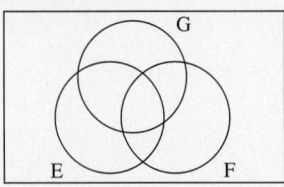

11. E ∩ F

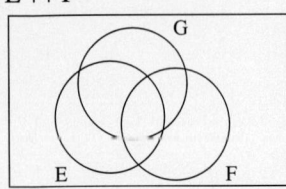

12. E ∪ F ∪ G

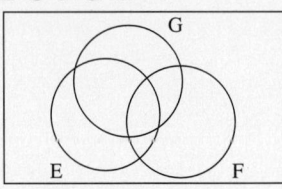

Short Answers: Answer the question asked or follow the directions given for each item.

13. Is the set of even numbers, i.e., {2, 4, 6, . . .}, finite or infinite? Why?

14. Is the set of digits, i.e., {0, 1, 2, 3, 4, 5, 6, 7, 8, 9}, finite or infinite? Why?

15. Use set notation to form the intersection of the natural numbers and the whole numbers.

16. Use set notation to form the union of the natural numbers and the whole numbers.

Use the following sets to determine whether the statements are true or false. Use appropriate set notation to change each false statement into a true one.

$$K = \{1, 2, 3\}, L = \{1, 2, 3, 4, 5\}$$
$$M = \{1, 3\}, \quad S = \{5, 4, 3, 2, 1\}$$
$$P = \{4\}$$

17. K ⊆ L _____

18. L ⊆ K _____

19. $M \subseteq K$ _____

20. $M \subseteq S$ _____

21. $L \subseteq S$ _____

22. $S \subseteq M$ _____

23. $M \not\subseteq S$ _____

24. $P \subseteq M$ _____

25. $2 \in K$ _____

26. $\emptyset \subseteq P$ _____

27. $S = L$ _____

SMALL GROUP ACTIVITY *Problem Solving with Venn Diagrams*

This semester all 200 first year college students at County Community College are enrolled in at least one of these three courses: Engl 101, Math 202, and Psych 303. One hundred fifty are enrolled in Engl 101, 125 in Math 202, and 83 in Psych 303. Five students are taking all three courses. One hundred five students are taking both Math 202 and Engl 101, and 33 students are taking both Engl 101 and Psych 303.

Use Venn diagrams to answer the following questions:

1. How many students are taking both Math 202 and Psych 303?
2. How many students are taking Engl 101 and Math 202, but not Psych 303?
3. How many students are taking Engl 101 and Psych 303, but not Math 202?
4. How many students are taking Math 202 and Psych 303, but not Engl 101?
5. How many students are taking only Engl 101?
6. How many students are taking only Math 202?
7. How many students are taking only Psych 303?

CHAPTER 1 REVIEW

Define the following terms.

1. Set _____

2. Member or element _____

3. Null or empty set _____

4. Finite set _____

5. Subset _____

6. Infinite set _____

7. Union of sets _____

8. Intersection of sets _____

9. Equal sets _____

10. Disjoint sets _____

11. Venn diagrams _____

12. Natural or counting numbers _____

13. Whole numbers _____

Express in words and use each symbol in an example.

14. $\in$ _____

15. $\notin$ _____

16. $\emptyset$ _____

17. $\cup$ _____

18. $\cap$ _____

19. $\subseteq$ _____

20. $\not\subset$ _____

21. { } _____

22. = _____

CHAPTER 1 PRACTICE TEST

List the elements in the following sets.

1. The set of natural numbers.

_____

2. The set of whole numbers.

_____

3. The set of the letters of the alphabet.

_____

Given the sets F = {1, 2, 3}, G = {3, 4, 5}, H = {1, 4}, form the following sets.

4. F ∩ G _____

5. F ∪ H _____

6. F ∪ (G ∩ H) _____

Using F, G, and H from Exercises 4 through 6, shade the indicated regions on the Venn diagrams.

7. F ∩ G

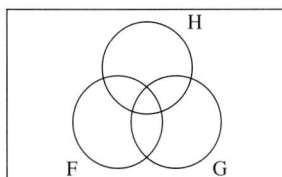

8. F ∪ H

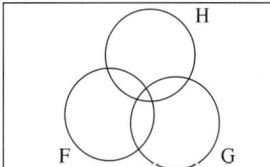

9. F ∪ (G ∩ H)

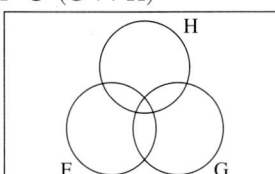

Using sets F, G, and H, label the following statements true or false. Then, use appropriate set notation to convert any false statements into true statements.

10. ∅ ⊆ F _____

11. G ⊄ F _____

12. H ⊆ H _____

13. 3 ∈ G _____

14. 2 ∈ H _____

15. 4 ∉ F _____

16. *Given: Z =* {*a, b, c, d*}, *list all the subsets of Z.*

_____

_____

_____

Consider the set of letters of the alphabet and the set of whole numbers.

17. *Classify each of them as finite or infinite.*

_____

_____

_____

_____

18. *State the reasons for your answers to 17.*

_____

_____

_____

_____

_____

_____

STUDY SKILLS TIP

Note Taking

As you prepared for the test on set theory, you probably found yourself referring to your class notes. Taking good notes during every class is essential. You'll want to organize them so you can find any topic you want to review. To do this, keep notes for your mathematics course in a separate notebook. Date the notes, and indicate the section being covered. Write on every other line, and leave wide margins for additions and corrections. All important information written on the board or shown on an overhead projector should be written in your notebook. Write any comments to the side of the notes. These will help you make more sense out of the notes later. Write down the information even if you do not completely understand what is being said, but put a large question mark in the margin so you can go back and make more sense out of the notes later. If the instructor uses vocabulary you do not understand, write down as much as you can and put a question mark in the margin. In a separate assignment notebook, be sure to record each assignment, the date it was given, and the date it is due.

Read and revise your notes as soon as possible after each class. Read them over to be sure you understand what you have written. Use the textbook to figure out what you had marked with a question mark. If you cannot find an explanation in the text, get help from your instructor.

If you are absent from a class, arrange to get the notes from a fellow student. Choose carefully when asking someone for his or her notes. Be sure to select a student who takes good notes. Make these arrangements as soon as you can, even before the absence occurs if possible. Pay to have the notes copied! You will definitely need them.

THE MATHEMATICS JOURNAL

Goals

Task: As you finish the first chapter and prepare to begin the second, take a few minutes to reflect on your goals for this course. Write a one-page (150-word) entry defining your goals.

Starters: What do you want to get out of this course? Discuss your plan for accomplishing these goals. What activities will you engage in? How will you know if you are accomplishing your goals?

2 *Whole Numbers*

2.1 Place Value and Expanded Notation
2.2 Rounding Whole Numbers
2.3 Addition and Subtraction of Whole Numbers
2.4 Multiplication and Division of Whole Numbers
2.5 Exponents and Roots
2.6 Order of Operations
2.7 Averages and Estimation

Introduction to Whole Numbers

The set of whole numbers is the union of the counting or natural numbers and the number zero. We know that the counting numbers date back to the Stone Age and had their origin in the piles of stones or sticks that the shepherd moved one-by-one each morning and evening to keep track of the sheep. Zero is of later origin, but we know it was in existence by the time the Egyptians developed geometry. The base ten place value system probably came about because we have ten fingers and ten toes. It's quite a leap from using a pile of stones to arriving at the abstract concept of number. Historically, we don't know how or when it happened, but we do know that our ability to form the concept of number sets us apart from other forms of life.

2.1 Place Value and Expanded Notation

OBJECTIVES

- Convert numbers from standard form to expanded form.
- Convert numbers from expanded form to standard form.

In our Hindu-Arabic number system, the number 349 is different from the number 493. This is because of place value—the value of a number depends on the position of each of its digits.

DEFINITION **Digits** = {0, 1, 2, 3, 4, 5, 6, 7, 8, 9}
All numbers in our Hindu-Arabic system are made up of these ten digits. Because there are ten of them, it is called a **base 10 system.**

DEFINITION **Natural numbers** or **counting numbers:** N = {1, 2, 3, . . .}

DEFINITION **Whole Numbers:** W = {0, 1, 2, 3, . . .}

Place Value

DEFINITION **Place value:** This term refers to the position of each digit of a number. (The position or place of a digit determines its value.) The system reads from right to left.

16

EXAMPLE 1   The number 123 means

Hundreds	Tens	Units
1	2	3

EXAMPLE 2   280 means

Hundreds	Tens	Units
2	8	0

Expanded Notation

Writing numbers in expanded notation is a good way to show the place value of each digit and to emphasize the advantage of the base 10 system.

DEFINITION   **Standard form or standard notation:** A number written in the base 10 place value system is said to be written in standard form.

EXAMPLE 3   The numbers 5, 76, and 103 are examples of numbers written in standard form.

DEFINITION   **Expanded form or expanded notation:** A number written as the sum of the products of its digits and the place values they hold.

EXAMPLE 4 $12 = 10 + 2$ or 1 ten and 2 units. In expanded notation, we write

$$1 \times 10 + 2 \times 1.$$

EXAMPLE 5   345 in expanded notation is written

$$3 \times 100 + 4 \times 10 + 5 \times 1.$$

EXAMPLE 6   Written in expanded notation,

$$2{,}520 = 2 \times 1{,}000 + 5 \times 100 + 2 \times 10 + 0 \times 1$$

EXAMPLE 7   Written in expanded notation,

$$600{,}034 = 6 \times 100{,}000 + 0 \times 10{,}000 + 0 \times 1{,}000 + 0 \times 100 + 3 \times 10 + 4 \times 1$$

EXAMPLE 8   Written in expanded notation,

$$739{,}469 = 7 \times 100{,}000 + 3 \times 10{,}000 + 9 \times 1{,}000 + 4 \times 100 + 6 \times 10 + 9 \times 1$$

Occasionally, in this book we will use a type of expanded notation we will call **modified expanded notation.**

EXAMPLE 9   In modified expanded notation,

$$245 = 200 + 40 + 5$$

Practice Problems

Write in standard notation.
1. $5 \times 100 + 0 \times 10 + 6 \times 1$
2. $7000 + 800 + 30 + 2$

Write in expanded notation.
3. 7432
4. 701

Write in modified expanded notation.
5. 639
6. 24,320

PROBLEM SET 2.1

Write in expanded form.

1. 23

2. 45

3. 789

4. 567

5. 1234

6. 9876

7. 10,004

8. 40,001

9. 478,291

10. 597,354

Write in standard notation.

11. $2 \times 10 + 1 \times 1$

12. $8 \times 10 + 3 \times 1$

13. $5 \times 100 + 6 \times 10 + 7 \times 1$

14. $2 \times 100 + 9 \times 10 + 8 \times 1$

15. $1 \times 10,000 + 0 \times 1000 + 1 \times 100 + 0 \times 10 + 1 \times 1$

16. $2 \times 10,000 + 0 \times 1000 + 0 \times 100 + 2 \times 10 + 0 \times 1$

Write in modified expanded form.

17. 101

18. 303

19. 457

20. 754

21. 4020

22. 8070

23. 10,000

24. 200,000

2.2 **Rounding Whole Numbers**

OBJECTIVE

● Round whole numbers.

Rounding a number is a way to approximate its value. There are several methods for rounding. Social scientists may use one method; physical scientists may use another. The method used will determine the resulting approximation. It is important to remember that not all approximations of a number will be equal. In this course, we will use the following method:

RULE **To round a number to a given place value:**

• Look at the digit to the immediate right of that place.
• If the digit is less than 5, that is, if it is a 0, 1, 2, 3, or 4, replace it with a 0 and replace all digits to the right of it with 0s.
• If the digit is 5 or greater, that is, if it is a 5, 6, 7, 8, or 9, add 1 to the digit to the left of it, then replace it with a 0, and finally, replace all digits to the right of it with 0s.

EXAMPLE 1 Round 24 to the nearest ten.

Tens	Units
2	4

⇑

The given position is tens. The digit to the right is 4, which is less than 5, so replace the 4 with a 0 and rewrite the number. 24 rounded to tens is 20.

On the number line, 24 lies between 20 and 30, but it is closer to 20. That's why 24 rounded to the nearest ten is 20.

DEFINITION   ≈ is the symbol for approximately equal to. So 24 ≈ 20.

Consider the number 482,153.

EXAMPLE 2   Round **137** to tens.

> **137 ≈ 140**
> The digit 3 is in the tens position. The digit immediately to the right of it is 7, which is greater than 5, so replace the 7 with 0 and add 1 to the 3, which is the tens digit. The result is 140. On the number line, 137 lies between 130 and 140, but it is closer to 140.

1. Round to tens.

EXAMPLE 3   Round 298 to tens.

> The digit 9 is in the tens position. The digit immediately to the right of it is 8, which is greater than 5, so replace the 8 with 0 and add 1 to the 9, which is the tens digit. Be careful, 9 + 1 = 10, so write the 0 in the tens position and add 1 to 2, which is the digit in the hundreds position. The result is 300.

2. Round to hundreds.

3. Round to thousands.

EXAMPLE 4   Round 96 to the nearest ten.

> In 96 the ∧ indicates the position we want to round to—the tens position. The digit in this position is 9. The digit immediately to the right of it is 6, which is greater than 5, so replace that 6 with 0, and add 1 to the 9 in the tens position. Since 9 + 1 = 10, write a 0 in the tens position and a 1 in the hundreds position. The result is 100. **96 ≈ 100**

4. Round to ten thousands.

EXAMPLE 5   Round 997 to the nearest ten.

> In 997, the ∧ indicates the position we want to round to—the tens position. The digit immediately to the right of it is a 7. Since 7 is greater than 5, replace it with a 0 and add 1 to the tens digit. Since 9 + 1 = 10, write down the 0 and carry the 1 to the hundreds position. The hundreds digit is also 9 and 9 + 1 = 10. **997 ≈ 1000**

5. Round to hundred thousands.

EXAMPLE 6   Round 49 to the nearest hundred.

> Written to hundreds, 49 is 049. Look to the right of the hundreds digit, 0. The number immediately to the right of it is 4. Since 4 is less than 5, replace it with a 0 and replace 9, the tens digit, with a 0 also. The result is 000 or 0.

> **49 ≈ 0**

EXAMPLE 7   Round 51 to the nearest hundred.

> **51 ≈ 100**   Why?

EXAMPLE 8   Round 902 to the nearest hundred.

> **902 ≈ 900**

EXAMPLE 9   Round 902 to the nearest ten.

> **902 ≈ 900**

Notice that 902 rounded to tens is the same as 902 rounded to hundreds.

EXAMPLE 10   Round 718 to the nearest ten.

> **718 ≈ 720**

EXAMPLE 11 Round 718 to the nearest hundred.

$$718 \approx 700$$

Notice that 718 rounded to tens is greater than 718 rounded to hundreds. Can you explain this?

APPLICATION 1   Jake's girlfriend sends him to the store to buy holiday napkins for a New Year's Eve party. She tells him she's expecting 78 people. He finds some nice napkins, but notices that they come in packs of 10. How many packs should he buy?   Round 78 to tens. $78 \approx 80$

Answer: Jake should buy 8 packs.   ◆

Note   Rounding, using the rules we've agreed to, isn't the only way to approximate, nor is it always the appropriate method. Consider this problem.

APPLICATION 2   For another party, his girlfriend sends Jake for napkins again, but this time she's expecting 82 guests. How many packs of 10 napkins each should he buy?

Answer: Jake should buy 9 packs. Rounding 82 to tens using the rules we've set up is not appropriate, since $82 \approx 80$ and that would leave two guests without napkins. In this case, we want to use 90 as the approximation for 82.   ◆

Problem Set 2.2

Round to the nearest ten.

1. 13

2. 74

3. 37

4. 48

5. 55

6. 65

Round to the nearest hundred.

7. 101

8. 202

9. 109

10. 208

11. 625

12. 465

13. 1999

14. 2999

Complete the table. Round each number to each place value listed.

	Number	*Ten thousands*	*Thousands*	*Hundreds*	*Tens*
15.	903	X			
16.	901	X			
17.	9084				
18.	9083				
19.	479	X	X		
20.	489	X	X		
21.	1005	X			
22.	3105	X			
23.	10,654				
24.	20,745				

Applications:

25. Jeff and Rick plan to go to Florida for spring break, a distance of 1153 miles. Estimate their trip to the nearest hundred miles.

26. Tammy and Brenda are planning a shopping trip to New York City over spring break, a distance of 521 miles. Estimate their trip to the nearest hundred mile.

27. If 1K of memory in a personal computer is equivalent to 1024 bytes, estimate the number of bytes to the nearest thousand.

28. Vanessa is doing her student teaching this semester. She wants to take a fruit treat to school for the 27 students in her class. She goes to the grocery store and decides to buy the oranges, which are in packages of 6. How many packages should she buy so that each of her students will have an orange?

29. Juan goes to the local landscape outlet to buy fertilizer for his lawn. He knows that he needs enough to cover 49,000 square feet. The label on a bag of the brand on sale states that one bag will cover 15,000 square feet. How many bags of that brand would he need to buy?

2.3   **Addition and Subtraction of Whole Numbers**

OBJECTIVES

- State the addition properties of real numbers.
- Solve application problems involving the addition and subtraction of whole numbers.

Addition

Recall from Chapter 1 that the set of whole numbers is a subset of the set of real numbers. Because of this relationship, all properties which are true for the real numbers also apply to the whole numbers.

DEFINITION **Addend:** Each of the numbers being added.
Sum: The result of performing the operation of addition.

Properties of the Real Numbers under Addition

Addition is a **binary** operation—we only add numbers two at a time. For all a, b, and $c \in R$, the following properties are true.

Associative Property of Addition: $a + (b + c) = (a + b) + c$

The Associative Property of Addition states that one real number plus the sum of a second and third is equal to the sum of the first and second added to the third.

EXAMPLE 1

$$2 + (3 + 4) = (2 + 3) + 4$$
$$2 + \quad 7 \quad = \quad 5 \quad + 4$$
$$9 \qquad\quad = \qquad 9$$

Identity Property of Addition: $a + 0 = a$ and $0 + a = a$

The Identity Property of Addition states that the sum of any real number and zero is the original real number and that the sum of zero and any real number is the real number.

EXAMPLE 2

$$5 + 0 = 5 \text{ and } 0 + 5 = 5$$

Commutative Property of Addition: $a + b = b + a$

The Commutative Property of Addition states that in adding two real numbers, either of them may be stated first.

EXAMPLE 3

$$8 + 9 = 9 + 8$$
$$17 = 17$$

APPLICATION   In preparing for his trip home at Thanksgiving, Joe fills up the gas tank in his car. When he stops for a soft drink, he fills the tank up again and notes that it takes 10 gallons. When he gets to the gas station at the corner of the street his parents live on, he fills it up again and notes that it takes 8 gallons. How many gallons of gas did Joe use to get home? $10 + 8 = 18$

Answer: Joe used 18 gallons of gas.   ◆

Subtraction

The operation subtraction takes its meaning from addition. The number 1 subtracted from 4 equals 3 because 3 plus 1 equals 4.

DEFINITION **Subtraction:** For all a, b, $c \in R$, $a - b = c$, if and only if $c + b = a$.

EXAMPLE 4

$$5 - 2 = 3 \text{ because } 3 + 2 = 5.$$

EXAMPLE 5 $6 - 4 \neq 4 - 6$ Subtraction is not commutative.

DEFINITION   **Difference:** The result of performing the operation of subtraction.

Note   When using the term *difference* the order in which the numbers are named matters.

EXAMPLE 6   Find the difference of 309 and 81.

$$309 - 81 = 228$$

APPLICATION   Jamie, a new freshman, has $531 left in her checking account after paying her tuition. She still needs to buy books which will cost $372. After buying her books, will she have enough money left to buy a $45 book pack she saw in the bookstore?

$$\begin{array}{r} \$531 \\ - \ \$372 \\ \hline \$159 \end{array}$$

Answer: Yes, she will have enough money left for the book pack.   ◆

PROBLEM SET 2.3

Write the name of the appropriate property of addition illustrated in each problem. Use the list at the right for reference.

1. $2 + 3 = 3 + 2$

2. $698 + 0 = 698$

3. $28 + (17 + 52) = (28 + 17) + 52$

4. $(9 + 10) + 11 = 11 + (9 + 10)$

5. $(15 + 0) + 8 = 15 + 8$

Associative Property of Addition

Commutative Property of Addition

Identity Property of Addition

Applications:

6. The first-string kicker on your football team scored 6 points in the first game this season, 7 points in the second, 11 points in the third, and 8 points in the fourth. How many points has he scored to date this season?

7. The leading rusher on your football team ran for 16 yards in the first game, 56 yards in the second, 80 yards in the third, and 110 yards in the fourth. How many yards is he credited with to date this season?

8. Sandy decides to redecorate her dorm room. She spends $52 for a new comforter, $29 for matching curtains, $25 for pillows, $16 for a new poster, and $9 for a frame. How much does it cost her to redecorate her room?

9. Kyle goes on a shopping trip to get some new clothes for the fall dance. He spends $145 for a sport coat, $35 for a pair of pants, $25 for a dress shirt, $48 for shoes, and $19 for a tie. How much did the clothes cost?

10. Catherine buys a new outfit for her installation as a student senator. She spends $85 for a business suit, $32 for shoes, $3 for stockings, $32 for a necklace, and $15 for matching earrings. How much did the outfit cost her?

11. The enrollment at State University was 21,729 last year. This year it is 20,982. How many fewer students are attending the university this year?

12. Sam's savings account has a balance of $1,532 the week before the semester begins. On the first day of classes, he withdraws $255 to pay for his books and supplies. If he makes no other withdrawals and no interest is credited, what is the balance in his account at the end of the first week of classes?

13. In last night's game of gin rummy, Hank scored 53 points during the first hand and lost 27 points on the second hand. After those two hands, what was his score?

2.4 Multiplication and Division of Whole Numbers

OBJECTIVES

- State the multiplication properties of real numbers.
- Apply the division algorithm.
- Classify a given whole number as prime or composite.
- Apply the divisibility rules for composite whole numbers.
- Determine the prime factors of composite whole numbers.
- Determine the greatest common factor of two or more whole numbers.

Multiplication

The operation of multiplication dates back to early Egyptian times. Its meaning then was the same as it is today. Multiplication is repeated addition. 3 multiplied by 2 means 3 + 3. Similarly, 5 + 5 + 5 + 5 can be thought of as 5 multiplied by 4. Multiplication is a binary operation, which means we multiply numbers two at a time.

DEFINITION **Factors:** The numbers being multiplied.

DEFINITION **Product:** The answer to a multiplication problem.

EXAMPLE 1 In $3 \times 2 = 6$, the numbers 3 and 2 are factors, and 6 is the product.

We indicate that two numbers are to be multiplied in several ways. We may place a multiplication symbol ($\times$ or $\cdot$) between them, or we may enclose one or both of them in parentheses.

SYMBOLS   For all $a, b \in$ R, the following symbols are all acceptable ways to indicate multiplication.

$$a \times b$$
$$a \cdot b$$
$$(a)(b)$$
$$a(b) \quad \text{or} \quad (a)b$$
$$a \cdot (b) \quad \text{or} \quad (a) \cdot b$$

Properties of the Real Numbers for Multiplication

For all a, b, and $c \in$ R, the following properties are true.

Associative Property of Multiplication: $a \cdot (b \cdot c) = (a \cdot b) \cdot c$

The Associative Property of Multiplication states that one real number multiplied by the product of a second and third is equal to the product of the first and second multiplied by the third.

EXAMPLE 2
$$2 \cdot (3 \cdot 4) = (2 \cdot 3) \cdot 4$$
$$2 \cdot \quad 12 \quad = \quad 6 \quad \cdot 4$$
$$24 \quad = \quad 24$$

Identity Property of Multiplication: $a \cdot 1 = a$ and $1 \cdot a = a$

The Identity Property of Multiplication states that the product of any real number and one is the original real number and that the product of one and any real number is the real number.

EXAMPLE 3 $5 \cdot 1 = 5$ and $1 \cdot 5 = 5$

Commutative Property of Multiplication: $a \cdot b = b \cdot a$

The Commutative Property of Addition states that in multiplying two real numbers, either of them may be stated first.

EXAMPLE 4
$$8 \cdot 9 = 9 \cdot 8$$
$$72 \quad = \quad 72$$

Distributive Property of Multiplication over Addition:
$a \cdot (b + c) = a \cdot b + a \cdot c$ and $(b + c) \cdot a = b \cdot a + c \cdot a$

The product of a real number and the sum of two other real numbers is equal to the sum of the products formed by multiplying the first real number by each of the other two real numbers.

EXAMPLE 5
$$2 \cdot (3 + 4) = 2 \cdot 3 + 2 \cdot 4$$
$$2 \quad \cdot \quad 7 \quad = \quad 6 \quad + \quad 8$$
$$14 \quad = \quad 14$$

Zero Product Property: $a \cdot 0 = 0$ and $0 \cdot a = 0$.

EXAMPLE 6 $3 \cdot 0 = 0$, since $0 + 0 + 0 = 0$. Remember, multiplication is a shortcut for addition.

EXAMPLE 7 $7 \cdot 0 = 0$

APPLICATION Emelio is collecting money for a soup kitchen sponsored by his fraternity. He is hoping that each freshman will contribute $2. If there are 2860 freshmen, how much money could he hope to collect?

$$2860 \times 2 = 5720$$

Answer: He can hope to raise $5720. ◆

Division

The operation of division as we know it today dates back to the Babylonians in the fourth century B.C.

DEFINITION **Division:** For all a, b, c, $\in$ R, $b \neq 0$, $c \div b = a$, if and only if $a \cdot b = c$.

In words, **c** divided by **b** equals **a,** if and only if **a** multiplied by **b** equals **c**. Since the whole numbers are a subset of the set R of real numbers, this definiton holds for the set of whole numbers also.

EXAMPLE 8 $42 \div 7 = 6$ because $6 \cdot 7 = 42$

SYMBOLS For all b, $c \in$ R, where $b \neq 0$, the following are all ways of indicating c divided by b.

$$c \div b$$

$$\frac{c}{b}$$

$$c/b$$

EXAMPLE 9 $6 \div 2 = 3, \dfrac{6}{2} = 3$, and $6/2 = 3$

DEFINITION **Dividend:** The number being divided.
In the expression $c \div b$, c is the dividend. In the previous example, 6 was the dividend.

Divisor: The number doing the dividing.
In the expression $c \div b$, b is the divisor. In the previous example, 2 was the divisor.

Quotient: The result of the division process.
In the expression $c \div b = a$, a is the quotient. In the previous example, 3 is the quotient.

Note Although the slash mark / is the symbol used for division by computer scientists, it can cause confusion especially if the divisor is an indicated product of two or more numbers. Avoid using it if possible.

Division by Zero

Division of nonzero numbers by zero is undefined. Why? Let's look at an example, say $5 \div 0$. What do you think the answer should be? Did you say **5**? Does it fit the *definition* of division? If $5 \div 0 = $ **5**, then **5** $\times$ 0 should equal 5, but we know that **5** $\times$ 0 = 0 and 0 $\neq$ 5. Try guessing other numbers as the quotient. You will never find one that works. **Division of nonzero numbers by 0 is undefined.**

The Division Algorithm

What is 7 $\div$ 3? 7 is *not evenly* divisible by 3. The quotient is 2, but there is also a remainder of 1. In notation, $7 = 2 \cdot 3 + 1$. This example reminds us that some whole numbers are *not evenly* divisible by other whole numbers, some have **remainders.** To carry out divisions which result in remainders, we will use the following property:

> **The Division Algorithm**
> Quotient times divisor plus remainder equals dividend. In notation, for all a, b, q, $r \in$ R, $b \neq 0$, if $q \times b + r = a$, then q is the quotient of $a \div b$ and r is the remainder.

EXAMPLE 10 $17 \div 3 = 5$ remainder 2 because $5 \times 3 + 2 = 17$

APPLICATION Sharon gets $50 from her aunt for her sixteenth birthday. She decides to buy as many CDs as possible and put the rest in the bank for college expenses. If CDs cost $12 each, how many can she buy? How much money will she have left to put in the bank?

$$50 \div 12 = 4 \text{ remainder } 2$$

Answer: She can buy 4 CDs and can bank $2. ◆

Prime Numbers and Prime Factorization

Certain numbers, because of their unique properties, have fascinated mathematicians since early times. Among them are prime numbers, whose history we can trace back to the ancient Greeks.

DEFINITION **Prime Number:** A whole number greater than 1 that has exactly two factors namely, itself and l.

EXAMPLE 11 2 has only itself and 1 as factors. It is prime.
   5 has only itself and 1 as factors. It is prime.

DEFINITION **Composite Number:** A whole number greater than 1 that is not prime. It has other factors in addition to itself and 1.

Note Composite numbers can always be factored in more than one way.

EXAMPLE 12 4 is composite because $4 = 4 \cdot 1$ and $4 = 2 \cdot 2 \cdot 1$.
15 is composite because $15 = 15 \cdot 1$ and $15 = 3 \cdot 5 \cdot 1$.

DEFINITION **Divisible:** Evenly divisible, i.e., no remainder.

> **RULES** Tests for divisibility
>
> A number is divisible by 2 if it is even, i.e., if its last digit is a 0, 2, 4, 6, or 8.
> A number is divisible by 3 if the sum of its digits is divisible by 3.
> A number is divisible by 5 if its last digit is a 5 or a 0.

EXAMPLE 13 Is 132 divisible by 3?

$1 + 3 + 2 = 6$, and 6 is divisible by 3, so 132 is divisible by 3. $132 \div 3 = 44$

EXAMPLE 14 Is 720 divisible by 2?

Yes, it ends with 0 and is, therefore, an even number.

EXAMPLE 15 Is 720 divisible by 3?

Yes, $7 + 2 + 0 = 9$ and 9 is divisible by 3; therefore, 720 is divisible by 3. Since 720 is divisible by 2 and by 3, it must also be divisible by $2 \cdot 3$, that is, it must be divisible by 6.

USING THE CALCULATOR

Check to see if 720 is divisible by 2, 3, and 6.

Enter: 720 ÷ 2 = **360** ← Display

so 720 is divisible by 2.

Enter: 720 ÷ 3 = **240** ← Display

so 720 is divisible by 3.

Enter: 720 ÷ 6 = **120** ← Display

so 720 is divisible by 6.

DEFINITION **Prime Factorization:** The process of expressing a number as a product of prime numbers.

There is only one set of prime factors for any number. When multiplied together, the product will be the original number. In order to find the prime factorization of a number, it is helpful to have a system to use. Generally, we start by trying to divide the number by the smallest prime, 2, then try 3, 5, 7, and continue dividing until the quotient equals 1.

EXAMPLE 16 Find the set of prime factors of 24.

$2\lfloor\underline{24}$ Divide by 2, if the number is even.
  12

$2\lfloor\underline{12}$ Keep dividing by 2 until the quotient is no longer an even
  6   number.

$2\lfloor\underline{\ 6}$ Now try 3, the next prime.
  3

$3\lfloor\underline{\ 3}$ Keep dividing by primes until the quotient equals 1.
  1

The divisors are the prime factors.

$$24 = 2 \cdot 2 \cdot 2 \cdot 3$$

USING THE CALCULATOR

Check the prime factorization of 24. Does $2 \cdot 2 \cdot 2 \cdot 3 = 24$?

Enter: 2 × 2 × 2 × 3 = **24** ← Display

The factors are all prime numbers and their product is 24, so $2 \cdot 2 \cdot 2 \cdot 3$ is the prime factorization of 24.

Practice Problems

Circle the prime numbers.

1. 9, 36, 41, 3, 99, 11, 1

Circle the composite numbers.

2. 2, 27, 19, 88, 45, 47, 1

Circle the numbers divisible by 2.

3. 44, 247, 742, 98, 12, 8, 41

Circle the numbers divisible by 5.

4. 505, 57, 670, 95, 40

Circle the numbers divisible by 3.

5. 57, 45, 73, 93, 79, 126, 86

6. What are the prime factors of 100, 48, 35, 255, 97?

Another method used to prime factor is a **factor tree.**

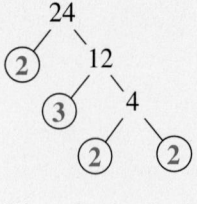

$$24 = 2 \cdot 2 \cdot 2 \cdot 3$$

Note   To find prime factors for any number, always start to divide by 2, the smallest prime, then try 3, 5, 7, and 11. Sometimes it is necessary to try larger primes, 13, 17, and 19, but be sure to try them in order from smallest to largest.

EXAMPLE 17   Find the prime factors of 36.

$$
\begin{array}{r|l}
2 & 36 \\ \hline
2 & 18 \\ \hline
3 & 9 \\ \hline
3 & 3 \\ \hline
  & 1
\end{array}
$$

Written as a product of primes, $36 = 2 \cdot 2 \cdot 3 \cdot 3$.

USING THE CALCULATOR

Check the prime factorization of 36.

Enter: 2 × 2 × 3 × 3 = **36** ← Display

The factors are all prime numbers and their product is 36, therefore, the factorization is correct.

EXAMPLE 18   Find the prime factors of 90

$$
\begin{array}{r|l}
2 & 90 \\ \hline
3 & 45 \\ \hline
3 & 15 \\ \hline
3 & 5 \\ \hline
5 & 1
\end{array}
$$

$$90 = 2 \cdot 3 \cdot 3 \cdot 5$$

EXAMPLE 19   Find the prime factors of 273.

$2 + 7 + 3 = 12$ and 12 is a multiple of 3, so 3 is a prime factor of 273.

$$
\begin{array}{r|l}
3 & 273 \\ \hline
7 & 91 \\ \hline
13 & 13 \\ \hline
  & 1
\end{array}
$$

Written as a product of primes, $273 = 3 \cdot 7 \cdot 13$.

DEFINITION **The Greatest Common Factor (GCF)** of two numbers is the largest number that divides into both numbers.

EXAMPLE 20 Find the greatest common factor of 18 and 30.

$18 = 2 \cdot 3 \cdot 3$	1. Write each number as a product of
$30 = 2 \cdot 3 \cdot 5$	prime factors.
$18 = ②\cdot③\cdot 3$	2. Circle each factor that appears in
$30 = ②\cdot③\cdot 5$	both factorizations.
GCF $(18, 30) = 2 \cdot 3$	3. The product of these factors is the
GCF $(18, 30) = 6$	greatest common factor.

Note In mathematical notation, the statement "the greatest common factor of 18 and 30 is 6" is written GCF $(18, 30) = 6$.

EXAMPLE 21 Find the GCF of 8 and 16.
Find the prime factors of both numbers.

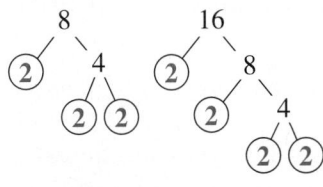

$$8 = ② \cdot ② \cdot ②$$

$$16 = ② \cdot ② \cdot ② \cdot 2$$

$$\text{GCF } (8, 16) = 2 \cdot 2 \cdot 2 = 8$$

EXAMPLE 22 Find the GCF of 21, 49, and 63.
Find the prime factors of the numbers.

 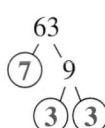

$$21 = 3 \cdot ⑦$$

$$49 = 7 \cdot ⑦$$

$$63 = 3 \cdot 3 \cdot ⑦$$

$$\text{GCF } (21, 49, 63) = 7$$

EXAMPLE 23 GCF $(15, 16) = ?$
$15 = 3 \cdot 5$
$16 = 2 \cdot 2 \cdot 2 \cdot 2$
15 and 16 have no prime factors in common, but
$15 = 15 \cdot 1$ and
$16 = 16 \cdot 1$ so
GCF $(15, 16) = 1$

Note Two or more numbers whose greatest common factor is 1 are called **relatively prime.** In the previous example, 15 and 16 are relatively prime.

ANSWERS TO PRACTICE PROBLEMS
1. Primes: 41, 3, 11
2. Composites: 27, 88, 45
3. Divisible by 2: 44, 742, 98, 12, 8
4. Divisible by 5: 505, 670, 95, 40
5. Divisible by 3: 57, 45, 93, 126
6. $100 = 2 \cdot 2 \cdot 5 \cdot 5$,
$48 = 2 \cdot 2 \cdot 2 \cdot 2 \cdot 3$, $35 = 5 \cdot 7$,
$255 = 3 \cdot 5 \cdot 17$, 97 is prime.

PROBLEM SET 2.4

Simplify. Use the Distributive Property.

1. 4(7 + 2)

2. 5(6 + 3)

3. (20 + 6)3

4. (5 + 10)4

Divide. Check by multiplying.

5. 10 ÷ 5   **6.** 12 ÷ 6   **7.** 15 ÷ 5   **8.** 18 ÷ 9

9. 0 ÷ 10   **10.** 0 ÷ 8   **11.** 22 ÷ 0   **12.** 36 ÷ 0

Write the name of the appropriate property of multiplication illustrated in each problem. Use the list at the right for reference.

13. 15 × 31 = 31 × 15

_____

Associative Property of Multiplication

Commutative Property of Multiplication

14. 56 × 1 = 56

_____

Identity Property of Multiplication

15. 28 × (17 × 52) = (28 × 17) × 52

_____

16. (21 × 8) × 16 = (8 × 21) × 16

_____

17. 216 = 216 × 1

_____

Circle the prime numbers. Prime factor the composite numbers.

18. 57   **19.** 37   **20.** 39   **21.** 91   **22.** 123

23. 41   **24.** 97   **25.** 19   **26.** 267   **27.** 65

28. 89   **29.** 30   **30.** 59   **31.** 51   **32.** 10

Factor the following numbers into a product of prime factors.

33. 50

34. 99

35. 200

36. 6

37. 12

38. 20

39. 144

40. 547

41. 310

42. 888

43. 424

44. 900

45. 122

46. 133

Find the Greatest Common Factor of the following sets of numbers.

47. GCF (9, 16) =

48. GCF (16, 18) =

49. GCF (24, 36) =

50. GCF (25, 35) =

51. GCF (72, 45) =

52. GCF (66, 132) =

53. GCF (78, 52) =

54. GCF (16, 36, 60) =

55. GCF (42, 98, 70) =

Applications:

56. The city decides to charge each of its 153,000 households $6 for a mandatory recycling bin. How much revenue should this action generate?

57. The record shop where Shellie works sold 150 CDs on Christmas Eve at $12 each. How much money did these CD sales generate?

58. Ricardo's Riviera gets an average of 18 miles per gallon of gasoline. He used 16 gallons of gas on his last trip. Approximately how many miles did he travel?

59. Eileen learns that she can save $5 per month on her electric bill by having a monitor installed. How much would she save in one year (12 months) if she has it installed?

60. Tim's grandfather leaves an estate worth $4,800,000. If the estate is to be divided equally among Tim, his sister, Lynn, and his brother, Lee, how much money will each receive?

61. Matt and his buddies Mark and Jose drove to Florida, a distance of 1150 miles, for spring break. If the trip took them 23 hours, what was the average number of miles they traveled per hour?

2.5 Exponents and Roots

OBJECTIVES
- Evaluate whole numbers raised to powers.
- Find the positive roots of selected whole numbers.

Exponents

Exponents are used to indicate repeated multiplication. Their modern day use is attributed to the seventeenth century French mathematician, René Descartes.

DEFINITION   In the expression a^n where a is a natural number and n is a whole number, a is called the **base** and n is called the **exponent.** The expression is read, "a raised to the nth power."

EXAMPLE 1 3^2 is read, "three to the second power."

DEFINITION $a^0 = 1$, for $a \neq 0$. Note: This is a special case. Learn it well!

EXAMPLE 2   **a.** $2^0 = 1$
b. $500^0 = 1$
c. $17^0 = 1$

DEFINITION a^1 means a used as a factor **1** time. The result is a.

EXAMPLE 3 $7^1 = 7$

DEFINITION a^2 means a used as a factor **2** times, or $a \cdot a$.

EXAMPLE 4   Evaluate: 3^2
$3^2 = 3 \cdot 3 = 9$

DEFINITION   In general, a^n means a used as a factor n times, or
$a \cdot a \cdot a \cdot a \cdot \cdots \cdot a$
⇑
There are n a's.

Practice Problems

Evaluate.

1. 3^0

2. 5^1

3. 8^2

4. 10^3

5. 2^4

Express in exponential form.

6. $36 \cdot 36$

7. 15

8. $8 \cdot 8 \cdot 8 \cdot 8 \cdot 8$

EXAMPLE 5   Evaluate: 5^3

$$5^3 = 5 \cdot 5 \cdot 5 = 125$$

EXAMPLE 6   Express in exponential form: $7 \cdot 7 \cdot 7 \cdot 7$

$$7 \cdot 7 \cdot 7 \cdot 7 = 7^4$$

USING THE CALCULATOR

Example A: Evaluate: 3^2 Enter: 3 x^2 **9** ← Display

Example B: Evaluate: 5^3 Enter: 5 y^x 3 = **125** ← Display

Note 1: Calculators differ. Not all label the key for raising numbers to powers y^x.

Note 2: When raising numbers to the second power, use either the x^2 key or the y^x key followed by 2.

APPLICATION   In 1991, China ranked first in the world in total population with more than 10^9 people. What was the approximate population of China?

$$10 \; y^x \; 9 \; = \; 1,000,000,000$$

Answer: 1,000,000,000 people ◆

Roots

Finding the root of a number is the inverse of raising a number to a power. In this section, we will limit our discussion to finding positive roots of selected natural numbers.

DEFINITION   In the expression $\sqrt[n]{b}$, where **b** and **n** are natural numbers, **b** is called the **radicand,** **n** is called the **index,** and the symbol $\sqrt{}$ is called the **radical sign.** The expression is read "the **n**th root of **b**."

Note   In the expression $\sqrt[2]{b}$, the index 2 is usually omitted and the expression is written $\sqrt{b}$.

DEFINITION   **Perfect square:** A number which can be expressed as some number raised to the second power or "squared."

EXAMPLE 7   The numbers 1, 4, 9, 16, etc., are perfect squares.

$$1 = 1^2, 4 = 2^2, 9 = 3^2, 16 = 4^2$$

DEFINITION   **Perfect cube:** A number which can be expressed as some number raised to the third power or cubed.

EXAMPLE 8   The numbers 1, 8, 27, 64, etc., are perfect cubes.

$$1 = 1^3, 8 = 2^3, 27 = 3^3, 64 = 4^3$$

Roots
For $a, n \in N$

$$\sqrt[n]{a^n} = a$$

Practice Problems

Evaluate. Use a calculator.

9. 2^5

10. 16^3

11. 5^7

12. 108^2

13. 17^0

EXAMPLE 9   Find $\sqrt{9}$

$$9 = 3^2$$
$$\text{so}\quad \sqrt{9} = \sqrt[2]{3^2}$$
$$\sqrt{9} = 3$$

EXAMPLE 10   Find $\sqrt{64}$

$$64 = 8^2$$
$$\text{so}\quad \sqrt{64} = \sqrt{8^2}$$
$$\sqrt{64} = 8$$

EXAMPLE 11   Find $\sqrt[3]{27}$

$$27 = 3^3$$
$$\text{so}\quad \sqrt[3]{27} = \sqrt[3]{3^3}$$
$$\sqrt[3]{27} = 3$$

RULE   **To find the positive *n*th roots of certain natural numbers**

1. Use prime factorization to express the number in exponential form.
2. Apply the formula: $\sqrt[n]{a^n} = a$.

EXAMPLE 12   Find $\sqrt{100}$

$$100 = 2 \cdot 2 \cdot 5 \cdot 5 = 10 \cdot 10 = 10^2$$
$$\sqrt{100} = \sqrt{10^2} = 10$$

EXAMPLE 13   Find $\sqrt[3]{125}$

$$125 = 5^3$$
$$\sqrt[3]{125} = \sqrt[3]{5^3} = 5$$

USING THE CALCULATOR

Example A: Find $\sqrt{49}$ Enter 49 $\sqrt{\ }$ **7** ← Display

Example B: Find $\sqrt[3]{8}$ Enter 8 $\sqrt[x]{y}$ 3 = **2** ← Display

---

Practice Problems

Find the roots.

14. $\sqrt{81}$

15. $\sqrt[3]{64}$

16. $\sqrt{144}$

17. $\sqrt[3]{216}$

ANSWERS TO PRACTICE PROBLEMS
1. 1 **2.** 5 **3.** 64 **4.** 1000
5. 16 **6.** 36^2 **7.** 15^1 **8.** 8^5
9. 32 **10.** 4096 **11.** 78,125
12. 11,664 **13.** 1 **14.** 9 **15.** 4
16. 12 **17.** 6

PROBLEM SET 2.5

Raise these numbers to the indicated powers. Do not use a calculator.

1. 2^0 **2.** 3^0 **3.** 5^1 **4.** 6^1

5. 4^2 **6.** 3^2 **7.** 10^3 **8.** 10^4

9. 2^3 **10.** 3^3

Write in exponential form.

11. $3 \cdot 3 \cdot 3 \cdot 3$

12. $8 \cdot 8 \cdot 8$

13. $5 \cdot 5 \cdot 5 \cdot 5$

14. $2 \times 2 \times 2 \times 2 \times 2 \times 2$

Use a calculator to raise these numbers to the indicated powers.

15. 15^3

16. 17^8

17. 32^2

18. 50^4

Use a calculator to find the positive roots of these selected natural numbers.

19. $\sqrt{121}$

20. $\sqrt{169}$

21. $\sqrt[3]{343}$

22. $\sqrt[3]{512}$

23. $\sqrt{225}$

24. $\sqrt{625}$

Application:

25. The number of persons in the United States purchasing group life insurance grew at an exponential rate of 2 (meaning the base raised to the second power) from 1940 to 1990. Approximately 700 people purchased group life insurance in 1940. Approximately how many people purchased it in 1990?

2.6 Order of Operations

OBJECTIVE

Follow the order of operations when simplifying an expression.

To avoid confusion in simplifying expressions that contain more than one operation, mathematicians have agreed to the following conventions:

RULES Order of operations

1. Perform all operations enclosed within **parentheses** or other grouping symbols.
2. Raise numbers to indicated powers (**exponents**).
3. Perform all **multiplications** and **divisions** in order from left to right.
4. Perform all **additions** and **subtractions** in order from left to right.

Note The four commonly accepted grouping symbols are braces { }, brackets [ ], parentheses ( ), and the fraction bar — when it is used for division.

To apply these rules:

P   1. Start with Rule 1. Look over the entire expression. Apply the rule wherever **parentheses** or other **grouping symbols** occur.

E   2. Move to Rule 2. Look over the entire expression. Apply the rule wherever **exponents occur.**

MD 3. Move to Rule 3. Look over the entire expression. Apply the rule wherever indicated **multiplications** or **divisions** occur.

AS   4. Move to Rule 4. There are only indicated **additions** and **subtractions** remaining. Apply the rule to them.

Note   Some students prefer to remember these rules by recalling the letters:

P
E
MD
AS

EXAMPLE 1   Simplify: $5 - 2 + 3$ There are no parentheses. Move to Rule 2. There are no exponents. Move to Rule 3. There are no multiplications or divisions. Move to Rule 4.

$= 5 - 2 + 3$ Perform the additions and subtractions in order from left to right.

$= 3 + 3$
$= 6$

EXAMPLE 2   Simplify: $3 \cdot 4 + 7$ No parentheses. No exponents.

$= 12 + 7$ Perform the multiplication.

$= 19$ Perform the addition.

EXAMPLE 3   Simplify: $5 + 6 \div 3 \cdot 7$ Move to Rule 3.

$= 5 + 2 \cdot 7$ Perform the division.

$= 5 + 14$ Perform the multiplication.

$= 19$ Perform the addition.

EXAMPLE 4   Simplify: $5 + 3^2 \cdot 8$ Move to Rule 2.

$= 5 + 9 \cdot 8$ Raise the number to the power.

$= 5 + 72$ Perform the multiplication.

$= 77$ Perform the addition.

EXAMPLE 5   Simplify: $(2 + 4) \cdot 8$ Remove the parentheses by adding the numbers within them.

$= 6 \cdot 8$

$= 48$ Perform the multiplication.

EXAMPLE 6   Simplify: $(7 + 6 \div 2) \div 5$ There are two operations enclosed in the parentheses. Remove them by performing the division followed by the addition.

$= (7 + 3) \div 5$

$= 10 \div 5$

$= 2$ Perform the division.

EXAMPLE 7 Simplify: $(8 - 5) \cdot (5 + 1)$

$$= 3 \cdot 6$$
$$= 18$$

EXAMPLE 8 Simplify: $(2 + 1)^3 - 4$ Perform the addition to remove the parentheses.

$$= 3^3 - 4$$ Next raise the number to the power.
$$= 27 - 4$$ Perform the subtraction.
$$= 23$$

EXAMPLE 9 Simplify: $[5 + (3 + 1)] \div 3$ Note the parentheses inside of the brackets. Remove the inner ones first; then remove the outer ones.

$$= [5 + 4] \div 3$$
$$= 9 \div 3$$
$$= 3$$

Notes **1.** The grouping symbols, { }, [ ], and ( ), may be nested within one another as well as within themselves. The usual format is $\{ [( )] \}$. As always, remove the innermost set first, then work outwards until all are removed.

  2. Parentheses, brackets, and braces may also be used to indicate multiplication. In such cases, follow the usual rules for the order of operations and perform the multiplications in order from left to right.

EXAMPLE 10 Simplify: $\dfrac{7 + 3}{5} - 2$ Here the fraction bar is used as both a grouping symbol and a symbol of division. Treat it as a grouping symbol first.

$$= \frac{10}{5} - 2$$
$$= 2 - 2$$
$$= 0$$

EXAMPLE 11 Simplify: $2\{5[1 + 2][3(1 + 3)] - 1\}$ Begin working in the inner most set, the ( ).

$$= 2\{5[1 + 2][3(4)] - 1\}$$ Simplify within the [ ].
$$= 2\{5[3][12] - 1\}$$ Simplify within the { }, following the order of operations. In this example, the { }, [ ], and ( ) all performed dual roles as grouping symbols and symbols of multiplication.
$$= 2\{15[12] - 1\}$$
$$= 2\{180 - 1\}$$
$$= 2\{179\}$$
$$= 358$$

EXAMPLE 12 Simplify: $\dfrac{169}{12 + 1} - [5(3 - 1)]$ Begin by adding 12 and 1 and subtracting 1 from 3.

$$= \frac{169}{13} - [5(2)]$$ Perform the multiplication and division.
$$= 13 - 10$$ Perform the subtraction.
$$= 3$$

Practice Problems

Simplify.

1. $4 - 3 + 6$

2. $3 \times 4 + 6$

3. $3 \cdot (4 + 6)$

4. $(3 + 4) \cdot 6$

5. $4 + 6 \div 2 + 1$

6. $4^2 \div 8 - 1$

Evaluate.

7. $2^2 + 3^2$

8. $(2 + 3)^2$

9. $\dfrac{3 + 11}{2} - 4$

10. $(1 + 2)^5$

11. $2 \times [13 + 7(1 + 3)]$

12. $[9 + (2 + 1)]6$

13. $11^2 - (6 + 1)(8 + 7)$

14. $\{[12 - (5 + 4)] \div (2 + 1)\}^2$

USING THE CALCULATOR

A scientific calculator with an algebraic operating system will automatically follow the order of operations, provided that parentheses or other grouping symbols do not take precedence.

Example A: Simplify $97 \cdot 3 - 100$

Enter 97 $\times$ 3 $-$ 100 $=$ **191** ← Display

Example B: Simplify $10 \cdot 4 \div (10 - 2)$

Enter 10 $\times$ 4 $\div$ (10 $-$ 2) $=$ **5** ← Display

Example C: Simplify $6 + 7(5 - 3)$

Enter 6 $+$ 7 $\times$ (5 $-$ 3) $=$ **20** ← Display

Example D: Simplify $3 \times 1 + 2[15 \div (6 - 1)] - 4$

Enter 3 $\times$ 1 $+$ 2 $\times$ (15 $\div$ (6 $-$ 1)) $-$ 4 $=$ **5** ←
Display

PROBLEM SET 2.6

Simplify. Do not use a calculator.

1. $3 + 4 \cdot 6$

2. $6 \cdot 3 + 2$

3. $8 + 12 - 6$

4. $4 \cdot 2 + 3 \cdot 4$

5. $7 + 2 \cdot 8 - 4$

6. $6 \cdot 2 + 9 \cdot 8$

7. $2(4 + 7)$

8. $(8 + 4)3$

9. $2[(3 + 6) \cdot 5]$

10. $7[(3 + 4) \cdot 6]$

11. $5 + 25 \div 5$

12. $7 + 18 \div 9$

13. $25 - 8(1 + 2)$

14. $9 + 6(3 + 4)$

15. $\dfrac{7(5)}{9 - 2} + 6$

16. $\dfrac{8(3 + 1)}{4(4)} + 15$

Simplify. Use a calculator to check answers.

17. $3 + 5(1 + 3^2)$

18. $2 + 7(2 + 4^2)$

19. $3 \cdot 2^2 - 1 + 4$

20. $2 \cdot 3^2 - 7 + 5$

21. $(7 + 1)^2 + 8 \div 4$

22. $(2 + 6)^2 + 10 \div 2$

23. $5(2 + 3^3) + 7$

24. $4(5 + 2^3) + 3$

Simplify. Use a calculator as appropriate.

25. $9 \times 14 + \{44 \div [19 - (5 + 3)]\}$

26. $7 \times 15 + \{55 \div [23 - (9 + 3)]\}$

27. $84 \div 7 - \{3 \times [8 - (3 \times 2)]\}$

28. $62 \div 2 - \{5 \times [7 - (6 \div 3)]\}$

29. $5 \times \{(100 - 50 \div 10) - [(33 + 2) \div 5] + 4 \times 3\}$

30. $7 \times \{(100 - 50 \div 10) - [(33 + 2) \div 5] + 4 \times 3\}$

2.7 Averages and Estimation

OBJECTIVES

- Find the average of a series of whole numbers.
- Estimate sums and differences of whole numbers.

Finding **averages** of whole numbers and **estimating** answers to problems will both provide us with opportunities to use the skills we've developed in this chapter as they each use more than one operation.

Averages

> **GENERAL RULE** To find the average of a subset of the whole numbers, find their sum and then divide it by the number in the subset.

EXAMPLE 1 Find the average of 6 and 8.

$$\frac{6 + 8}{2} = \frac{14}{2} = 7 \qquad \text{The average of 6 and 8 is 7.}$$

This result corresponds with our common notion of average as being in the middle. The number 7 is exactly midway between 6 and 8.

EXAMPLE 2 Find the average of 3, 8, and 10.

$$(3 + 8 + 10) \div 3$$
$$= (21) \div 3$$
$$= 7 \qquad \text{The average of 3, 8, and 10 is 7.}$$

EXAMPLE 3   Find the average of {3, 8, 15, 2, 7}.

$$(3 + 8 + 15 + 2 + 7) \div 5$$
$$= (35) \div 5$$
$$= 7 \qquad \text{The average of } \{3, 8, 15, 2, 7\} \text{ is 7.}$$

 USING THE CALCULATOR

We will use the previous example to illustrate finding averages on a calculator.

Enter: 3 [+] 8 [+] 15 [+] 2 [+] 7 [=] 35 [÷] 5 [=] 7

                   ↑   ↑

               Display   Display

Notes

- In this section, we are working only with averages that are whole numbers.
- In statistics, the average is called the **arithmetic mean** or simply, the **mean.**

APPLICATION   Alex wants to get an A in his math class. To get one, he needs to have an average test score of 90 or above. To date, he has received the following test scores: 100, 91, 83, 97, and 89. Is he maintaining an A average?

Answer: Yes *(100 + 91 + 83 + 97 + 89) ÷ 5 = 92* ◆

Estimation

Estimation is an important mathematical skill. It has become especially important in our modern technological society because it is the skill we use to judge the reasonableness of the results we get when using calculators and computers.

When we're in the supermarket, we use estimation to decide whether the cost of the items we're placing in the shopping cart is not more than the money in our pockets. When we get our monthly phone bill, it is one of the skills we use to decide whether the amount due is reasonable and whether we have enough money in our checking account to pay it immediately.

In this section, we will discuss estimates of sums and differences.

RULE To estimate the sum of two or more whole numbers:

1. Decide on the accuracy desired for the sum. (Should it be accurate to tens, hundreds, etc.?)
2. Round all addends to that place value.
3. Add the rounded numbers.

EXAMPLE 4   Estimate the sum of 105, 14, 27 to the nearest ten.

$$105 \approx 110$$
$$14 \approx 10$$
$$\underline{+\ 27 \approx \ \ 30}$$
$$\qquad 150 \text{ is the estimate of the sum.}$$

Practice Problems

Find the averages.

1. 6, 8, and 7

2. 12, 14, 16, 18

3. 92, 90, 74, 80

4. 16 and 0

Estimate these results to tens.

5. 15 + 37

6. 121 − 10

7. 459 + 672

8. 1005 − 268

RULE To estimate the difference of two whole numbers:

1. Decide on the accuracy desired for the result.
2. Round the numbers to that place value.
3. Perform the subtraction on the rounded numbers.

EXAMPLE 5   Subtract 6592 from 8411. Estimate the answer to hundreds.

$$8411 \approx 8400$$
$$-6592 \approx 6600$$
1800 is the estimate.

PROBLEM SET 2.7

Find the averages of these numbers. Use a calculator.

1. 37 and 49

2. 43 and 89

3. 111, 222, 46, and 1

4. 360, 945, 524, and 867

5. 999, 2, 52, 101, and 46

6. 788, 1, 47, 64, and 100

7. 1111, 2222, 3333, and 6666

8. 4444, 5555, 7777, and 8888

9. 12, 581, 0, 75, 17, and 587

10. 35, 782, 43, 0, 249, and 13

Estimate the results by rounding as indicated.

11. 74 + 22 + 58 + 90 (to tens)

12. 83 + 17 + 38 + 65 (to tens)

13. 235 + 117 + 89 + 22 (to hundreds)

14. 391 + 219 + 77 + 30 (to hundreds)

15. 10,000 − 2500 (to thousands)

16. 50,000 − 7500 (to thousands)

17. 25,659 + 31,743 (to ten thousands)

18. 72,556 + 37,921 (to ten thousands)

19. 123 + 456 + 789 (to thousands)

20. 879 + 564 + 321 (to thousands)

Applications:

21. This season, Willy Joe, your football team's quarterback, completed three passes in the first game, seven in the second game, and two in the third game. Find the average number of passes he completed in these games.

22. This season, Eric Harrison, the leading rusher on the football team, ran 3 yards in the first game, 5 yards in the second game, and 100 yards in the third game. Find the average number of yards he ran in these games.

23. According to the payroll office of ADV Advertising, Jim earns $942 per month; Agnes, $1098; Mike, $1458; Kay, $1830; Patti $2376; and Gary, $3660.
A. What is the average monthly salary of this group?

24. Robbi, Beth, Patricia, and Geraldine are all employees of Brite State University. Robbi earns $28,500 per year; Beth, $30,258; Patricia, $31,580; and Geraldine, $22,490.
A. What is the average yearly salary of these employees?

B. Check the reasonableness of your answer by estimating it to hundreds.

B. Check the reasonableness of your answer by estimating their average yearly salaries to thousands.

SMALL GROUP ACTIVITY *Eratosthenes's Sieve*

Eratosthenes was a geographer and mathematician who lived during the third century B.C. He used a method he called a sieve to find the prime numbers from 2 to 100. First, he wrote down the numbers from 2 to 100. Next, he punched out the numbers which were multiples of 2—the even numbers, except 2 itself. After that, he punched out all the multiples of 3, except 3 itself. He continued in the same manner, looking for multiples of 5, then 7, etc., until he had eliminated all the composite numbers. When he finished, he noted that the results looked like a sieve and that the numbers remaining were prime.

Activity 1
Make your own sieve for the primes from 2 to 100.

Activity 2
List the prime numbers you found.

Activity 3
After completing the sieve, discuss these questions.
   a. Why did you only need to search for multiples of the prime numbers? Why not multiples of 6, 10, etc.?
   b. Note that 11 was the largest prime number you needed to find multiples of. Why?
   c. If you were asked to find the prime numbers from 2 to 150, what is the largest number you would need to find multiples of? Why?

CHAPTER 2 REVIEW

The terms listed below are among those used in this chapter. Write a brief definition for each term. (Refer to the appropriate section in the chapter as needed.)

1. Natural numbers _____

2. Whole numbers _____

3. Place value _____

4. Standard form _____

5. Expanded form _____

6. Subtraction _____

7. Factors _____

8. Product _____

9. Exponent _____

10. Dividend _____

11. Divisor _____

12. Quotient _____

Construct an example illustrating each of the following properties of real numbers. (Refer to the appropriate section for assistance.)

13. Associative Property of Addition _____

14. Identity Property of Addition _____

15. Commutative Property of Addition _____

16. Identity Property of Multiplication _____

17. Associative Property of Multiplication _____

18. Commutative Property of Multiplication _____

19. Distributive Property of Multiplication over Addition _____

20. Zero Product Property _____

In this chapter, you used several rules and procedures. State the following ones. (Refer to the appropriate section as needed.)

21. Order of operations _____

22. Finding an average _____

23. Rounding _____

24. Estimating a sum _____

25. Findinig the prime factors of a number _____

26. Finding the GCF of two or more numbers _____

After completing Chapter 2, you should have mastered the objectives listed at the beginning of the chapter. Test yourself by working the following problems. (Refer to the section indicated as needed.)

• *Convert numbers from expanded form to standard form (2.1).*

27. Write $3 \times 100 + 1 \times 10 + 8 \times 1$ in standard form

28. Write $5 \times 1000 + 7 \times 100 + 9 \times 10 + 1 \times 1$ in standard form.

29. Write $9 \times 100 + 0 \times 10 + 0 \times 1$ in standard form.

30. Write $8000 + 200 + 20 + 4$ in standard form.

31. Write $6 \times 1000 + 0 \times 100 + 2 \times 10 + 7 \times 1$ in standard form.

• *Convert numbers from standard form to expanded form (2.1).*

32. Write 57 in expanded form.

33. Write 203 in expanded form.

34. Write 8502 in expanded form.

35. Write 18,511 in expanded form.

36. Write 400,746 in expanded form.

• *Round whole numbers to ten. (2.2).*

37. 67

38. 504

39. 2369

40. 899

41. 1,000,007

• *Perform the indicated operations (2.3–2.5).*

42. 0 ÷ 8

43. 8 ÷ 0

44. 1^0

45. 1^1

46. 10^5

47. 27^3

• *Classify a given whole number as prime or composite (2.4). Determine the prime factors of composite whole numbers (2.4). Write the prime factors of the composite numbers. Circle the prime numbers.*

48. 97

49. 17

50. 285

51. 300

52. 41

53. 72

54. 27

55. 2

56. 383

57. 500

58. 112

59. 231

• *Determine the Greatest Common Factor to two or more whole numbers (2.4).*

60. GCF (26,40)

61. GCF (50, 110)

62. GCF (18, 25)

63. GCF (8, 20, 44)

• *Find the positive roots of selected whole numbers. (2.5)*

64. $\sqrt{64}$

65. $\sqrt[3]{64}$

• *Follow the order of operations when simplifying an expression (2.6).*

66. (2 + 7)(8 − 3)

67. 2(6 − 4) + 7

68. (6 − 2)(1 + 5)

69. $(2 + 3)^3 ÷ 5$

70. $\dfrac{12 + 2}{8 - 1} + 5^2$

71. [9 ÷ (2 + 1)] + 7

72. $5^2 + 6^2$

73. 9 + 7 · 8

74. Use the Distributive Property to simplify: $3(15 + 8)$.

75. Write in exponential form: $6 \cdot 6 \cdot 6 \cdot 6$.

• *Find the average of each series of whole numbers (2.7).*

76. {8, 12, 10}

77. {27, 33, 24}

78. {15, 20, 22, 18, 13, 20}

79. {101, 336, 267, 212}

80. {202, 75, 12, 7}

• *Estimate sums and differences of whole numbers (2.7).*

81. Estimate by rounding to tens: $354 + 729$.

82. Estimate by rounding to hundreds: $267 + 505$.

83. Estimate by rounding to tens: $385 - 72$.

84. Estimate by rounding to hundreds: $1467 - 983$.

Applications:

85. Erin makes $436 per month at her part-time job. She sees an ad in the paper for an "almost new" bike for $98. She'd like to buy it, but is afraid she won't have enough money left to pay her bills. This month she owes $100 for rent, $68 for tuition, $25 for phone, $75 for food, $20 on a credit card, $60 on a loan from her boyfriend, and $75 on an outfit on layaway that she wants to wear to a party this weekend. She also has $70 in a savings account that she can use if needed. Help her decide whether she can afford to buy the bike.

CHAPTER 2 PRACTICE TEST

Match the name of the appropriate property of real numbers with each example.

_____ **1.** $3 \cdot 0 = 0$

A. Associative Property of Addition

_____ **2.** $17 + 14 = 14 + 17$

B. Identity Property of Addition

_____ **3.** $18 \cdot 1 = 18$

C. Commutative Property of Addition

_____ **4.** $2(3 + 19) = 2 \cdot 3 + 2 \cdot 19$

D. Associative Property of Multiplication

_____ **5.** $8 + (9 + 10) = (8 + 9) + 10$

E. Identity Property of Multiplication

_____ **6.** $3 + 0 = 3$

F. Commutative Property of Multiplication

_____ **7.** $2 \cdot 13 = 13 \cdot 2$

G. Distributive Property

_____ **8.** $7(10 \cdot 11) = (7 \cdot 10)11$

H. Zero Product Property

Write in standard or expanded form, as indicated.

9. Write $7000 + 100 + 50 + 2$ (standard form). _____

10. Write $3 \times 100 + 0 \times 10 + 5 \times 1$ (standard form). _____

11. Write 10,704 (expanded form). _____

Round or estimate, as requested.

12. Round 7483 to hundreds _____

13. Round 7483 to thousands _____

14. Round 7483 to tens _____

15. Round 7483 to ten thousands _____

16. Estimate the difference of 10,050 and 4149 to hundreds _____

Perform the indicated operations.

17. 15^1

18. 6^3

19. 160^4

20. $17 + (8 - 2)$

21. $(7^2 + 3) \div 4$

22. $2^4 + 4^2$

23. $2 + \{[5 - 1][6(3)] \cdot 2\}$

24. Use the distributive property to simplify: 6(5 + 4)

25. Write in exponential form: 5 · 5 · 5

26. Write 106 as an indicated product of prime factors.

27. List the prime numbers between 60 and 70.

28. Find the GCF (35, 140).

29. Find the $\sqrt{144}$.

30. Find the $\sqrt[3]{216}$.

31. Find the average of 30, 55, 60, 25, and 75.

Application:

32. Karen is a recent transfer student who has junior status. She'd like to get an apartment, but needs to decide how much rent she can afford to pay. Her monthly income is $625, and her other monthly expenses—not including entertainment—are $300 per month. She's found the "perfect" apartment for $300 per month including utilities. She also met two girls who share a house and are looking for a third housemate. She looked at the house and everything seemed in order. Her share of the rent would be $100. She would need to have her own phone installed ($40) and pay her own phone bill ($26 per month plus long distance charges). How would you advise her financially? State the reasons for the advice you'd give.

STUDY SKILLS TIP

Managing Time for Study
Part I: Homework

As you are already learning in this course, mathematics homework must be done each day. Doing your homework soon after class will help reinforce the learning that took place that day.

Before starting your homework, be sure you have a set of notes for the day that you understand. Reviewing your notes, and then clarifying them using the textbook, will prepare you to do the assigned problems.

Begin your study session by writing each new vocabulary word and each new formula on its own 3 × 5 index card. Keep these cards on your desk for reference as you do your assignment.

As you work the problems, write out each step so that you can easily find any mistakes if they occur. Take your time. Refer to the textbook or to your notes if you have any difficulty. If there are problems you cannot solve by referring to the text and your notes, mark them with a red pen or highlighter. Be sure to have your instructor or a tutor explain them to you as soon as possible. Never let any problem remain a mystery. To gain mastery in mathematics, you'll want to know how to do every problem. Before you end your study session, make it a habit to read through all the 3 × 5 index cards you have accumulated.

A final cautionary note: Never put off doing your mathematics homework for even one day. Each day's assignment is meant to reinforce what you learned that day.

Managing Time for Study
Part II: Time Management

As a college student, you are faced with the need to manage your own time. In high school, your parents and teachers supervised much of your time. In college, you are on your own. At the beginning of each semester or quarter, you should develop a study schedule to accommodate the courses you are studying, as well as time for everything else in your life. How can you go about organizing your time so that you make the most productive use of it? Make a *plan*. Use the accompanying chart. *Be realistic!* Mark off obvious activities such as sleeping, eating, personal hygiene time, work time, travel time (to and from work, school, and home), class time, and social time. Once you have all your daily activities marked on the chart, block out your study times. Remember, the rule of thumb is two hours of study time outside of class for every hour in class. Count the hours. If you are taking 12 credit hours, have you charted 24 study hours per week—that is, about $3\frac{1}{2}$ hours a day? If not, you need to rework your schedule. It is extremely important that you have an adequate amount of time to study; otherwise, you will begin to feel overwhelmed within a few weeks. If you still can't fit in enough study time, you may have to reevaluate the load you're carrying.

Once you have allotted yourself enough time for study, decide which classes will be studied at which times. Be sure to put mathematics study time as soon after each mathematics class as possible.

Preparing your time-management plan is important; using it and keeping to it are equally important. Train yourself to follow the plan. Hang it in a prominent place over your desk. Monitor your progress, and be honest with yourself. If some time slots are not working out, revise the plan. Remember, *you* made the plan to help you organize *your* time efficiently. Make it fit your style of life and learning; make it work for *you*. Reward yourself when you have kept your schedule.

Now that you know when you are going to study mathematics, the next step is to find a place to study. It is not necessary to have absolute silence, but effective study cannot be accomplished with the TV distracting you or with constant interruptions from the telephone or other people. Take some time to think about a suitable place. Choose what you think is an ideal location and try it out. If it doesn't work, change it. Finally, settle on a spot and get into the habit of studying there.

Pacing yourself during your study time is the key to making effective use of it. Break up your study time into 50-minute intervals, and take 10-minute breaks between them. These breaks will refresh you and help you use your time productively. Don't make the mistake of trying to study for three or four hours without a break. You will become bored, tired, and restless. Do, however, get into the habit of getting back to studying after your 10-minute breaks. You may find this hard to do, but keep to your schedule. The secret to being a successful student is to learn the art of managing your time.

THE MATHEMATICS JOURNAL

Time Management

Task: After reading the Study Skills Tip "Managing Time for Study," use this schedule to develop a time-management plan for each day of the week.

	Sun.	Mon.	Tue.	Wed.	Thurs.	Fri.	Sat.
7:00							
8:00							
9:00							
10:00							
11:00							
12:00							
1:00							
2:00							
3:00							
4:00							
5:00							
6:00							
7:00							
8:00							
9:00							
10:00							
11:00							
12:00							

3 Introduction to Algebra

3.1 Simplifying Algebraic Expressions
3.2 Exponents
3.3 Evaluating Algebraic Expressions
3.4 Translating Words into Algebraic Expressions

Introduction to Algebra

The word *algebra* is derived from the title of a ninth century Arabic work on mathematics by Al-khwarizmi called *Al-jabr wa'l maqabalah* (the science of reuniting). Algebraic methods were used by the Babylonians about 1900–1600 B.C. Progress was slow until the seventeenth century when Descartes, a French philosopher and mathematician, introduced the modern system of notation.

Algebra is a part of mathematics. When using algebra, letters are used to hold the place of unknown numbers. In the following example $x + 6 = 10$, the letter x represents the unknown. Letters are also used to express general relationships between numbers as shown in the following example, $a + b = b + a$, where the equation holds true for all real number values of a and b. Algebra also allows us to find the values of variables in equations and inequalities, such as $x + 6 = 9$ or $7x < 96$. In this chapter, you will start to learn basic algebraic concepts.

3.1 Simplifying Algebraic Expressions

OBJECTIVES

- Recognize an algebraic expression.
- Name the terms of an algebraic expression.
- Recognize a variable, the coefficient of the variable, a monomial, and a constant.
- Combine like terms.
- Use the Distributive Property to simplify an expression.

An algebraic expression is a sum of terms. To simplify means to make as simple as possible.

DEFINITION **Term:** A number or variable or a product of numbers and or variables.

DEFINITION **Variable:** A symbol, usually a lower case letter of the English alphabet, used to represent an unknown number.

We know that multiplication is repeated addition: $3 + 3$ is equal to $2 \cdot 3$. Also, $5 + 5 = 2 \cdot 5$. We can make this into a general statement by using symbols $\bigcirc + \bigcirc = 2\, \bigcirc$.

Using a letter as a symbol, this would read $x + x = 2x$.

$x + x = 2x$, substitute 0 for x, $0 + 0 = 2(0) = 0$. $x + x = 2x$ is true for 0. In fact, it is true for any number. Try some!

DEFINITION **Constant** term: A number with no variable.

EXAMPLE 1 In $3x + 7 + 2a$, the middle term, 7, is a constant, and the coefficient of 7 is 7.

DEFINITION **Monomial:** A term of an algebraic expression. It is a real number, variable, or a product of a real number and one or more variables.
  A monomial in one variable can be written as ax^n, where a is a real number, x is a variable, and n is a whole number.

EXAMPLE 2 The following are monomials: $5x$, $3xy$, 9, d, $3x^2$, $7a^2b^3$.

Note These are *not* monomials: $3x + 6$, $x \div y^2$, $8 - 4a$. Monomials cannot involve addition or subtraction or division by a variable.

DEFINITION **Degree of a monomial:** The degree of a monomial is the whole number used as the exponent of the variable.
  A constant has a degree of 0. Why?
  Recall $b^0 = 1$ for $b \in$ R, $b \neq 0$. Since $c = c \cdot 1$ for $c \in$ R, and $x^0 = 1$ for any variable x where $x \neq 0$, we may think of any constant c as cx^0 ($c = c \cdot 1 = cx^0$).

EXAMPLE 3 $3x^2$ has a degree of 2.

EXAMPLE 4 $7y$ has a degree of $1 (y = y^1)$.

EXAMPLE 5 $4z^5$ has a degree of 5.

EXAMPLE 6 12 has a degree of 0. $(c = c \cdot 1 = cx^0)$

DEFINITION **Algebraic expression:** A sum of terms. Each term of an algebraic expression, including the constant term, is a monomial.

EXAMPLE 7 $4x^2 + 6x + 7$ is an algebraic expression which is the sum of the terms $4x^2$, $6x$, and 7. The terms $4x^2$ and $6x$ contain variables, in these cases x^2 and x. The 7 is a constant.

DEFINITION **Coefficient of the variable:** The numerical factor of the term.

EXAMPLE 8 The 4 in $4x^2$ and the 6 in $6x$ are both coefficients of their variables. The coefficient and the variable are multiplied. $7y$ means $7 \cdot y$ or $7(y)$. The multiplication symbol is omitted between the coefficient and the variable.

Note A variable which is written without a coefficient actually has an understood coefficient of 1.

EXAMPLE 9 The term x is the same as $1 \cdot x$; the coefficient of x is 1. The term ab is the same as $1 \cdot ab$; the coefficient of ab is 1.

Associative Property (See section 2.4.)

EXAMPLE 10 $6x$ is $6 \cdot x$. To multiply $6x$ by 3, $6x(3)$ or $3(6x)$, use the Associative Property. We have $(3 \cdot 6) \cdot x = 18x$.

EXAMPLE 11 $4(2y)$. Use the Associative Property: $(4 \cdot 2) \cdot y = 8y$.

EXAMPLE 12 $3x(2y) = 3 \cdot x \cdot 2 \cdot y$, use the Commutative Property: $3 \cdot 2 \cdot x \cdot y = 6xy$.

EXAMPLE 13 $5x(3x) = 5 \cdot x \cdot 3 \cdot x$, use the Cummutative Property: $5 \cdot 3 \cdot x \cdot x = 15x^2$.

Distributive Property

As you learned in Chapter 2, some expressions involve both addition and multiplication, for example $5(4 + 7)$. There are two different ways to simplify this expression.

METHOD 1

$5(4 + 7)$ Add the numbers inside the parentheses.

$= 5(11)$

$= 55$ Find the product of 5 and 11.

Alternatively, we could use the Distributive Property.

> **RULE** **Distributive property**
> for all real numbers a, b, and c
> $$a(b + c) = ab + ac$$
> $$(b + c)a = ab + ac$$
> and $a(b - c) = ab - ac$ (See section 2.4.)

METHOD 2 $5(4 + 7)$ Multiply 5 by the first number in the parentheses. Add to that the product of 5 and the second number in the parentheses.

$= 5(4) + 5(7)$ Find the product of 5 and 4, and the product of 5 and 7.

$= 20 + 35$ Find the sum of 20 and 35.

$= 55$

The Distributive Property also works when the numbers inside the parentheses are subtracted, in which case the products are subtracted.

EXAMPLE 14 $2(8 - 3)$

$= 2(8) - 2(3)$

$= 16 - 6 = 10$

Expressions containing only constants can be simplified either way, but expressions containing variables and constants inside the parentheses can only be simplified by using the Distributive Property. (Constants cannot be combined with variables.) $5 + x$ and $5 - x$ are in their simplest forms. Remember, $5x$ is $5 \cdot x$ *not* $5 + x$.

EXAMPLE 15 $5(x + 3)$ We cannot simplify x + 3.

$= 5(x) + 5(3)$ Use the Distributive Property.

$= 5x + 15$ This expression is in simplest form.

EXAMPLE 16 $8(x - 4)$ We cannot simplify x − 4

$= 8(x) - 8(4)$ Use the Distributive Property.

$= 8x - 32$ This expression is in simplest form.

Practice Problems

List the terms of the following expressions.

1. $6a^3 + 3a + 11$

2. $7x^3 + 7x + 2xy + 7$

List the coefficients of the following terms.

3. $5x^2$

4. $15xy$

5. $23t$

6. a

Multiply.

7. $6(3x)$

8. $10a(9b)$

9. $4x(12x)$

Use the Distributive Property to rewrite the following.

10. $3(y + 7)$

11. $(6 + a)6$

12. $2(x - 3)$

13. $(y - 5)4$

14. $2x(x + 4)$

EXAMPLE 17 $(3x + 2y)4x$ We cannot simplify $3x + 2y$.

  $= (3x)4x + (2y)4x$ Use the Distributive Property.

  $= 12x^2 + 8xy$ This expression is in simplest form.

EXAMPLE 18 $2x(x + 4)$ We cannot simplify $x + 4$

  $= 2x(x) + 2x(4)$ Use the Distributive Property

  $= 2x^2 + 8x$ This expression is in simplest form.

Combining Like or Similar Terms

The next step to simplify an algebraic expression is to combine the like or similar terms.

DEFINITION   **Like or similar terms:** Terms with the same variable to the same exponent, or constant terms.

The Distributive Property can be used to combine like terms.

$$ab + ac = a(b + c)$$

EXAMPLE 19 $2x + 3x$

  $= x(2) + x(3)$ Use the Commutative Property.

  $= x(2 + 3)$ Use the Distributive Property.

  $= x(5)$ $2 + 3 = 5$

  $= 5x$ Use the Commutative Property.

  Therefore, $2x + 3x = 5x$.

EXAMPLE 20 $6y - 4y$

  $= y(6) - y(4)$ Use the Commutative Property.

  $= y(6 - 4)$ Use the Distributive Property.

  $= y(2)$ $6 - 4 = 2$

  $= 2y$ Use the Commutative Property.

  Therefore, $6y - 4y = 2y$.

EXAMPLE 21 $4c + c$ Remember a variable with no coefficient actually has a coefficient of 1.

  $= c(4) + c(1)$

  $= c(4 + 1)$

  $= c(5)$

  $= 5c$

Often we shortcut this procedure by saying that we keep the variable part of the term and add or subtract the coefficients.

EXAMPLE 22 $3x - 2x = 1x$ or x

EXAMPLE 23 $23y - 7y = 16y$

EXAMPLE 24 $4a + 19a = 23a$

$7x$, $4x$, $23x$, and $57x$ are all like terms and can be combined.
$4ab$, $79ab$, $3ab$, and $24ab$ are all like terms and can be combined.

$3y^2$, $67y^2$, and $12y^2$ are all like terms and can be combined.
$3x$, $5x^2$, and $7x^3$ are *not* like terms because even though they all contain the variable x, it is *not* to the same power in each term. As a result, they *cannot* be combined.

Note Constants *cannot* be combined with variables.

EXAMPLE 25 $5 + x$ and $5 - x$ are in their simplest form.

Note Remember, $5x$ is $5 \cdot x$ not $5 + x$.

> **RULE** **To simplify an algebraic expression**
> **1.** Remove parentheses by using the Distributive Property.
> **2.** Combine like terms.

EXAMPLE 26 $5x + 2 + 3x + 5$ There are no parentheses, so use the Commutative Property and combine terms.
$= 5x + 3x + 2 + 5$
$= 8x + 7$ This expression is in simplest form.

EXAMPLE 27 $3(6 + x) + 5$ Use the Distributive Property.
$= 3(6) + 3(x) + 5$ Simplify the products.
$= 18 + 3x + 5$ Use the Commutative Property of Addition to combine like terms.
$= 3x + 18 + 5$
$= 3x + 23$ This expression is in simplest form.

Practice Problems

Simplify.

15. $2a + 4 + 5a + 8$

16. $x + y + 8y + 7$

17. $4x^2 + 5y + x^2 + 3$

18. $3(a + 7) + 4$

19. $9x + (4x + 3)2$

ANSWERS TO PRACTICE PROBLEMS
1. $6a^3$, $3a$, 11 2. $7x^3$, $7x$, $2xy$, 7
3. 5 4. 15 5. 23 6. 1 7. $18x$
8. $90ab$ 9. $48x^2$ 10. $3(y) + 3(7)$
11. $6(6) + a(6)$ 12. $2(x) - 2(3)$
13. $y(4) - 5(4)$ 14. $2x(x) + 2x(4)$
15. $7a + 12$ 16. $x + 9y + 7$
17. $5x^2 + 5y + 3$ 18. $3a + 25$
19. $17x + 6$

●PROBLEM SET 3.1

List the terms of the following expressions.

1. $6x + 7y$

2. $3x^3 + 2x^2 + 6x + 7$

3. $4a + 5b + 6c + 7d$

4. $6xy + 5xy^2 + 4x + 6y$

5. $2a^2b + 3a^2 + 4b$

6. $9a + 7y$

7. $4xy + 7y + 8x$

8. $9d + 8e + 7g + 2z$

9. $9x^2 + 6x + 7$

10. $4x^2 + 9x + 7y + 4$

List the coefficients of the following terms.

11. x **12.** $8xy$ **13.** $19x^2y^2$

14. z **15.** $53y$ **16.** 9

17. $8ab$ **18.** $3cd$ **19.** $4x^2y^2z^2$

20. $4fgh$

Simplify by multiplying the following.

21. $7(10a)$ **22.** $15(4x)$ **23.** $7y(6)$ **24.** $2x(27)$

25. $9(3ab)$ **26.** $6(4x^2)$ **27.** $4x(x)$ **28.** $2(3y)$

29. $9(9xy)$ **30.** $7(x^2)$ **31.** $9a(ab)$ **32.** $4c(2d)$

Use the Distributive Property to simplify the following expressions.

33. $7(x + 4)$ **34.** $a(b + 6)$

35. $(x + 4)7$ **36.** $3x(4x + 6y)$

37. $x(3 + y)$ **38.** $2a(b - 9)$

39. $8(4 - x)$ **40.** $(3a - 2)7$

41. $(3x + 4y)4$ **42.** $9(8 + 2a)$

43. $9(a + 2b)$ **44.** $7(2c + d)$

45. $(3a + 2b)a$ **46.** $(3z + y)2a$

47. $3b(c + 2d)$ **48.** $4x(x + 2y)$

49. $3z(2x + 3y)$ **50.** $17(4a + 8b)$

51. $15(9z + 7x)$ **52.** $(8a + 6b)14$

Simplify by combining like terms.

53. $2x + 3x + 10x - 3x$ **54.** $6a - 5a + 10a$

55. $12y + y - 11y$

56. $4xy + 5xy - xy$

57. $3a^2 + 4a + 2a - a$

58. $2y^2 + 6y + 4xy$

59. $3x^2 + 7x^2 - 2x^2 + 5x + 2x + x$

60. $2y^2 + 4y^2 + 6xy + 3x^2 + 2x^2$

61. $18t + 23t + t$

62. $b + 6b + 19b + 21b$

63. $4a^2 + 3a^2 + 5a + 7a$

64. $9b + 7b + 7b + 4b$

65. $19x^2 + 9x + 9x^2 + 14x$

66. $5z^2 + 9y + 12z^2$

67. $15c + 5cd + 8cd$

68. $2m + 9m + 8m + 4m^2$

69. $16jk + 9j + 14k + 4jk$

70. $13x + 19y + 14z$

71. $23x + 14x + 9x + 2x$

72. $15x + 7y + 8x + 12y$

Simplify these expressions using the Distributive Property and combining like terms.

73. $4(a + 6) + 6$

74. $y + 6y + 3(y + 4)$

75. $4 + 3(x + 7)$

76. $4(x + 3) + 8(x + 6)$

77. $6 + (x + 3)3$

78. $3 + 4(x + 7) + 6$

79. $7x + 6y + 3(x + 4)$

80. $12(x + y) + 4y + 3x$

81. $2x + 7(3x + 1) + 6x$

82. $a(7 + b) + 3ab$

Try these more challenging problems. Simplify.

83. $4a(b + a) + 3b(a + 4) + 7(b)$

84. $3x(x + 3y + 9) + 7x(9)$

85. $19(a + 3b + c) + 8c(4)$

86. $13(3j + 2k) + 8jk$

87. $9c(8c + 5d) + 3(4d + 5)$

88. $8(3x + 2x) + 9(3x + y)$

89. $(8a + 3b)4 + 7(a + 4b)$

90. $5(2m + 3n) + 6(8n + 4m)$

91. $7x(x + 3) + 2x(x + 4)$

92. $9y(8 + y) + 5y(3y + 8)$

3.2 Exponents

OBJECTIVE

● Evaluate whole numbers and variables raised to powers.

In Section 2.5, we defined a^n where the base a was a natural number and the exponent n was a whole number. Now we will extend this definition to include variables as the replacement set for the base a and the set of whole numbers as the replacement set for the exponent n.

DEFINITIONS $a^0 = 1$, for $a \neq 0$

$$a^1 = a$$
$$a^2 = a \cdot a$$
$$a^3 = a \cdot a \cdot a$$

In general,

$$a^n = [a \cdot a \cdot a \cdot a \cdot \cdots \cdot a]$$

There are n a's

EXAMPLE 1 $5^3 = 5 \cdot 5 \cdot 5 = 125$

EXAMPLE 2 $x^3 = x \cdot x \cdot x$

EXAMPLE 3 $3^0 = 1$

EXAMPLE 4 $y^0 = 1$, for $y \neq 0$

EXAMPLE 5 $1^3 = 1;\ 1^6 = 1;\ 1^{92} = 1$

Laws of exponents: For all a, $b \in$ whole numbers and all m, $n \in$ whole numbers, the following statements are true:

1. $a^m \cdot a^n = a^{m+n}$
2. $(a^m)^n = a^{mn}$
3. $(ab)^m = a^m \cdot b^m$

These laws, together with the definitions in Section 3.1, can be used to simplify algebraic expressions. The following examples will illustrate the use of each law.

Law 1: $a^m \cdot a^n = a^{m+n}$

EXAMPLE 6 $2^3 \cdot 2^4 = (2 \cdot 2 \cdot 2)(2 \cdot 2 \cdot 2 \cdot 2)$ Definition of exponents
                      $= 2^7$
              $2^3 \cdot 2^4 = 2^{3+4} = 2^7$ Law 1 for exponents

EXAMPLE 7 $x^3 \cdot x^2 = x^{3+2} = x^5$

Law 2: $(a^m)^n = a^{mn}$

EXAMPLE 8 $(3^2)^4 = 3^2 \cdot 3^2 \cdot 3^2 \cdot 3^2$
$= (3 \cdot 3) \cdot (3 \cdot 3) \cdot (3 \cdot 3) \cdot (3 \cdot 3)$
$= 3^8$ Definition of exponents.
$(3^2)^4 = 3^{2 \cdot 4}$ Law 2 for exponents.
$= 3^8$

EXAMPLE 9 $(6^2)^4 = 6^{2 \cdot 4} = 6^8$
or 1,679,616   Use a calculator.

EXAMPLE 10 $(x^5)^3 = x^{5 \cdot 3} = x^{15}$

Law 3: $(ab)^m = a^m \cdot b^m$

EXAMPLE 11 $(3x)^2 = (3x)(3x)$
$= (3 \cdot 3)(x \cdot x)$
$= 9x^2$ Definition of exponents.
$(3x)^2 = 3^2 \cdot x^2$
$= 9x^2$ Law 3 of exponents.

EXAMPLE 12 $(2xy)^3 = 2^3 \cdot x^3 \cdot y^3 = 8x^3y^3$

EXAMPLE 13 $(2y^3)^5 = 2^5 \cdot (y^3)^5$ $(ab)^m = a^m \cdot b^m$
$= 32y^{15}$ $(a^m)^n = a^{nm}$

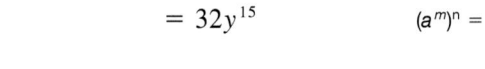

USING THE CALCULATOR

A calculator can be very useful in applying the laws of exponents to the *numerical* parts of an expression. It will not, however, help in simplifying the *variables* in an expression. You may find that it is convenient to apply one or more of the laws of exponents before using the calculator.

$(10^2)^3$ Enter 10 | x^2 | y^x | 3 | $=$ | **1000000** ← Display

or apply the law $(a^m)^n = a^{mn}$ first then use the calculator. $(10^2)^3 = 10^{2 \cdot 3}$

Enter 10 | y^x | (| 2 | × | 3 |) | $=$ | **1000000** ← Display

or $10^{2 \cdot 3} = 10^6$

Enter 10 | y^x | 6 | $=$ | **1000000** ← Display

Practice Problems

Evaluate.

1. 4^2

2. 9^3

3. 1^{10}

4. 2^3

5. 10^4

Simplify.

6. $6^2 \cdot 6^{10}$

7. $(17^2)^8$

8. $(6x^2y)^3$

ANSWERS TO PRACTICE PROBLEMS
1. 16 **2.** 729 **3.** 1 **4.** 8
5. 10,000 **6.** 6^{12} **7.** 17^{16}
8. $216x^6y^3$

PROBLEM SET 3.2

Express in exponential form.

1. $3 \cdot 3$

2. $4 \cdot 4 \cdot 4$

3. $9 \cdot 9 \cdot 9 \cdot 9$

4. $x \cdot x \cdot x \cdot x \cdot x$

5. $7 \cdot 7 \cdot 7 \cdot 7$

6. $23 \cdot 23 \cdot 23 \cdot 23 \cdot 23$

7. $4y \cdot 4y \cdot 4y \cdot 4y$

8. $19 \cdot 19 \cdot 19 \cdot 19 \cdot 19 \cdot 19$

9. $ab \cdot ab \cdot ab$

10. $3pq \cdot 3pq$

11. 9

12. 16

13. 81

14. 32

15. 125

Evaluate by using the Laws of Exponents. Check by calculator.

16. 3^0

17. 2^0

18. 15^1

19. 27^1

20. 1^7

21. 1^1

22. 5^4

23. 3^6

24. $(4x)^2$

25. $(5y)^3$

26. 9^3

27. 10^4

28. 1^{45}

29. 7^3

30. 6^5

Use the Laws of Exponents to simplify these expressions.

31. $a^2 \cdot a^6$

32. $7^2 \cdot 7^3$

33. $x^5 \cdot x$

34. $4^2 \cdot 4^3$

35. $12^2 \cdot 12^2$

36. $y^4 \cdot y^7$

37. $b \cdot b$

38. $c^6 \cdot c^7$

39. $z \cdot z^2$

40. $(2^2)^6$

41. $(5^2)^3$

42. $(1^9)^{16}$

43. $(a^8)^5$

44. $(x^3)^4$

45. $(x^7)^7$

46. $(2y^2)^6$

47. $(z^5)^4$

48. $(d^4)^7$

49. $(b^3)^5$

50. $(ab)^2$

51. $(xy)^5$

52. $(fg)^6$

53. $(3a)^8$

54. $(4x)^4$

55. $(7ab)^2$

56. $(2xy)^5$

57. $(xyz)^4$

58. $(6pq)^2$

59. $(9x)^3$

60. $(5y)^5$

61. $(a^3)^7$

62. $x^4 \cdot x^7$

63. $(y^8)^7$

64. $(jk)^{23}$

65. $s^8 \cdot s^7$

66. $(x^{12})^{10}$

67. $(x^2y^4)^4$

68. $(y^6 \cdot y^4)^5$

69. $(a^4b^6)^9$

70. $(b^4 \cdot b^8)^{11}$

3.3 **Evaluating Algebraic Expressions**

OBJECTIVE

● Evaluate an algebraic expression with a given value for the variable.

The variable in an expression can have any value. By assigning a value to the variable in an expression, we can evaluate the expression to find its value.

EXAMPLE 1   Find the value of the expression $8x + 5$ given the fact that $x = 2$.

$$8(x) + 5 \qquad \text{Substitute for the } x.$$
$$= 8(2) + 5 \qquad \text{Multiply.}$$
$$= 16 + 5 \qquad \text{Then add.}$$
$$= 21$$

The value of the expression $8x + 5$ when $x = 2$ is 21.

EXAMPLE 2   Evaluate the same expression $8x + 5$, if $x = 6$.

$$8x + 5$$
$$= 8(x) + 5 \qquad \text{Substitute 6 for the } x.$$
$$= 8(6) + 5 \qquad \text{Multiply.}$$
$$= 48 + 5 \qquad \text{Then add.}$$
$$= 53$$

The value of the expression $8x + 5$ when $x = 6$ is 53.

▦   **USING THE CALCULATOR**

Enter 8 $\times$ (substitute 6 for the x) 6 $+$ 5 $=$ **53** ← Display

EXAMPLE 3   Evaluate $x^2 + 5x + 4$, if $x = 3$.

$$= (x)(x) + 5(x) + 4$$
$$= (3)(3) + 5(3) + 4 \qquad \text{Substitute 3 for each } x.$$
$$= 9 + 15 + 4 \qquad \text{Multiply.}$$
$$= 28 \qquad \text{Then add.}$$

The value of the expression $x^2 + 5x + 4$ when $x = 3$ is 28.

EXAMPLE 4   Evaluate $x^2 + 5x + 4$, if $x = 0$.

$$= (x)(x) + 5(x) + 4$$
$$= (0)(0) + 5(0) + 4$$
$$= 0 + 0 + 4$$
$$= 4$$

The value of the expression $x^2 + 5x + 4$ when $x = 0$ is 4.

EXAMPLE 5   Evaluate x^2y^2, if $x = 1$ and $y = 2$.

$$= (x)(x)(y)(y)$$
$$= (1)(1)(2)(2)$$
$$= 4$$

The value of the expression x^2y^2, when $x = 1$ and $y = 2$, is 4.

 USING THE CALCULATOR

Enter 1 $\boxed{x^2}$ $\times$ 2 $\boxed{x^2}$ $\boxed{=}$ **4** ← Display

EXAMPLE 6   Evaluate $xy + 2y$, if $x = 5$ and $y = 8$.

$$= (x)(y) + (2)(y)$$
$$= (5)(8) + (2)(8) = 40 + 16 = 56$$

The value of the expression $xy + 2y$, when $x = 5$ and $y = 8$, is 56.

Evaluating an expression is a useful way to check that an algebraic simplifi-cation is correct. Substitute the same value for the variable in the original expres-sion and in the answer to the simplification. If the expressions have the same value, you simplified correctly.

EXAMPLE 7   Simplify $2(x + 5) = 2x + 10$.
To check your simplification, evaluate both expressions using $x = 3$.

$2(x + 5)$	$2x + 10$
$2(3 + 5)$	$2(x) + 10$
$2[8]$	$2(3) + 10$
16	$6 + 10$
	16

EXAMPLE 8   To check your simplification, evaluate both expressions using $x = 2$.

$3x + 6x^2 + 2x + x^2$	$7x^2 + 5x$
$3x + 6x^2 + 2x + x^2$	$7x^2 + 5x$
$3(x) + 6(x)(x) + 2(x) + (x)(x)$	$7(x)(x) + 5(x)$
$3(2) + 6(2)(2) + 2(2) + (2)(2)$	$7(2)(2) + 5(2)$
$6 + 24 + 4 + 4$	$28 + 10$
38	38

Practice Problems

Evaluate.

1. $4x + 6$, if $x = 4$

2. $3y + 9$, if $y = 0$

3. $2a - 4$, if $a = 9$

4. $3x + 4y$ if $x = 3$ and $y = 0$

5. x^2y, if $x = 2$ and $y = 7$

Simplify; then check your sim-plification by evaluating both expressions.

6. $4(3x + 7)$, if $x = 5$

7. $3a^3 + 5a^2 + 7a^3 + 9a$, if $a = 8$

8. $3x + 4(x + 8)$, if $x = 7$

ANSWERS TO PRACTICE PROBLEMS
1. 22   **2.** 9   **3.** 14   **4.** 9   **5.** 28
6. 88   **7.** 5512   **8.** 81

PROBLEM SET 3.3

Evaluate the following algebraic expressions using the given value of the variable.

1. $2a + 6$, if $a = 3$

2. $7x - 8$, if $x = 4$

3. $4a + 9$, if $a = 5$

4. $9x + 9$, if $x = 9$

5. $7b + 2$, if $b = 3$

6. $10y - 5$, if $y = 2$

7. $2a + 6$, if $a = 0$

8. $7x - 8$, if $x = 8$

9. $4a + 9$, if $a = 10$

10. $9x + 9$, if $x = 1$

11. $7b + 2$, if $b = 0$

12. $10y - 5$, if $y = 12$

13. $x^2 + 2x - 4$, if $x = 4$

14. $7x + 9$, if $x = 6$

15. $9y - 6$, if $y = 2$

16. $3a + 4$, if $a = 0$

17. $2m + 5$, if $m = 9$

18. $q + 9$, if $q = 17$

19. $29 + r$, if $r = 8$

20. $t^2 + 2t + 4$, if $t = 1$

21. $a^2 + 2a + 4$, if $a = 8$

22. $j^2 + 6j + 4$, if $j = 7$

23. $x^2 + 8$, if $x = 6$

24. $y^2 + y$, if $y = 5$

25. $2m^2 + m + 9$, if $m = 0$

26. $3x^2 + 4x + 7$, if $x = 3$

27. $x + xy + y$, if $x = 4$ and $y = 6$

28. $5a + ab + 4b$, if $a = 5$ and $b = 9$

29. $7a^2 + 9a - 7$, if $a = 8$

30. $4x^3 + 6x^2 + 9x + 5$, if $x = 4$

31. $5x + 7xy + 7y$, if $x = 6$ and $y = 7$

32. $6x + 7xy + 4y$, if $x = 3$ and $y = 4$

33. $5a^3 + 2a^2 + a + 6$, if $a = 9$

34. $3x^2 + 4x + 6$, if $x = 8$

35. $9x^3 + 16x^2 + 4x$, if $x = 0$

36. $3m^2 + 2n^2 + 6n + 9$, if $m = 3$ and $n = 4$

37. $7j + 9jk + 6k$, if $j = 2$ and $k = 7$

38. $3c^2 + 2d + 5cd$, if $c = 1$ and $d = 0$

39. $5x + 7y + 9xy$, if $x = 3$ and $y = 8$

40. $12a + 16b + 12ab + 3a^2$, if $a = 1$ and $b = 3$

3.4 Translating Words into Algebraic Expressions

OBJECTIVE

● Translate phrases into expressions using algebraic symbols.

To be successful at working with word problems, it is necessary that you can translate words into algebraic expressions. To apply the operations you have learned so far, you need to be familiar with the following words and phrases used to express these operations.

Words or phrases meaning add
Add
Find the sum
Addition
More than
Increased by
Plus
Total

Words or phrases meaning subtract
Subtract
Decreased by
Less than
Subtracted from
Find the difference
Diminished by
Reduced by
Minus

Words or phrases meaning multiply
Multiply
Find the product
Times as much
Three times
Double
Of
Twice means multiply by 2
Triple means multiply by 3

Words or phrases indicating divide
Divided by
Divided into
Find the quotient

RULE **Translating from words into algebraic expressions**

1. Find the unknown quantity, frequently "the number," and assign it a variable.
2. Decide which operation is indicated using the previous list of terms.
3. Translate the phrase into an algebraic expression.

Examples of word phrases and their algebraic equivalents follow:

Words	Algebra
The sum of 3 and a number.	$3 + x$
Six plus a number.	$6 + y$
Ten more than a number	$x + 10$
The total of two different numbers.	$a + b$

Note In the previous example, notice that two different numbers must be represented by two different variables.

The difference between a number and 14.	$x - 14$
The difference between 14 and a number.	$14 - x$
Seven less than a number.	$x - 7$
A number less than 7.	$7 - x$

Note In the previous examples, note that in subtraction, the order of the terms is vital, while in addition, order is not important because addition is commutative.

Ten decreased by any number.	$10 - n$
Subtract a number from 8.	$8 - x$
From 10 subtract 3.	$10 - 3$
Subtract 8 from 15.	$15 - 8$

Note In the previous examples, notice the order of the numbers.

The difference between two numbers.	$x - y$
Twice a number.	$2x$
Five times a number.	$5x$
The product of two numbers.	ab
The quotient of a number and 6.	$y \div 6$
Forty-two divided by a number.	$42 \div a$
Forty-two divided into a number.	$a \div 42$
A number divided by 42.	$a \div 42$

Note In the previous example, notice that when using division, order is important. If you reverse the numbers, you change the problem.

PROBLEM SET 3.4

Translate the following English words into algebraic expressions.

1. The quotient of 14 and the number.

2. Seventeen times the number.

3. The number increased by 20.

4. Nine less than the number.

5. The difference of 4 and the number.

6. Triple the number.

7. Eight more than the number.

   8. The number less than 21.

   9. The number divided by 2.

10. The product of the number and 6.

11. Thirteen decreased by the number.

12. The sum of 28 and the number.

13. One minus the number.

14. The number decreased by 1.

15. Seventeen divided by the number.

16. The difference between the number and 24.

17. Twice the number.

18. Fifty subtracted from the number.

19. The number multiplied by 10.

20. The number subtracted from 50.

21. Double the number.

22. The total of 34 and the number.

23. Twenty-five divided into the number.

24. Forty-five reduced by the number.

25. Six times as much as the number.

26. The number diminished by 76.

27. Sixty-seven more than the number.

28. Find the product of a number and 54.

29. The number plus 43.

30. The quotient of 76 and a number.

31. Subtract 9 from 19.

32. From x subtract 10.

Write algebraic expressions that represent the following phrases.

33. How many books does Jose have if he has 10 more books than Maria?

34. Joe has twice as many baseballs as James. How many baseballs does Joe have?

35. Keesha scored 10 more points than Jane on the test. How many points did Keesha score?

36. LeTisha grew 20 fewer tomatoes than Mike. How many tomatoes did LeTisha grow?

37. The cost of Jeff's textbooks this semester was reduced by $19 from last semester. How much will his textbooks cost this semester?

38. DarRita's salary was decreased by $9 this month. What is her new salary?

39. Pedro scored 8 times as many points on the test as Maria. What is Pedro's score?

40. Find the sum of Vanessa's and Amanda's scores on the test.

41. Multiply Ian's salary by Fiona's salary.

42. How many daisies and buttercups are there in the field?

43. If there are x buttercups in one-tenth of the field, how many buttercups are there in the field?

44. If it costs x dollars to clean 100 ft^2, how much does it cost to clean 1 ft^2?

45. A cat eats x cans of food in a week. How many cans does he eat in a day? How many cans does he eat in 6 weeks? How many cans does he eat in a year of 52 weeks?

46. Ten cars pass through the intersection on each light change. How many cars will pass through the intersection during x light changes?

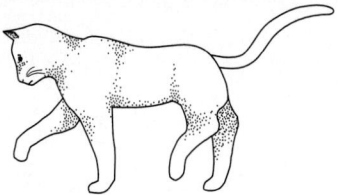

47. A truck can carry x tons of sand per load. How many truck loads will be needed to move 150 tons of sand?

48. An English phone number has 11 digits, a Norwegian phone number has x fewer digits. How many digits in a Norwegian phone number?

49. A nursing home has 30 patients during a week when no one is admitted, but x patients go home and y patients die. How many patients does the home have now?

50. A nursing home has 50 patients. x patients are admitted and y patients go home. How many patients does the home have now?

51. Your computer has x megabytes of memory. The new computer you are considering buying has 10 times more memory. How much memory does the new computer have?

52. Seven hundred students take a developmental mathematics class, x students withdraw, and y students fail. How many students pass the class?

53. Keith is 6 years older than Eleanor. How old is Keith?

54. Kateeka is 3 years younger than Osam. How old is Kateeka?

55. Elliott has twice as many marbles as Dawayne. How many marbles do they have together?

56. There are 27 passengers riding on a train. At the first stop, x passengers got off, and y passengers got on. How many passengers are on the train now?

57. Tim is twice as old as Fred. How old is Tim?

58. LaFawn is 4 years younger than Todd. How old is LaFawn?

SMALL GROUP ACTIVITY *Practicing Evaluating Algebraic Expressions*

Give each group of students ten index cards. Tell each group to write an algebraic expression on each index card. The expression should contain no more than three terms, in one variable. The terms should be added; for example, $7x^2 + 9x + 14$. On the same side of the index card, write three values for the variable used in the expression. On the back of the index card write the solutions for the three values. When each group has finished these tasks, exchange cards with another group and check their solutions. This activity could be repeated using more than one variable.

CHAPTER 3 REVIEW

The terms listed below are among those used in the chapter. Write a brief definition for each term. (Refer to the appropriate section in the chapter as needed.)

1. Variable _____

2. Term _____

3. Algebraic expression _____

4. Constant _____

5. Like terms _____

6. Associative Property of Multiplication _____

7. Distributive Property _____

8. Mononial _____

9. Exponent _____

10. List the three Laws of Exponents _____

After completing Chapter 3, you should be able to apply the following concepts. Test yourself by working the following problems. (Refer to the indicated section as needed.)

• *Name the terms of an algebraic expression (3.1).*

11. $3x + 2y + 92$ _____

12. $4x^2 + 5x + 7$ _____

13. $6a + 4ab + 7b$ _____

14. $4x + 6y$ _____

15. $9a + 7$ _____

16. $14b + 3a$ _____

17. $2c + 3d + 4cd$ _____

18. $9x + 7y + 46$ _____

19. $3jk + 2j + 7k + 6$ _____

• *Recognize a variable, the coefficient of the variable, a monomial, and a constant (3.1).*

  Circle the variables in the following expressions.

20. $6x^3 + 4x^2 + 5x + 6xy + xy + 3y^2 + y + 8$ _____

21. $3a^2 + 4a + 6ab + 7b^2 + b + 9$ _____

22. $4c^2 + c + cd + 5d + 7$ _____

List the coefficients of the variables in the previous expressions.

23. _____

24. _____

25. _____

Circle the constant terms in the following expressions.

26. $9 + 4x + 7y + 3x^2 + 46 + 179x + 86$ _____

27. $4a + 7b + 19 + 3c + 4ab + 2$ _____

28. $7m + 4n + 15 + 6mn + 0$ _____

• *Combine like terms (3.1).*

29. $6x + x + 9x$ _____

30. $3y + 18y - y + 4x - 3x$ _____

31. $2a + 4a + ab + 3ab + 6b - b$ _____

32. $8y - 2y$ _____

33. $3x^2 + 5x + 2x^2 + 6x$ _____

34. $3x + 2x + 7x$ _____

35. $7x^2 + 9x + 3x^2 + 4x^2$ _____

36. $ab + 3a + 7b$ _____

37. $y + 3y + 9y + 4y^2$ _____

38. $3m^2 + 6m^2 + 9m + 5m$ _____

39. $7c + 9d + 3cd$ _____

40. $4x^2 + 0 + 9x^2 + 6$ _____

41. $4x + x + 2xy$ _____

42. $3ab + 6ab + 96a$ _____

43. $7x + 9x + 3y + 9y + 6xy + 2xy$ _____

- *Use the Distributive Property to simplify an expression (3.1).*

44. $7(x + 3)$ _____

45. $(a + 5)8$ _____

46. $(y - 8)9y$ _____

47. $x(y - 1)$ _____

48. $7(a + b)$ _____

49. $3(2x + 4)$ _____

50. $(2y + 9)4$ _____

51. $4x(3 + y)$ _____

52. $5(x + 4)$ _____

53. $6(y - 3)$ _____

- *Use the Distributive Property to simplify and combine like terms (3.1).*

54. $4 + 6(3 + x)$ _____

55. $x + 3(x + 9)$ _____

56. $(y + 4)5 + 3$ _____

57. $2(a + 4) + 3(2 + a)$ _____

58. $7 + 3(x + 6)$ _____

59. $y + 4(y + 3)$ _____

60. $(2a + 4)5 + 6$ _____

61. $(2c + d)3 + 2d$ _____

62. $4(2a + 2b) + 3(a + b)$ _____

63. $9(2x + 3y) + (3x + 4y)6$ _____

- *Evaluate whole numbers and variables raised to powers (3.2).*

Evaluate by using the laws of exponents.

64. 7^2 _____

65. 5^3 _____

66. x^9 _____

67. 93^0 _____

68. 1^{56} _____

69. $a^5 \cdot a^7$ _____

70. $(x^6)^9$ _____

71. $(gh)^6$ _____

72. $(y^2 \cdot y^4)^5$ _____

73. $(y^4z^2)^4$ _____

74. $(4x^5)^3$ _____

• *Evaluate an algebraic expression with a given value for the variable (3.3).*

Evaluate.

75. $3x + 6$, if $x = 4$ _____

76. $7a + 3$, if $a = 6$ _____

77. $2xy + x^2$, if $x = 3$ and $y = 4$ _____

78. $3x^2 + 2x + 4$, if $x = 0$ _____

79. $2a + 6$, if $a = 3$ _____

80. $4b + 9$, if $b = 5$ _____

81. $7x + 6$, if $x = 0$ _____

82. $9y + 1$, if $y = 1$ _____

83. $8z + z$, if $z = 8$ _____

84. $4x + 5y + 9$, if $x = 2$ and $y = 3$ _____

85. $9c + 3cd + 4$, if $c = 4$ and $d = 4$ _____

86. $2m^2 + 3m + 9$, if $m = 5$ _____

87. $2m + 3mn + 7n$, if $m = 6$ and $n = 2$ _____

88. $3x^2 + 6x + 19$, if $x = 7$ _____

89. $4y^2 + 4y + 16$, if $y = 8$ _____

• *Translate phrases into expressions using algebraic symbols (3.4).*

90. A number increased by 7. _____

91. The difference of two numbers. _____

92. Double a number. _____

93. A number divided into 9. _____

94. The sum of two numbers. _____

95. The product of a number and 72. _____

96. The quotient of 15 and a number. _____

97. Nineteen reduced by a number. _____

98. Six plus a number. _____

99. The difference between 14 and 7. _____

100. Nine less than a number. _____

101. A number subtracted from x. _____

102. Seven less than 3. _____

103. Subtract 7 from a number. _____

104. The sum of 9 and 10. _____

105. Twenty-one less than a number. _____

106. The product of 3 and a number. _____

107. The quotient of two numbers. _____

108. The sum of a number and 19. _____

109. Thirty-six divided by a number. _____

110. Mary's new salary is $9 a month less than her old salary. What is her new salary? _____

111. Serenity has x cards, but Krystal has twice as many. How many cards does Krystal have? _____

112. The journey from Kent to Youngstown on the freeway is 50 miles. The journey driven on the back roads is x miles farther. How far is it from Kent to Youngstown on the back roads? _____

113. There are 43 people on the train. If x number of people get off and y number of people get on, how many people are now on the train? _____

CHAPTER 3 PRACTICE TEST

Circle the terms of this expression.

1. $6x + 9xy + 7y$

List the coefficients of the following terms.

2. y **3.** $7y$ **4.** $81xy$ **5.** $27a$

Simplify by combining like terms.

6. $10a - 4a$ _____

7. $5y - y$ _____

8. $4ab + 5ab$ _____

9. $3x + 8x + 7y + 4x + x - y$ _____

10. $4x^3 + 3x^2 + 2x^2 + 5x + x + 9 - 2$ _____

Use the Distributive Property to simplify.

11. $4(x + 7)$ _____

12. $(3 + x)9x$ _____

13. $(x - 8)5$ _____

14. $3y(2y - 4)$ _____

15. $6(4x + 3) - 10 + x$ _____

16. $(a + 4)6 + 2(3 + a)$ _____

17. $3x + 4x + 7(x + 9)$ _____

18. Evaluate: $8x + 1$, if $x = 6$ _____

19. Evaluate: $3x^2 + 5x + 10$, if $x = 9$ _____

20. Evaluate: $3x^2 + 4x + 9$, if $x = 0$ _____

Translate the following into algebraic expressions.

21. The sum of a number and 10. _____

22. Seventeen decreased by a number. _____

23. One number multiplied by another number. _____

24. The quotient of any two different numbers. _____

25. A number divided by 4. _____

26. Six reduced by a number. _____

27. Eight times as much as a number. _____

28. Six less than a number. _____

29. A number divided by 7. _____

30. The sum of 23 and a number. _____

31. Evaluate: $(7x^2)^3$ _____

32. Evaluate: $a^4 \cdot a^{17}$ _____

STUDY SKILLS TIP

Tests
Part I: Preparation

By this time in the semester, you will most likely be getting ready for a test. Begin preparing at least a week ahead of time. Do not make the common mistake of cramming all the study time into six hours the day and night before the test. Such behavior is counterproductive because you will arrive at the test tired, anxious, and underprepared. Beginning at least a week ahead, studying for two to three hours a day in 50-minute segments with 10-minute breaks throughout is a more efficient and more productive way to prepare for a test.

As you begin to think about the upcoming test, aim to get 100% of the problems correct. You cannot achieve a high grade unless you aim high. If you are aiming to get 100% on the test, you must completely understand every topic to be tested. Start your preparation with your perfect score in mind. When preparing, you must cover well all of the material to have complete understanding. To gain mastery of any skill—be it athletics, music, or mathematics—you need to practice. Concepts should be understood and skills then practiced to achieve mastery.

There is too much material in a one-semester mathematics course to try to memorize each series of steps needed to solve particular kinds of problems. You must understand the concepts, not merely memorize specific steps.

To determine exactly what you need to master for a test, make a list of every topic or concept covered. You can accomplish this by using the notes you have taken in class. Check the list by asking your instructor.

Once you have the list of topics, begin to review the topics one by one. You must understand the topics before you can practice the skills associated with them. To master a concept, you need to understand completely why you are following a particular sequence of steps to solve problems which use that concept. *Do not* memorize the steps without understanding them. Instead, comprehension should be your focus. If there are topics on the list you cannot master using your notes and textbook, *do not give up on them.* Remember, your aim is to get 100% on the test. Get help from your instructor or a tutor.

It is important to practice these newly acquired skills. You practiced with your homework problems; remember that in those exercises, you were given sets of similar problems. Tests are not like that. On a test, the problems are not always arranged by type, so you need to be able to determine the type of problem before you can work it. To prepare for the test, you can work through practice tests provided by your instructor, tutor, or textbook; however, if you want even more practice, construct your own practice tests. Choose three or four problems from the assigned homework for each topic on your list. (Be sure you have the correct answers to them.) Include both easy and more-difficult problems. Put each problem on a 3×5 index card. Include the section and problem numbers so that you can check each answer later. Shuffle the cards. Now you have a set of practice problems. Work each problem on the reverse side of the card. Now each card has a problem on one side and a solution on the other. When you have completed about 20 problems, check your answers. Try to find your mistakes. Analyze them. Review those concepts, then work a few more practice problems using those concepts.

You, like many students, may find it helpful to prepare for tests by forming a small study group with a few of your classmates. Studying in a group gives you an opportunity to share practice tests, to get assistance if needed, and to help others. Explaining concepts to others enables you to clarify your own thinking about these concepts. Someone in your group may be able to explain a concept in a way that makes sense to you. Your study group can become a real "support group" for you.

Remember the keys to good test preparation: set the goal of getting a perfect score, start your preparation early, and practice to make that score a reality.

Tests
Part II: Top Ten Test-Taking Strategies

These strategies will help you actually take the test.

1. Get a good night's sleep (7 to 8 hours) the night before the test. You should then be alert and better able to concentrate on the test.
2. Come to the test a few minutes early. Come prepared with several sharpened pencils—and a working calculator if permitted. It can be very disturbing to break a pencil lead in the middle of a test and not have another pencil to use.
3. Before you look over the test, write all the formulas and other information you think might be helpful on the back of it.
4. Write your name on each page of the test. Look over the entire test. As you are doing this, see whether there are other formulas or information needed for the test. If so, turn to the back of the test and note the needed information.
5. Start at the beginning since problems usually start out easy and get more difficult toward the end of the test.
6. *Read directions carefully*. Sometimes only an equation or proportion is required. Do only what you are directed to do.
7. Skip initially any problem that puzzles you. Finish the problems you know how to do first, then return to and rethink the skipped ones.
8. Watch your time and pace yourself. Divide the total time by the number of problems on the test to determine the average time you can afford to spend on a problem. Glance at the clock at periodic intervals to check your pace. Leave sufficient time to review the whole test, checking in particular that you did not make any unintentional errors. If you still have time remaining, return to any problems you did not know how to do, or review the entire test again.
9. *Never* hand in the test before the time is finished. Use all the time you are allowed.
10. After the test has been graded and returned to you, check your thought processes and calculations for each problem you missed. Do not simply correct the problems—determine also *why* you were wrong. If you can do this, you are less likely to make the same mistake again. If you cannot figure out why you went wrong and you have all your calculations for the problem, take them to your instructor. He or she can probably help you determine where you made the error. *Learn from tests!*

	THE MATHEMATICS JOURNAL	
	Study Strategies	
	Task: Reflect on the study strategies you are using for this course. Write a one-page (150-word) entry on your reflections.	
	Starters: Have you developed any new study strategies? Discuss them. Have you changed or discarded any old ones? Discuss them. Have your study strategies improved? In what ways? Why? What caused the improvements? Are you satisfied with your present strategies? Why or why not? Do you have a plan for future improvements? Discuss it.	

CUMULATIVE REVIEW: CHAPTERS 1, 2, and 3

DIRECTIONS *Test your mastery of the topics covered in chapters 1, 2, and 3 by taking this practice test. All items are worth 2 points, unless indicated otherwise. The total number of possible points is 100. Remember, this is a review, so check your answers and then refer to the appropriate chapter and section to get assistance in correcting any mistakes.*

COMPLETION *Choose the appropriate word or phrase from the list below.*

Empty set
Associative Property of Multiplication
Distributive Property
Identity Property of Addition
Identity Property of Multiplication
Finite set
Composite number
Variable

Commutative Property of Addition
Constant
Subset
Commutative Property of Multiplication
Natural numbers
Whole numbers
Prime number
Coefficient

1. $3 \times 4 = 4 \times 3$ is an example of the _____

2. The symbol $\subseteq$ is used to represent that one set is a _____ of another set.

3. The _____ states that, for all $a, b, c \in$ Reals, $(a \cdot b) \cdot c = a \cdot (b \cdot c)$

4. The _____ states that, for all $a, b \in$ Reals, $a + b = b + a$.

5. The _____ states that, for all $a, b, c \in$ Reals, $a(b + c) = a \cdot b + a \cdot c$.

6. The set of _____ $= \{1, 2, 3, \ldots\}$.

7. The set of _____ $= \{0, 1, 2, 3, \ldots\}$.

8. In the expression $35x$, x is a _____ .

9. A whole number greater than 1 whose only factors are itself and 1 is a _____ .

10. In the expression $3y^2 + 7$, the number 7 is a _____ .

Simplify by performing the indicated operations.

11. $15(x + 2y) =$ _____

12. $4(a + 17) + 12 =$ _____

13. $31j - 5j =$ _____

14. $6p + 2r + 7p - r =$ _____

Follow the directions for each item.

15. Write in exponential form: $5 \cdot 5 \cdot 5 =$ _____

16. Round 76,082 to the nearest ten _____

17. Round 76,082 to the nearest hundred _____

18. Round 76,082 to the nearest thousand _____

19. Round 76,082 to the nearest ten thousand _____

20. $12(5) =$ _____

21. $\sqrt{81} =$ _____

22. 6^3 _____

23. $5 \cdot 0$ _____

24. 7^0 _____

25. $8 + 2 \cdot 13 =$ _____

26. $34 + \sqrt{9} =$ _____

27. $\dfrac{16}{2 + 2} =$ _____

28. $\dfrac{41}{0} =$ _____

29. $(x^2 \cdot x^4)^5 =$ _____

30. $(3y^4)^3 =$ _____

31. $\sqrt[3]{64} =$ _____

32. Simplify: $10ab + 7ab - ab =$ _____

33. Simplify: $5(3x + 4) - 10 + x =$ _____

34. Simplify: $5 \cdot 6 + 45 \div 9 =$ _____ (3 pts.)

35. Simplify: $28 \cdot 7 - 6 \cdot 8 =$ _____ (3 pts.)

36. Simplify: $31 - [2(4 + 5) + 8] =$ _____ (3 pts.)

37. Simplify: $\{24 - [5 - (4 - 2 \div 2)]\} \div (3 + 16 \div 2) =$ _____ (3 pts.)

38. Estimate the sum by rounding each number to the nearest hundred. _____ (3 pts.)

$$9782$$
$$+\ 2934$$

39. Evaluate: $4y^2 - y + 3$, when $y = 3$ _____ (3 pts.)

40. Evaluate: $\dfrac{7 - h}{3 + h}$, when $h = 7$ _____ (3 pts.)

Translate these phrases into algebraic expressions.

41. The sum of $2l$ and $2w$ _____

42. Forty less than the three times r _____

Set theory. (3 pts. each)

43. List the subsets of $\{x, y, z\}$ _____

44. Shade the indicated region.

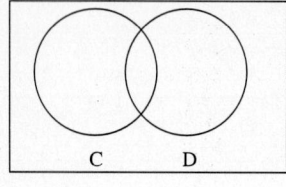

$C \cap D$

45. Shade the indicated region.

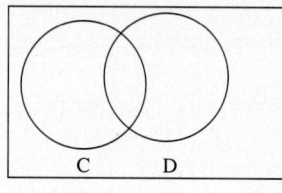

$C \cup D$

4 Integers

4.1 The Set of Integers
4.2 Absolute Value
4.3 Addition of Integers
4.4 Subtraction of Integers
4.5 Multiplication of Integers
4.6 Division of Integers
4.7 Order of Operations and Averages Using Integers
4.8 Using the Addition Property of Equality to Solve Equations

Introduction to Integers

You already learned in Chapter 2 that the counting numbers (also called positive integers or natural numbers) and the whole numbers (natural numbers and zero) can be traced back to the Stone Age. **Negative integers,** however, have a more recent history. There is evidence that Hindus used negative numbers by the seventh century. By the ninth century, Hindus were indicating that a number was negative by circling it. We still use this notation today when we refer to a card player with a negative score as being "in the hole." It was also the Hindus who first used negative numbers to represent debts.

There is evidence that the Chinese used negative numbers as early as the twelfth century. Chinese mathematicians of that day were not in complete agreement on how to refer to these numbers, however. Red and black bamboo rods were used for computing, with one mathematician using red and black ink to write the numbers—red for the negatives and black for the positives. That tradition endures to this day in the business world, especially in banking. Some printing calculators have a ribbon that is red (for subtraction) and one that is black (for addition). It is also common to refer to a profitable business as being "in the black" and to an unprofitable one as being "in the red."

The $+$ and $-$ signs we use today are European in origin. They seem to have come from the Germans and probably date back to the fifteenth century. Gerome Cardan, a sixteenth-century mathematician, was the first European to treat negative numbers seriously. By the seventeenth century, René Descartes, a French philosopher and mathematician, was using negative and positive numbers extensively to represent opposite directions.

In today's world, the use of negative numbers is commonplace. On a bitter cold winter day, we may refer to the temperature as minus $35[-35°$ F], meaning 35 degrees below zero on the Fahrenheit scale. As we wait for a space shuttle to lift off, we watch negative numbers appear on the corner of our TV screens and hear a voice from NASA count, "T minus $5[-5]$, T minus $4[-4]$, ... With such widespread use of negative numbers today, it's hard to believe that it wasn't until the nineteenth century that negative numbers became widely accepted.

4.1 The Set of Integers

OBJECTIVES

● Recognize that integers have direction as well as distance from zero.

● Order integers.

In Chapter 1, the sets of numbers were introduced. In Chapter 2, we studied the set of whole numbers. We are now ready to begin our study of the set of integers.

DEFINITION   **The set of integers:** {. . . , −3, −2, −1, 0, 1, 2, 3, . . .} is a subset of the set of real numbers.

The set of whole numbers {0, 1, 2, 3, . . .} and the set of natural numbers {1, 2, 3, . . .} are both subsets of the set of integers.

Unlike the other sets of numbers we have considered, the set of integers has direction.

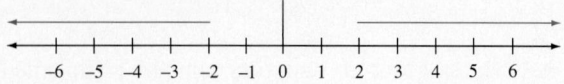

Negative numbers are marked with a **negative sign.** The numbers −3 and −6 are negative while positive numbers could be marked +3 and +6, or 3 and 6. If a number has no sign, it is positive.

USING THE CALCULATOR

To enter a negative number into your calculator, enter the number followed by the +/− key. For example, to enter −3, enter [3] [+/−]. To enter a positive number, just press the number key. For example, to enter +3, enter [3].

Comparison of Integers

Positive and negative numbers can be ordered or compared by using the greater than (>), less than (<), and equal to (=) symbols. We call this the **Principle of Trichotomy.**

EXAMPLE 1   Consider the numbers 4 and 6.

    6 is greater than 4;
        in symbols: $6 > 4$

    4 is less than 6;
        in symbols: $4 < 6$

Note   The smaller part of the inequality symbol always points to the smaller number.

Principle of Trichotomy

Given any two real numbers a and b

a is equal to b	$a = b$
or	or
a is greater than b	$a > b$
or	or
a is less than b	$a < b$

For all real numbers, exactly one of these statements is true.

$$a > b \text{ or } a < b \text{ or } a = b$$

In terms of a number line:

> $a > b$, a is greater than b if a is located to the right of b on the number line.
> $a < b$, a is less than b if a is located to the left of b on the number line.
> $a = b$, a is equal to b if a and b have the same location on the number line.

As you can see on the number line below, the numbers increase in value as you move from left to right. This is movement in the positive direction. Any number to the left is smaller than the numbers to the right. Negative numbers are always smaller than positive numbers.

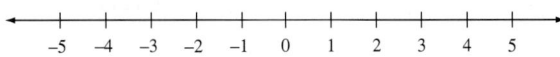

EXAMPLE 2

If we compare -7 and 2, -7 is smaller than 2 because it is to the left of 2 on the number line. $-7 < 2$. Conversely, 2 is larger than -7 because it is to the right of -7 on the number line. $2 > -7$.

EXAMPLE 3

If we compare -5 and -3, -5 is smaller than -3 because it is to the left of -3 on the number line. $-5 < -3$. Conversely, -3 is larger than -5 because it is to the right of -5 on the number line. $-3 > -5$.

PROBLEM SET 4.1

Write the following statements using words for the inequality symbols.

1. $4 > 2$

2. $-6 > -16$

3. $4 < 8$

4. $4 > -8$

5. $0 > -4$

6. $47 < 49$

7. $-43 > -44$

8. $-194 < 94$

9. $14 > 0$

10. $5 > -9$

Replace the words in the following statements using the proper inequality symbols.

11. 9 is greater than 5   **12.** −9 is less than −2

13. 15 is less than 17   **14.** 19 is greater than −4

15. 0 is greater than −4   **16.** −9 is less than 0

17. −64 is less than −63   **18.** 1 is less than 10

19. −1 is greater than −10   **20.** 1 is greater than −10

Fill in >, <, = symbols between the following numbers to make the resulting statements true.

21. 7   16   **22.** 29   −30

23. −7   16   **24.** 7   0

25. 7   −16   **26.** −9   −8

27. −7   −16   **28.** −4   −4

29. 0   −3   **30.** 93   −100

31. 30   30   **32.** −13   −3

33. 24   25   **34.** 13   3

35. −24   −25   **36.** 27   −28

4.2 Absolute Value

OBJECTIVES
- Recognize the absolute value of a number.
- Evaluate the absolute value of a number.
- Determine the opposite or additive inverse of a number.
- Note the difference between finding the absolute value of a number and removing parentheses.

We will use absolute value when we add, subtract, multiply, and divide integers.

DEFINITION **Absolute value** of a number: The distance the number is from zero (the origin) on the number line.

EXAMPLE 1 The number 4 is 4 units from zero. The number -4 is also 4 units from zero. The absolute value of any real number a is written $|a|$.

$$|4| = 4 \text{ and } |-4| = 4$$

EXAMPLE 2 $|7| = 7$

EXAMPLE 3 $|-3| = 3$

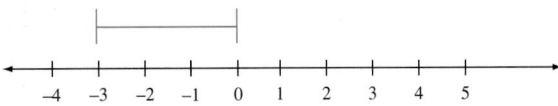

Opposite or Additive Inverse

The opposite of a number is a number that is the same distance from zero but in the opposite direction on the number line. Mathematicians call the opposite of a number its **additive inverse.**

EXAMPLE 4 -3 is the opposite of 3.
-3 is the additive inverse of 3.

EXAMPLE 5 24 is the opposite of -24.
24 is the additive inverse of -24.

The opposite of every negative number is a positive number. The opposite of every positive number is a negative number. Zero is its own opposite. A negative sign outside the absolute value symbol means to take the opposite (additive inverse) of the absolute value.

EXAMPLE 6 $-|4|$ is read: the opposite of the absolute value of 4. The absolute value of 4 is 4. The opposite of 4 is -4. Therefore, $-|4| = -4$.

EXAMPLE 7 $-|-9|$ is read: the opposite of the absolute value of -9. The absolute value of -9 is 9. The opposite of 9 is -9. Therefore, $-|-9| = -9$.

A negative sign outside a parentheses indicates the opposite or additive inverse of the number inside the parentheses.

$-(4)$ reads the opposite or additive inverse of 4.
$-(9)$ reads the opposite or additive inverse of 9.
$-(-1)$ reads the opposite or additive inverse of -1.
$-(-4)$ reads the opposite or additive inverse of -4.

Therefore,
$$-(4) = -4$$
$$-(9) = -9$$
$$-(-1) = 1$$
$$-(-4) = 4$$

Note Notice the differences between removal of parentheses and the removal of absolute value symbols. They are sometimes confused.

EXAMPLE 8

$$-(4) = -4$$
$$-(-4) = 4$$
$$-|4| = -4$$
$$-|-4| = -4$$

PROBLEM SET 4.2

Evaluate the following absolute values.

1. $|-4|$ **2.** $|15|$ **3.** $|0|$ **4.** $|-5|$

5. $|17|$ **6.** $|27|$ **7.** $|-27|$ **8.** $|8|$

9. $|734|$ **10.** $|-32|$ **11.** $|27|$ **12.** $|1|$

13. $|-9|$ **14.** $|-731|$ **15.** $|3|$ **16.** $|-11|$

17. $|1369|$ **18.** $|-2|$ **19.** $|-6|$ **20.** $|75|$

Evaluate these absolute values.

21. $-|-8|$ **22.** $|23|$ **23.** $|-1|$

24. $|-6|$ **25.** $-|4683|$ **26.** $-|9|$

27. $-|15|$ **28.** $-|-94|$

29. $|5|$ **30.** $-|-12|$

Simplify.

31. $-(3)$ **32.** $-(26)$ **33.** $-(1)$

34. (-4) **35.** (9) **36.** $-(-1)$

37. $-(-17)$ **38.** $-(-73)$ **39.** $-(19)$

40. (23) **41.** (14) **42.** $-(-64)$

43. (-23) **44.** (78) **45.** (-304)

46. $-(23)$ **47.** (-93) **48.** $-(42)$

49. $-(-23)$ **50.** (-2)

Simplify each of the following. Notice the difference between simplifying absolute value symbols or removing parentheses.

51. (-10) **52.** (8)

53. $|-7|$ **54.** $-(8)$

55. $-(-5)$ **56.** $-|4|$

57. $-|-13|$ **58.** (93)

59. $|4|$ **60.** $-|67|$

61. $|1|$ **62.** $-(5)$

63. (-99) **64.** $|-11|$

65. $-|-2|$ **66.** $-(77)$

67. $-|83|$ **68.** (14)

69. $|55|$ **70.** (-6)

4.3 **Addition of Integers**

OBJECTIVE

● Add integers.

To add integers we use distance from zero, absolute value, and the direction of the numbers, positive or negative. In the following examples, we will illustrate them using a number line.

EXAMPLE 1 To add 4 + 5

Always begin counting at 0, the origin. Since the addends (numbers to be added together) are positive integers, count 4 units in a positive direction (to the right) and from that point continue for 5 more units in a positive direction (to the right). The sum is 9 units, (9 units from 0).

$$4 + 5 = 9$$

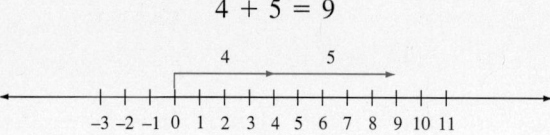

To test the Commutative Property of 4 + 5, does 4 + 5 = 5 + 4? This time, begin with positive 5, and from that point count 4 more in a positive direction.

$$5 + 4 = 9$$

▦ **USING THE CALCULATOR**

Enter 5 + 4 = 9 ← Display

EXAMPLE 2 To add (−5) + (−4)

Always begin counting at 0, the origin. Since the addends are negative integers, count 5 units in a negative direction; from that point, count 4 units more in a negative direction.

$$-5 + (-4) = -9$$

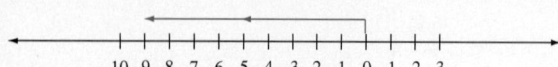

To test the Commutative Property, does −5 + (−4) = −4 + (−5)? To add two negative numbers, count 4 in a negative direction; from that point, count 5 more in a negative direction.

$$-4 + (-5) = -9$$

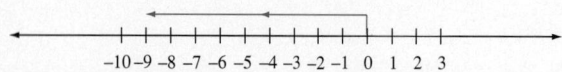

▦ **USING THE CALCULATOR**

Enter 4 +/- + 5 +/- = −9 ← Display

EXAMPLE 3 To add 8 + (−3)

If the addends have different signs, remember to begin at 0 and move 8 units in a positive direction, then add (−3) by moving 3 units in a negative direction.

$$8 + (-3) = 5$$

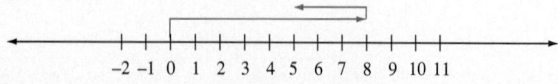

This problem is more difficult to understand because the numbers have different directions.

▦ *USING THE CALCULATOR*

Enter 8 <kbd>+</kbd> 3 <kbd>+/−</kbd> <kbd>=</kbd> 5 ← Display

---

1. $7 + 4$

Again, to check the Commutative Property, does $8 + (-3) = (-3) + 8$? Begin at 0 and count 3 units in a negative direction, then add 8 by moving 8 units in a positive direction.

$$-3 + 8 = 5$$

2. $-3 + (-6)$

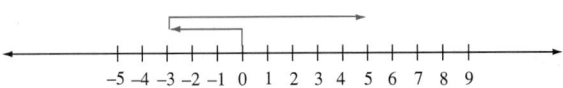

3. $5 + (-3)$

EXAMPLE 4   To add $5 + (-8)$

Begin at 0 and count 5 units in a positive direction. From that point add -8 by counting 8 units in a negative direction.

$$5 + (-8) = -3$$

4. $-5 + 3$

5. $-15 + (-20)$

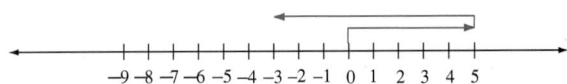

Again, to check the Commutative Property, does $5 + (-8) = -8 + 5$?

Begin at 0 and count 8 units in a negative direction, then add 5 by moving 5 units in a positive direction.

6. $14 + (-8)$

$$-8 + 5 = -3$$

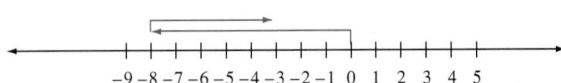

▦ *USING THE CALCULATOR*

Enter 8 <kbd>+/−</kbd> <kbd>+</kbd> 5 <kbd>=</kbd> −3 ← Display

---

> **RULE Addition of integers**
>
> **1.** To add integers that have like signs, add the absolute values of the integers. The sum has the same sign as the original integers.
> **2.** To add integers that have unlike signs, subtract the smaller absolute value from the larger absolute value. The sum has the sign of the integer with the greater absolute value.

The Associative Property of Addition

$(a + b) + c = a + (b + c)$, which holds true for all real numbers holds true for the integers which are a subset of the set of real numbers. (See Section 2.3.)

EXAMPLE 5 To test the Associative Property more than two addends are needed.

Does $[2 + (-5)] + 4 = 2 + [(-5) + 4]$? To simplify $[2 + (-5)] + 4$, begin by working inside the brackets. As always, begin at 0. Add 2 and -5. Count 2 units in a positive direction, and to add a (-5) count 5 units in a negative direction to reach (-3). Add 4 by moving 4 units in a positive direction.

$$[2 + (-5)] + 4$$
$$(-3) + 4 = 1$$

$$-5\ -4\ -3\ -2\ -1\ \ 0\ \ 1\ \ 2\ \ 3\ \ 4\ \ 5$$

Use the Associative Property to regroup the addends:

To simplify $2 + [(-5) + 4]$, add the numbers inside the brackets first. Count 5 units in a negative direction, and add 4 by moving 4 units in a positive direction to (-1). Add 2 by moving 2 units in a positive direction.

$$2 + [(-5) + 4]$$
$$2 + (-1) = 1$$

$$-5\ -4\ -3\ -2\ -1\ \ 0\ \ 1\ \ 2\ \ 3\ \ 4\ \ 5$$

DEFINITION The **identity element of addition** is zero because zero added to any number leaves the number identical to what it was. (See Section 2.3.) In symbols, we say $a + 0 = a$. We can test the property using integers.

EXAMPLE 6 $$8 + 0 = 8$$
$$\text{or}$$
$$-8 + 0 = -8$$

EXAMPLE 7 Adding opposites always equals 0. This is illustrated in Example 7. This is the property of additive inverses. See Section 4.2.

EXAMPLE 8 $5 + (-5) = 0$

APPLICATION 1

Leslie owed Sarah $10, Maria $5, and Sasha $15. How much money does Leslie owe?

$$-10 + (-5) + (-15) = -30$$

Answer: Leslie owes a total of $30. ◆

APPLICATION 2

As a birthday gift, you were given $50, but you owed $17 to your mother and $18 to your girlfriend. How much do you have left to spend after you pay your debts? $50 + (-17) + (-18) = 15$.

Answer: You have $15 to spend. ◆

Practice Problems

Simplify.

7. $-6 + (-2) + (-3)$

8. $[7 + 4] + (-8)$

9. $(-8) + 4 + (-4)$

APPLICATION 3

In a football game, suppose the quarterback passes the ball 10 yards forward, then he is tackled 5 yards behind the line of scrimmage. Next, he throws a 17 yard pass. How far has the ball been advanced? In mathematical symbols, the yard gain would be $10 + (-5) + 17 = 22$.

Answer: The ball has been advanced 22 yards down the field. ◆

APPLICATION 4

Look at the thermometer. Use it as a numberline if it will help you.
  a. If the temperature is 10° F and it increases by 7° F, mathematically you have $10 + 7 = 17$ or 17° F.
  b. If the temperature is 15° F and it drops by 5° F, mathematically you have $15 + (-5) = 10$ or 10° F.
  c. If the temperature is 10° F below zero and increases by 10° F, mathematically you have $-10 + 10 = 0$ or 0° F.
  d. If the temperature is −1° F and it drops 12° F, mathematically you have $-1 + (-12) = -13$ or −13° F. ◆

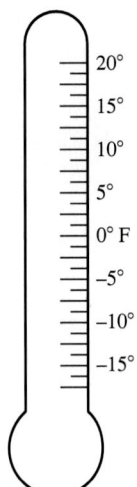

ANSWERS TO PRACTICE PROBLEMS
1. 11 2. −9 3. 2 4. −2 5. −35
6. 6 7. −11 8. 3 9. −8

PROBLEM SET 4.3

Use the number line to add the following numbers.

-10 -9 -8 -7 -6 -5 -4 -3 -2 -1 0 1 2 3 4 5 6 7 8 9 10

1. $4 + 3$ **2.** $9 + (-5)$

3. $5 + (-2)$ **4.** $-7 + 3$

5. $-8 + 4$ **6.** $-4 + (-1)$

7. $-6 + (-4)$ **8.** $-9 + 0$

9. $-3 + (-5)$ **10.** $5 + (-5)$

Add the following numbers using rules for addition of integers. Check your answers with a calculator.

11. $(+5) + (+9)$

12. $(-5) + (-9)$

13. $(+5) + (-9)$

14. $(-5) + (+9)$

15. $(-8) + (-6)$

16. $9 + (-9)$

17. $9 + (-16)$

18. $-1 + (-16)$

19. $-16 + 12$

20. $9 + (-19)$

21. $(-12) + (-18)$

22. $25 + 85$

23. $93 + (-102)$

24. $0 + (-150)$

25. $-117 + 286$

26. $-51 + 36$

27. $-78 + 49 + 78$

28. $-2 + 5 + (-3)$

29. $97 + 3 + (-200)$

30. $(-92) + 16 + (-35)$

31. $63 + 36 + (-59)$

32. $(-23) + (-37) + 43$

33. $93 + (-45) + (-15) + 27$

34. $(-87) + (-4) + 9 + (-6)$

35. $427 + (-384) + 749 + (-257)$

36. $-8 + 3$

37. $0 + (-8)$

38. $8 + (-7)$

39. $20 + [(-14) + (-10)]$

40. $(-6) + (-9) + (-5)$

41. $-4 + 5 + (-2)$

42. $(-11) + (-9) + 7$

43. $[8 + (-8)] + (-7)$

44. $18 + (-15) + 17$

45. $35 + (-50) + 10 + (-5)$

46. $-6 + (-11)$

47. $15 + (-32)$

48. $(-8) + (12)$

49. $(-11) + 11$

50. $0 + (-21)$

Use mathematical symbols to express the following and simplify.

51. Find the sum of (-6) and $-(8)$.

52. What is (-12) increased by (-3)?

53. Find the sum of (-18) and 16.

54. What number do you add to 9 to get 4?

55. What number do you add to (-4) to get 7?

56. What number do you add to (-6) to get (-10)?

57. What is 17 increased by (-7)?

58. Find the sum of (-8), (-9), and 27.

Applications:

59. If the temperature is 2° F below zero and it drops 12° F overnight, what is the temperature in the morning?

60. The morning temperature is 6° F below zero. During the day, the temperature rises 56° F. What is the new temperature?

61. I owe Kissee $26, Isa $41, and Pete $12. How much money do I need to repay my debts?

62. The quarterback threw the ball for a gain of 42 yards. On the next play, the running back ran the ball for 3 yards. On the third play, the quarterback was tackled and lost 5 yards. How many yards were gained on those plays?

63. On the coldest day last winter, the temperature was $-24°$ F. On the hottest day last summer, the temperature was 116° F warmer. How hot was the hottest day last summer?

64. The quarterback threw the ball for a gain of 32 yards. On the next play, the running back ran the ball for 8 yards. On the third play, the quarterback was tackled and lost 10 yards. How many yards were gained on those plays?

4.4 Subtraction of Integers

OBJECTIVE

● Subtract integers.

Subtraction is the inverse or opposite of addition. A subtraction problem can be rewritten as an addition problem. For example, $13 - 8 = 5$ can be rewritten $8 + 5 = 13$. Subtraction is defined in terms of addition.

EXAMPLE 1 $6 - 2 = 4$ because $4 + 2 = 6$

To show this using integers, let us look at some examples:

$$4 - 3 = [\ ]$$ As an addition problem it would be:

$$[\ ] + 3 = 4$$ What number fills both boxes? 1.

Therefore, $4 - 3 = 1$ and $1 + 3 = 4$.

EXAMPLE 2 $4 - (-1) = [\ ]$ As an addition problem this would be:

$$[\ ] + (-1) = 4$$ What number must be added to (-1) to make 4? 5.

Therefore, $5 + (-1) = 4$ and $4 - (-1) = 5$.

EXAMPLE 3   To explain why subtraction works:

$$3 - 2 = 1$$
$$3 + (-2) = 1$$

Therefore, it would be true to say that

$$3 - 2 = 3 + (-2).$$

Subtracting 2 is the same as adding the opposite of 2, which is (-2). This is true for all real numbers.

RULE Subtraction of integers

For all real numbers a and b, $a - b = a + (-b)$.

To subtract integers, use the definition of subtraction to change the problem from subtraction to addition.

$$3 - 2 = 3 + (-2)$$

Then use the rules for addition of integers and add.

EXAMPLE 4 $6 - 2$ can be changed to $6 + (-2) = 4$

$$6 - 2 = 4$$

Subtracting 2 is equivalent to adding (-2).

EXAMPLE 5 $-7 - 3$ can be changed to: -7 plus the opposite of 3, which is (-3). Subtracting 3 is the same as adding (-3).

$$-7 + (-3) = -10$$
$$\text{so}$$
$$-7 - 3 = -10$$

USING THE CALCULATOR

Enter 7 +/− − 3 = −10 ← Display

Note When entering an integer subtraction problem into the calculator you *do not* have to convert the problem to addition.

EXAMPLE 6 −8 − 4 is −8 plus the opposite of 4, which is (−4), so

$$-8 + (-4) = -12, \text{ so } -8 - 4 = -12.$$

EXAMPLE 7 13 − (−5) is 13 plus the opposite of (−5), which is 5, so

$$13 + 5 = 18, \text{ so } 13 - (-5) = 18.$$

Note The symbol − can be interpreted three different ways.

1. 6 − 4 is read 6 minus 4 and means subtract.
2. −(3) is read the opposite of 3.
3. −9 is read negative nine.

The way the symbol is interpreted has no effect on the result, which will be the same no matter which interpretation is used.

USING THE CALCULATOR

Enter 13 − 5 +/− = 18 ← Display

EXAMPLE 8 −15 − (−20)
  −15 plus the opposite of (−20) which is 20
  −15 + 20 = 5, so −15 − (−20) = 5

USING THE CALCULATOR

Enter 15 +/− − 20 +/− = 5 ← Display

Note Notice that in all these problems the first integer *does not* change. The subtraction sign becomes an addition sign, and the opposite of the second integer is taken.

EXAMPLE 9 6 − (−5) − [4 − 5]
Follow the rules for order of operations (Section 2.6). Work inside the brackets first.

6 − (−5) − [4 − 5]
6 − (−5) − [4 + (−5)]
6 − (−5) − [−1]   Use the definition of subtraction
6 + (+5) + (+1)   to change to an addition problem.
6 + 5 + 1 = 12

USING THE CALCULATOR

Enter 6 − 5 +/− − (4 − 5) = 12 ← Display

Practice Problems

Subtract the following integers.

1. 19 − 7

2. 15 − 20

3. −8 − 6

4. −4 − (−2)

5. 17 − (−3)

EXAMPLE 10 $(6 + 4) - (7 - 4)$

$(10) - (3)$ Work inside the parentheses first.

$10 - 3 = 7$

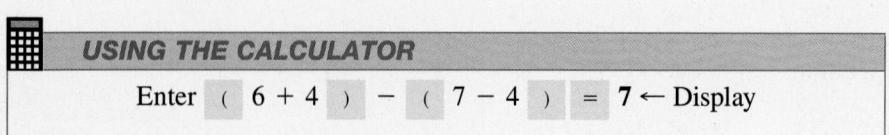

📷 **USING THE CALCULATOR**

Enter   (6 + 4   )   −   (7 − 4   )   =   **7** ← Display

Review the vocabulary Section 3.4 before trying these problems.

EXAMPLE 11 Find the difference between 9 and (-4).

$$9 - (-4) = 9 + 4 = 13$$

APPLICATION 1

Find the difference or the actual distance, between a plane flying at 30,000 feet above sea level and a submarine 267 feet below sea level. First, subtract the number of feet below sea level, which would be negative. Mathematically, this would be

$$30,000 - (-267) = 30,000 + 267 = 30,267.$$

Answer: 30,267 feet between the plane and the submarine. ◆

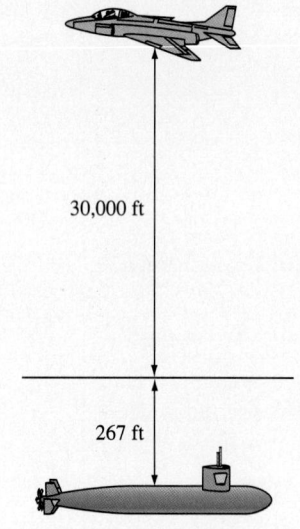

30,000 ft

267 ft

APPLICATION 2

Find the difference between 1992 A.D. and 18 B.C. How many years separate these two dates? B.C. would be negative, so

$$1992 - (-18) = 1992 + 18 = 2010 \text{ years.}$$

500 BC   0   500 AD   1000 AD   1500 AD   2000 AD

18 BC   1992 AD

Answer: 2010 years. ◆

PROBLEM SET 4.4

Subtract the following numbers by changing each problem to an addition problem. Check each answer with a calculator.

1. $6 - 4$

2. $6 - (-4)$

3. $-6 - (-4)$

4. $-6 - 4$

5. $7 - (-7)$

6. $0 - 2$

7. $3 - 7$

8. $40 - (-25)$

9. $0 - (-5)$

10. $-12 - 7$

11. $-8 - (-15)$

12. $-8 - (-8)$

13. $-9 - 14$

14. $-3 - (-5)$

15. $(-6) - 7$

16. $4 - 7$

17. $-2 - (-7)$

18. $86 - 77$

19. $-137 - (-142)$

20. $(+5) - (-6)$

21. $-42 - (-42)$

22. $3 - 10 + 7 - 4 - 5 + 9$

23. $-15 - 3$

24. $-9 - (-3)$

25. $-10 - 6 + 3$

26. $-(-3) - 5$

27. $6 - 9$

28. $-15 - 6$

29. $5 - (-4) - [3 - 4]$

30. $4 - (5 + 2)$

31. $-(-2) - [4 - 6]$

32. $-86 - 77$

33. $49 - 69$

34. $24 - (-24)$

35. $-2 - 7$

36. $(5 + 3) - (6 - 2)$

37. $-16 - (-16)$

38. $-11 - 4 - 15$

39. $-217 - 13$

40. $-36 - (-4)$

Use mathematical symbols to express the following and simplify.

41. Find the difference between 8 and (-3).

42. What is (-6) decreased by (-4)?

43. Subtract (-9) from 10.

44. Subtract 3 from (-7)

45. Find the difference between (-8) and 7.

46. What is (-27) decreased by 10?

47. Subtract (-9) from the sum of 7 and (-3).

48. Find the difference between 6 and (-3).

49. Subtract (-2) from the sum of (-12) and 9.

50. Find the difference between (-9) and (-3).

Applications:

51. Find the difference between a plane flying at 28,340 feet above sea level and a submarine at 271 feet below sea level.

52. How many years separate 1873 A.D. from 37 B.C.?

4.5 **Multiplication of Integers**

OBJECTIVE

● Multiply integers.

As you learned in Chapter 2, multiplication is a quick way to calculate repeated addition.

EXAMPLE 1   3 · 4 means adding 3 sets of 4

$$4 + 4 + 4 = 12 \text{ so } 3 \cdot 4 = 12$$

or 4 · 3 means adding 4 sets of 3.

$$3 + 3 + 3 + 3 = 12 \text{ so } 4 \cdot 3 = 12$$

We can use a number line to count out 3(4), which is adding 3 sets of 4:

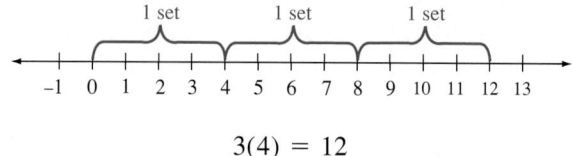

$$3(4) = 12$$

EXAMPLE 2   3(−4) means adding 3 sets of (−4)

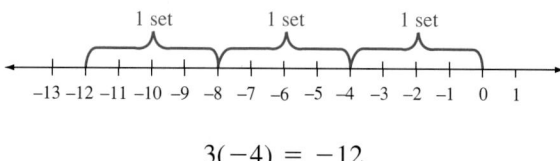

$$3(-4) = -12$$

EXAMPLE 3   −3(4). As you cannot count out negative sets, use the Commutative Property of Multiplication to change −3(4) to 4(−3), which means adding 4 sets of −3.

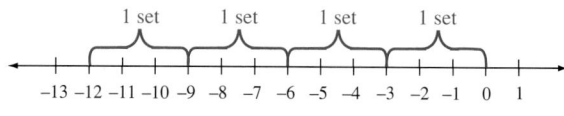

$$4(-3) = -12 \quad \text{and} \quad -3(4) = -12$$

To show that (−3)(−4) = 12, we can look at the patterns from the previous examples. We have determined that

$$4(-3) = -12$$
$$3(-3) = -9$$
$$2(-3) = -6$$
$$1(-3) = -3 \qquad \text{Any number multiplied by 1 maintains its identity.}$$
$$0(-3) = 0 \qquad \text{Any number multiplied by 0 is 0.}$$

Examine the pattern of the numbers in the list above. As the changing factor becomes 1 smaller, the product gets 3 larger.

−9 is 3 larger than −12.

−6 is 3 larger than −9.

−3 is 3 larger than −3.

Therefore, to maintain the pattern 3 more than 0 is 3.

$$(-1)(-3) = 3$$
$$(-2)(-3) = 6$$
$$(-3)(-3) = 9$$
$$(-4)(-3) = 12$$

Multiply the following integers.

1. (−4)(8)

2. 7(9)

3. (−6)(8)

4. 9 (−6)

5. 15(−5)

6. (−18)(−4)

7. (−6)14

8. 17(23)

USING THE CALCULATOR

Try multiplying $(-4)(-3)$. Enter 4 +/− × 3 +/− = **12** ← Display

RULE **Multiplication of integers**

Multiply the absolute values. If the numbers have the same sign, the product is positive. If the numbers have different signs, the product is negative.

$$(2)(6) = 12$$

Same sign, product is positive.

$$(-2)(-6) = 12$$

Same sign, product is positive.

$$(-2)(6) = -12$$

Different signs, product is negative.

$$(2)(-6) = -12$$

Different signs, product is negative.

APPLICATION 1

A car manufacturing company retooled, and on the first 100 cars built, the company lost $78 on each car. How much would the company lose?

Answer: $(-78)(100) = -7800$, which means the company lost $7800. ◆

APPLICATION 2

A running back lost 5 yards on each of 6 plays. How many yards did he lose?

Answer: $-5(6) = -30$. He lost 30 yards. ◆

APPLICATION 3

The university library has approximately 357 books stolen each semester. Approximately how many books have been stolen in the last 10 semesters?

Answer: $(-357)(10) = -3570$. The library lost approximately 3570 books. ◆

PROBLEM SET 4.5

Multiply using the rules for multiplication of integers. Check your answers with a calculator.

1. 6(4)　　　**2.** 6(−4)　　　**3.** (−6)(−4)

4. (−6)(4)　　　**5.** (−7)(−3)　　　**6.** 8(−4)

7. (9)7　　　**8.** (−6)(−7)　　　**9.** 12(−3)

10. (−11)(4)　　　**11.** 4(−5)(1)　　　**12.** 4(−5)(−1)

13. $2(-3)(-1)(4)$ **14.** $5(5)(-3)$ **15.** $6(-9)$

16. $(-1)(24)(0)$ **17.** $7(-7)(2)$ **18.** $9(-9)$

19. $(-4)(-3)(-2)(4)$ **20.** $11(-11)$ **21.** $5(-5)(5)$

22. $4(2)6$ **23.** $-21 \cdot 4$ **24.** $-3(-33)$

25. $[-16 \cdot 9](-2)$ **26.** $(-24)(0)$ **27.** $(25)(11)$

28. $(-4)(-9)(-6)$ **29.** $0(-5)(-4)$ **30.** $[-39][-29]$

31. $(-5)(45)$ **32.** $(-25)(12)(1)$ **33.** $2(-3)(-33)$

34. $[-102][12]$ **35.** $(6)(7)(-4)$ **36.** $(0)(8)(96)$

37. $(-3)(-107)(0)$ **38.** $(64)(4 \cdot 2)$ **39.** $19 \cdot 10 \cdot 8$

40. $(1)(-1)(1)(-1)$ **41.** $(8)(-3)(-4)$ **42.** $(5)(5)(4)$

43. $(100)(1)(0)$ **44.** $(13)(2)(-1)$

Use mathematical symbols to express the following and simplify.

45. Find the product of -9 and -8.

46. Multiply 7 by -3.

47. Find the product of 12 and -3.

48. Find the product of -11 and 10.

49. Multiply the sum of 3 and (-5) by 7.

50. Multiply the difference of (-8) and (-2) by 27.

Applications:

51. If a toy manufacturer lost $7 on each bike sold, how much would be lost if 297 bikes were sold?

52. A running back lost 8 yards on each of 6 plays. How many yards did he lose all together?

4.6 Division of Integers

OBJECTIVE

● Divide integers.

Division can be defined in terms of multiplication. Division is the inverse of multiplication.

In multiplication, factor · factor = product.
In division, product ÷ factor = factor.

EXAMPLE 1 12 ÷ 3 = [ ], therefore, 3 · [ ] = 12
What number multiplied by 3 will give us 12? 4.
Therefore, 12 ÷ 3 = 4.

EXAMPLE 2 (−12) ÷ 3 = [ ], therefore, 3 · [ ] = −12
What number multiplied by 3 will give (−12)? (−4). Therefore, (−12) ÷ 3 = (−4).

USING THE CALCULATOR
Enter 12 +/− ÷ 3 = −4 ← Display

EXAMPLE 3 (−12) ÷ (−3) = [ ], therefore, (−3) · [ ] = −12
What number multiplied by (−3) will give (−12)? 4.
Therefore, (−12) ÷ (−3) = 4.

USING THE CALCULATOR
Enter 12 +/− ÷ 3 +/− = 4 ← Display

EXAMPLE 4 12 ÷ (−3) = [ ], therefore, (−3) · [ ] = 12
What number multiplied by (−3) will give 12? (−4).
Therefore, 12 ÷ (−3) = −4.

Note Looking at these examples, you will see that if you divide two numbers with the same sign, the quotient is positive. If you divide two numbers with different signs, the quotient is negative.

RULE Division of integers
Divide the absolute values. If the numbers have the same signs, the quotient is positive. If the numbers have different signs, the quotient is negative.

8 ÷ 2 = 4   Same signs, positive quotient.
−8 ÷ −2 = 4   Same signs, positive quotient.
−8 ÷ 2 = −4   Different signs, negative quotient.
8 ÷ (−2) = −4   Different signs, negative quotient.

Note Division by 0 is undefined. −9 ÷ 0 = undefined.
Division into 0 is 0. 0 ÷ −9 = 0.

Practice Problems

Divide the following integers.

1. 21 ÷ 7

2. (−84) ÷ 12

3. 63 ÷ (−7)

4. −14)42

5. −36)−72

6. −18/9

7. −50 ÷ 5

8. −28 ÷ (−7)

9. 54 ÷ (−6)

ANSWERS TO PRACTICE PROBLEMS
1. 3 2. −7 3. −9 4. −3 5. 2
6. −2 7. −10 8. 4 9. −9

PROBLEM SET 4.6

Find the following quotients.

1. $12 \div 3$

2. $-12 \div 3$

3. $12 \div (-3)$

4. $-12 \div (-3)$

5. $\dfrac{-27}{3}$

6. $\dfrac{72}{-9}$

7. $\dfrac{-42}{-6}$

8. $5\overline{)45}$

9. $-4\overline{)-36}$

10. $8\overline{)-32}$

11. $\dfrac{49}{-7}$

12. $(-63) \div (-9)$

13. $11\overline{)121}$

14. $90 \div (-5)$

15. $\dfrac{-56}{-7}$

16. $9 \div (-1)$

17. $\dfrac{-63}{0}$

18. $\dfrac{0}{-1}$

19. $144 \div 12$

20. $\dfrac{-28}{4}$

21. $-17\overline{)187}$

22. $\dfrac{-625}{-25}$

23. $288 \div (-16)$

24. $\dfrac{225}{-15}$

25. $3\overline{)147}$

26. $-19\overline{)-418}$

27. $\dfrac{710}{-710}$

28. $-13\overline{)169}$

29. $-400 \div 8$

30. $\dfrac{329}{0}$

Use mathematical symbols to express the following and simplify.

31. Find the quotient of -54 and -9.

32. Divide -3 into 36.

33. How many times does 6 divide into 48?

34. Find the quotient of 55 and -11.

35. Find -60 divided by -12.

36. Divide 72 by -9.

37. Find the quotient of -99 and -11.

38. How many times does -13 divide into -143?

39. Divide -204 by -17.

40. Divide -19 into -361.

4.7 **Order of Operations and Averages Using Integers**

OBJECTIVE

● Use the order of operations rules to simplify problems containing more than one operation.

Combining operations requires following the order of operations discussed and practiced in Section 2.6. Review this section before working with combinations of integers.

RULE Order of operations

1. Perform all operations enclosed within parentheses or other grouping symbols.
2. Exponents: raise to indicated powers, or find the indicated root.
3. Perform all multiplication and division in order from left to right.
4. Perform all addition and subtraction in order from left to right.

Refer back to Sections 2.5 and 3.2 and review exponents. Exponents using integers as bases are simplified following the same rules.

EXAMPLE 1 $(-3)^2 = (-3)(-3) = 9$

EXAMPLE 2 $(-2)^5 = (-2)(-2)(-2)(-2)(-2) = -32$

Remember that the exponent refers only to the letter or number directly to its left. In the term $4x^2$, the exponent refers only to the x.

$$4x^2 = 4(x)(x) = 4x^2$$

In the term -4^2, the exponent refers only to the 4.

$$-4^2 = -(4)(4) = -16$$

However, an exponent outside a parentheses raises the entire quantity inside the parentheses to the power of the exponent.

EXAMPLE 3 $(4x)^2 = (4x)(4x) = 16x^2$

EXAMPLE 4 $(-6)^2 = (-6)(-6) = 36$

EXAMPLE 5 $(3 + 2)^3 = (5)^3 = (5)(5)(5) = 125$

EXAMPLE 6 $(10 - 12)^2 = (-2)^2 = (-2)(-2) = 4$

Note Be careful in working with exponents when the base has a negative sign. The exponent affects only the number immediately before it.

Note A negative number raised to an even power will always result in a positive number.

EXAMPLE 7 $(-5)^2 = (-5)(-5) = +25$
For $x > 0$, $(-x)^2 = (-x)(-x) = x^2$

Note A negative number raised to an odd power will always result in a negative number.

EXAMPLE 8 $(-4)^3 = (-4)(-4)(-4) = -64$
$(-2)^5 = (-2)(-2)(-2)(-2)(-2) = -32$
For $y > 0$, $(-y)^3 = (-y)(-y)(-y) = -y^3$

Note Note the difference between -4^2 and $(-4)^2$.

$$-4^2 = -(4)(4) = -16$$
$$(-4)^2 = (-4)(-4) = +16$$

$-x^2$ cannot be simplified any further.

$$(-x)^2 = (-x)(-x) = +x^2$$

EXAMPLE 9 $(4 - 10)^2 = (-6)^2 = (-6)(-6) = 36$

USING THE CALCULATOR

(4 − 10) x^2 = **36** ← Display

Practice Problems

Evaluate these exponents.

 1. -3^2

 2. $(-8)^2$

 3. -2^3

 4. $(8 - 3)^2$

 5. $(7 - 13)^2$

EXAMPLE 10 $7 - 2^2 = 7 - (2)(2) = 7 - 4 = 3$

EXAMPLE 11 $5 + 3^2 - 4 = 5 + (3)(3) - 4 = 5 + 9 - 4 = 10$
Other combinations. Use the order of operations.

EXAMPLE 12 $7(2 - 6) - 4$
$\qquad$ $7(-4) - 4$
$\qquad$ $-28 - 4$
$\qquad$ $-28 + (-4)$
$\qquad$ -32

USING THE CALCULATOR

7 $\boxed{\times}$ $\boxed{(}$ 2 $\boxed{-}$ 6 $\boxed{)}$ $\boxed{-}$ 4 $\boxed{=}$ $\;$ **−32** ← Display

Note   Recall that any nonzero number raised to the zero power equals the number 1.

EXAMPLE 13 To work with nested parentheses, work the innermost parentheses first, and work out to the outer parentheses last.

$\qquad -4[3 + 5^0(4 - 9)]$
$\qquad -4[3 + 1(-5)]$
$\qquad -4[3 - 5]$
$\qquad -4[-2] = 8$

USING THE CALCULATOR

To work nested parentheses, work the inside parentheses first and store the answer.

5 $\boxed{+/-}$ $\boxed{y^x}$ 0 $\boxed{=}$ $\boxed{\times}$ $\boxed{(}$ 4 $\boxed{-}$ 9 $\boxed{)}$ $\boxed{=}$ $\boxed{\text{STO}}$

4 $\boxed{+/-}$ $\boxed{\times}$ $\boxed{(}$ 3 $\boxed{+}$ $\boxed{\text{RCL}}$ $\boxed{)}$ $\boxed{=}$ **8** ← Display

EXAMPLE 14 $\dfrac{4(-7) + (-1)5}{-2 - 1} = \dfrac{-28 + (-5)}{-2 + (-1)} = \dfrac{-33}{-3} = 11$

USING THE CALCULATOR

4 $\boxed{\times}$ 7 $\boxed{+/-}$ $\boxed{+}$ 1 $\boxed{+/-}$ $\boxed{\times}$ 5 $\boxed{=}$ **−33** ← display

$\boxed{(}$ 2 $\boxed{+/-}$ $\boxed{-}$ 1 $\boxed{)}$ $\boxed{=}$ **11** ← Display

Notice that the numbers in the denominator need to be enclosed in parentheses.

Averages

Averages of integers are found in the same way as averages of whole numbers (see Section 2.7). To find the average of a set of numbers, find the sum of the elements of the set, then divide that sum by the number of elements in the set.

Practice Problems

Simplify.

6. $8 + 2^2 - 9$

7. $4(4 - 9) + 4$

8. $8[2 + 3(4^2 + 2)]$

9. $\dfrac{-4(3) + (-2)(4)}{-3 - 2}$

EXAMPLE 15 Find the average of $\{-16, 3, -13, 6\}$.

$$\frac{-16 + 3 + (-13) + 6}{4}$$ The sum of the elements divided by the number of elements.

$$\frac{-20}{4} = -5$$

The average of the set $\{-16, 3, -13, 6\}$ is -5.

USING THE CALCULATOR

Enter 16 +/- + 3 + 13 +/- + 6 =

$-20 \leftarrow$ display $\div$ 4 = **−5** $\leftarrow$ Display

EXAMPLE 16 Find the average of the set $\{7, 9, 0, (-8)\}$.

$$\frac{7 + 9 + 0 + (-8)}{4}$$

$$\frac{8}{4} = 2$$

Find the sum and divide by 4. The average of the set $\{7, 9, 0, (-8)\}$ is 2.

Note Zero is just as much a number as any other integer.

APPLICATION 1

For 5 days, the heating engineer monitored the deviations from the desired heating setting, which were $-2°$ F, $-20°$ F, $19°$ F, $-3°$ F, and $1°$ F. What was the average temperature deviation?

$$\frac{-2 + (-20) + 19 + (-3) + 1}{5}$$

$$\frac{-5}{5} = -1$$

Answer: Therefore, there was a 1° F average temperature deviation. ◆

APPLICATION 2

On three consecutive days, a stock price changed up $1, up $6, and down $4. What is the average change in the stock price?

$$\frac{1 + 6 + (-4)}{3} = \frac{3}{3} = 1$$

Answer: On average, there was a $1 rise in the stock price each day. ◆

● *PROBLEM SET 4.7*

Simplify the following expressions using the rules for order of operations. Work the problems using paper and pencil, then check your answers with a calculator.

1. $3 + 7[9 - (3 - 5)]$

2. $5[-6 + 9(-3 + 2)] - 8$

3. $4[9 - (7 + 4)]$

4. $6 - 4[15 - 4(5 - 7)]$

5. $7 + [(9 - 7) + (-2 - 3)]$

6. $9[(3 - 5)(-4)]$

7. $3\{4[-30 - 5(-9 + 3)]\}$

8. $7 + 6[3(-5) + 4]$

9. $[6 - 3(5 - 9)]7$

10. $9 - 2[(7 - 10) - 2]$

11. $-10[(-2^2 + 7) + (-4 + (-10))]$

12. $4[(3 + 8)(-5 + 5)]$

13. $-2 - [(8 - 1)(5)]$

14. $3 - [6 - 3(5 - 9)]$

15. $17(-4)$

16. $(-5)(-2)$

17. $(-48) \div 4$

18. $(-24) \div (-6)$

19. $(-5) + (-13)$

20. $-5 - 13$

21. $-5 - (-13)$

22. $-8 + 2$

23. $8 + (-2)$

24. $8 - 2$

Expand the following exponents and then multiply. For example:
$(-4)^3 = (-4)(-4)(-4) = -64$

25. $(-2)^2$

26. $(-9)^2$

27. $(-1)^3$

28. $(-4)^2$

29. $(-7)^3$

30. $(-8)^3$

31. $(-3)^4$

32. $(-4)^2(3)^2$

33. $(-2)^2(-3)^2$

34. $(6)^2(-2)^3$

35. $(-4)(-2)(-1)(2)(-2)$

36. $\dfrac{-36}{0}$

37. $-14 - 2 - 3 + 7 + 41$

38. $6 - 13 + 5 - 7 - 9 - 10$

39. $2(5 - 6)^2 + 7 - 3$

40. $[3 + 7(2^2 - 3)] - 6$

41. $[3 + 7(2 - 3)^2]6$

42. $-6[3 + 7(2 - 3)]$

43. $\dfrac{4(5 + 3) + 3}{2(3) + 1}$

44. $\dfrac{5(-2) - 3(4)}{-2[3 - (-2)] - 1}$

45. $(-3)(-15)$

46. $13(-4)^2$

47. $(-54) \div (-9)$

48. $\dfrac{72}{-8}$

49. $-3 + 13$

50. $-3 - (-13)$

51. $-12 + (-10)$

52. $-12 - 10$

53. $4 + (-11)$

54. $4 - 11$

55. $5(-3)(4)(-1)(-2)$

56. $\dfrac{0}{-36}$

57. $5 - 13 - 19 + 7 - (8)^2$

58. $9 - (3)^2 - 19 + 7 - 8$

59. $12(-2^2) + [-3 + (-4)]$

60. $-5[2 + 9(3 - 6)]$

61. $[2 + 9(3 - 6)] - 5$

62. $[2 + 9(3 - 6)]5$

63. $\dfrac{8(4 + 2) - 6}{-8 - 5(-3)}$

64. $\dfrac{7(5) - 6(-4) - (-1)}{5[-2 - 4]}$

65. $9 - 3^2$

66. $9 + (-3)^2$

67. $4^2 - 4^2$

68. $3^3 - 14$

69. $-3^3 - 14$

70. $9^1 - 5^2$

71. $18° - 6^2$

72. $4(3^2 - 3^2)$

73. $5 + 2^2(3^2 + 4)$ **74.** $3^2 + 8 - 3^2$

Applications: Use mathematical symbols to express the following and simplify.

75. If a stock lost $4 on Monday and $7 on Tuesday, gained $12 on Wednesday and $9 on Thursday, then lost $5 on Friday, what was the average gain or loss for the week?

76. The temperature was $-2°$ F on Friday, $15°$ F on Saturday, and $-1°$ F on Sunday. What was the average temperature during the weekend?

77. The temperature was $-18°$ F at 7:00 A.M., by noon it was $8°$ F, and by 7:00 P.M. it was $-2°$ F. What was the average temperature that day?

78. A stock lost $7 on Monday, gained $2 on Tuesday, gained another $11 on Wednesday, but lost $9 Thursday and $12 Friday. What was the average gain or loss for the week?

79. For the last six flights sold by Ugo Airlines, their computers were down, causing an overbooking. As a result, Flight 1 was overbooked 20 seats, Flight 2 was overbooked 15 seats, Flight 3 had 10 empty seats, Flight 4 was overbooked 3 seats, Flight 5 had 19 empty seats, and Flight 6 was overbooked 3 seats. What was the average mistake in bookings?

80. The top scorers in the golf championships scored: first place 10 below par; second place 8 below par; third place 5 below par; fourth place 4 below par; fifth place 3 below par; sixth place even par; seventh place 2 above par; and eighth place 4 above par. What was the average score of the top eight players? (Hint: par is zero, above par is positive, and below par is negative.)

4.8 **Using the Addition Property of Equality to Solve Equations**

OBJECTIVE

● Use the Addition Property of Equality to solve equations.

Some people think that solving equations is all there is to algebra. This is not true, but working with equations can be a very satisfying experience—one that appeals to our sense of order.

DEFINITION   **Equation:** A statement that two algebraic expressions are equal.

Equations have their own set of properties.

| **Identity Property of Equality** |
| For all a, b, and $c \in$ R. $a = a$ |

EXAMPLE 1
$$5 = 5$$
$$-37 = -37$$
$$3(1 + 17) = 3(1 + 17)$$

| **Symmetric Property of Equality** |
| For all a, b, and $c \in$ R.   If $a = b$, then $b = a$. |

EXAMPLE 2   If $2 + 3 = 5$,   then $5 = 2 + 3$.
          If $-1(8 + 7) = -15$, then $-15 = -1(8 + 7)$

> **Transitive Property of Equality**
> For all a, b, and $c \in$ R.   If $a = b$, and $b = c$, then $a = c$.

EXAMPLE 3   If $1 + 2 = 3$, and $3 = 7 - 4$, then $1 + 2 = 7 - 4$.

> **Addition Property of Equality**
> For all a, b, and $c \in$ R.   If $a = b$, then $a + c = b + c$

The addition property of equality states that if two real numbers are equal, and if the same real number is added to both of them, then the sums formed will also be equal.

Note   This property also holds true when a and b represent expressions.

EXAMPLE 4   Since $5 = 3 + 2$

$$5 + 1 = 3 + 2 + 1 \quad \text{Add 1 to both sides.}$$
$$6 = 6 \quad \text{The sums are equal.}$$

DEFINITION   **Solution set:** The subset of the real numbers that can be substituted for the variable in an equation to make the equation a true statement.

EXAMPLE 5 $x + (-2) = 5$, where $x \in$ R.
In this example, we want to find the real numbers we can substitute for x to make the statement true. We can guess that the solution set is $\{7\}$, since $7 + (-2) = 5$. We can find the value of x in a more organized way by using the properties for real numbers and the properties for equalities.

If $\quad x + (-2) = 5,$
then $x + (-2) + 2 = 5 + 2,$
then $\qquad x + 0 = 5 + 2$, since $(-2) + 2 = 0$
then $\qquad x \quad = 5 + 2$, since $x + 0 = x$
so $\qquad x \quad = 7$

We say that $x = 7$ is the solution or that $\{7\}$ is the solution set. When we have found the values of the variable that make the statement true, we say we have "solved" the equation.

Note   In this book, we are interested in equations that contain only one variable and are first degree. By first degree, we mean the variable is raised to the first power; in other words, it has an implied or expressed exponent of 1. (See Section 3.1.)

Always check your answer. This is particularly easy in solving equations since we are looking for the number we can substitute for the variable to make the equation true.
From Example 5, check:

$x + (-2) = 5$ Write the original equation.
$7 + (-2) = 5$ Replace the variable with a number from the solution set.
$5 = 5$ Simplify each side of the equation.

If the result is a true statement, the number checked is a member of the solution set.

EXAMPLE 6   Solve for y: $y + 3 = 8$

$$y + 3 + (-3) = 8 + (-3) \qquad \text{Add } -3 \text{ to both sides.}$$
$$y + 0 = 5$$
$$y = 5$$

$$\text{Check: } y + 3 = 8$$
$$5 + 3 \ ? \ 8$$
$$8 = 8$$

Note   While the format used in these two examples clearly illustrates the properties being used, some students prefer to show their solutions in the following format.

EXAMPLE 7   Solve for x: $x + (-2) = 5$

$$\underline{ +2 \quad +2}$$
$$x + 0 = 7$$
$$x = 7$$

EXAMPLE 8   Solve for y: $y + 3 = 8$

$$\underline{ -3 \quad -3}$$
$$y + 0 = 5$$
$$y = 5$$

Use the method that seems easier for you or makes you more comfortable.

The solution set of a linear equation may consist of *one real number, all real numbers,* or *no real numbers.* So far, we have considered examples in which the solution set was one real number. Now let's look at the other two possibilities.

EXAMPLE 9 $y = y + 7$

$$y + (-y) = y + 7 + (-y)$$
$$0 = 7$$

But $0 \neq 7$. Because of order on the number line, $0 < 7$. Because of the principle of trichotomy, 0 cannot both equal 7 and be less than 7.

Solution Set: $\emptyset$ There are no real numbers, y which make the statement $y = y + 7$ true. This equation has *no solutions.*

EXAMPLE 10 $x + 3 = x + 5 - 2$ Simplify.

$$x + 3 = x + 3$$
$$x + 3 + (-3) = x + 3 + (-3) \qquad \text{Addition Property of Equality}$$
$$x + 0 = x + 0$$
$$x = x \qquad\qquad \text{This identity is true for all real numbers.}$$

Solution Set: All real numbers. The equation has *infinitely many* solutions.

Note   When solving equations, do so vertically. Have only one equal sign on a line, and write each equal sign directly under the one on the previous line. Check your work.

Practice Problems

Solve for the variable.

1. $x + (-5) = 8$

2. $y + 15 = 25$

3. $s - 5 = s$

4. $q + (-8) = -10$

5. $t - 3 = t + 2 - 5$

EXAMPLE 11

$$2x - x = 6 \quad \text{Simplify.}$$
$$2x - 1x = 6$$
$$x = 6$$

EXAMPLE 12

$$6 = 2x - x \quad \text{Simplify.}$$
$$6 = 2x - 1x$$
$$6 = x$$

EXAMPLE 13

$$-90 + x = 0$$
$$\underline{\quad 90 \quad} = \underline{90}$$
$$x = 90$$

EXAMPLE 14

$$g - (-5) = -18$$
$$g + 5 = -18$$
$$g + 5 + (-5) = -18 + (-5)$$
$$g = -23$$

Because of the Symmetric Property of Equality, the solutions to Examples 11 and 12 are equivalent.

The equations we have discussed in this section may vary in appearance, but the same procedure is used for solving them.

RULE **Solving an equation**

1. Simplify each side.
2. Apply the Addition Property of Equality.

APPLICATION

Yesterday, it took 18 gallons of gasoline to fill the 25 gallon tank in your roommate's truck. How many gallons were in the tank before you filled it?

Let x = The number of gallons of gas originally in the tank.

$$x + 18 = 25 \qquad \text{Translate the sentence into an equation.}$$
$$x + 18 + (-18) = 25 + (-18) \qquad \text{Solve the equation.}$$
$$x = 7$$

Refer to the original problem. Answer the question asked.

Answer: The tank had 7 gallons of gas in it. ◆

PROBLEM SET 4.8

Identify the property of equality illustrated.

1. $7 = 7$ _____

2. If $13 + 1 = 14$, then $14 = 13 + 1$. _____

3. If $8 + 17 = 25$ and $25 = 5 \cdot 5$, then $8 + 17 = 5 \cdot 5$. _____

Solve for the variable. Check your work.

4. $x + 5 = 8$

5. $y + 7 = 12$

6. $x - 98 = 100$

7. $s - 76 = 92$

8. $z + 8 = 6$

9. $w + 6 = 2$

10. $y + 5 = -2$

11. $x + 4 = -6$

12. $5 = y + 13$

13. $8 = p + 12$

14. $x - (-2) = -8$

15. $y - (-5) = -3$

16. $0 = y + 34$

17. $0 = x + 43$

18. $0 = a - 7$

19. $0 = c - 23$

20. $q + 50 = q$

21. $z - 70 = z$

22. $a + 37 = a$

23. $d - 87 = 3d - 2d$

24. $3 + s = s + 8 - 5$

25. $4 + r = r + 48 \div 12$

26. $12 - 3 - x = 9 - x$

27. $22 - x = 32 - 10 - x$

28. $19 + a = 6$

29. $13 + 12b - 11b = 3$

30. $15 + c = -5$

31. $9f - 8f - 11 = 33$

32. $z + 2 = 41 - 27$

33. $4q + (-3q) + 8 = 42$

34. $x + 17 = -5 + (-20)$

35. $2b - b - 10 = 6$

Use an equation to answer the question asked.

36. Northwest Central College is on the quarter system. If three days were added to each quarter, the result would be 48 days per quarter. How many days per quarter does the school currently have?

37. If I earn $15 in cash today, spend no money, and have twice as much money in my pocket returning from work as I did going to work, how much money did I have going to work?

SMALL GROUP ACTIVITY   *Understanding Integer Numbers*

Give each group the same set of four numbers, and have them see how many ways the numbers can be arranged using parentheses, addition, subtraction, and multiplication. For example: Give the numbers 2, 10, -4, and 1. Possible arrangements could be

$2 \cdot 10 + (-4) + 1 = 17$

$2(10) + (-4)(1) = 16$

$10(2)(-4)(1) = -80$

$10 - 2 - (-4) - 1 = 11$

See which group can make the most arrangements. Which group can make the largest and/or smallest answer. Investigate how the largest and/or smallest answer is made.

CHAPTER 4 REVIEW

The terms listed below are among those used in this chapter. Write a brief definition for each term. (Refer to the appropriate section in the chapter as needed.)

1. Set of integers _____

2. Principle of Trichotomy _____

3. Absolute value _____

4. Opposite or additive inverse _____

5. Equation _____

6. Identity Property of Equality _____

7. Symmetric Property of Equality _____

8. Transitive Property of Equality _____

9. Addition Property of Equality _____

10. Solution set _____

Write the rules for the following.

11. Simplifying an absolute value _____

12. Simplifying a set of parentheses _____

13. Addition of integers _____

14. Subtraction of integers _____

15. Multiplication of integers _____

16. Division of integers _____

After completing Chapter 4, you should be familiar with the following concepts and be able to apply them.

• *Recognize that integers have direction as well as distance from zero (4.1).*

• *Order integers (4.1)*

Fill in <, >, or = .

17. −3   −23   **18.** −3   −3

19. 0   −3   **20.** 19   19

21. −92   27   **22.** −4   4

23. −10   6   **24.** −5   −10

25. 0   15   **26.** 0   −15

• *Recognize the meaning of absolute value and evaluate the absolute value of a number (4.2).*

Simplify.

27. $|3|$ **28.** $|-54|$

29. $|-6|$ **30.** $|-86|$

31. $|342|$ **32.** $|-732|$

33. $|693|$ **34.** $|67|$

35. $|93|$ **36.** $|0|$

• *Determine the opposite or additive inverse of a number (4.2).*

Write the opposite of each of these numbers.

37. 7   **38.** −63

39. −14

40. 84

41. 0

• *Note the difference between finding the absolute value of a number and removing parentheses (4.2).*

Simplify.

42. $-|-9|$

43. $-(-76)$

44. $-|23|$

45. $-(53)$

46. $-|6|$

47. $-(-6)$

48. $-|-17|$

49. $-(27)$

50. (19)

51. $|-3|$

• *Add integers (4.3).*

52. $8 + 6$

53. $13 + (-6)$

54. $-8 + 6$

55. $(-7) + (-9)$

56. $8 + (-6)$

57. $(-6) + 16$

58. $-8 + (-6)$

59. $(-14) + (-18)$

60. $-3 + (-5) + 7 + (-8)$

61. $15 + (-15)$

• *Subtract integers (4.4).*

62. $8 - 6$

63. $17 - (-8)$

64. $-8 - 6$

65. $(-3) - 3$

66. $8 - (-6)$

67. $14 - 2$

68. $-8 - (-6)$

69. $-13 - (-13)$

70. $7 - (-3) - [4 - 6]$

71. $-18 + 15 - [6 - 9]$

72. $-7 - 14 - (-2)$

73. $13 - 4 - (-3)$

74. $-18 - (-18) - 4$

75. $7 - 8 - (2 - 1)$

76. $(2 - 8) - 6 - 9$

• *Multiply integers (4.5)*

77. $8 \cdot 9$

78. $3(0)(-6)$

79. $(-8)(-9)$

80. $3(-1)(-6)$

81. $8(-9)$

82. $-3(-6)(-9)$

83. $(-8)9$

84. $2(7)(9)$

85. $(-3)(-6)(-1)(10)$

86. $[-7 \cdot 11](-2)$

87. $64(-8)(0)$

88. $8(1)(-4)(-3)$

89. $-(8)(5)(2)$

90. $5 \cdot (6)(-3)$

91. $(-4)(3 \cdot 9)$

• *Divide integers (4.6).*

92. $35 \div 7$

93. $-72 \div 9$

94. $\dfrac{-35}{-7}$

95. $99 \div (-11)$

96. $7\overline{)-35}$

97. $-323 \div (-17)$

98. $35 \div (-7)$

99. $-100 \div 10$

100. $\dfrac{-6}{0}$

101. $102 \div (-3)$

102. $-434 \div 7$ **103.** $115 \div (-5)$

104. $-88 \div (-8)$ **105.** $-98 \div 7$

• *Use the order of operations rules to simplify problems containing more than one operation (4.7).*

106. -7^2 _____

107. $(-7)^2$ _____

108. $(8 - 6)^2$ _____

109. $8 - 6^2$ _____

110. $2 + 6[4^2 - (2 - 4)]$ _____

111. $[8 + 2(3 - 6)]6$ _____

112. $4 + [(8 - 2^2) + (-6 - 8)]$ _____

113. $\dfrac{4(3 + 2) + 4}{-4 - 2}$ _____

114. $(5^2 - 25)0$ _____

115. $27 + 3^2 - 6(-5)$ _____

116. $(17 - 3)(3^2 - 9)$ _____

117. $3(8 - 4) - 7(2^2 + 4)$ _____

118. $4(3 - 2^2) - 4(3 - 2)^2$ _____

119. $3 + 4[2^2 - (6 - 8)]$ _____

120. $\dfrac{7(4 + 2) + 3}{-3 - 2}$ _____

121. Find the average of the set $\{-29, -35, 7\}$. _____

122. Find the average of the set $\{5, -19, 0, -12, 6\}$. _____

123. Find the average of the set $\{9, -15, 8, 2\}$. _____

124. Find the average of the set $\{6, -7, 9, 0\}$ _____

125. Find the average of the set $\{19, -9, 2\}$. _____

• *Use the Addition Property of Equality to solve equations (4.8).*

Solve for the variable.

126. $x + 6 = 10$

127. $7y + (-6y) - 4 = 4$

128. $-4a + 3 + 5a = 6$

129. $10q = 10q + 3$

130. $7 + t = 3 + t + 4$

CHAPTER 4 PRACTICE TEST

1. Which number is greater, -13 or 0? _____

2. Which absolute value is greater $|-6|$ or $|4|$? _____

3. Which is smaller $-|-7|$ or $-(-7)$? _____

4. Which number is greater $-(-47)$ or $-(47)$? _____

5. Find the product $(-7)(-8)$ _____

6. Add: $-4 + 17 + (-23)$ _____

7. Find the sum: $15 + (-15)$ _____

8. Find the sum: $-15 + (-15)$ _____

9. Divide: $-72 \div 9$ _____

10. Find the product: $[6 \cdot 5](-4)$ _____

11. Find the quotient: $-23 \div 0$ _____

12. Find the difference: $6 - 12$ _____

13. Find the difference: $-6 - 12$ _____

14. Multiply: $(-1)(7)(-6)(-2)(-3)$ _____

15. Subtract: $22 - (-22)$ _____

16. Find the average of the set $\{15, 0, -12\}$ _____

17. Simplify: -6^2 _____

18. Simplify: $(-6)^2$ _____

19. Simplify: $6[-5 - (-5)](-34)$ _____

20. $6 + 2[3^2 - (3 - 7)]$ _____

21. $5 + [7^2 - 5(8)]3$ _____

22. $5 + [7^2 - 5(8)] - 3$ _____

23. $\dfrac{7(9 - 5)}{-2^2}$ _____

24. $6 + [(3 - 17) + (-2 - 5)]$ _____

25. Solve for x: $x + 9 = 17$ _____

26. Solve for y: $y - 9 = 9$ _____

27. Solve for a: $2a = 2a + 7$ _____

28. Solve for t: $6t = 9 + 5t$ _____

29. Solve for c: $c - 23 = -13$ _____

30. Solve for x: $x - 8 = 2x$ _____

		THE MATHEMATICS JOURNAL
		Analysis of Test Preparation
	Task I:	Did you achieve your goal on the test? [ ] yes [ ] no
		If no, did you come close? [ ] yes [ ] no
	Task II:	Reflect on the strategies you used to prepare for this test. Write
		a one-page (150-word) entry on your reflections.
	Starters:	Did the strategies you used work for you? Which ones would you use
		again? If you were preparing for the test again, what would you
		do differently?

5

Fractions and Mixed Numbers

5.1 The Set of Rational Numbers as Fractions
5.2 Using the Identity Property of Multiplication with Fractions
5.3 Multiplication of Fractions and Mixed Numbers
5.4 Division of Fractions and Mixed Numbers
5.5 Addition and Subtraction of Fractions and Mixed Numbers
5.6 Order of Operations and Complex Fractions
5.7 Solving Equations Using the Multiplication Property of Equality

Introduction to Fractions and Mixed Numbers

The word *fraction* is derived from the latin word *frangere* meaning "to break." That is what a fraction is, a whole broken into equal pieces.

The first people known to use fractions were the Egyptians who used them in their systems of weights and measures. Egyptian fractions were composed only of numerators of 1, except for the fractions $\frac{3}{4}$ and $\frac{2}{3}$. The fraction $\frac{5}{8}$ would be expressed as $\frac{1}{2} + \frac{1}{8}$. The symbol for a fractional number was ⬯ which was placed over the denominator.

$$\frac{1}{8} = \overset{\bigcirc}{\underset{\text{||||}}{\text{||||}}} \qquad \frac{1}{21} = \overset{\bigcirc}{\cap\cap|} \qquad \frac{1}{100} = \overset{\bigcirc}{\underset{\bigcirc}{\bigcirc}}$$

The Babylonian system of fractions was easier to use than the Egyptian system. The Babylonians used a sexagesimal system, using a base of 60 for fractions as they did for whole numbers. In the Babylonian system, all fractions had denominators that were factors or multiples of 60, while the numerators could be numbers other than 1.

Fractions as we use them today, stacked with the numerator over the denominator, can be traced to Hindu mathematicians. The fraction bar dividing the numerator and denominator was used extensively by Fibonacci, a Florentine mathematician who lived from 1170 to 1250, but it was not generally used until the sixteenth century.

5.1 The Set of Rational Numbers as Fractions

OBJECTIVE

● Recognize the numerator and denominator of a fraction.

In this chapter, we will consider rational numbers expressed as fractions. This is an important chapter for an algebra student because you will need to be comfortable with fractions in order to be able to understand algebra.

DEFINITION **Rational number:** A real number which can be expressed in the form $\frac{a}{b}$, where both a and b are integers and $b \neq 0$.

$$\frac{a}{b} = \frac{\text{numerator}}{\text{denominator}}$$

DEFINITION In this fraction, *b* is the **denominator.** It denotes the number of sections into which the unit is divided.

Note Recall from Section 2.4 that division by zero is undefined. A fraction indicates division. Any number divided by zero or any fraction with a denominator of zero is undefined. Zero cannot be a denominator.

DEFINITION The **numerator** is *a*. It denotes the number of sections used from the number of sections into which the unit is divided. The numerator can be any integer including zero.

DEFINITION The line between the numerator and denominator, the **fraction bar,** means to divide. The fraction, $\frac{a}{b}$, means $a \div b$.

EXAMPLE 1 This circle is divided into 6 sections, so the denominator would be 6.

EXAMPLE 2 This circle was divided into 6 sections, but only 5 of the sections are pictured. This diagram represents the fraction $\frac{5}{6}$. The fraction $\frac{5}{6}$ means $5 \div 6$.

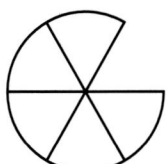

EXAMPLE 3 In the fraction $\frac{2}{3}$, the unit is divided into 3 sections, and 2 of those sections are used. The numerator is 2; 3 is the denominator.

EXAMPLE 4 In the fraction $\frac{4}{x}$, the unit is divided into *x* sections, and 4 of those sections are used. The numerator is 4; *x* is the denominator.

If a unit is divided into 6 sections, the denominator is 6. If 6 of those sections are used, the numerator is 6. The fraction is written $\frac{6}{6}$ and $\frac{6}{6} = 1$. Similarly, a fraction with a denominator of 10 in which 10 sections are used is written $\frac{10}{10}$ and $\frac{10}{10} = 1$.

EXAMPLE 5 $1 = \frac{1}{1} = \frac{9}{9} = \frac{7}{7} = \frac{204}{204} = \frac{x}{x} = \frac{4x}{4x}$, for $x \neq 0$.

Any fraction with the same numerator and denominator is equal to 1.

Expressing Integers as Fractions

Every integer can be expressed as a fraction with a denominator of 1. If the unit is divided into 1 section and 6 of these sections are used, the denominator is 1, and the numerator is 6.

EXAMPLE 6 $\frac{6}{1} = 6, \frac{9}{1} = 9, \frac{a}{1} = a.$

Note   When working with integers and fractions, it sometimes helps to convert the integer into a fraction. This can easily be done by giving the integer a denominator of 1. A fraction with an integer as a numerator and a denominator of 1 is an integer.

In fractions with 0 for the numerator, for example, $\frac{0}{6}$, the unit is divided into 6 sections (the denominator), but 0 sections are used. $\frac{0}{6} = 0$. Regardless of how many sections the unit is divided into, if 0 sections are used, the result is 0.

EXAMPLE 7 $\frac{0}{9} = 0; \frac{0}{x} = 0$, for $x \neq 0$

Any fraction with a zero numerator is equal to zero.

Types of Fractions

Three types of fractions will be defined in this section and used in this chapter.

DEFINITION   **Proper fraction:** A fraction that represents less than one unit. A proper fraction is a fraction in which the absolute value of the numerator is less than the absolute value of the denominator.

EXAMPLE 8 $\frac{1}{6}, \frac{2}{3}, \frac{4}{5}, \frac{7}{9}, -\frac{3}{4}, -\frac{4}{9}, \frac{6}{14}$ are all proper fractions.

DEFINITION   **Improper fraction:** A fraction in which the absolute value of the numerator is equal to or greater than the absolute value of the denominator.

Note   There is nothing wrong or "improper" about these fractions. In fact, they are used extensively in algebra.

EXAMPLE 9 $\frac{7}{4}, \frac{13}{2}, \frac{11}{5}, -\frac{9}{2}, -\frac{7}{3}$ are improper fractions. Improper fractions can also be written as mixed fractions.

DEFINITION   **Mixed number:** The sum of an integer and a proper fraction. $2\frac{1}{4}$ is an example of a mixed number. It means $2 + \frac{1}{4}$. Improper fractions can be expressed as mixed numbers.

EXAMPLE 10   Express $\frac{7}{4}$ as a mixed number.

$$\frac{7}{4} = \frac{4+3}{4}$$
$$= \frac{4}{4} + \frac{3}{4}$$
$$= 1 + \frac{3}{4}$$
$$= 1\frac{3}{4}$$

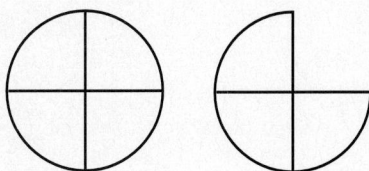

EXAMPLE 11 Express $\frac{13}{2}$ as a mixed number.

$$\frac{13}{2} = \frac{2 + 2 + 2 + 2 + 2 + 2 + 1}{2}$$

Express 13 as the sum of six 2s and one 1.

$$= \frac{2}{2} + \frac{2}{2} + \frac{2}{2} + \frac{2}{2} + \frac{2}{2} + \frac{2}{2} + \frac{1}{2}$$

Write each term with its own denominator of 2.

$$= 1 + 1 + 1 + 1 + 1 + 1 + \frac{1}{2}$$

Simplify. $\frac{2}{2} = 1$

$$= 6\frac{1}{2}$$

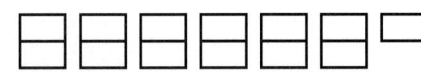

$6\frac{1}{2}$ is a mixed number equivalent of $\frac{13}{2}$.

EXAMPLE 12 Expresss $\frac{11}{5}$ as a mixed number.

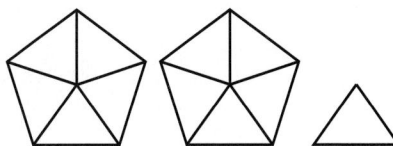

$2\frac{1}{5}$ is the mixed number equivalent of $\frac{11}{5}$.

The following are some examples of converting mixed numbers to improper fractions.

EXAMPLE 13 Express $2\frac{1}{2}$ as an improper fraction.

Two whole units equal four half units.

$$\frac{4}{2} + \frac{1}{2} = \frac{5}{2}$$

$\frac{5}{2}$ is equivalent to $2\frac{1}{2}$.

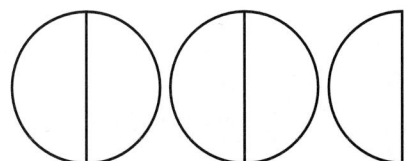

EXAMPLE 14 Express $3\frac{3}{4}$ as an improper fraction.

Three (3) whole units equal twelve fourths of a unit $\left(\frac{12}{4}\right)$.

$$\frac{12}{4} + \frac{3}{4} = \frac{15}{4}$$

$\frac{15}{4}$ is equivalent to $3\frac{3}{4}$.

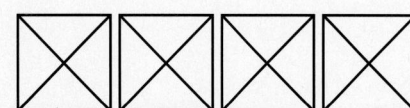

EXAMPLE 15 Express $3\frac{1}{6}$ as an improper fraction.

Three (3) whole units equal eighteen sixths of a unit $\left(\frac{18}{6}\right)$.

$$\frac{18}{6} + \frac{1}{6} = \frac{19}{6}.$$

$\frac{19}{6}$ is equivalent to $3\frac{1}{6}$.

USING THE CALCULATOR

Convert $\frac{17}{3}$ to a mixed number.

Enter 17 [a$^{b/c}$] 3 [=] **5 _ 2 ⌋ 3** ← Display

Convert $5\frac{2}{3}$ to an improper fraction.

Enter 5 [a$^{b/c}$] 2 [a$^{b/c}$] 3 [$^{d/c}$] **17⌋ 3** ← Display

Note When negative mixed numbers are changed to improper fractions, or negative improper fractions are changed to mixed numbers, the signs are maintained.

EXAMPLE 16 $-1\frac{1}{2} = -\frac{3}{2}$ and $-\frac{10}{3} = -3\frac{1}{3}$

Positive and Negative Fractions

Since the fraction bar indicates division, the rules for division of integers hold.

RULES For all integers, *a* and *b*, where *b* ≠ 0

$\dfrac{-a}{b} = -\dfrac{a}{b}$	A negative numerator divided by a positive denominator equals a negative fraction.
$\dfrac{a}{-b} = -\dfrac{a}{b}$	A positive numerator divided by a negative denominator equals a negative fraction.
$\dfrac{-a}{-b} = \dfrac{a}{b}$	A negative numerator divided by a negative denominator equals a positive fraction.

EXAMPLE 17

$$\frac{-3}{5} = -\frac{3}{5}$$

$$\frac{3}{-5} = -\frac{3}{5}$$

$$\frac{-3}{-5} = \frac{3}{5}$$

ANSWERS TO PRACTICE PROBLEMS

1. Proper fractions: $\frac{3}{4}, \frac{19}{38}, \frac{1}{8}$;

Improper fractions; $\frac{5}{2}, \frac{7}{7}$

2. $2\frac{1}{4}, 2\frac{2}{3}, 2, -3\frac{1}{2}, 3\frac{4}{5}, -2\frac{1}{6}$

3. $\frac{13}{4}, \frac{14}{3}, \frac{-17}{2}, \frac{7}{5}, -\frac{65}{9}$ 4. –

5. – 6. + 7. + 8. +

PROBLEM SET 5.1

Name the denominator in each of the following fractions.

1. $\frac{2}{3}$ **2.** $\frac{4}{5}$ **3.** $\frac{7}{1}$ **4.** $\frac{8}{3}$ **5.** $\frac{1}{8}$

6. $\frac{-2}{5}$ **7.** $\frac{10}{3}$ **8.** $\frac{12}{1}$ **9.** $\frac{9}{13}$ **10.** $\frac{1}{9}$

Name the numerator in each of the following fractions.

11. $\frac{3}{4}$ **12.** $\frac{7}{3}$ **13.** $\frac{2}{5}$ **14.** $\frac{8}{1}$ **15.** $\frac{0}{6}$

16. $\frac{-2}{3}$ **17.** $\frac{20}{9}$ **18.** $\frac{13}{1}$ **19.** $\frac{5}{9}$ **20.** $\frac{-5}{8}$

Circle the proper fractions, and change the improper fractions to mixed numbers.

21. $\frac{1}{6}$ **22.** $\frac{20}{6}$ **23.** $\frac{6}{1}$ **24.** $\frac{22}{9}$ **25.** $\frac{8}{9}$

26. $\frac{8}{5}$ **27.** $\frac{5}{8}$ **28.** $\frac{55}{55}$ **29.** $\frac{57}{7}$ **30.** $\frac{40}{19}$

Are the following fractions positive or negative?

31. $\frac{-2}{3}$ **32.** $\frac{4}{-5}$ **33.** $\frac{-3}{-8}$ **34.** $\frac{7}{8}$ **35.** $\frac{-3}{3}$

36. $\frac{6}{-6}$ **37.** $\frac{-9}{-9}$ **38.** $\frac{1}{-6}$ **39.** $\frac{-1}{6}$ **40.** $\frac{0}{-6}$

Change the following improper fractions to mixed numbers.

41. $\dfrac{17}{5}$ **42.** $\dfrac{28}{7}$ **43.** $\dfrac{11}{3}$ **44.** $\dfrac{19}{2}$ **45.** $\dfrac{76}{7}$ **46.** $\dfrac{14}{3}$

47. $\dfrac{-10}{3}$ **48.** $\dfrac{10}{4}$ **49.** $\dfrac{32}{-3}$ **50.** $\dfrac{82}{9}$ **51.** $\dfrac{106}{3}$ **52.** $\dfrac{-67}{6}$

Change the following mixed numbers to improper fractions.

53. $7\dfrac{3}{4}$ **54.** $9\dfrac{7}{8}$ **55.** $7\dfrac{2}{9}$ **56.** $8\dfrac{1}{2}$ **57.** $4\dfrac{2}{3}$ **58.** $3\dfrac{9}{13}$

59. $3\dfrac{4}{5}$ **60.** $10\dfrac{7}{8}$

5.2 **Using the Identity Property of Multiplication with Fractions**

OBJECTIVES

- Recognize that equivalent fractions look different but represent the same part of the whole.
- Calculate equivalent fractions with a given denominator.
- Reduce a fraction to lowest terms by prime factoring the numerator and denominator.

We have used the Identity Property of Multiplication in Sections 2.4 and 4.5. In this section, we will use it to work with fractions.

DEFINITION   **Equivalent fractions:** Fractions that look different but name the same number.

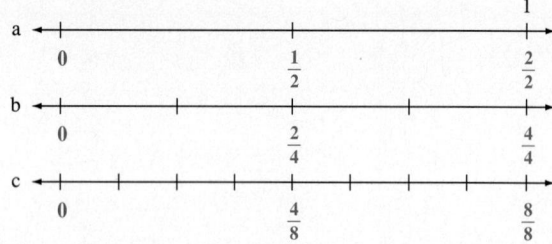

Consider the equivalent fractions represented on these three number lines. Notice that the number 1 is represented as $\frac{2}{2}$ on number line a; as $\frac{4}{4}$ on number line b; and as $\frac{8}{8}$ on number line c.

$$\frac{2}{2} = \frac{4}{4} = \frac{8}{8} = 1$$

$\frac{2}{2}, \frac{4}{4}$, and $\frac{8}{8}$ are equivalent fractions.

Notice that $\frac{1}{2}, \frac{2}{4}$, and $\frac{4}{8}$ all name the same position on the three number lines.

$$\frac{1}{2} = \frac{2}{4} = \frac{4}{8}$$

$\frac{1}{2}, \frac{2}{4}$, and $\frac{4}{8}$ are equivalent fractions.

When working with fractions, it is often necessary to use equivalent fractions. Recall the Identity Property of Multiplication from Section 2.4.

For all real numbers a, $a \cdot 1 = a$.

In words, the product of a real number and 1 is the original real number. When multiplied by 1, the number remains the same. Because $a \cdot 1 = a$, it is also true that $a \div a = 1$. In fraction notation, $\frac{a}{a} = 1$. We use this property to find a fraction equivalent to a given one.

RULE To find a fraction equivalent to a given fraction, multiply the original fraction by the number 1 in the form of $\frac{a}{a}$.

EXAMPLE 1 Convert $\frac{1}{2}$ to a fraction with a denominator of 6.

Multiply $\frac{1}{2}$ by 1 in the form of $\frac{3}{3}$.

$$\frac{1}{2} \cdot \frac{3}{3} = \frac{3}{6}$$

Therefore, $\frac{1}{2}$ is equivalent to $\frac{3}{6}$.

EXAMPLE 2 Convert $\frac{1}{2}$ to a fraction with a denominator of 4.

$$\frac{1}{2} \cdot \frac{2}{2} = \frac{2}{4}$$

Therefore, $\frac{1}{2}$ is equivalent to $\frac{2}{4}$.

EXAMPLE 3 Convert $\frac{2}{3}$ to a fraction with a denominator of 12.

$$\frac{2}{3} \cdot \frac{4}{4} = \frac{8}{12}$$

Therefore, $\frac{2}{3}$ is equivalent to $\frac{8}{12}$.

> **The Identity Element used with Fractions**
> If a, b, and c are integers and $b \neq 0$ and $c \neq 0$, then
>
> $$\frac{a}{b} = \frac{ac}{bc}.$$
>
> Similarly, the numerator and denominator of a fraction may be divided by the same nonzero integer to obtain an equivalent fraction.
>
> $$\frac{a}{b} = \frac{a \div c}{b \div c}$$

EXAMPLE 4   Convert $\frac{5}{10}$ to a fraction with a denominator of 2.

$$\frac{5}{10} = \frac{5 \div 5}{10 \div 5} = \frac{1}{2}$$

$$\text{so } \frac{5}{10} = \frac{1}{2}$$

EXAMPLE 5   Change $\frac{2}{5}$ to a fraction with a denominator of 10.

$$\frac{2}{5} = \frac{2 \cdot 2}{5 \cdot 2} = \frac{4}{10}$$

$$\text{so } \frac{2}{5} = \frac{4}{10}$$

EXAMPLE 6   Convert $\frac{15}{20}$ to a fraction with a denominator of 4.

$$\frac{15}{20} = \frac{15 \div 5}{20 \div 5} = \frac{3}{4}$$

$$\text{so } \frac{15}{20} = \frac{3}{4}$$

EXAMPLE 7   Write $\frac{4}{10}$ as a fraction with a denominator of 5.

$$\frac{4}{10} = \frac{4 \div 2}{10 \div 2} = \frac{2}{5}$$

$$\text{so } \frac{4}{10} = \frac{2}{5}$$

EXAMPLE 8   Express $\frac{3}{8}$ as a fraction with a denominator of 40.

$$\frac{3}{8} = \frac{3}{8} \cdot \frac{?}{?} = \frac{?}{40}$$

$$\frac{3}{8} = \frac{3}{8} \cdot \frac{5}{5} = \frac{15}{40}$$

$$\text{so } \frac{3}{8} = \frac{15}{40}$$

EXAMPLE 9   Express $\frac{2}{3x}$ as a fraction with a denominator of $6x$.

$$\frac{2}{3 \cdot x} = \frac{2 \cdot ?}{3 \cdot 2 \cdot x}$$

$$\frac{2}{3x} = \frac{2 \cdot 2}{3 \cdot 2 \cdot x} = \frac{4}{6x}$$

$$\text{so } \frac{2}{3x} = \frac{4}{6x}$$

Reducing Fractions to Lowest Terms

Reducing fractions to lowest terms makes the fractions as simple as possible.

DEFINITION   A fraction is in **lowest terms** when the numerator and denominator are relatively prime, i.e., when their greatest common factor is 1.

Note   Any fraction that has 1 for either the numerator or denominator is in lowest terms.

EXAMPLE 10   Some examples of fractions in lowest terms are

$$\frac{1}{2}, \frac{5}{6}, \frac{7}{1}, \frac{9}{4}, -\frac{1}{4}, -\frac{2}{3}.$$

To reduce a fraction to lowest terms:

Method 1

1. Find the prime factors of the numerator and denominator.
2. Match the factors from the denominator and numerator which are equal. Replace them with a 1.
3. Carry out the multiplication.

EXAMPLE 11 $\dfrac{9}{15} = \dfrac{3 \cdot 3}{3 \cdot 5} = \dfrac{3}{3} \cdot \dfrac{3}{5} = 1 \cdot \dfrac{3}{5} = \dfrac{3}{5}$

EXAMPLE 12 $\dfrac{8}{16} = \dfrac{2 \cdot 2 \cdot 2}{2 \cdot 2 \cdot 2 \cdot 2} = \dfrac{2}{2} \cdot \dfrac{2}{2} \cdot \dfrac{2}{2} \cdot \dfrac{1}{2} = 1 \cdot 1 \cdot 1 \cdot \dfrac{1}{2} = \dfrac{1}{2}$

EXAMPLE 13 $\dfrac{21}{49} = \dfrac{3 \cdot 7}{7 \cdot 7} = \dfrac{3}{7} \cdot \dfrac{7}{7} = \dfrac{3}{7} \cdot 1 = \dfrac{3}{7}$

Method 2

1. Find the GCF of the numerator and denominator.
2. Divide both the numerator and denominator by the GCF.

EXAMPLE 14 $\frac{9}{15}$ GCF $(9, 15) = 3$

$$\frac{9}{15} = \frac{9 \div 3}{15 \div 3} = \frac{3}{5}$$

EXAMPLE 15 $\frac{8}{16}$ GCF $(8, 16) = 8$

$$\frac{8}{16} = \frac{8 \div 8}{16 \div 8} = \frac{1}{2}$$

Practice Problems

Express these fractions as equivalent fractions with denominators of 12.

1. $\dfrac{1}{4}$

2. $\dfrac{45}{60}$

3. $\dfrac{2}{3}$

4. $\dfrac{12}{36}$

5. $\dfrac{x}{2}$

EXAMPLE 16 $\frac{21}{49}$ GCF $(21, 49) = 7$

$$\frac{21}{49} = \frac{21 \div 7}{49 \div 7} = \frac{3}{7}$$

USING THE CALCULATOR

Use the $a^{b/c}$ key. Reduce $\frac{9}{15}$ to lowest terms.

Enter: 9 $a^{b/c}$ 15 = **3⌐5** ← Display

EXAMPLE 17 Reduce $\dfrac{x^2y}{xy}$

Method 1

$$\frac{x^2y}{xy} = \frac{x \cdot x \cdot y}{x \cdot y} = \frac{x}{x} \cdot \frac{y}{y} \cdot \frac{x}{1} = 1 \cdot 1 \cdot \frac{x}{1} = \frac{x}{1} = x$$

Method 2

$$\text{GCF } (x^2y, xy) = xy$$
$$\frac{x^2y}{xy} = \frac{x^2y \div xy}{xy \div xy} = \frac{x}{1} = x$$

EXAMPLE 18 Reduce $\dfrac{6xy}{3x^2}$

Method 1

$$\frac{6xy}{3x^2} = \frac{2 \cdot 3 \cdot x \cdot y}{3 \cdot x \cdot x} = \frac{2 \cdot y}{x} \cdot \frac{3}{3} \cdot \frac{x}{x}$$
$$= \frac{2y}{x} \cdot 1 \cdot 1 = \frac{2y}{x}$$

Method 2

$$\text{GCF } (6xy, 3x^2) = 3x$$
$$\frac{6xy \div 3x}{3x^2 \div 3x} = \frac{2y}{x}$$

Practice Problems

Reduce to lowest terms.

6. $\dfrac{30}{35}$

7. $\dfrac{25}{120}$

8. $\dfrac{3x}{3y}$

9. $\dfrac{3x^2}{4xy}$

ANSWERS TO PRACTICE PROBLEMS

1. $\frac{3}{12}$ 2. $\frac{9}{12}$ 3. $\frac{8}{12}$ 4. $\frac{4}{12}$ 5. $\frac{6x}{12}$
6. $\frac{6}{7}$ 7. $\frac{5}{24}$ 8. $\frac{x}{y}$ 9. $\frac{3x}{4y}$

PROBLEM SET 5.2

Express these fractions as equivalent fractions with denominators of 6.

1. $\dfrac{1}{3}$ 2. $\dfrac{1}{2}$ 3. $\dfrac{2}{12}$ 4. $\dfrac{8}{24}$

Express these fracitons as equivalent fractions with denominators of 10.

5. $\dfrac{1}{5}$ 6. $\dfrac{1}{2}$ 7. $\dfrac{6}{30}$ 8. $\dfrac{18}{20}$

Express these fractions as equivalent fractions with denominators of 12x.

9. $\dfrac{1}{4}$　　　　　**10.** $\dfrac{2}{2x}$　　　　　**11.** $\dfrac{8}{24x}$　　　　　**12.** $\dfrac{7}{12}$

Find the greatest common factors.

13. GCF $(3y, 3y^2)$　　　　　　　　**14.** GCF $(7x, 14x)$

15. GCF $(2ab, 2a^2)$　　　　　　　　**16.** GCF (x^2y, y^2)

Reduce each fraction to lowest terms. Check the answers on a calculator.

17. $\dfrac{6}{12}$　　　　　　　　　　**18.** $\dfrac{12}{6}$

19. $\dfrac{10}{15}$　　　　　　　　　　**20.** $\dfrac{7}{8}$

21. $\dfrac{20}{25}$　　　　　　　　　　**22.** $\dfrac{27}{72}$

23. $\dfrac{11}{99}$　　　　　　　　　　**24.** $\dfrac{21}{49}$

25. $\dfrac{28}{56}$　　　　　　　　　　**26.** $\dfrac{19}{38}$

27. $\dfrac{13}{52}$　　　　　　　　　　**28.** $\dfrac{45}{54}$

29. $\dfrac{36}{25}$　　　　　　　　　　**30.** $\dfrac{24a^2b}{12ab^2}$

31. $\dfrac{x^2}{xy}$　　　　　　　　　　**32.** $\dfrac{45x}{40x^2}$

5.3 **Multiplication of Fractions and Mixed Numbers**

OBJECTIVE

● Multiply fractions and mixed numbers.

You learned in Sections 2.4 and 4.5 that multiplication is repeated addition. In this section, we will apply this definition to fractions.

EXAMPLE 1 $3\left(\frac{1}{4}\right)$ means add 3 sets of $\frac{1}{4}$; $3\left(\frac{1}{4}\right) = \frac{3}{4}$.

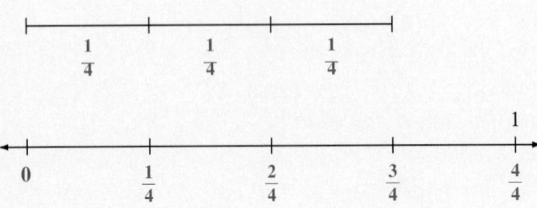

EXAMPLE 2 $6\left(\frac{2}{3}\right)$ means add 6 sets of $\frac{2}{3}$; $6\left(\frac{2}{3}\right) = 4$.

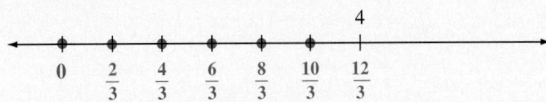

EXAMPLE 3 To test the Commutative Property of Multiplication, consider $\frac{2}{3}$ of a set of 6. (See Section 3.4; *of* indicates multiplication.) $\frac{2}{3} \cdot (6) = 4$. When the set of 6 is divided into 3 equal parts, each part has a size of 2, so 2 of the parts, that is $\frac{2}{3}$, would have a size of 4.

EXAMPLE 4 Another way to look at multiplication of fractions is with diagrams. The large rectangle represents one unit.

$$\frac{1}{2} \cdot \frac{1}{3} \text{ is represented as } \frac{1}{2} \text{ of a set of } \frac{1}{3};$$

$$\frac{1}{2} \cdot \frac{1}{3} = \frac{1}{6}.$$

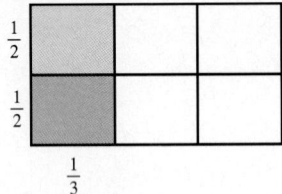

EXAMPLE 5 $\frac{3}{4} \cdot \frac{2}{3}$ is represented as $\frac{3}{4}$ of a set of $\frac{2}{3}$, $\frac{3}{4} \cdot \frac{2}{3} = \frac{6}{12} = \frac{1}{2}$.

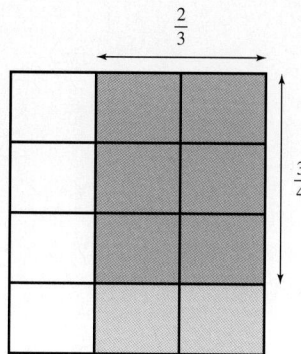

Look carefully at these examples. We have actually multiplied the numerators together and multiplied the denominators together.

RULE Multiplying fractions

For all integers, a, b, c, and d, where $b \neq 0$ and $d \neq 0$,

$$\frac{a}{b} \cdot \frac{c}{d} = \frac{a \cdot c}{b \cdot d}$$

In words, the product of two fractions is the product of their numerators divided by the product of their denominators.

Note Before multiplying fractions, reduce each to lowest terms. This will make the numbers smaller and easier to work with.

USING THE CALCULATOR

A. Find $\frac{2}{3} \cdot 6$

Enter 2 $a^{b/c}$ 3 $\times$ 6 $=$ **4** ← Display

B. Find $\frac{1}{3} \cdot \frac{1}{2}$

Enter 1 $a^{b/c}$ 3 $\times$ 1 $a^{b/c}$ 2 $=$ **1⌋6** ← Display

C. Find $\frac{3}{4} \cdot \frac{2}{3}$

Enter 3 $a^{b/c}$ 4 $\times$ 2 $a^{b/c}$ 3 $=$ **1⌋2** ← Display

---

EXAMPLE 6 Find $\frac{15}{16} \cdot \frac{4}{5}$

$$\frac{15}{16} \cdot \frac{4}{5} = \frac{60}{80} = \frac{6}{8} = \frac{3}{4}$$

Another way is to express the numerators and denominators as products of prime factors and to *reduce* before carrying out the multiplication. (This method is often called *canceling*.)

$$\frac{15}{16} \cdot \frac{4}{5} = \frac{3 \cdot \overset{1}{\cancel{5}} \cdot \overset{1}{\cancel{2}} \cdot \overset{1}{\cancel{2}}}{2 \cdot 2 \cdot \underset{1}{\cancel{2}} \cdot \underset{1}{\cancel{2}} \cdot \underset{1}{\cancel{5}}} = \frac{3}{4}$$

Multiplying Positive and Negative Fractions

For all positive integers a and b, $\frac{-a}{b} = \frac{a}{-b} = -\frac{a}{b}$. For example,

$$\frac{-2}{3} = \frac{2}{-3} = -\frac{2}{3}.$$

The rules for multiplying positive and negative integers (see Section 4.5) hold for all real numbers including fractions. The product of two positive fractions is a positive fraction. The product of two negative fractions is a positive fraction. The product of a positive fraction and a negative fraction is a negative fraction.

EXAMPLE 7
$$\left(-\frac{1}{2}\right)\left(\frac{4}{5}\right) = -\frac{2}{5}$$

EXAMPLE 8
$$\left(-\frac{3}{4}\right)\left(-\frac{2}{3}\right) = \frac{1}{2}$$

EXAMPLE 9
$$\left(\frac{4}{7}\right)\left(-\frac{1}{3}\right) = -\frac{4}{21}$$

EXAMPLE 10
$$\left(\frac{x}{y}\right)\left(\frac{x}{y^2}\right) = \frac{x^2}{y^3}$$

EXAMPLE 11
$$\frac{x^2}{y^2} \cdot \frac{y^3}{x} = \frac{x \cdot \overset{1}{\cancel{x}} \cdot y \cdot \overset{1}{\cancel{y}} \cdot \overset{1}{\cancel{y}}}{\underset{1}{\cancel{y}} \cdot \underset{1}{\cancel{y}} \cdot \underset{1}{\cancel{x}}} = xy$$

🖩 USING THE CALCULATOR

A. Find $\left(-\frac{1}{2}\right)\left(\frac{4}{5}\right)$

    Enter 1 $a^{b/c}$ 2 $+/-$ $\times$ 4 $a^{b/c}$ 5 $=$ **− 2⌋5** ← Display

B. Find $\left(-\frac{3}{4}\right)\left(-\frac{2}{3}\right)$

    Enter 3 $a^{b/c}$ 4 $+/-$ $\times$ 2 $a^{b/c}$ 3 $+/-$ $=$ **1⌋2** ← Display

---

Multiplying Mixed Numbers

Remember that a mixed number is the sum of an integer and a proper fraction.

$$2\frac{1}{2} = 2 + \frac{1}{2}$$

RULE To multiply mixed numbers, first convert them to improper fractions. Reduce them as much as possible and then multiply.

EXAMPLE 12

$$\left(2\frac{1}{2}\right)\left(3\frac{1}{4}\right) = \left(\frac{5}{2}\right)\left(\frac{13}{4}\right) = \frac{65}{8} = 8\frac{1}{8}$$

USING THE CALCULATOR

Find $\left(2\frac{1}{2}\right)\left(3\frac{1}{4}\right)$

     Enter 2 $\boxed{a^{b/c}}$ 1 $\boxed{a^{b/c}}$ 2 $\boxed{\times}$ 3 $\boxed{a^{b/c}}$ 1 $\boxed{a^{b/c}}$ 4 $\boxed{=}$ **8_1⌋8** ← Display

Note: It is not necessary to convert mixed numbers to improper fractions when using a calculator to multiply them.

EXAMPLE 13 Find $\frac{3}{5}\left(5\frac{1}{3}\right)$

$$\frac{3}{5}\left(5\frac{1}{3}\right) =$$

$$\frac{3}{5} \cdot \frac{16}{3} = \qquad \text{Change } 5\frac{1}{3} \text{ to an improper fraction.}$$

$$\frac{3}{5} \cdot \frac{16}{3} = \qquad \text{Reduce.}$$

$$\frac{16}{5} \text{ or } 3\frac{1}{5}$$

USING THE CALCULATOR

Find $\frac{3}{5}\left(5\frac{1}{3}\right)$

     Enter 3 $\boxed{a^{b/c}}$ 5 $\boxed{\times}$ 5 $\boxed{a^{b/c}}$ 1 $\boxed{a^{b/c}}$ 3 $\boxed{=}$ **3 _1⌋5** ← Display

---

EXAMPLE 14 $2\left(4\frac{7}{8}\right) =$

$$\frac{2}{1} \cdot \frac{39}{8} = \qquad \text{Change to improper fractions.}$$

$$\frac{2}{1} \cdot \frac{39}{8} = \qquad \text{Reduce.}$$

$$\frac{39}{4} \text{ or } 9\frac{3}{4}$$

USING THE CALCULATOR

Find $2\left(4\frac{7}{8}\right)$

     Enter 2 $\boxed{\times}$ 4 $\boxed{a^{b/c}}$ 7 $\boxed{a^{b/c}}$ 8 $\boxed{=}$ **9_3⌋4** ← Display

---

Practice Problems

Multiply. Reduce first.

1. $\dfrac{1}{3} \cdot \dfrac{3}{8}$

2. $-16 \cdot \dfrac{1}{8}$

3. $\left(\dfrac{2}{3}\right)(-6)$

4. $-\dfrac{4}{5}\left(-\dfrac{15}{4}\right)$

5. $\dfrac{32}{33}\left(\dfrac{11}{8}\right)$

6. $\left(\dfrac{a^2 b}{xy}\right)\left(\dfrac{xy}{ab}\right)$

7. $\left(\dfrac{1}{2}\right)^2\left(\dfrac{2}{9}\right)^2$

APPLICATION 1

A cook needs to double the brownie recipe. The original recipe calls for $\frac{1}{2}$ cup of sugar, $\frac{3}{4}$ cup of flour, and $\frac{1}{8}$ cup of cocoa. How much sugar, flour, and cocoa will be needed? (To double means to multiply by 2.)

Sugar: $2 \cdot \frac{1}{2} = \frac{2}{1} \cdot \frac{1}{2} = 1$ cup

Flour: $2 \cdot \frac{3}{4} = \frac{2}{1} \cdot \frac{3}{4} = \frac{3}{2}$ or $1\frac{1}{2}$ cups

Cocoa: $2 \cdot \frac{1}{8} = \frac{2}{1} \cdot \frac{1}{8} = \frac{1}{4}$ cup

Answer: The cook will need 1 cup of sugar, $1\frac{1}{2}$ cups of flour, and $\frac{1}{4}$ cup of cocoa. ◆

APPLICATION 2

How many tons could a truck haul in $4\frac{1}{2}$ loads if the truck can haul $\frac{3}{4}$ ton of gravel in each load?

$$\left(4\frac{1}{2}\right)\left(\frac{3}{4}\right) = \frac{9}{2} \cdot \frac{3}{4} = \frac{27}{8} \text{ or } 3\frac{3}{8}$$

Answer: The truck could haul $3\frac{3}{8}$ tons. ◆

APPLICATION 3

At EGU, $\frac{1}{3}$ of the entering students never graduate. If the freshman class has 3057 students, how many will never graduate?

$$\frac{1}{3} \text{ of } 3057 = \frac{1}{3} \cdot \frac{3057}{1} = 1019$$

Answer: 1019 students will not graduate. ◆

APPLICATION 4

In a mathematics class, $\frac{2}{9}$ of the students earn grades of A. If 216 students are enrolled in the class, how many will get A's?

$$\frac{2}{9} \text{ of } 216 = \frac{2}{9} \cdot \frac{216}{1} = 48$$

Answer: 48 students will get A's. ◆

ANSWERS TO PRACTICE PROBLEMS

1. $\frac{1}{8}$ 2. -2 3. -4 4. 3 5. $\frac{4}{3}$ 6. a 7. $\frac{1}{81}$

PROBLEM SET 5.3

Find the product.

1. $\frac{2}{3} \cdot \frac{4}{5}$

2. $\frac{4}{5} \cdot \frac{1}{4}$

3. $\frac{2}{5} \cdot \frac{3}{7}$

4. $\frac{4}{9} \cdot \frac{9}{4}$

5. $5 \cdot -\frac{1}{5}$

6. $7 \cdot \frac{3}{7}$

7. $-3 \cdot \dfrac{5}{6}$

8. $\left(\dfrac{2}{5}\right)\left(-\dfrac{2}{3}\right)\left(\dfrac{4}{5}\right)$

9. $\left(-\dfrac{1}{8}\right)\left(-\dfrac{2}{3}\right)(-8)$

10. $\dfrac{9}{20} \cdot \dfrac{4}{3}$

11. $-\dfrac{27}{40} \cdot \dfrac{8}{3}$

12. $\left(-\dfrac{3}{5}\right)\left(-\dfrac{3}{4}\right)$

13. $\left(-\dfrac{2}{3}\right)\left(\dfrac{1}{5}\right)$

14. $\left(1\dfrac{5}{9}\right)\left(-\dfrac{6}{7}\right)$

15. $\left(-3\dfrac{1}{2}\right)\left(-1\dfrac{1}{7}\right)$

16. $\left(-\dfrac{9}{11}\right)\left(-7\dfrac{1}{3}\right)$

17. $\left(1\dfrac{7}{9}\right)\left(\dfrac{27}{32}\right)$

18. $\left(-\dfrac{42}{7}\right)\left(-\dfrac{15}{36}\right)\left(-\dfrac{6}{7}\right)$

19. $\left(3\dfrac{3}{5}\right)\left(-\dfrac{5}{36}\right)$

20. $\left(-\dfrac{7}{8}\right)\left(\dfrac{1}{14}\right)16$

21. $\left(3\dfrac{3}{7}\right)\left(-2\dfrac{5}{8}\right)\left(-\dfrac{1}{9}\right)$

22. $\left(-\dfrac{14}{45}\right)\left(\dfrac{3}{56}\right)$

23. $\left(-\dfrac{15}{32}\right)\left(-1\dfrac{3}{5}\right)$

24. $10 \cdot \dfrac{1}{100}\left(-\dfrac{1}{10}\right)$

Expand and simplify.

25. $\left(\dfrac{1}{2}\right)^2$

26. $\left(\dfrac{2}{3}\right)^3$

27. $\left(-\dfrac{1}{4}\right)^2$

28. $\left(-\dfrac{3}{5}\right)^2$

29. $\left(-\dfrac{1}{2}\right)^3\left(-\dfrac{2}{3}\right)^2$

30. $\left(\dfrac{3}{4}\right)^2\left(-\dfrac{2}{3}\right)^2$

31. $\dfrac{a^2 b}{c} \cdot \dfrac{c^3}{ab^2}$

32. $\dfrac{xy^2}{z^2} \cdot \dfrac{z}{x^2 y}$

33. $\left(2\dfrac{1}{2}\right)\left(4\dfrac{2}{3}\right)$

34. $\left(5\dfrac{1}{4}\right)\left(-1\dfrac{2}{3}\right)$

35. $\left(-7\dfrac{1}{2}\right)\left(-2\dfrac{1}{5}\right)(2)$

36. $\left(-3\dfrac{5}{9}\right)\left(2\dfrac{5}{8}\right)$

37. $\left(\dfrac{3}{5}\right)^2\left(-1\dfrac{2}{9}\right)^2$

38. $\dfrac{3ab}{2c^2}\cdot\dfrac{8c}{6b^2}$

39. $\dfrac{x}{y}\cdot\dfrac{y}{x}$

40. $\dfrac{64x^2}{9y^2}\cdot\dfrac{72y}{16x}$

41. $\dfrac{63ab}{21c}\cdot\dfrac{7c^2}{35b^2}$

42. $\dfrac{55m}{19n}\cdot\dfrac{38mn}{11m^2}$

Perform the indicated operations.

43. Find the product of $\dfrac{1}{2}$ and $\dfrac{7}{8}$.

44. Find $\dfrac{1}{4}$ of 44.

45. Find $\dfrac{4}{5}$ of $\dfrac{3}{4}$.

46. Find $\dfrac{3}{5}$ of $\dfrac{1}{3}$ of 15.

Applications:

47. A cook needs to double a recipe that uses $\dfrac{7}{8}$ cup of flour. How much flour will be needed?

48. If the same cook were to take half of the same recipe in Problem 47, how much flour would be needed?

49. In a mathematics class, $\dfrac{3}{4}$ of the students passed, and $\dfrac{1}{3}$ of those who passed received a grade of A. What fraction of the class received a grade of A?

50. Statistics reveal that 1 woman in 9 will develop cancer during her lifetime. What fraction is this? In a college freshman class of 1737 women, how many can expect to develop cancer?

51. A truck can carry $\dfrac{3}{4}$ ton of gravel. If the driver delivered $\dfrac{1}{2}$ of the load, how much gravel was left in the truck?

52. Last night it snowed $\dfrac{3}{4}$ inch. As the temperature rose, $\dfrac{1}{8}$ of the snow melted. How much snow is still on the ground?

53. A cook needs to triple a recipe that uses $2\frac{1}{2}$ cups of flour. How much flour will be needed?

54. If the same cook from Problem 53 were to take $\frac{3}{4}$ of that same recipe using $2\frac{1}{2}$ cups of flour, how much flour would be needed?

55. In a mathematics class, $\frac{7}{8}$ of the students passed, and $\frac{1}{4}$ of those who passed received a grade of A. What fraction of the class received a grade of A?

56. Statistics reveal that 1 man in 9 will develop cancer during his lifetime. What fraction is this? In a college freshman class of 3663 men, how many can expect to develop cancer?

57. A trucker delivered $\frac{1}{10}$ of the truck's load of gravel. How much gravel was left in the truck if it can carry $\frac{7}{8}$ ton of gravel?

58. Last night it snowed $2\frac{7}{8}$ inches. As the temperature rose, $\frac{1}{3}$ of the snow melted. How much snow is still on the ground?

59. The state wildlife agency stocked one of the Portage Lakes with 15,246 fish. How many fish could you expect to find in $\frac{3}{22}$ of the lake?

60. A biologist counted 108 fish in $\frac{1}{3}$ of Milton Lake. How many fish could you expect to find in the whole lake?

5.4 **Division of Fractions and Mixed Numbers**

OBJECTIVE

● Use reciprocals in dividing fractions and mixed numbers.

Before we can look at division of fractions, we need to look at reciprocals. Consider these multiplication problems.

$$7 \cdot \frac{1}{7} = 1 \qquad \frac{3}{4} \cdot \frac{4}{3} = 1 \qquad \left(-\frac{2}{3}\right)\left(-\frac{3}{2}\right) = 1$$

DEFINITION   When the product of two numbers is 1, the numbers are called **reciprocals.**

The reciprocal of 3 is $\frac{1}{3}$; the reciprocal of $\frac{1}{3}$ is 3.

The reciprocal of $\frac{3}{7}$ is $\frac{7}{3}$.

The reciprocal of $-\frac{4}{5}$ is $-\frac{5}{4}$.

A positive number has a positive reciprocal.
A negative number has a negative reciprocal.

Note   The reciprocal of a mixed number can only be found by changing the mixed number into an improper fraction.

To find the reciprocal of $1\frac{1}{2}$, first find the equivalent improper fraction, which is $\frac{3}{2}$. Then, form the reciprocal, which is $\frac{2}{3}$.

To find the reciprocal of $-3\frac{2}{3}$, first find the equivalent improper fraction, which is $-\frac{11}{3}$. Then, form the reciprocal, which is $-\frac{3}{11}$.

Note All real numbers, except 0, have reciprocals. Since division by 0 is undefined, 0 has no reciprocal.

Since $12 \div 3 = 4$ and $12 \cdot \frac{1}{3} = 4$, $12 \div 3 = 12 \cdot \frac{1}{3}$.

Since $35 \div 7 = 5$ and $35 \cdot \frac{1}{7} = 5$, $35 \div 7 = 35 \cdot \frac{1}{7}$.

These examples illustrate that dividing by a number is equivalent to multiplying by the reciprocal of the number.

Dividing 12 by 3 is equivalent to multiplying 12 by $\frac{1}{3}$, the reciprocal of 3.

Dividing 35 by 7 is equivalent to multiplying 35 by $\frac{1}{7}$, the reciprocal of 7.

RULE To divide one fraction by another
For all real numbers a, b, c, d, where $b \neq 0$, $c \neq 0$, and $d \neq 0$,

$$\frac{a}{b} \div \frac{c}{d} = \frac{a}{b} \cdot \frac{d}{c} = \frac{a \cdot d}{b \cdot c}$$

In words, to divide, multiply the dividend by the reciprocal of the divisor.

EXAMPLE 1 $\dfrac{3}{4} \div \dfrac{1}{2} = \dfrac{3}{4} \cdot \dfrac{2}{1} = \dfrac{3}{2}$

EXAMPLE 2 $\dfrac{2}{3} \div \dfrac{8}{9} = \dfrac{2}{3} \cdot \dfrac{9}{8} = \dfrac{3}{4}$

EXAMPLE 3 $\dfrac{2}{7} \div \dfrac{7}{4} = \dfrac{2}{7} \cdot \dfrac{4}{7} = \dfrac{8}{49}$

USING THE CALCULATOR

Find $\dfrac{3}{4} \div \dfrac{1}{2}$

Enter 3 a^b/c 4 $\div$ 1 a^b/c 2 $=$ **1_1⌋2** ← Display

To divide by fractions using a calculator, it is not necessary to use the reciprocal.

Notes Before reducing fractions in a division problem, change the problem to multiplication.

To divide a mixed number by another mixed number, first change both mixed numbers to improper fractions. Then, multiply by the reciprocal of the divisor.

EXAMPLE 4

$$\left(9\frac{3}{5}\right) \div \left(3\frac{1}{5}\right) =$$

$$\frac{48}{5} \div \frac{16}{5} = \qquad \text{Change to improper fractions.}$$

$$\frac{48}{5} \cdot \frac{5}{16} = \qquad \text{Multiply by the reciprocal.}$$

$$\frac{3}{1} = 3$$

USING THE CALCULATOR

Find $9\frac{3}{5} \div 3\frac{1}{5}$.

Enter 9 a^b/c 3 a^b/c 5 $\div$ 3 a^b/c 1 a^b/c 5 $=$ **3** $\leftarrow$ Display

To divide mixed numbers using a calculator, it is not necessary to change the mixed numbers to improper fractions.

EXAMPLE 5

$$\left(3\frac{2}{7}\right) \div 2 =$$

$$\frac{23}{7} \div \frac{2}{1} = \qquad \text{Change to improper fractions.}$$

$$\frac{23}{7} \cdot \frac{1}{2} = \qquad \text{Multiply by the reciprocal.}$$

$$\frac{23}{14} \text{ or } 1\frac{9}{14}$$

Note The rules for division of signed integers also apply to division of negative and positive fractions. See Section 4.6.

EXAMPLE 6

$$-\frac{2}{3} \div \frac{1}{6} = -\frac{2}{3} \cdot \frac{6}{1} = -4$$

EXAMPLE 7

$$-\frac{4}{5} \div \left(-\frac{16}{15}\right) = -\frac{4}{5} \cdot \left(-\frac{15}{16}\right) = \frac{3}{4}$$

EXAMPLE 8

$$\frac{7}{8} \div \left(-\frac{3}{16}\right) = \frac{7}{8} \cdot \left(-\frac{16}{3}\right) = -\frac{14}{3} \text{ or } -4\frac{2}{3}$$

EXAMPLE 9

$$\frac{x^2}{y} \div \frac{x}{y^2} = \frac{x^2}{y} \cdot \frac{y^2}{x} = xy$$

USING THE CALCULATOR

Find $\left(-\frac{2}{3}\right) \div \frac{1}{6}$

Enter 2 a^b/c 3 +/- $\div$ 1 a^b/c 6 $=$ **−4** $\leftarrow$ Display

Find $\left(\frac{7}{8}\right) \div \left(-\frac{3}{16}\right)$

Enter 7 a^b/c 8 $\div$ 3 a^b/c 16 +/- $=$ **−4_2⌋3** $\leftarrow$ Display

Practice Problems

Find the quotients.

6. $-3 \div \left(-\frac{4}{5}\right)$

7. $\frac{6}{7} \div \frac{3}{14}$

8. $12 \div \left(-\frac{4}{3}\right)$

9. $-\frac{5}{6} \div \left(6\frac{2}{3}\right)$

10. $\frac{3}{19} \div \frac{15}{38}$

11. $\frac{x}{y^2} \div \frac{x^3}{y^3}$

12. $\left(4\frac{1}{2}\right) \div \left(3\frac{3}{5}\right)$

APPLICATION 1

It takes $\frac{3}{4}$ of a yard of fabric to make a scarf. How many scarves can be made from 6 yards of fabric? (How many sets of $\frac{3}{4}$ are there in 6 yards?)

$$6 \div \frac{3}{4} = 6 \cdot \frac{4}{3} = 8$$

Answer: 8 scarves.   ◆

APPLICATION 2

A recipe calls for $\frac{3}{4}$ cup of flour. If the cook only has an $\frac{1}{8}$-cup measuring cup, how many $\frac{1}{8}$ cups will be needed to measure $\frac{3}{4}$ cup of flour? (How many sets of $\frac{1}{8}$ are there in $\frac{3}{4}$?)

$$\frac{3}{4} \div \frac{1}{8} = \frac{3}{4} \cdot \frac{8}{1} = 6$$

Answer: 6 measures of $\frac{1}{8}$ cup each.   ◆

APPLICATION 3

How many bottles containing $\frac{3}{5}$ pint of liquid can be filled from a 21 pint container? (How many sets of $\frac{3}{5}$ are there in 21?)

$$21 \div \frac{3}{5} = 21 \cdot \frac{5}{3} = 35$$

Answer: Thirty-five bottles can be filled.   ◆

ANSWERS TO PRACTICE PROBLEMS
1. $\frac{5}{2}$ 2. $\frac{9}{4}$ 3. $-\frac{8}{3}$ 4. $-\frac{1}{8}$ 5. $\frac{6}{17}$
6. $\frac{15}{4}$ 7. 4 8. -9 9. $-\frac{1}{8}$
10. $\frac{2}{5}$ 11. $\frac{y}{x^2}$ 12. $\frac{5}{4}$

PROBLEM SET 5.4

Divide. Work the problems with paper and pencil, then check the answers on a calculator.

1. $\frac{1}{2} \div \frac{3}{5}$ 2. $\frac{2}{3} \div \frac{5}{7}$ 3. $\frac{1}{4} \div \left(-\frac{3}{8}\right)$

4. $-\frac{4}{5} \div \frac{10}{2}$ 5. $\frac{1}{3} \div (-6)$ 6. $-\frac{4}{5} \div (-4)$

7. $\frac{7}{8} \div \left(-\frac{7}{8}\right)$ 8. $-\frac{25}{36} \div \frac{5}{6}$ 9. $\frac{3}{4} \div (-3)$

10. $\frac{14}{3} \div \frac{16}{9}$ 11. $-1\frac{1}{14} \div \frac{5}{2}$ 12. $-\frac{3}{2} \div \left(-\frac{3}{2}\right)$

13. $\frac{19}{45} \div \left(-5\frac{7}{10}\right)$

14. $\frac{17}{13} \div \left(-\frac{17}{13}\right)$

15. $1\frac{1}{14} \div \left(-1\frac{17}{28}\right)$

16. $3\frac{3}{7} \div \left(-\frac{8}{21}\right)$

17. $\frac{13}{9} \div \frac{13}{9}$

18. $\frac{17}{20} \div 13\frac{3}{5}$

19. $-\frac{3}{14} \div \left(-\frac{6}{7}\right)$

20. $\frac{1}{4} \div \frac{1}{9}$

21. $\frac{x^2 y}{z} \div \frac{z^2}{xy}$

22. $\frac{x}{y^2} \div \frac{y}{x^2}$

23. $\frac{ab^2}{cd} \div \frac{c^2 d^2}{ab^2}$

24. $\frac{a^2}{b^2} \div \frac{a}{b}$

25. $\left(3\frac{3}{5}\right) \div \left(2\frac{7}{10}\right)$

26. $\left(-1\frac{4}{7}\right) \div \left(7\frac{1}{3}\right)$

27. $\dfrac{3\frac{3}{4}}{-2\frac{5}{8}}$

28. $\left(-1\frac{5}{7}\right) \div 24$

Perform the indicated operations.

29. Find the quotient of $\frac{7}{8}$ and $\frac{7}{16}$.

30. What is the quotient of $\frac{5}{6}$ and $\frac{15}{36}$.

31. Divide $\frac{4}{5}$ by $\frac{16}{25}$.

Applications:

32. If it takes $\frac{3}{8}$ yard of fabric to make a pillow, how many pillows can be from $3\frac{3}{4}$ yards of fabric?

33. A shirt requires $2\frac{2}{5}$ meters of material. How many shirts can be made from 84 meters of material?

34. How many $\frac{5}{8}$ pint glasses of pop can be filled from a 15 pint barrel?

35. How many $\frac{1}{6}$ cup measures are there in $1\frac{2}{3}$ cups?

36. If it takes $\frac{5}{8}$ yard of fabric to make a pillow, how many pillows can be made from $3\frac{1}{8}$ yards of fabric?

37. A shirt requires $3\frac{1}{8}$ meters of material. How many shirts can be made from 100 meters of material?

38. How many $\frac{4}{5}$ pint glasses of pop can be filled from a 12 pint barrel?

39. How many $\frac{1}{8}$ cup measures are there in $2\frac{3}{4}$ cups?

40. After painting his house for 20 hours, José has $\frac{4}{5}$ of it painted. How long will it take him to finish painting the house?

5.5 Addition and Subtraction of Fractions and Mixed Numbers

OBJECTIVE

● Add and subtract fractions and mixed numbers.

Fractions can be added and subtracted only if they have common denominators.

> **RULE** To add or subtract fractions with the same denominators
>
> **1.** Keep the denominator.
> **2.** Add or subtract the **numerators.**
> **3.** Reduce the resulting fraction to lowest terms.

EXAMPLE 1 $\quad \dfrac{1}{5} + \dfrac{3}{5} = \dfrac{4}{5}$

Add only the **numerators.** Think of 1 bat plus 3 bats equal 4 bats. Therefore, one-fifth plus three-fifths equal four-fifths.

Let each pentagon represent one unit.

This diagram represents $\dfrac{1}{5}$.   This diagram represents $\dfrac{3}{5}$.

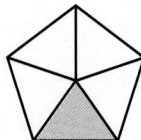 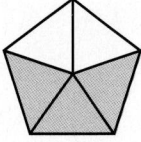

Add the shaded regions. How many fifths are there? $1 + 3 = 4$. There are four fifths $\left(\frac{4}{5}\right)$.

USING THE CALCULATOR

Find $\frac{1}{5} + \frac{3}{5}$

Enter 1 $a^{b/c}$ 5 $+$ 3 $a^{b/c}$ 5 $=$ **4⌋5** ← Display

EXAMPLE 2
$$\frac{5}{8} - \frac{4}{8} = \frac{1}{8}$$

This diagram represents $\frac{5}{8}$.   This diagram represents $\frac{4}{8}$.

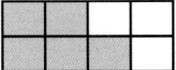   

If we subtract the 4 shaded regions in the second diagram from the 5 shaded regions in the first diagram, how many shaded regions do we have left? $5 - 4 = 1$. There is one-eighth left.

$$\frac{5}{8} - \frac{4}{8} = \frac{1}{8}$$

USING THE CALCULATOR

Find $\frac{5}{8} - \frac{4}{8}$

Enter 5 $a^{b/c}$ 8 $-$ 4 $a^{b/c}$ 8 $=$ **1⌋8** ← Display

DEFINITION   The **lowest common denominator (LCD)** of two or more fractions is the smallest number that is a multiple of each denominator.

Note   The lowest common denominator (LCD) is the least common multiple of the denominators of a set of fractions.

RULE To add or subtract fractions with different denominators

1. Find the lowest common denominator of the denominators.
2. Express each fraction as an equivalent fraction having the lowest common denominator as its denominator.
3. Add or subtract the **numerators** of the newly formed fractions.
4. Reduce the resulting fraction.

EXAMPLE 3 $\dfrac{1}{2} + \dfrac{1}{4}$ The lowest common denominator of 2 and 4 is 4.

$$= \dfrac{2}{4} + \dfrac{1}{4} \qquad \dfrac{1}{2} \cdot \dfrac{2}{2} = \dfrac{2}{4}$$

$$= \dfrac{2 + 1}{4} \qquad \text{Add the numerators.}$$

$$= \dfrac{3}{4}$$

EXAMPLE 4 $\dfrac{1}{2} - \dfrac{1}{4}$

$$= \dfrac{2}{4} - \dfrac{1}{4} \qquad \dfrac{1}{2} \cdot \dfrac{2}{2} = \dfrac{2}{4}$$

$$= \dfrac{2 - 1}{4} \qquad \text{Subtract the numerators.}$$

$$= \dfrac{1}{4}$$

In order to add or subtract fractions, the fractions must be renamed or converted into equivalent fractions which have like denominators.

EXAMPLE 5 $\dfrac{2}{3} + \dfrac{5}{6}$

$$= \dfrac{4}{6} + \dfrac{5}{6} \qquad \dfrac{2}{3} \cdot \dfrac{2}{2} = \dfrac{4}{6}$$

$$= \dfrac{4 + 5}{6}$$

$$\dfrac{9}{6} = \dfrac{3}{2} \text{ or } 1\dfrac{1}{2} \qquad \text{Express the answer in lowest terms.}$$

EXAMPLE 6 $\dfrac{5}{6} - \dfrac{2}{3}$

$$= \dfrac{5}{6} - \dfrac{4}{6} \qquad \dfrac{2}{3} \cdot \dfrac{2}{2} = \dfrac{4}{6}$$

$$= \dfrac{5 - 4}{6} \qquad \text{Subtract the numerators.}$$

$$= \dfrac{1}{6}$$

EXAMPLE 7 $\dfrac{1}{4} + \dfrac{1}{3}$ Since 4 is not a multiple of 3, use 12 as the common denominator.

$$= \dfrac{1}{4} \cdot \dfrac{3}{3} + \dfrac{1}{3} \cdot \dfrac{4}{4}$$

$$= \dfrac{3}{12} + \dfrac{4}{12}$$

$$= \dfrac{7}{12}$$

Note It is always possible to find a common denominator by multiplying the denominators together, but it may not be the lowest common denominator and doing so can result in large, often unwieldy calculations.

EXAMPLE 8 $\dfrac{1}{6} + \dfrac{1}{8}$ Find the Lowest Common Denominator

Method 1: 2 is a common factor of 6 and 8, so divide both 6 and 8 by 2.

$$2\overline{)\,6\quad 8\,}$$
$$\quad\;\; 3\quad 4 \qquad \text{3 and 4 have no common factors, so multiply the divisor by}$$
$$\qquad\qquad\qquad\qquad \text{both dividends}$$

$$\text{LCD}(6, 8) = 2 \cdot 3 \cdot 4 = 24$$

Method 2:   6   8   Write each denominator as a product of primes.

$$6 = 2 \cdot 3 \qquad 8 = 2 \cdot 2 \cdot 2 = 2^3$$
$$\text{LCD}(6, 8) = 2^3 \cdot 3 = 24$$

The LCD is the product of all factors of each denominator raised to their highest powers.

$$\dfrac{1}{6} + \dfrac{1}{8} \qquad\qquad \text{LCD}(6, 8) = 24$$

$$= \dfrac{1}{6} \cdot \dfrac{4}{4} + \dfrac{1}{8} \cdot \dfrac{3}{3} \qquad 6 \cdot 4 = 24 \text{ and } 8 \cdot 3 = 24$$

$$= \dfrac{4}{24} + \dfrac{3}{24}$$

$$= \dfrac{7}{24}$$

EXAMPLE 9 $\dfrac{5}{12} + \dfrac{5}{18}$ First find the LCD

Method 1 $2\overline{)\,12\quad 18\,}$
$$\quad\; 3\overline{)\,\;6\quad\; 9\,}$$
$$\qquad\quad\; 2\quad\;\; 3$$

$$\text{LCD}(12, 18) = 2 \cdot 2 \cdot 3 \cdot 3$$

Method 2

$$LCD(12, 18) = 2^2 \cdot 3^2 = 36$$

$$\frac{5}{12} + \frac{5}{18}$$ Express each fraction as an equivalent fraction with the LCD.

$$\frac{5}{12} \cdot \frac{3}{3} + \frac{5}{18} \cdot \frac{2}{2}$$

$$\frac{15}{36} + \frac{10}{36}$$

$$= \frac{25}{36}$$

EXAMPLE 10 $\dfrac{2}{9} - \dfrac{9}{15}$ First find the LCD.

Method 1

$$3\overline{)9 \quad 15}$$
$$3 \quad 5 \quad \text{so LCD}(9, 15) = 3 \cdot 3 \cdot 5 = 45$$

Method 2

$$9 = 3^2$$
$$15 = 3 \cdot 5 \quad \text{so LCD}(9, 15) = 3^2 \cdot 5 = 45$$

Now use the LCD to perform the subtraction.

$$\frac{2}{9} - \frac{9}{15}$$

$$= \frac{2}{9} \cdot \frac{5}{5} - \frac{9}{15} \cdot \frac{3}{3}$$

$$= \frac{10}{45} - \frac{27}{45}$$

$$= \frac{10 - 27}{45} = \frac{-17}{45} \text{ or } -\frac{17}{45}$$

EXAMPLE 11 $-\dfrac{9}{25} - \dfrac{1}{20}$

Method 1

$$5\overline{)25 \quad 20}$$
$$5 \quad 4 \quad \text{so LCD}(25, 20) = 5 \cdot 5 \cdot 4 = 100$$

Method 2

$$25 = 5^2$$
$$20 = 2^2 \cdot 5 \quad \text{so LCD}(25, 20) = 2^2 \cdot 5^2 = 100$$

$$-\frac{9}{25} - \frac{1}{20}$$

$$= \frac{-9}{25} \cdot \frac{4}{4} - \frac{1}{20} \cdot \frac{5}{5}$$

$$= \frac{-36}{100} - \frac{5}{100} = \frac{-36 - 5}{100} = \frac{-41}{100} \text{ or } -\frac{41}{100}$$

EXAMPLE 12

$$\frac{1}{9} + \frac{1}{5} + \frac{1}{6}$$

Method 1

$$3\overline{)9 \quad 5 \quad 6}$$
$$3 \quad 5 \quad 2$$

Note 3 does not divide evenly into 5, so just bring down the 5.

$$LCD(9, 5, 6) = 3 \cdot 3 \cdot 5 \cdot 2 = 90$$

Method 2

$$9 = 3^2$$
$$5 = 5$$
$$6 = 2 \cdot 3 \qquad \text{so } LCD(9, 5, 6) = 2 \cdot 3^2 \cdot 5 = 90$$

$$\frac{1}{9} + \frac{1}{5} + \frac{1}{6}$$

$$= \frac{1}{9} \cdot \frac{10}{10} + \frac{1}{5} \cdot \frac{18}{18} + \frac{1}{6} \cdot \frac{15}{15}$$

$$= \frac{10}{90} + \frac{18}{90} + \frac{15}{90}$$

$$= \frac{10 + 18 + 15}{90} = \frac{43}{90}$$

USING THE CALCULATOR

Find $\frac{1}{9} + \frac{1}{5} + \frac{1}{6}$

Enter 1 $\boxed{a^{b/c}}$ 9 $\boxed{+}$ 1 $\boxed{a^{b/c}}$ 5 $\boxed{+}$ 1 $\boxed{a^{b/c}}$ 6 $\boxed{=}$ **43⌋90** ← Display

---

EXAMPLE 13

$$2 - \frac{1}{4} \qquad \text{Recall that } 2 = \frac{2}{1}.$$

$$= \frac{2}{1} - \frac{1}{4} \qquad LCD(1, 4) = 4$$

$$= \frac{2}{1} \cdot \frac{4}{4} - \frac{1}{4}$$

$$= \frac{8}{4} - \frac{1}{4}$$

$$= \frac{8 - 1}{4} = \frac{7}{4} \text{ or } 1\frac{3}{4}$$

Addition and Subtraction of Mixed Numbers

When adding and subtracting mixed numbers, remember that mixed numbers are sums of integers and proper fractions.

$$3\frac{1}{4} + 4\frac{2}{3} \text{ is really } 3 + \frac{1}{4} + 4 + \frac{2}{3}$$

In adding mixed numbers we are actually adding integers and fractions. One method for adding them is to first convert them to improper fractions and then to add the improper fractions.

EXAMPLE 14 $3\frac{1}{4} + 4\frac{2}{3}$

Method 1 $3\frac{1}{4} + 4\frac{2}{3}$ Change the mixed numbers to improper fractions.

$$= \frac{13}{4} + \frac{14}{3}$$

$$= \frac{13}{4} \cdot \frac{3}{3} + \frac{14}{3} \cdot \frac{4}{4} \qquad \text{LCD(4, 3)} = 12$$

$$= \frac{39}{12} + \frac{56}{12}$$

$$= \frac{39 + 56}{12}$$

$$= \frac{95}{12} \text{ or } 7\frac{11}{12}$$

Method 2 $3\frac{1}{4} = 3\frac{3}{12}$ Align the numbers vertically.

$$\underline{4\frac{2}{3} = 4\frac{8}{12}} \qquad \text{LCD(4, 3)} = 12$$

$$7\frac{11}{12}$$

USING THE CALCULATOR

Find $3\frac{1}{4} + 4\frac{2}{3}$

Enter 3 $\boxed{a^{b}/_{c}}$ 1 $\boxed{a^{b}/_{c}}$ 4 $\boxed{+}$ 4 $\boxed{a^{b}/_{c}}$ 2 $\boxed{a^{b}/_{c}}$ 3 $\boxed{=}$ **7_11⌋12** ← Display

To add or subtract mixed numbers on a calculator, it is not necessary to convert them to improper fractions.

EXAMPLE 15 $5\frac{3}{4} + 3\frac{5}{6}$

Method 1 $5\frac{3}{4} + 3\frac{5}{6}$

$$= \frac{23}{4} + \frac{23}{6}$$

$$= \frac{23}{4} \cdot \frac{3}{3} + \frac{23}{6} \cdot \frac{2}{2}$$

$$= \frac{69}{12} + \frac{46}{12}$$

$$= \frac{69 + 46}{12}$$

$$= \frac{115}{12} \text{ or } 9\frac{7}{12}$$

Method 2
$$5\frac{3}{4} = 5\frac{9}{12}$$
$$+\ 3\frac{5}{6} = 3\frac{10}{12}$$
$$8\frac{19}{12} = 8 + \frac{19}{12} = 8 + \left(1 + \frac{7}{12}\right) = 9 + \frac{7}{12} = 9\frac{7}{12}$$

EXAMPLE 16 $3\frac{9}{10} - 1\frac{3}{10}$

Method 1
$$3\frac{9}{10} - 1\frac{3}{10}$$
$$= \frac{39}{10} - \frac{13}{10}$$
$$= \frac{39 - 13}{10}$$
$$= \frac{26}{10} = \frac{13}{5} \qquad \text{Reduce to lowest terms.}$$
$$\text{or } 2\frac{3}{5}$$

Method 2
$$3\frac{9}{10}$$
$$-1\frac{3}{10}$$
$$2\frac{6}{10} = 2\frac{3}{5}$$

USING THE CALCULATOR

Find $3\frac{9}{10} - 1\frac{3}{10}$

Enter 3 a^{b}/c 9 a^{b}/c 10 $-$ 1 a^{b}/c 3 a^{b}/c 10 $=$ **2_3⌋5** ← Display

EXAMPLE 17 $2\frac{1}{7} - 3\frac{1}{2}$

This problem is easier to work using Method 1.
$$2\frac{1}{7} - 3\frac{1}{2}$$
$$= \frac{15}{7} - \frac{7}{2}$$
$$= \frac{15}{7} \cdot \frac{2}{2} - \frac{7}{2} \cdot \frac{7}{7}$$
$$= \frac{30}{14} - \frac{49}{14}$$
$$= -\frac{19}{14} \text{ or } -1\frac{5}{14}$$

Practice Problems

Add or subtract the following fractions.

1. $\frac{5}{6} + \frac{3}{8}$

2. $2\frac{1}{4} + 1\frac{2}{9}$

3. $\frac{3}{10} - \frac{8}{15}$

4. $-\frac{3}{4} - \frac{3}{8}$

5. $3\frac{1}{10} - 2\frac{4}{5}$

EXAMPLE 18 $3\frac{1}{8} - 1\frac{3}{4}$

To use Method 2, first rewrite $\frac{3}{4}$ as $\frac{6}{8}$, then rewrite $3\frac{1}{8}$ as $2\frac{9}{8}$.

Note $3\frac{1}{8} = 3 + \frac{1}{8} = (2 + 1) + \frac{1}{8} = 2 + \left(\frac{8}{8} + \frac{1}{8}\right)$

Method 2 $3\frac{1}{8} = \quad 3\frac{1}{8} = \quad 2\frac{9}{8}$

$\underline{-1\frac{3}{4} = -1\frac{6}{8} = -1\frac{6}{8}}$

$\qquad\qquad\qquad\qquad 1\frac{3}{8}$

APPLICATION 1

In a mathematics class requiring a C to pass, $\frac{2}{9}$ of the students received a grade of A; $\frac{1}{3}$ received a grade of B; and $\frac{3}{18}$ received a grade of C. What fraction of the class passed? The total passing will be the sum of the As, Bs, and Cs. Add the fractions.

$$\frac{2}{9} + \frac{1}{3} + \frac{3}{18} \qquad \text{LCD}(9, 3, 18) = 18$$

$$= \frac{2}{9} \cdot \frac{2}{2} + \frac{1}{3} \cdot \frac{6}{6} + \frac{3}{18}$$

$$= \frac{4}{18} + \frac{6}{18} + \frac{3}{18}$$

$$= \frac{4 + 6 + 3}{18}$$

$$= \frac{13}{18}$$

Answer: A total of $\frac{13}{18}$ of the class passed. ◆

APPLICATION 2

John runs $\frac{7}{8}$ of a mile, Joe runs $\frac{1}{4}$ of a mile, and Amy runs $\frac{7}{12}$ of a mile. Find the sum of the distances they ran.

$$\frac{7}{8} + \frac{1}{4} + \frac{7}{12} =$$

$$\frac{7}{8} \cdot \frac{3}{3} + \frac{1}{4} \cdot \frac{6}{6} + \frac{7}{12} \cdot \frac{2}{2} =$$

$$\frac{21}{24} + \frac{6}{24} + \frac{14}{24} = \frac{21 + 6 + 14}{24} = \frac{41}{24} = 1\frac{7}{24}$$

Answer: They ran $1\frac{7}{24}$ miles. ◆

APPLICATION 3

Amy runs $\frac{8}{9}$ of mile each morning, while Joe runs $\frac{15}{18}$ of a mile. How much farther does Amy run each day? In one week, how much farther would Amy run if they each ran 5 days a week?

$$\frac{8}{9} - \frac{15}{18} = \frac{8}{9} \cdot \frac{2}{2} - \frac{15}{18} = \frac{16}{18} - \frac{15}{18} = \frac{1}{18}$$

Amy runs $\frac{1}{18}$ of a mile farther each day.

$$\left(5 \cdot \frac{1}{18}\right) = \frac{5}{8}$$

Answer: In a week, Amy runs $\frac{5}{18}$ of a mile farther. ◆

PROBLEM SET 5.5

Find the following sums and differences. Reduce answers to lowest terms.

1. $\frac{2}{5} + \frac{3}{5}$ **2.** $\frac{4}{6} + \frac{1}{6}$ **3.** $\frac{6}{7} - \frac{3}{7}$

4. $\frac{3}{4} - \frac{1}{4}$ **5.** $\frac{1}{3} - \frac{2}{3}$ **6.** $-\frac{4}{9} - \frac{2}{9}$

7. $\frac{1}{6} + \frac{1}{3}$ **8.** $\frac{2}{5} + \frac{7}{10}$ **9.** $\frac{2}{5} - \frac{3}{7}$

10. $\frac{6}{5} - \frac{9}{15}$

Find LCD and then add or subtract. Reduce answers to lowest terms.

11. $\frac{4}{9} + \frac{1}{3}$ **12.** $\frac{3}{10} + \frac{2}{5}$ **13.** $\frac{5}{6} - \frac{1}{12}$

14. $\frac{2}{3} + \frac{1}{4}$ **15.** $\frac{2}{3} - \left(-\frac{1}{3}\right)$ **16.** $-2 - \frac{2}{3}$

17. $\frac{4}{3} + \frac{5}{2}$ **18.** $-\frac{3}{10} + \frac{4}{15}$ **19.** $\frac{4}{5} + \left(-\frac{5}{6}\right)$

20. $\dfrac{7}{3} - \dfrac{13}{4}$

21. $\dfrac{5}{8} - \dfrac{11}{16}$

22. $-\dfrac{5}{9} - \dfrac{5}{6}$

23. $\dfrac{10}{3} - \dfrac{7}{6}$

24. $\dfrac{13}{8} - \dfrac{17}{6}$

25. $-\dfrac{9}{4} + \left(-\dfrac{14}{3}\right)$

26. $\dfrac{11}{28} + \dfrac{5}{21}$

27. $-\dfrac{4}{3} - \left(-\dfrac{4}{5}\right)$

28. $\dfrac{7}{8} - \dfrac{5}{7}$

29. $\dfrac{9}{14} + \left(-\dfrac{5}{21}\right)$

30. $\dfrac{9}{4} + \left(-\dfrac{3}{2}\right)$

31. $\dfrac{23}{5} + \left(-\dfrac{4}{15}\right)$

32. $\dfrac{4}{15} + \left(-\dfrac{17}{45}\right)$

Perform the indicated operations.

33. Find the sum of $\dfrac{2}{3}, \dfrac{4}{5}$, and $\dfrac{11}{15}$.

34. What is the sum of $\dfrac{5}{6}, \dfrac{7}{12}$, and $\dfrac{1}{24}$?

35. What is the difference between $\dfrac{9}{10}$ and $\dfrac{2}{15}$?

36. Find the difference between $\dfrac{5}{6}$ and $\dfrac{2}{9}$.

Applications:

37. A recipe calls for $\dfrac{3}{4}$ cup of sugar in the cake and $\dfrac{3}{8}$ cup of sugar in the frosting. How many cups of sugar are called for in this recipe?

38. John runs $\dfrac{1}{2}$ mile on Monday, $\dfrac{3}{4}$ mile on Tuesday, $\dfrac{2}{3}$ mile on Wednesday, $\dfrac{5}{6}$ mile on Thursday, and $\dfrac{3}{4}$ mile on Friday. How far did he run this week?

39. A family spends $\frac{3}{10}$ of its income on rent and $\frac{2}{5}$ on food. What fraction of its income is left?

40. A recipe calls for $\frac{7}{8}$ cup of flour and $\frac{4}{5}$ cup of sugar. How many cups of flour and sugar are listed in this recipe?

41. Jane runs $\frac{7}{8}$ mile Monday, $\frac{2}{3}$ mile Tuesday, $\frac{5}{8}$ mile Wednesday, $\frac{3}{4}$ mile Thursday, and $\frac{2}{3}$ mile Friday. How far did she run this week?

42. A family spends $\frac{1}{3}$ of its income on rent and $\frac{1}{6}$ on food. What fraction of its income is left?

43. The stock of the Midwestern Company rose $\frac{5}{8}$ points on Monday, $\frac{3}{8}$ points on Tuesday, $\frac{5}{8}$ points on Wednesday, $\frac{7}{8}$ points on Thursday, and $1\frac{1}{8}$ points Friday. What was the total rise for the week?

44. During a football game, $\frac{1}{3}$ of the total points were scored during the first quarter, $\frac{1}{4}$ of the total points were scored during the second quarter, and $\frac{3}{8}$ of the total points were scored during the third quarter. What fraction of the total points in the game were scored in the final quarter?

45. The distance around a triangle is called the perimeter. To find the perimeter of a triangle, add the lengths of all the sides. Find the perimeter of the triangle below.

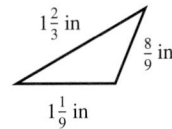

$1\frac{2}{3}$ in $\frac{8}{9}$ in $1\frac{1}{9}$ in

46. The distance around a rectangle is called the perimeter. To find the distance around a rectangle, add the length of all the sides. Find the perimeter of the rectangle below.

$4\frac{3}{4}$ in

$2\frac{3}{8}$ in $2\frac{3}{8}$ in

$4\frac{3}{4}$ in

5.6 Order of Operations and Complex Fractions

OBJECTIVE

• Use the rules for order of operations to simplify expressions containing more than one operation.

Before considering the order of operations for expressions containing fractions, look back to Sections 2.6 and 4.7. The order of operations we used for expressions containing integers is also used for fractions and mixed numbers.

> **RULE: Order of operations**
>
> 1. Perform all operations enclosed in parentheses or other grouping symbols.
> 2. Raise numbers to indicated powers or find roots.
> 3. Perform all multiplication and division in order from left to right.
> 4. Perform all addition and subtraction in order from left to right.

EXAMPLE 1 Simplify $3 + \left(\frac{7}{8}\right)\left(\frac{4}{7}\right)$

$$= 3 + \frac{7}{8} \cdot \frac{4}{7} \qquad \text{Multiply before adding.}$$

$$= 3 + \frac{1}{2} \qquad \text{Add. (Write as a mixed number.)}$$

$$= 3\frac{1}{2}$$

USING THE CALCULATOR

Simplify $3 + \left(\frac{7}{8}\right)\left(\frac{4}{7}\right)$

Enter 3 $\boxed{+}$ 7 $\boxed{a^{b/c}}$ 8 $\boxed{\times}$ 4 $\boxed{a^{b/c}}$ 7 $\boxed{=}$ **3_1⌋2** ← Display

Enter the numbers and operations in order as they appear in the expression; the calculator will automatically follow the order of operations.

EXAMPLE 2 Simplify $7\frac{1}{4} + \left(1\frac{1}{2}\right)\left(1\frac{2}{3}\right)$ Convert to improper fractions and multiply.

$$= 7\frac{1}{4} + \frac{3}{2} \cdot \frac{8}{3}$$

$$= 7\frac{1}{4} + 4 \qquad \text{Add.}$$

$$= 11\frac{1}{4}$$

EXAMPLE 3 Simplify $\left(\frac{1}{2} + \frac{2}{3}\right)\left(\frac{1}{7} + \frac{5}{7}\right)$ Add within the parentheses.

$$= \left(\frac{3}{6} + \frac{4}{6}\right)\left(\frac{6}{7}\right)$$

$$= \frac{7}{6} \cdot \frac{6}{7} \qquad \text{Multiply.}$$

$$= 1$$

USING THE CALCULATOR

Simplify $\left(\frac{1}{2} + \frac{2}{3}\right)\left(\frac{1}{7} + \frac{5}{7}\right)$

Enter $\boxed{(}$ 1 $\boxed{a^{b/c}}$ 2 $\boxed{+}$ 2 $\boxed{a^{b/c}}$ 3 $\boxed{)}$ $\boxed{\times}$ $\boxed{(}$ 1 $\boxed{a^{b/c}}$

7 $\boxed{+}$ 5 $\boxed{a^{b/c}}$ 7 $\boxed{)}$ $\boxed{=}$ **1** ← Display

Note: Remember to enter the parentheses so that the calculator will add before multiplying. Remember to enter the multiplication symbol. The calculator does not recognize writing parentheses next to each other as a symbol for multiplication.

EXAMPLE 4   Simplify $2\frac{1}{2} + \frac{1}{3}(6 + 3)$

Method 1

$$= 2\frac{1}{2} + \frac{1}{3} \cdot 6 + \frac{1}{3} \cdot 3 \qquad \text{Use the Distributive Property.}$$

$$= 2\frac{1}{2} + 2 + 1$$

$$= 2\frac{1}{2} + 3 \qquad\qquad\qquad \text{Add.}$$

$$= 5\frac{1}{2}$$

Method 2

$$= 2\frac{1}{2} + \frac{1}{3}(6 + 3)$$

$$= 2\frac{1}{2} + \frac{1}{3} \cdot 9 \qquad\qquad \text{Add within the parentheses.}$$

$$= 2\frac{1}{2} + 3 \qquad\qquad\qquad \text{Multiply.}$$

$$= 5\frac{1}{2} \qquad\qquad\qquad\qquad \text{Add.}$$

USING THE CALCULATOR

Simplify $2\frac{1}{2} + \frac{1}{3}(6 + 3)$

Enter 2 a^{b}/c 1 a^{b}/c 2 + 1 a^{b}/c 3 × (6 + 3) = **5_1⌋2** ← Display

---

EXAMPLE 5   Simplify $\left(\frac{2}{3}\right)^2 + \left(\frac{1}{4}\right)^2$

$$= \left(\frac{2}{3}\right)\left(\frac{2}{3}\right) + \left(\frac{1}{4}\right)\left(\frac{1}{4}\right)$$

$$= \frac{4}{9} + \frac{1}{16}$$

$$= \frac{64}{144} + \frac{9}{144}$$

$$= \frac{73}{144}$$

Complex Fractions

Simplifying complex fractions is not complex at all. Just use the order of operations you already know.

DEFINITION   **Complex fraction:** A fraction in which the numerator and/or the denominator are themselves fractions or combinations of fractions.

EXAMPLE 6   Find $\dfrac{\frac{3}{4}}{\frac{1}{3}}$ The fraction bar indicates division.

Method 1

$= \dfrac{3}{4} \div \dfrac{1}{3}$ Rewrite as a division problem.

$= \dfrac{3}{4} \cdot \dfrac{3}{1}$ Multiply by the reciprocal.

$= \dfrac{9}{4}$

Method 2

$\dfrac{\frac{3}{4}}{\frac{1}{3}}$ Find the LCD of the two denominators.

   LCD (3, 4) = 12

$= \dfrac{\frac{3}{4} \cdot 12}{\frac{1}{3} \cdot 12}$ Multiply both fractions by the LCD.

$= \dfrac{\frac{3}{\cancel{4}} \cdot \overset{3}{\cancel{12}}}{\frac{1}{\cancel{3}} \cdot \overset{4}{\cancel{12}}}$ Simplify.

$= \dfrac{9}{4}$

EXAMPLE 7   Find $\dfrac{\frac{1}{2} + \frac{2}{3}}{\frac{3}{4} - \frac{1}{6}}$ ← The fraction bar indicates division.

Method 1

$= \left(\dfrac{1}{2} + \dfrac{2}{3}\right) \div \left(\dfrac{3}{4} - \dfrac{1}{6}\right)$ Rewrite as a division problem.

$= \left(\dfrac{3}{6} + \dfrac{4}{6}\right) \div \left(\dfrac{9}{12} - \dfrac{2}{12}\right)$ Rewrite with LCDs.

$= \left(\dfrac{7}{6}\right) \div \left(\dfrac{7}{12}\right)$ Simplify.

$= \dfrac{7}{6} \cdot \dfrac{12}{7}$ Multiply by the reciprocal.

$= 2$

Method 2

$$\frac{\dfrac{1}{2} + \dfrac{2}{3}}{\dfrac{3}{4} - \dfrac{1}{6}}$$ Find LCD of all the denominators.

LCD (2, 3, 4, 6) = 12

$$= \frac{\dfrac{1}{2} + \dfrac{2}{3}}{\dfrac{3}{4} - \dfrac{1}{6}} \cdot \frac{12}{12}$$ Multiply both the numerator and denominator by the LCD.

$$= \frac{\dfrac{1}{2} \cdot \dfrac{12}{1} + \dfrac{2}{3} \cdot \dfrac{12}{1}}{\dfrac{3}{4} \cdot \dfrac{12}{1} - \dfrac{1}{6} \cdot \dfrac{12}{1}}$$ Use the Distributive Property.

$$= \frac{6 + 8}{9 - 2}$$ Simplify.

$$= \frac{14}{7}$$ Reduce to lowest terms.

$$= 2$$

USING THE CALCULATOR

Find $\dfrac{\dfrac{1}{2} + \dfrac{2}{3}}{\dfrac{3}{4} - \dfrac{1}{6}}$

Enter (1 $a^b/_c$ 2 + 2 $a^b/_c$ 3) ÷ (3 $a^b/_c$ 4 −

1 $a^b/_c$ 6) = **2** ← Display

Remember to enclose the numerator and the denominator of a complex fraction in parentheses.

PROBLEM SET 5.6

Simplify these complex fractions. Reduce to lowest terms.

1. $\dfrac{\dfrac{7}{8}}{\dfrac{3}{4}}$ 2. $\dfrac{\dfrac{3}{10}}{\dfrac{1}{2}}$ 3. $\dfrac{\dfrac{1}{3}}{\dfrac{5}{6}}$

4. $\dfrac{\dfrac{3}{4}}{\dfrac{2}{3}}$

Simplify using the order of operations.

5. $\frac{2}{3} + 4\left(\frac{1}{2} + \frac{3}{2}\right)$

6. $5 + 12\left(\frac{1}{6} + \frac{1}{8}\right)$

7. $\left(\frac{2}{3} + \frac{2}{9}\right)\left(\frac{1}{2} - \frac{2}{5}\right)$

8. $\left(\frac{3}{4} - \frac{4}{5}\right)\left(\frac{3}{7} - \frac{1}{14}\right)$

9. $\frac{4}{7} \div \left(-\frac{2}{7}\right)\left(-\frac{3}{8}\right)$

10. $\frac{5}{9} \div \left(-\frac{35}{39}\right)\left(\frac{14}{13}\right)$

11. $\left(\frac{3}{4}\right)\left(\frac{1}{2}\right)^2 + \frac{1}{8}$

12. $\left(2\frac{1}{2}\right)^2 + \left(1\frac{1}{2}\right)^2$

13. $\left(\frac{2}{3}\right)^2\left(\frac{3}{4}\right) - \frac{2}{5}$

14. $\frac{2}{3}\left(\frac{3}{4} + 3\right) + \left(\frac{1}{3}\right)^2$

15. $\frac{7}{8} + \frac{2}{3}\left(\frac{1}{2} + \frac{6}{5}\right)$

Simplify. Reduce to lowest terms.

16. $\dfrac{\frac{15}{6}}{9}$

17. $\dfrac{\frac{2}{5} + \frac{1}{6}}{\frac{3}{4} - \frac{7}{8}}$

18. $\dfrac{\frac{1}{8} + \frac{3}{4}}{\frac{1}{2} - \frac{1}{3}}$

19. $\dfrac{\frac{3}{16} + 5}{6 - \frac{7}{8}}$

20. $\dfrac{-\frac{4}{3} + \frac{5}{6}}{-\frac{3}{4} + \frac{2}{9}}$

21. $\dfrac{\frac{2}{3} + \frac{3}{2}}{\frac{1}{3} - \frac{5}{6}}$

22. $\dfrac{\frac{3}{5} - \frac{7}{10}}{\frac{3}{20} - \frac{1}{5}}$

23. $\dfrac{5 - \frac{3}{4}}{2 - \frac{1}{3}}$

24. $\dfrac{\frac{4}{3} - \frac{5}{6}}{\frac{3}{4} + \frac{2}{5}}$

25. $\dfrac{\frac{4}{3} + \frac{5}{6}}{\frac{2}{3} - \frac{3}{2}}$

26. $\dfrac{\frac{13}{18} - \frac{11}{24}}{\frac{5}{12} - \frac{7}{36}}$

27. $\dfrac{\frac{2}{5} + \frac{7}{2}}{\frac{12}{3} - \frac{6}{5}}$

5.7 Solving Equations Using the Multiplication Property of Equality

OBJECTIVES

● Apply the Multiplication Property of Equality.

● Solve first degree equations in one variable by using the Multiplication Property of Equality.

In this section, we will use the **Multiplication Property of Equality** to solve first degree equations in one variable. It states that if two real numbers are equal and are multiplied by the same real number, the products formed also will be equal.

Multiplication Property of Equality: For all real numbers a, b, and c,

$$\text{If } a = b, \text{ then } a \cdot c = b \cdot c$$

Note This property will also hold true when a and b represent expressions.

EXAMPLE 1 Since $3 = 2 + 1$

$$5 \cdot 3 = 5(2 + 1) \qquad \text{Multiply both sides by 5.}$$
$$15 = 5(3)$$
$$15 = 15 \qquad \text{The products are equal.}$$

Using the Multiplication Property of Equality to Solve Equations

In general, multiply both sides of the equation by the reciprocal of the coefficient of the variable. This will result in getting the coefficient to be 1.

$$\left(\text{Remember, } a \cdot \frac{1}{a} = 1 \text{ for all } a \in \text{R}, a \neq 0 \right)$$

EXAMPLE 2 $3x = 6$

$$\frac{1}{3} \cdot 3x = 6 \cdot \frac{1}{3} \qquad \text{Multiply both sides of the equation by } \frac{1}{3} \text{ to obtain a}$$
$$1 \cdot x = 2 \qquad \text{coefficient of 1 for the variable. } \left(\frac{1}{3} \cdot 3 = 1 \right)$$
$$x = 2 \qquad \text{Recall: } 1 \cdot a = a.$$

Check: $\quad 3x = 6$
$$3 \cdot 2 \ ? \ 6$$
$$6 = 6$$

EXAMPLE 3 $5y = -30$

$$\frac{1}{5} \cdot 5y = \frac{1}{5} \cdot (-30) \qquad \text{Multiply both sides by } \frac{1}{5}, \text{ the reciprocal of 5.}$$
$$1 \cdot y = -6$$
$$y = -6$$

Check: $\quad 5y = -30$
$$5(-6) \ ? \ -30$$
$$-30 = -30$$

EXAMPLE 4 $-2x = 18$

$$-\frac{1}{2} \cdot (-2)x = \overset{9}{\cancel{18}} \cdot \left(-\frac{1}{\underset{1}{\cancel{2}}}\right)$$ Multiply both sides by $-\frac{1}{2}$, the reciprocal of -2.

$$1 \cdot x = -9$$
$$x = -9$$

EXAMPLE 5 $\frac{1}{3}x = 12$

$$3 \cdot \frac{1}{3}x = 12 \cdot 3$$ Multiply both sides by 3, the reciprocal of $\frac{1}{3}$.

$$1 \cdot x = 36$$
$$x = 36$$

EXAMPLE 6 $-2y = -12$

$$\left(-\frac{1}{2}\right)(-2y) = \left(-\frac{1}{2}\right)(-12)$$ Multiply both sides by $-\frac{1}{2}$.

$$1y = 6$$
$$y = 6$$

Note   While the format used in these examples clearly illustrates the use of the Multiplication Property of Equality, some students may prefer to show their solutions in the following format.

$$3x = 6$$

$$\frac{\overset{1}{\cancel{3}}x}{\underset{1}{\cancel{3}}} = \frac{\overset{2}{\cancel{6}}}{\underset{1}{\cancel{3}}}$$ Divide both sides by 3, the coefficient of *x*.

$$1x = 2$$
$$x = 2$$

Since dividing by a number is equivalent to multiplying by its reciprocal, this format is acceptable. (Recall: $a \div b = a \cdot \frac{1}{b}$ for all real numbers, where $b \neq 0$.)

Use the method that seems easier or more comfortable to you. Before using the multiplication property of equality, be sure that only one term contains the variable and that it is "isolated" on one side of the equation.

EXAMPLE 7 $-y = 22$

$$-1 \cdot y = 22$$ Recall: $-1 \cdot a = -a$.

$$\frac{-1 \cdot y}{-1} = \frac{22}{-1}$$

$$y = -22$$

Note   Each of the following examples has a fraction as the coefficient of the variable. Recall that for all real numbers, $a, b \neq 0$,

$$\frac{a}{b} \cdot \frac{b}{a} = 1.$$

In other words, $\frac{b}{a}$ is the reciprocal of $\frac{a}{b}$. To obtain a coefficient of 1 for the variable, multiply both sides of the equation by the reciprocal of the original coefficient.

Practice Problems

Solve for the variable.

1. $2x = 8$

2. $-3x = 39$

3. $-7y = -56$

4. $-36 = 4q$

5. $15 = -w$

6. $\frac{1}{2}x = 3$

7. $-\frac{2}{5}x = 8$

8. $\frac{3y}{7} = -12$

EXAMPLE 8 $\frac{5}{6}y = 20$

$$\frac{6}{5} \cdot \frac{5}{6}y = 20 \cdot \frac{6}{5}$$

$$y = 24$$

EXAMPLE 9 $\frac{2}{3}x = \frac{3}{4}$

$$\frac{3}{2} \cdot \frac{2}{3}x = \frac{3}{4} \cdot \frac{3}{2} \qquad \text{Multiply both sides by } \frac{3}{2}, \text{ the reciprocal of } \frac{2}{3}.$$

$$1x = \frac{3 \cdot 3}{4 \cdot 2}$$

$$x = \frac{9}{8} \quad \text{or} \quad 1\frac{1}{8}$$

EXAMPLE 10 $-\frac{3}{8}r = 15$

$$\left(-\frac{8}{3}\right)\left(-\frac{3}{8}\right)r = \frac{15}{1} \cdot \left(-\frac{8}{3}\right) \qquad \text{Multiply both sides by } -\frac{8}{3}, \text{ the reciprocal of } -\frac{3}{8}.$$

$$\left(-\frac{8}{3}\right)\left(-\frac{3}{8}\right)r = \frac{\overset{5}{\cancel{15}}}{1} \cdot \left(-\frac{8}{\underset{1}{\cancel{3}}}\right)$$

$$1r = -40$$

$$r = -40$$

EXAMPLE 11 $\dfrac{-2x}{9} = \dfrac{1}{27}$ The coefficient of x is $-\frac{2}{9}$.

$$-\frac{2}{9}x = \frac{1}{27} \qquad \text{Rewrite the equation to emphasize it.}$$

$$\left(-\frac{9}{2}\right)\left(-\frac{2}{9}\right)x = \left(-\frac{9}{2}\right) \cdot \frac{1}{27} \qquad \text{Multiply both sides by } -\frac{9}{2}.$$

$$1x = -\frac{\overset{1}{\cancel{9}}}{2} \cdot \frac{1}{\underset{3}{\cancel{27}}} \qquad \text{Simplify}$$

$$x = -\frac{1}{6}$$

Note   When an equation expresses the equality of two fractions (as these examples do), the equation is called a proportion.

Note   Convert any mixed numbers to improper fractions before solving the equation. (See the following example.)

EXAMPLE 12

$$5x = 1\frac{3}{10}$$

$$5x = \frac{13}{10}$$ Convert the mixed number to a fraction.

$$\frac{1}{5} \cdot 5x = \frac{13}{10} \cdot \frac{1}{5}$$ Multiply both sides by $\frac{1}{5}$.

$$1x = \frac{13 \cdot 1}{10 \cdot 5}$$ Simplify.

$$x = \frac{13}{50}$$

APPLICATION

If three pounds of apples cost 93 cents, how much does one pound cost?

Let x = the cost in cents of 1 lb apples.

$$3x = 93$$

$$\frac{1}{3} \cdot 3x = 93 \cdot \frac{1}{3}$$

$$1x = \frac{\overset{31}{\cancel{93}}}{1} \cdot \frac{1}{\underset{1}{\cancel{3}}}$$

$$x = 31$$

Answer: The apples cost 31 cents a pound. ◆

Practice Problems

Work these word problems.

9. Shellie lives 12 miles from campus, which is twice as far as her lab partner lives. How far from campus does her lab partner live?

10. If one-half dozen eggs cost 49 cents, how much does one dozen cost?

ANSWERS TO PRACTICE PROBLEMS
1. $x = 4$ 2. $x = -13$ 3. $y = 8$
4. $q = -9$ 5. $w = -15$ 6. $x = 6$
7. $x = -20$ 8. $y = -28$ 9. Her lab partner lives 6 miles from campus. 10. A dozen eggs cost 98 cents.

PROBLEM SET 5.7

Solve for the variable. Show all work. Check.

1. $3x = 15$
2. $2x = 28$
3. $5q = 25$
4. $7r = 35$
5. $60 = 6z$
6. $42 = 6y$
7. $-2x = 14$
8. $-3x = 21$
9. $8y = -64$
10. $9y = -63$
11. $-4m = 12$
12. $-3n = 18$
13. $6s = -24$
14. $7t = -28$
15. $-2z = -26$
16. $-5w = -45$
17. $\frac{1}{3}x = 4$
18. $\frac{1}{2}x = 5$
19. $-\frac{1}{4}y = 2$
20. $-\frac{1}{5}y = 3$
21. $\frac{2}{7}x = 8$

22. $\frac{3}{5}x = 9$

23. $-\frac{3}{11}y = 12$

24. $-\frac{2}{13}y = 26$

25. $\frac{3x}{4} = 27$

26. $\frac{2x}{5} = 32$

27. $7x = \frac{29}{14}$

28. $9x = \frac{19}{10}$

29. $\frac{8}{5}y = -31$

30. $\frac{16}{7}y = -22$

Applications: Show all work.

31. If you burn 687 calories running 3 miles, how many calories do you burn running 1 mile?

32. It takes Iveliz 5 minutes to walk from the dorm to class. If her roommate walks twice as fast as she does, how long does it take her roommate to walk to class?

SMALL GROUP ACTIVITY   *Using Egyptian Fractions*

In the introduction to this chapter, it was noted that the ancient Egyptians wrote all their fractions with numerators of 1. For example, $\frac{5}{8}$ could be written as $\frac{1}{8} + \frac{1}{2}$, and $\frac{11}{14}$ could be written as $\frac{1}{2} + \frac{1}{7} + \frac{1}{14} + \frac{1}{14}$ or as $\frac{1}{7} + \frac{1}{7} + \frac{1}{7} + \frac{1}{7} + \frac{1}{7} + \frac{1}{14}$.

Activity 1
Rewrite the following fractions using only fractions with numerators of 1. $\frac{7}{8}, \frac{5}{9}, \frac{13}{20}$.

Activity 2
Choose a few other fractions, and rewrite them using only fractions with numerators of 1.

CHAPTER 5 REVIEW

The terms listed below are among those used in this chapter. Write a brief definition for each term. (Refer to the appropriate section in the chapter as needed.)

1. Rational numbers _____

2. Denominator _____

3. Numerator _____

4. Proper fraction _____

5. Improper fraction _____

6. Mixed number _____

7. Equivalent fractions _____

8. Lowest terms _____

9. Reciprocal _____

10. Lowest common denominator _____

11. Complex fraction _____

12. Identity element for multiplication _____

13. Multiplication Property of Equality _____

After completing Chapter 5, you should have mastered the objectives listed at the beginning of each section. Test yourself by working the following problems. (Refer to the section indicated as needed.)

• *Recognize the numerator and denominator of a fraction (5.1).*

Name the numerator and denominator of these fractions.

14. $\dfrac{1}{6}$ **15.** $\dfrac{23}{4}$ **16.** $\dfrac{7}{9}$ **17.** $\dfrac{9}{8}$

18. $\dfrac{12}{7}$ **19.** $\dfrac{0}{5}$ **20.** $\dfrac{7}{1}$ **21.** $\dfrac{1}{12}$

22. $\dfrac{5}{2}$ **23.** $\dfrac{37}{86}$

• *Recognize that equivalent fractions look different but represent the same part of the whole (5.2).*

  • *Calculate equivalent fractions with a given denominator (5.2).*

Compute the following numerators.

24. $\dfrac{1}{2} = \dfrac{?}{18}$ **25.** $\dfrac{3}{8} = \dfrac{?}{24}$ **26.** $\dfrac{2}{9} = \dfrac{?}{54}$ **27.** $\dfrac{3}{4} = \dfrac{?}{8x}$

28. $\dfrac{15}{36} = \dfrac{?}{12}$ **29.** $\dfrac{3}{9} = \dfrac{?}{3}$ **30.** $\dfrac{35}{30} = \dfrac{?}{6}$ **31.** $\dfrac{36}{72} = \dfrac{?}{2}$

32. $\dfrac{55}{110} = \dfrac{?}{10}$ **33.** $\dfrac{55}{110} = \dfrac{?}{2}$

• *Reduce a fraction to lowest terms by prime factoring the numerator and denominator (5.2).*

Reduce to lowest terms.

34. $\dfrac{19}{38}$

35. $\dfrac{12}{36}$

36. $\dfrac{30a^2b}{40ab^2}$

37. $\dfrac{15x}{16y}$

38. $\dfrac{72x^3}{9x^2}$

39. $\dfrac{4x^2}{8x}$

40. $\dfrac{100}{200}$

41. $\dfrac{35}{63}$

• *Multiply fractions and mixed numbers (5.3).*

Multiply.

42. $\left(\dfrac{3}{4}\right)\left(-\dfrac{8}{9}\right)(-4)$

43. $\left(5\dfrac{1}{2}\right)\left(3\dfrac{1}{3}\right)$

44. $\left(2\dfrac{1}{2}\right)\left(4\dfrac{6}{7}\right)$

45. $\left(\dfrac{20x}{30}\right)\left(\dfrac{15y}{88}\right)$

46. $-\dfrac{5}{6} \cdot \dfrac{6}{5}$

47. $\left(\dfrac{9}{12}\right)\left(\dfrac{10}{15}\right)$

48. $\dfrac{x^2}{y} \cdot \dfrac{y^2}{x}$

49. $\dfrac{7}{8} \cdot \dfrac{8}{21}$

50. $\left(-\dfrac{3}{18}\right)\left(-\dfrac{9}{10}\right)\left(\dfrac{5}{14}\right)\left(-\dfrac{8}{21}\right)$

51. Find the product of $\dfrac{9}{10}$ and $\dfrac{5}{18}$. _____

52. At a state university, $\dfrac{3}{4}$ of the entering freshmen will graduate in 4 years. If the freshman class has 2024 students, how many of them will graduate in 4 years? _____

53. Find the product of $6\frac{3}{7}$ and $4\frac{1}{11}$. _____

54. A book was priced to sell for $6. During a sale, the price was reduced by $\frac{1}{3}$. By how many dollars was the price reduced? _____

• *Use reciprocals in dividing fractions and mixed numbers (5.4).*

Divide.

55. $\frac{2}{5} \div \frac{15}{4}$

56. $\frac{9}{8} \div \frac{3}{4}$

57. $-\frac{7}{8} \div \frac{24}{14}$

58. $\frac{2}{5} \div \frac{3}{5}$

59. $-2\frac{4}{5} \div \left(-\frac{7}{10}\right)$

60. $\frac{6}{7} \div \frac{3}{14}$

61. $\frac{1}{6} \div \left(-3\frac{2}{3}\right)$

62. $\frac{1}{96} \div \left(-\frac{3}{50}\right)$

63. Find the quotient of $\frac{2}{3}$ and $-\frac{4}{21}$. _____

64. How many cans of pop can be poured into a bottle which holds $7\frac{1}{2}$ cups? Note: Each can holds $\frac{2}{3}$ cup.

_____

65. Find the quotient of $\frac{8}{9}$ and $\frac{8}{3}$. _____

66. George has $\frac{7}{8}$ lb of cheese. If an omelette recipe calls for $\frac{1}{16}$ lb of cheese, how many omelettes can he make?

_____

• *Add and subtract fractions and mixed numbers (5.5).*

Simplify.

67. $\frac{5}{6} + \frac{5}{12}$

68. $\frac{1}{3} + \frac{1}{4} + \frac{1}{12}$

69. $-1\frac{1}{2} + 3\frac{1}{4}$

70. $\frac{1}{2} + \frac{3}{4} + \left(-\frac{3}{16}\right)$

71. $\frac{5}{7} - \frac{4}{5}$

72. $\frac{1}{6} + \left(-\frac{3}{8}\right) + \frac{1}{2}$

73. $-1\frac{3}{4} - \frac{11}{12}$

74. $\frac{5}{12} + \left(-\frac{7}{16}\right) + \left(-\frac{1}{24}\right)$

75. $\frac{3}{4} + \frac{2}{3}$

76. $\frac{11}{20} + \frac{2}{5} + \frac{13}{40}$

77. Find the sum of $\frac{9}{4}$, $\frac{5}{6}$, and $\frac{1}{12}$. _____

78. Find the difference between $\frac{7}{3}$ and $\frac{7}{4}$. _____

79. A family spends $\frac{2}{5}$ of its income on rent and $\frac{4}{15}$ on food. What fraction of the family's income is left _____

80. Find the sum of $\frac{1}{2}$, $\frac{1}{3}$, and $\frac{1}{3}$. _____

81. Find the sum of $\frac{1}{2}$, $\frac{3}{10}$, and $\frac{3}{5}$. _____

• *Use the rules for order of operations to simplify expressions containing more than one operation (5.6).*

Simplify.

82. $\dfrac{\frac{5}{9}}{\frac{3}{4}}$

83. $\dfrac{\frac{1}{2} + \frac{2}{3}}{\frac{3}{4} - \frac{1}{6}}$

84. $\dfrac{3 + \frac{1}{2}}{2 - \frac{3}{4}}$

85. $5 + \left(1\frac{1}{3}\right)\left(\frac{3}{4}\right)$

86. $\frac{2}{9} + \frac{1}{3}\left(1\frac{1}{2} + 1\frac{1}{2}\right)$

87. $\dfrac{12\frac{1}{3}}{6\frac{2}{3}}$

88. $\dfrac{5 - \frac{3}{4}}{2 + \frac{3}{4}}$

89. $\dfrac{\frac{3}{4} + \frac{1}{3}}{\frac{2}{3} + \frac{1}{6}}$

90. $\dfrac{1 + \frac{2}{3}}{1 - \frac{2}{3}}$

- *Solve first-degree equations in one variable by using the Multiplication Property of Equality (5.7).*

91. $2y = 36$

92. $-6x = 102$

93. $7t = -28$

94. $\dfrac{1}{4}z = 40$

95. $-\dfrac{2}{3}w = 20$

96. $\dfrac{2}{5}x = -\dfrac{3}{10}$

CHAPTER 5 PRACTICE TEST

1. Name the denominator in each of the following fractions.

   a. $\dfrac{8}{3}$ **b.** $\dfrac{2}{9}$ **c.** $\dfrac{4}{17}$ **d.** 6 **e.** $\dfrac{1}{4}$

2. Name the numerator in each of the following fractions.

   a. $\dfrac{0}{4}$ **b.** $\dfrac{19}{2}$ **c.** $\dfrac{3}{4}$ **d.** $\dfrac{5}{9}$ **e.** 7

3. Are the following fractions positive or negative?

   a. $\dfrac{-3}{4}$ **b.** $\dfrac{-5}{-3}$ **c.** $\dfrac{7}{9}$ **d.** $\dfrac{1}{-6}$

4. Find the missing number: $\dfrac{7}{8} = \dfrac{?}{64}$ _____

5. Find the missing number: $\dfrac{5}{6} = \dfrac{?}{72x}$ _____

6. Reduce to lowest terms: $\dfrac{56}{63}$. _____

7. Reduce to lowest terms: $\dfrac{38}{76}$. _____

8. Reduce to lowest terms: $\dfrac{8x^2y}{64xy^2}$ _____

9. Find the reciprocal of $\dfrac{3}{5}$. _____

10. Find the reciprocal of $-2\dfrac{5}{6}$. _____

11. Add: $\dfrac{7}{8} + \dfrac{1}{6}$ _____

12. Multiply: $\left(\dfrac{17}{19}\right)\left(\dfrac{38}{51}\right)$ _____

13. Subtract: $\dfrac{1}{5} - \dfrac{4}{7}$ _____

14. Find the sum: $\dfrac{3}{4} + \left(-\dfrac{2}{5}\right)$ _____

15. Divide: $-3\dfrac{1}{3} \div \dfrac{3}{5}$ _____

16. Find the product: $-4\dfrac{1}{6} \cdot \dfrac{2}{5}$ _____

17. Subtract: $-\dfrac{1}{4} - \left(-\dfrac{11}{12}\right)$ _____

18. Divide: $\dfrac{-30x}{25} \div \dfrac{-6x}{15}$ _____

19. Simplify: $\left(\dfrac{2}{3} + \dfrac{3}{4}\right) + \dfrac{1}{6}$ _____

20. Simplify: $\dfrac{\dfrac{2}{5} + \dfrac{1}{10}}{\dfrac{1}{2} - \dfrac{4}{5}}$ _____

21. Find the quotient of $\dfrac{9}{8}$ and $\dfrac{27}{62}$. _____

22. What is the sum of $\dfrac{3}{8}, \dfrac{5}{6}$, and $1\dfrac{7}{12}$? _____

23. Find the difference between $\dfrac{4}{15}$ and $2\dfrac{4}{5}$. _____

24. What is the product of $\left(-\dfrac{1}{2}\right)$ and $\left(-\dfrac{2}{3}\right)$? _____

25. In a math class, $\dfrac{2}{7}$ of the students earned an A, $\dfrac{1}{4}$ earned a B, and $\dfrac{2}{14}$ earned a C. What fraction of students

earned an A, B, or C? _____

26. In another math class, $\dfrac{3}{8}$ of the students earned a grade of A. If there are 232 student in the class, how many

earned As? _____

27. How many bottles containing $\frac{2}{7}$ pint of liquid can be filled from a 24 pint container? _____

28. A family spends $\frac{1}{3}$ of its income on rent and $\frac{1}{4}$ on food. What fraction of its income is left? _____

Solve the following equations.

29. $\frac{2}{3}x = 8$ _____

30. $9 = 3x$ _____

31. $\frac{3}{4}x = \frac{4}{3}$ _____

32. $\frac{9}{14} = \frac{1}{7}x$ _____

THE MATHEMATICS JOURNAL

Explaining a Procedure I

Task: Choose a method for solving a problem you learned in this chapter. (Make sure it's one you understand thoroughly.) Work a sample problem using it. Imagine that you are explaining it to a new student who just entered the class. Provide a step-by-step solution to the problem, and write your step-by-step explanation next to the solution.

6

Algebraic Expressions

6.1 Exponents (Revisited)
6.2 Polynomials
6.3 Simplifying, Multiplying, and Dividing Algebraic Fractions
6.4 Addition and Subtraction of Algebraic Fractions
6.5 Evaluating Algebraic Expressions

Introduction to Algebraic Expressions

Although algebra dates back to the early Babylonian times (2000 B.C. to 600 B.C.), it is a branch of mathematics that has attracted much interest in recent centuries, particularly the nineteenth and twentieth. So much work has been done in this field that we must now speak of algebras rather than algebra. We can think of the algebra we are using in this text as both a system and a set of symbols. We have been working in an algebraic system from the beginning of the text as we developed the definitions and properties that would serve as guides for our study. In Chapter 3, we began to consider algebra as a set of symbols—a language to help us express mathematics. In that chapter, we used the symbols of algebra with the set of whole numbers. In this chapter, we will extend our use of the symbols of algebra to the integers and rational numbers as we continue our study of algebra as a system.

6.1 Exponents (Revisited)

OBJECTIVE

● Apply the laws of exponents.

We began our discussion of exponents in Section 2.5 where we defined the terms and, limiting the base of an exponential expression to one of the natural numbers, introduced and applied three of the Laws of Exponents. We continued our discussion in Section 3.2 where we applied these three Laws of Exponents to exponential expressions which had variables as bases. In Section 4.7, we enlarged the replacement set for the base of an exponential expression to the set of nonzero integers. In this chapter, we will extend the replacement set for an exponent to the set of integers, add one definition, and two new laws. We'll begin by reviewing a few key concepts.

DEFINITION $a^0 = 1$, for $a \in$ rational numbers, $a \neq 0$

$$a^1 = a$$
$$a^2 = a \cdot a$$
$$a^3 = a \cdot a \cdot a$$

In general

$$a^n = [a \cdot a \cdot a \cdot a \cdot \cdots \cdot a], \text{ for } n \in \text{ whole numbers}$$

There are n a's.

EXAMPLE 1 $\left(\dfrac{2}{3}\right)^0 = 1$

$\left(\dfrac{c}{d}\right)^0 = 1$, for $c, d \in$ integers and $c, d \neq 0$.

$\left(\dfrac{2}{5}\right)^3 = \dfrac{2}{5} \cdot \dfrac{2}{5} \cdot \dfrac{2}{5} = \dfrac{8}{125}$

$\left(\dfrac{x}{y}\right)^2 = \dfrac{x}{y} \cdot \dfrac{x}{y} = \dfrac{x^2}{y^2}$, for $x, y \in$ integers and $y \neq 0$.

DEFINITION $\qquad a^{-n} = \dfrac{1}{a^n}; a \neq 0$

EXAMPLE 2 $\qquad 6^{-1} = \dfrac{1}{6^1} = \dfrac{1}{6}$

$z^{-1} = \dfrac{1}{z^1} = \dfrac{1}{z}$

$3^{-2} = \dfrac{1}{3^2} = \dfrac{1}{3 \cdot 3} = \dfrac{1}{9}$

$x^{-3} = \dfrac{1}{x^3}$

Note Be aware of the difference between multiplying a number by a negative number and raising a number to a negative power.

EXAMPLE 3 $(-1) \cdot 3 = -3$ $\qquad$ The result is the additive inverse of 3.

$3^{-1} = \dfrac{1}{3}$ $\qquad$ The result is the reciprocal or multiplicative inverse of 3.

EXAMPLE 4 Express in exponential form.

$$\dfrac{1}{27} = \dfrac{1}{3 \cdot 3 \cdot 3} = \dfrac{1}{3^3} \text{ or } 3^{-3}$$

$$\dfrac{1}{z^5} = z^{-5}$$

Note When simplifying expressions containing exponents, you may be asked to express the results with positive exponents.

USING THE CALCULATOR

Evaluate 8^{-3}

$$8^{-3} = \dfrac{1}{8^3}$$

Enter: 8 y^x 3 $=$ **512** ← Display

so, $8^{-3} = \dfrac{1}{512}$

Express with positive exponents

1. 7^{-2}

2. x^{-5}

3. $-(z^{-2})$

4. $\dfrac{1}{c^{-2}}$

5. $\dfrac{49}{100}$

6. $x \cdot x \cdot x \cdot x \cdot x$

7. $\dfrac{1}{y \cdot y \cdot y}$

Evaluate. Use a calculator when appropriate.

8. $(-6)^5$

9. $4^{(-7)}$

10. $(-3)^{-4}$

11. $-(3)^4$

Laws of Exponents:
For all $a, b \in$ integers, $a, b \neq 0$ and all $m, n \in$ integers, the following statements are true.

1. $a^m \cdot a^n = a^{m+n}$
2. $(a^m)^n = a^{mn}$
3. $(ab)^m = a^m \cdot b^m$

4. $\dfrac{a^m}{a^n} = a^{m-n}$, for $a \neq 0$

5. $\left(\dfrac{a}{b}\right)^m = \dfrac{a^m}{b^m}$, for $b \neq 0$

These laws, together with the definitions stated earlier, can be used to simplify algebraic expressions. The following examples will illustrate the use of each law.

Law 1: $a^m \cdot a^n = a^{m+n}$

EXAMPLE 5
$$(3^{-1}) \cdot (3^{-2}) = 3^{[(-1)+(-2)]} = 3^{-3} = \frac{1}{3^3} \text{ or } \frac{1}{27}$$

$$y^{-4} \cdot y^3 = y^{-4+3} = y^{-1} = \frac{1}{y^1} \text{ or } \frac{1}{y}$$

Law 2: $(a^m)^n = a^{mn}$

EXAMPLE 6
$$(10^2)^{-3} = 10^{2(-3)} = 10^{-6} = \frac{1}{10^6} \text{ or } \frac{1}{1,000,000}$$

$$(y^3)^{-4} = y^{3(-4)} = y^{-12} = \frac{1}{y^{12}}$$

Law 3: $(ab)^m = a^m \cdot b^m$

EXAMPLE 7
$$(3x)^2 = 3^2 \cdot x^2 = 9x^2$$

Law 4: $\dfrac{a^m}{a^n} = a^{m-n}$, for $a \neq 0$

EXAMPLE 8
$$\frac{5^6}{5^3} = 5^{6-3} = 5^3 \text{ or } 125$$

$$\frac{z^5}{z^3} = z^{5-3} = z^2$$

$$\frac{2^6}{2^8} = 2^{6-8} = 2^{-2} = \frac{1}{2^2} \text{ or } \frac{1}{4}$$

$$\frac{y^9}{y^{12}} = y^{9-12} = y^{-3} = \frac{1}{y^3}$$

$$\frac{(-2)^{10}}{(-2)^7} = (-2)^{10-7} = (-2)^3 = -8$$

$$\frac{(-x)^4}{(-x)^6} = (-x)^{4-6} = (-x)^{-2} = \frac{1}{(-x)^2} = \frac{1}{x^2}$$

Practice Problems

Simplify.

12. $2^{-3} \cdot 2^{-4}$

13. $[(-8)^3]^2$

14. $(3y)^5$

15. $\left(\dfrac{9}{10}\right)^2$

16. $y \cdot y^2$

17. $(x^9)^7$

18. $\left(\dfrac{z}{3}\right)^3$

19. $(2c^7)^3$

20. $(12xy^2z^3)^2$

Law 5: $\left(\frac{a}{b}\right)^m = \frac{a^m}{b^m}$, for $b \neq 0$

Use a calculator where appropriate.

EXAMPLE 9

$$\left(\frac{2}{3}\right)^4 = \frac{2^4}{3^4} = \frac{16}{81}$$

$$\left(\frac{1}{9}\right)^3 = \frac{1^3}{9^3} = \frac{1}{729}$$

21. $(-7)^5$

22. $(5^2)^5$

Recall that $-\frac{a}{b} = \frac{-a}{b} = \frac{a}{-b}$, for all $a, b \in$ real numbers, $b \neq 0$.

23. $\left(-\frac{2}{3}\right)^{11}$

EXAMPLE 10

$$\left(-\frac{1}{4}\right)^3 = \left(\frac{-1}{4}\right)^3 = \frac{(-1)^3}{4^3} = \frac{-1}{64} = -\frac{1}{64}$$

$$\left(\frac{y}{6}\right)^2 = \frac{y^2}{6^2} = \frac{y^2}{36}$$

24. $(6x^2)^{-3}$

25. $(27y)^0$

Note $\left(\frac{a}{b}\right)^{-1} = \left(\frac{b}{a}\right)^1$, for $a, b \in$ integers, $a, b \neq 0$

26. $\frac{(5x)^4}{(3y)^4}$

Using the definitions and Laws of Exponents, we can show why this is true.

$$\left(\frac{a}{b}\right)^{-1} = \frac{1}{\dfrac{a}{b}} \qquad \text{By definition: } a^{-n} = \frac{1}{a^n}$$

$$= 1 \div \frac{a}{b}$$

$$= 1 \cdot \frac{b}{a}$$

$$= \frac{b}{a}$$

EXAMPLE 11

$$\left(\frac{x^2}{5}\right)^3 = \frac{(x^2)^3}{5^3} \qquad \left(\frac{a}{b}\right)^m = \frac{a^m}{b^m}$$

$$= \frac{x^6}{125} \qquad (a^m)^n = a^{mn}$$

USING THE CALCULATOR

A calculator can be very useful in applying the Laws of Exponents to the *numerical* parts of an expression. it will not, however, help in simplifying the *variables* in an expression

You may find that it is more convenient to apply one or more of the Laws of Exponents before using the calculator.

Evaluate: $\frac{5^6}{5^2}$

Enter 5 [y^x] 6 [=] [÷] 5 [y^x] 2 [=] **625** ← Display

or $\frac{5^6}{5^2} = 5^{6-2} = 5^4$

Enter 5 [y^x] 4 [=] **625** ← Display

Note   On some "fraction" calculators, using the y^x key in conjunction with the fraction key will result in decimal answers. To avoid this, apply Law 5: $\left(\dfrac{a}{b}\right)^n = \dfrac{a^n}{b^n}$, before using the calculator.

PROBLEM SET 6.1

Express in exponential form.

1. 9

2. 16

3. 81

4. 32

5. $\dfrac{4}{9}$

6. $\dfrac{9}{4}$

7. $\dfrac{16}{81}$

8. $\dfrac{25}{36}$

9. $\dfrac{1}{x \cdot x \cdot x \cdot x}$

10. $\dfrac{1}{y \cdot y \cdot y}$

11. $a^2b \cdot a^2b$

12. $x^2y \cdot x^2y$

13. $\dfrac{3pq \cdot 3pq}{z^3 \cdot z^3}$

14. $\dfrac{7st \cdot 7st}{r^4 \cdot r^4}$

Evaluate by using the Laws of Exponents. Check numerical answers by calculator.

15. 3^0

16. $(2)^0$

17. $(-15)^0$

18. $(-12)^0$

19. $(-5)^4$

20. $(-3)^6$

21. $-(12)^6$

22. $-(9)^4$

23. $(-14)^5$

24. $(-11)^3$

25. $-(8^3)$ **26.** $-(6^5)$

27. $-c^2$ **28.** $-q^4$

29. $(-a)^4$ **30.** $(-b)^2$

31. $(-f)^3$ **32.** $(-z)^5$

33. $\left(\dfrac{2}{3}\right)^3$ **34.** $\left(\dfrac{3}{4}\right)^3$

35. $-\left(\dfrac{5}{6}\right)^2$ **36.** $-\left(\dfrac{7}{8}\right)^2$

37. $\left(-\dfrac{1}{2}\right)^2$ **38.** $\left(-\dfrac{1}{3}\right)^2$

39. 6^{-3} **40.** 5^{-4}

Use the Laws of Exponents to simplify these expressions. Express results with positive exponents.

41. x^{-2} **42.** y^{-4}

43. $\left(\dfrac{1}{2}q\right)^3$ **44.** $\left(\dfrac{1}{3}r\right)^2$

45. $(18x^2yz^3)^2$ **46.** $(16a^2bc^3)^2$

47. $\dfrac{(5q^2)^4}{10q^4}$ **48.** $\dfrac{(4p^3)^4}{16p^6}$

Evaluate. Use a calculator when appropriate.

49. $\left(\dfrac{7}{8}\right)^{10}$ **50.** $\left(\dfrac{5}{6}\right)^8$

51. $\left(-\dfrac{2}{3}\right)^5$

52. $\left(-\dfrac{3}{4}\right)^7$

53. $(9^5)^{-2}$

54. $(7^{-1})^3$

55. $3^{-4} \cdot 3^{-8}$

56. $4^{-7} \cdot 4^{-2}$

6.2 Polynomials

OBJECTIVES

- Recognize polynomials.
- Add and subtract polynomials.
- Multiply polynomials.
- Multiply binomials using the FOIL method.
- Find the common monomials factors of polynomials.

In Section 3.1, we began our study of algebraic expressions with a discussion of monomials. In this section, we will discuss polynomials as we continue our study of algebraic expressions.

DEFINITION **Polynomial:** An algebraic expression whose terms are monomials.

EXAMPLE 1 $3x^2 + 7$ is a polynomial.
$11y - 8$ or $11y + (-8)$ is a polynomial.
$-2z^5 + 17z^3 + 8z$ is a polynomial.

DEFINITION **Degree of a polynomial in one variable:** The degree of the term that has the highest degree.

EXAMPLE 2 $3x + 2$ has degree 1 because $3x$ has degree 1 and 2 has degree 0.

$-17y^2 + 5y - \dfrac{1}{2}$ has degree 2 because $-17y^2$ has degree 2, $5y$ has degree 1, and $\dfrac{1}{2}$ has degree 0.

$t^{12} + 2t^6 + 3$ has degree 12 because t^{12} has degree 12, $2t^6$ has degree 6, and 3 has degree 0.

Two types of polynomials used frequently have special names.

DEFINITION **Binomial:** A polynomial with two terms.

EXAMPLE 3 $2x + 7$ is a binomial.

DEFINITION **Trinomial:** A polynomial with three terms.

EXAMPLE 4 $15y^2 - 8y + \dfrac{1}{2}$ is a trinomial.

Note If a polynomial has only one variable, it is customary to arrange the terms according to their degrees—from highest to lowest or lowest to highest. In this text, we will arrange them from highest to lowest. This is called descending order.

Addition and Subtraction of Polynomials

Just as we can find the sums and differences of real numbers, so too we can indicate the sums and differences of polynomials by adding and subtracting their similar terms.

> **RULE** **To add two polynomials**
> Add their similar or like terms.

> **RULE** **To subtract one polynomial from another**
> Subtract the terms of the one from the similar terms of the other.

We will illustrate these operations with the polynomials arranged two ways.

EXAMPLE 5 Add $(3y + 6)$ and $(18y + 5)$.

Method 1 Add: $3y + 6$ Line up the similar terms under each
$$\underline{18y + 5}$$ other, and add their coefficients.
$$21y + 11$$

Method 2 $(3y + 6) + (18y + 5$ Use the Associative and Commutative
$= (3y + 18y) + (6 + 5)$ Properties of Addition to group the
$= 21y + 11$ similar terms together.
    Add.

EXAMPLE 6 Add: $27z^5 + 8z^4 - 12z^3 + 15z^2 - 2z + 11$
$$\underline{13z^5 - 6z^4 + 20z^3 + 8z^2 - 3z - 42}$$
$$40z^5 + 2z^4 + 8z^3 + 23z^2 - 5z - 31$$

To subtract, remember to add the opposite or additive inverse of each term of the polynomial being subtracted.

EXAMPLE 7 Subtract: Rewrite as:

$3x^3 + 7x^2 - 5x + 2$ $3x^3 + 7x^2 - 5x + 2$
$\underline{-2x^3 + 9x^2 - 12x - 3)}$ $\underline{+(-2x^3 - 9x^2 + 12x + 3)}$
                            $1x^3 - 2x^2 + 7x + 5$
                  or $x^3 - 2x^2 + 7x + 5$

EXAMPLE 8 $(4x^4 + 3x^2 + 7x) + (6x^4 + 3x^3 + 2x + 9)$

Note Each of these polynomials has missing terms.

Method 1 Rewrite each polynomial, replacing the missing terms with zeros.

$$4x^4 + 0x^3 + 3x^2 + 7x + 0$$
$$\underline{6x^4 + 3x^3 + 0x^2 + 2x + 9}$$
$$10x^4 + 3x^3 + 3x^2 + 9x + 9$$

Practice Problems

Identify the type and degree of each of the following polynomials.

1. $3x^2 - 24x + 11$

2. $7q - \dfrac{5}{6}$

3. $15y$

4. 5280

5. $\dfrac{1}{2}z^7 + 25$

Method 2   Rewrite each polynomial, leaving spaces for the missing terms.

$$\begin{array}{r} 4x^4 \qquad\quad + 3x^2 + 7x \\ 6x^4 + 3x^3 \qquad\quad + 2x + 9 \\ \hline 10x^4 + 3x^3 + 3x^2 + 9x + 9 \end{array}$$

EXAMPLE 9 $(3x^4 - 2x^2 + 7) - (5x^4 + 3x^3 - 5x)$

$$\begin{array}{rl} 3x^4 \qquad - 2x^2 \qquad + 7 & \text{Leave spaces for missing terms.} \\ -5x^4 - 3x^3 \qquad + 5x & \text{Rewrite as an addition problem.} \\ \hline -2x^4 - 3x^2 - 2x^2 + 5x + 7 & \end{array}$$

Multiplication of Polynomials

We have already multiplied monomials by monomials when we discussed the Laws of Exponents in Section 6.1.

EXAMPLE 10 $(3x^2)(-6x) = -18x^3$

Now we'll consider multiplying a polynomial by a monomial.

RULE To multiply a polynomial by a monomial

1. Apply the Distributive Property.
2. Use the Laws of Exponents to multiply each term of the polynomial by the monomial.
3. Simplify.

EXAMPLE 11 $3x(x^2 + 5x + 7)$

$$\begin{aligned} &= 3x(x^2) + 3x(5x) + 3x(7) \qquad && \text{Apply the Distributive Property.} \\ &= 3x^3 + 15x^2 + 21x && \text{Use the Laws of Exponents and simplify.} \end{aligned}$$

EXAMPLE 12 $\quad -6y(-2y^4 + 8y^2 - 5)$

$$\begin{aligned} &= -6y(-2y^4) + (-6y)(8y^2) + (-6y)(-5) \\ &= 12y^5 + (-48y^3) + 30y \\ &= 12y^5 - 48y^3 + 30y \end{aligned}$$

Besides the horizontal pattern illustrated in the previous examples, a vertical pattern can also be used.

EXAMPLE 13

$$\begin{array}{r} x^2 + 5x + 7 \\ \times \quad 3x \\ \hline 3x^3 + 15x^2 + 21x \end{array}$$

EXAMPLE 14
Method 1

$$\begin{aligned} &\quad (2m^2 + 7mn + 13)(5m) \\ &= 2m^2(5m) + 7mn(5m) + 13(5m) \\ &= 10m^3 + 35m^2n + 65m \end{aligned}$$

Practice Problems

Add or subtract these polynomials. (Use either method.)

6. $(-3x + 4) + (7 + 5x)$

7. $(-3x + 4)(- (7 + 5x)$

8. $(5y + z) - (y + 3z)$
$\qquad\qquad\qquad - (y - 2z)$

9. Add: $\begin{aligned} -3t^2 - 7t + 2 \\ 4t^2 + 6t + 3 \end{aligned}$

10. Subtract: $\begin{aligned} -3t^2 - 7t + 2 \\ 4t^2 + 6t + 3 \end{aligned}$

11. Add: $\begin{aligned} r^2s^2 + 7rs - 20 \\ -3r^2s^2 - 2rs + \;\; 6 \end{aligned}$

12. $(-4x^3 - 4)$
$\qquad + (4x^2 + 6x - 9)$

13. $(12y^3 - y^2 + 4y)$
$\qquad\qquad - (12y^2 + 18)$

14. $(a^3 - 8a + 1)$
$\qquad + (a^3 + 2a^2 - 1)$

Method 2

$$2m^2 + 7mn + 13$$
$$\underline{\times\ 5m}$$
$$10m^3 + 35m^2n + 65m$$

15. $2z(z^2 + 8z - 11)$

Multiplication of Binomials

We'll examine two commonly used methods for multiplying binomials.

16. $-pq(5p - 6q + 1)$

Generalized Distributive Property

17. $y(3y^2 + 8yz - 7)$

EXAMPLE 15 Multiply $(x + 2)$ by $(x + 3)$.
   Let $a = x + 3$, then $(x + 2)(x + 3)$.

18. $(15x^2 - 5x + 21)(-3x)$

$= (x + 2)a$	Substitute a for $x + 3$.
$= x \cdot a + 2 \cdot a$	Apply the Distributive Property.
$= x(x + 3) + 2(x + 3)$	Replace a with $x + 3$
$= x \cdot x + x \cdot 3 + 2 \cdot x + 2 \cdot 3$	Use the Distributive Property.
$= x^2 + 3x + 2x + 6$	Combine like terms.
$= x^2 + 5x + 6$	

We can omit the use of another variable to represent one of the binomials and apply the Distributive Property immediately if we care to.

EXAMPLE 16 $(x + 3)(x + 2)$

$= (x + 3)x + (x + 3)2$	Apply the Distributive Property.
$= x \cdot x + 3x + x \cdot 2 + 3 \cdot 2$	Apply it again.
$= x^2 + 3x + 2x + 6$	Combine like terms.
$= x^2 + 5x + 6$	

We can extend this generalized Distributive Property to multiply any polynomial by another polynomial.

FOIL Method for Multiplying Binomials

Mathematicians have also developed a shorter method to work just with binomials. It's called **FOIL**, and this is how it works.

Note Each binomial has two terms—a first (F) and a last (L). When written next to one another, two binomials have a pair of inside (I) terms and a pair of outside (O) terms.

RULE **To multiply two binomials using the FOIL method**

1. Write the binomials next to each other.
2. Find the product of the **F**irst terms.
3. Find the product of the **O**utside terms.
4. Find the product of the **I**nside terms.
5. Find the product of the **L**ast terms.
6. Add the resulting similar terms.

EXAMPLE 17 Multiply $(x + 3)$ by $(x + 2)$ using the FOIL method:

$(x + 3)(x + 2)$

F $(x + 3)(x + 2)$ $x \cdot x = x^2$ **1.** Write the binomials next to each other.

O $(x + 3)(x + 2)$ $x \cdot 2 = +2x$ **2.** Find the product of the first terms.

I $(x + 3)(x + 2)$ $3 \cdot x = +3x$ **3.** Find the product of the outside terms.

L $(x + 3)(x + 2)$ $3 \cdot 2 = +6$ **4.** Find the product of the inside terms.

$x^2 + 3x + 2x + 6 =$ **5.** Find the product of the last terms.

$x^2 + 5x + 6$ **6.** Add the resulting terms.

So $(x + 3)(x + 2) = x^2 + 5x + 6$

EXAMPLE 18 $(y - 5)(2y + 1)$ Remember, $y - 5$ means $y + (-5)$

$(\mathbf{y} - \mathbf{5})(\mathbf{2y} + \mathbf{1})$ $2y^2$ F

$(\mathbf{y} - \mathbf{5})(\mathbf{2y} + \mathbf{1})$ $+1y$ O

$(\mathbf{y} - \mathbf{5})(\mathbf{2y} + \mathbf{1})$ $-10y$ I

$(\mathbf{y} - \mathbf{5})(\mathbf{2y} + \mathbf{1})$ -5 L

$2y^2 - 9y - 5$ Take note of the signs.

EXAMPLE 19 $(a + 5b)(4a + 6b) =$

     F   O   I   L

$a \cdot 4a + a \cdot 6b + 5b \cdot 4a + 5b \cdot 6b =$

$4a^2 + 6ab + 20ab + 30b^2 =$

$4a^2 + 26ab + 30b^2$

Finding Common Monomial Factors

Besides being used to multiply, the Distributive Property can be used to factor — for example, $3x + 3y = 3(x + y)$. In Section 2.4, we discussed finding the Greatest Common Factor (GCF) of two or more integers. In Section 5.2, we extended this discussion to include monomials. In this section, we'll find the Greatest Common Factor of the terms of a polynomial and then use the Distributive Property to express the polynomial as a product.

EXAMPLE 20 Factor: $x^2 + 7x$

$x \cdot x + 7 \cdot x$ **1.** Write each term in factored form.

$x \cdot \widehat{x} + 7 \cdot \widehat{x}$ **2.** Find the GCF. GCF $(x^2, 7x) = x$.

$x(x + 7)$ **3.** Apply the Distributive Property.

EXAMPLE 21 $26a^2 + 6a - 12$

$= 2 \cdot 13 \cdot a \cdot a + 2 \cdot 3 \cdot a - 2 \cdot 2 \cdot 3$

$= \textcircled{2} \cdot 13 \cdot a \cdot a + \textcircled{2} \cdot 3 \cdot a - \textcircled{2} \cdot 2 \cdot 3$

$= 2(13a^2 + 3a - 6)$

Practice Problems

Use FOIL to multiply.

19. $(x + 9)(x + 10)$

20. $(3q + 2)(q - 1)$

21. $(7y - 8)(y - 2)$

22. $(12 - 3z)(6 + z)$

23. $(3x + y)(x + 2y)$

Factor.

24. $5x^2 + 10x$

25. $36m^2 + 24m + 12$

26. $8x^3y^2 - 6x^2y^3$

27. $-3z^2 - 15z$

EXAMPLE 22 $16x^3 + 32x^2 - 40x$

$$= 2^4x^3 + 2^5x^2 - 2^3 \cdot 5 \cdot x$$

$$= \boxed{2^3} \cdot 2 \cdot x^2 \cdot \boxed{x} + \boxed{2^3} \cdot 2^2 \cdot \boxed{x} \cdot x - \boxed{2^3} \cdot 5 \cdot \boxed{x}$$

$$= \mathbf{2^3 x}(2x^2 + 2^2 x - 5)$$

$$= \mathbf{2^3 x}(2x^2 + 4x - 5)$$

or $8x(2x^2 + 4x - 5)$

EXAMPLE 23 $9x^2 + 12x + 3$

$$= 3 \cdot 3 \cdot x^2 + 2^2 \cdot 3 \cdot x + 3 \cdot 1 \qquad \text{Factor 3 as } 3 \cdot 1.$$

$$= \boxed{3} \cdot 3 \cdot x^2 + 2^2 \cdot \boxed{3} \cdot x + \boxed{3} \cdot 1 \qquad \text{GCF} = 3.$$

$$= \mathbf{3}(3x^2 + 2^2 x + 1) \qquad \text{Don't forget the 1.}$$

$$= 3(3x^2 + 4x + 1)$$

EXAMPLE 24 $-2xy - 6x^2$

$$= \boxed{(-2)} \cdot \boxed{x} \cdot y + \boxed{(-2)} \cdot 3 \cdot \boxed{x} \cdot x$$

$$= [(-2)x](\qquad y + \qquad 3 \cdot \qquad x)$$

$$= -2x(y + 3x)$$

PROBLEM SET 6.2

Identify the coefficient and the degree of each monomial.

1. $-x^2$

2. $-y^3$

3. $\dfrac{2}{3}x^3$

4. $\dfrac{3}{4}y^2$

5. 27

6. 86

7. $\dfrac{15}{31}$

8. $\dfrac{11}{40}$

9. $3z$

10. $7q$

Identify the type and degree of the polynomial.

11. $7x + 8$

12. $6y - 2$

13. $2x^2 + 15x - 3$

14. $3y^2 - 9y + 12$

15. $25c^3 - 18c$

16. $49r^5 + 2r$

Add or subtract these polynomials as indicated.

17. $(3a + 17) + (14 + 2a)$

18. $(21 + 5b) + (6b + 7)$

19. $(21 + 5b) - (7 + 6b)$

20. $(3a + 17) - (2a + 14)$

21. $(2x + y) - (7x + 3y) + (x + y)$

22. $(t + s) + (2t + 5s) - (7t + 3s)$

23. Add:
$$t^2 + 3t + 6$$
$$3t^2 + 5t - 1$$

24. Add:
$$3x^2 + 2x - 5$$
$$2x^2 - 6x + 7$$

25. Add:
$$6a^4 + a^3 + 2a^2 - 5a + 7$$
$$a^4 \qquad\qquad + 2a + 1$$

26. Add:
$$3c^4 + 6c^3 + c^2 - 2c + 11$$
$$c^3 \qquad - 5c + 7$$

27. Subtract
$$12t^2 + 6t + 7$$
$$t^2 + 2t + 3$$

28. Subtract:
$$13r^2 + 4r + 3$$
$$r^2 + 3r + 1$$

29. Subtract:
$$5x^4 - 2x^2 + 1$$
$$-2x^4 + 7x^2 - 3$$

30. Subtract:
$$4y^3 - 7y + 2$$
$$-3y^3 + 10y - 5$$

31. $(4a^3 - a + 3) - (3a^3 + 6a^2 + 6a - 1)$

32. $(7x^2 - 4x - 9) - (-8x^3 - 4x + 4)$

33. $(3x + 6) + (4x^2 - 9) - (2x + 4)$

34. $(7y + 5) + (-y^2 + 3) - (3y^2 - 6y)$

Multiply.

35. $(3x^2)(5x^3)$

36. $(2x^3)(4x^2)$

37. $(-2m)^3(6m)$

38. $(-3n)^3(5n)$

39. $(3z^2)(-3z)^2$

40. $(2y^2)(-2y)^2$

41. $q(q + 7)$

42. $s(s + 4)$

43. $-3z^2(z + 1)$

44. $-2w^2(w + 3)$

45. $x(3x^2 + 7x + 5)$

46. $2x(x^2 + 3x + 1)$

47. $-3y(2y^2 + 7y + 1)$

48. $-4y(3y^2 + 2y + 10)$

49. $-a(ab^2 + b - 3)$

50. $-b(bc^2 + 2b - 8)$

51. $(22x^2 + 17x + 5)(-2x)$

52. $(19y^2 + 15y + 6)(-3y)$

Multiply these binomials using FOIL.

53. $(x + 5)(x + 6)$

54. $(y + 3)(y + 5)$

55. $(y + 1)(2y + 7)$

56. $(x + 1)(3x + 2)$

57. $(3r + 5)(6r + 2)$

58. $(4s + 5)(3s + 8)$

59. $(7 - y)(2 + y)$

60. $(4 - x)(3 + x)$

61. $(a - 2b)(3a - b)$

62. $(p - 3q)(2p - q)$

63. $(3t + 4)(2t - 5)$

64. $(2r + 9)(3r - 10)$

65. $(q + 2)(q - 2)$

66. $(w + 5)(w - 5)$

67. $\left(x + \dfrac{1}{2}\right)\left(x + \dfrac{1}{4}\right)$

68. $\left(y + \dfrac{1}{5}\right)\left(y + \dfrac{1}{10}\right)$

Supply the missing terms for each factor.

69. $3x^2 + 12x = 3x(\underline{} + \underline{})$

70. $4y^2 + 16y = 4y(\underline{} + \underline{})$

Factor. Find the common monomial factors, and rewrite each expression as the product of a monomial and a polynomial.

71. $6z^2 + 18z + 4$

72. $4v^2 + 24v + 6$

73. $12a^2b + 2b$

74. $10c^2d + 5d$

75. $5ab^2 + 15a^2b - 35ab$

76. $4g^2h - 16gh^2 + 24gh$

77. $-3x^2 + 6x - 24$

78. $-2y^2 + 18y - 30$

79. $27 - 9h^2$

80. $35 - 7k^2$

6.3 Simplifying, Multiplying, and Dividing Algebraic Fractions

OBJECTIVES
- Recognize algebraic fractions.
- Reduce algebraic fractions to lowest terms.
- Multiply algebraic fractions.
- Divide algebraic fractions.

DEFINITION **Algebraic fractions** are indicated divisions (fractions) that contain variables. They are also referred to as **rational expressions.**

EXAMPLE 1 $\frac{x}{4}, \frac{6}{y}$, and $\frac{2y^2}{7x}$ are examples of algebraic fractions.

EXAMPLE 2 $3\frac{1}{2}x$ and $15z$ are also examples of algebraic fractions, since $3\frac{1}{2}x = \frac{7x}{2}$ and $15z = \frac{15z}{1}$.

Simplifying Algebraic Fractions

In Section 5.2, we discussed reducing fractions to lowest terms. We can simplify algebraic fractions by following the same procedure. The only difference is that now the numerators and denominators contain variables.

> **RULE** To reduce an algebraic fraction to lowest terms
>
> 1. Express both the numerator and the denominator as products of prime factors.
> 2. Use the Properties of Real Numbers and the Laws of Exponents to simplify or reduce the fraction.

Recall from Section 5.3 that $\frac{ac}{bd} = \frac{a}{b} \cdot \frac{c}{d}$, for all $a, b, c,$ and $d \in$ R, $b, d \neq 0$; from Section 2.4 that $\frac{a}{a} = 1$, for $a \neq 0$; and from Section 6.1 that $\frac{a^m}{a^n} = a^{m-n}$, for $m, n \in$ integers, $a \neq 0$.

EXAMPLE 3

$$\frac{3x^2}{6x} = \frac{3}{2 \cdot 3} \cdot \frac{x^2}{x} = \frac{\overset{1}{\cancel{3}}}{2 \cdot \underset{1}{\cancel{3}}} \cdot x = \frac{1}{2} \cdot x = \frac{x}{2}$$

EXAMPLE 4

$$\frac{-15y^2}{3y^5} = \frac{-5 \cdot \overset{1}{\cancel{3}}}{\underset{1}{\cancel{3}}} \cdot \frac{y^2}{y^5} = \frac{-5}{1} \cdot \frac{y^2}{y^3 \cdot y^2} = \frac{-5}{y^3} \text{ or } -\frac{5}{y^3}$$

EXAMPLE 5

$$\frac{a^7 b^6}{a^3 b^8} = \frac{a^7}{a^3} \cdot \frac{b^6}{b^8} = \frac{a^{7-3}}{1} \cdot \frac{b^6}{b^6 \cdot b^2} = \frac{a^4}{b^2}$$

Note We have worked these examples so that we could illustrate several variations of the method to be used. Through practice, get comfortable with one or two variations, and use them as needed.

Multiplication and Division of Algebraic Fractions

In Sections 5.3 and 5.4, we discussed the multiplication and division of fractions. In this section, we will use the procedures we discussed then, together with the knowledge we now have of algebra, to multiply and divide algebraic fractions. Specifically, we will use the following definitions, properties, and Laws of Exponents:

For all a, b, c, and $d \in$ R, $b, d \neq 0$,

$$\frac{a}{b} \cdot \frac{c}{d} = \frac{ac}{bd}$$

$$a \div b = a \cdot \frac{1}{b}, \text{ for } b \neq 0$$

$$\frac{a}{a} = 1, \text{ for } a \neq 0$$

$$\frac{a^m}{a^n} = a^{m-n}, \text{ for } m, n \in \text{ integers}$$

$$\left(\frac{a}{b}\right)^{-1} = \frac{1}{\dfrac{a}{b}} = \frac{b}{a}, \text{ for } a, b \in \text{ rational numbers, } a, b \neq 0$$

Multiplication of Algebraic Fractions

Just as we can multiply arithmetic fractions, so too we can multiply algebraic fractions.

> **RULE** To multiply algebraic fractions
>
> 1. Form a new algebraic fraction whose numerator is the product of the original numerators and whose denominator is the product of the original denominators.
> 2. Reduce it to lowest terms.

EXAMPLE 6

$$\frac{x}{y} \cdot \frac{3}{y} = \frac{3x}{y \cdot y} = \frac{3x}{y^2}$$

Many times it is convenient to prime factor the algebraic fractions before multiplying them. Writing them as the product of prime factors will help us recognize any common factors in the numerators and denominators (those of the form $\frac{a}{a}$, which equal 1) we can reduce.

Reduce these algebraic fractions.

1. $\dfrac{y^2}{xy}$

2. $\dfrac{24c^2 d}{12cd^2}$

3. $\dfrac{-45x}{40x^2}$

4. $\dfrac{30a^3 b^2}{-5ab^3}$

Multiply.

5. $\dfrac{4a}{b} \cdot \dfrac{2}{b}$

6. $\dfrac{3x}{y^3} \cdot \dfrac{5x^2}{y}$

7. $\dfrac{c}{4d} \cdot \dfrac{2d^4}{7c^3}$

8. $\dfrac{-3h}{4k^2} \cdot \dfrac{2k^3}{15h^3}$

EXAMPLE 7 $\dfrac{z}{x^2} \cdot \dfrac{x}{az^2} = \dfrac{\overset{1}{\cancel{z}}}{\underset{x}{\cancel{x^2}}} \cdot \dfrac{\overset{1}{\cancel{x}}}{\underset{az}{\cancel{az^2}}} = \dfrac{1 \cdot 1}{x \cdot a \cdot z} = \dfrac{1}{axz}$

EXAMPLE 8 $\dfrac{4a}{5} \cdot \dfrac{b}{12a^2} = \dfrac{\overset{1}{\cancel{4}} \cdot \overset{1}{\cancel{a}}}{5} \cdot \dfrac{b}{3 \cdot \underset{1}{\cancel{4}} \cdot \underset{a}{\cancel{a^2}}} = \dfrac{b}{15a}$

Division of Algebraic Fractions

> **RULE** To divide one algebraic fraction by another
>
> **1.** Form the reciprocal of the divisor.
> **2.** Multiply the first fraction by that reciprocal.
> **3.** Reduce the product to lowest terms.

Note Recall from Section 2.4 that division by zero is undefined. In this section, it is assumed that any variables used as denominators will not equal zero.

EXAMPLE 9 $\dfrac{x}{6y} \div \dfrac{y}{6x}$ Multiply the first fraction by the reciprocal of the second fraction.

$= \dfrac{x}{6y} \cdot \dfrac{6x}{y}$

$= \dfrac{x}{\underset{1}{\cancel{6}}y} \cdot \dfrac{\overset{1}{\cancel{6}}x}{y}$

$= \dfrac{x^2}{y^2}$

EXAMPLE 10 $\dfrac{x^2}{10} \div \dfrac{x}{12} = \dfrac{x^2}{10} \cdot \dfrac{12}{x} = \dfrac{\overset{x}{\cancel{x^2}}}{\underset{5}{\cancel{10}}} \cdot \dfrac{\overset{6}{\cancel{12}}}{\underset{1}{\cancel{x}}} = \dfrac{6x}{5}$

EXAMPLE 11 $\dfrac{22pq^2}{5p} \div \dfrac{11q^2}{10p} = \dfrac{22pq^2}{5p} \cdot \dfrac{10p}{11q^2}$

$= \dfrac{\overset{2}{\cancel{22}} \cdot \overset{1}{\cancel{p}} \cdot \overset{1}{\cancel{q^2}}}{\underset{1}{\cancel{5}} \cdot \underset{1}{\cancel{p}}} \cdot \dfrac{\overset{2}{\cancel{10}}p}{\underset{1}{\cancel{11}} \cdot \underset{1}{\cancel{q^2}}} = \dfrac{4p}{1} = 4p$

Combining Operations

Recall from Sections 2.6 and 5.6 that the order of operations states that multiplication and division are performed in order from left to right.

EXAMPLE 12 $\dfrac{v^2}{3} \div \dfrac{v}{9} \cdot \dfrac{6}{v}$

$= \dfrac{v^2}{3} \cdot \dfrac{9}{v} \cdot \dfrac{6}{v}$

$= \dfrac{\overset{1}{\cancel{v}} \cdot \overset{1}{\cancel{v}}}{\underset{1}{\cancel{3}}} \cdot \dfrac{\overset{3}{\cancel{9}}}{\underset{1}{\cancel{v}}} \cdot \dfrac{6}{\underset{1}{\cancel{v}}}$

$= \dfrac{18}{1} = 18$

PROBLEM SET 6.3

Reduce these algebraic fractions.

1. $\dfrac{xy}{x^2}$

2. $\dfrac{cd}{d^2}$

3. $\dfrac{15ab^2}{10a^2b}$

4. $\dfrac{12x^2y}{20xy^2}$

5. $\dfrac{-30w}{3wx}$

6. $\dfrac{-28q}{4qr}$

7. $\dfrac{3x^2y}{12x^2y}$

8. $\dfrac{25cd^2}{5cd^2}$

Multiply and/or divide as indicated. Reduce results.

9. $\dfrac{4a}{b^2} \cdot \dfrac{3b}{2a^2}$

10. $\dfrac{6d}{f^2} \cdot \dfrac{2f}{3d^2}$

11. $\dfrac{x^3}{y} \cdot \dfrac{y^5}{4x^2}$

12. $\dfrac{4x^3}{y^2} \cdot \dfrac{y^3}{x^5}$

13. $\dfrac{3y}{2x^2} \div \dfrac{6y^2}{4x}$

14. $\dfrac{4z}{3w^2} \div \dfrac{8z^2}{6w}$

15. $\dfrac{-12t^4}{36s^3} \div \dfrac{-t}{s}$

16. $\dfrac{-15s^3}{45t^4} \div \dfrac{-s}{t}$

17. $\dfrac{2x+4}{9y} \cdot \dfrac{27y^2}{2}$

18. $\dfrac{3y+6}{10x} \cdot \dfrac{25x^2}{3}$

19. $\dfrac{2ab}{7c} \cdot \dfrac{14c^2}{a^2b^2} \div \dfrac{2c}{4b}$

20. $\dfrac{3xy}{5z} \cdot \dfrac{15z^2}{x^2y^2} \div \dfrac{6z}{2y}$

6.4 Addition and Subtraction of Algebraic Fractions

OBJECTIVE

● Add and subtract algebraic fractions.

Adding and subtracting algebraic fractions will be similar to adding and subtracting fractions in arithmetic as we discussed in Section 5.5.

Adding and Subtracting Algebraic Fractions with the Same Denominators

Recall from Section 5.5 that $\frac{a}{b} + \frac{c}{b} = \frac{a+c}{b}$, for all $a, b, c \in R$, $b \neq 0$.

$$\text{and } \frac{a}{b} - \frac{c}{b} = \frac{a-c}{b}$$

> **RULE** To add or subtract two algebraic fractions with the same denominator.
>
> 1. Keep the denominator.
> 2. Add or subtract the numerators.
> 3. Simplify. (Reduce the resulting fraction.)

EXAMPLE 1

$$\frac{7}{y} + \frac{2}{y} \qquad \text{Keep the denominator.}$$

$$= \frac{7+2}{y} \qquad \text{Add the numerators.}$$

$$= \frac{9}{y}$$

EXAMPLE 2

$$\frac{5}{x} - \frac{3}{x} \qquad \text{Keep the denominator.}$$

$$= \frac{5-3}{x} \qquad \text{Subtract the numerators.}$$

$$= \frac{2}{x}$$

EXAMPLE 3

$$\frac{3y}{5z} + \frac{2y}{5z} = \frac{3y+2y}{5z} = \frac{\overset{1}{\cancel{5}y}}{\underset{1}{\cancel{5}z}} = \frac{y}{z}$$

EXAMPLE 4

$$\frac{5x}{2y} - \frac{x}{2y} = \frac{5x-x}{2y}$$

$$= \frac{4x}{2y} = \frac{\overset{2}{\cancel{4}x}}{\underset{1}{\cancel{2}y}} = \frac{2x}{y}$$

Practice Problems

Add or subtract as indicated.

1. $\dfrac{15}{t} - \dfrac{7}{t}$

2. $\dfrac{2x}{3a} + \dfrac{5x}{3a}$

3. $\dfrac{2y+2}{5x} + \dfrac{y-8}{5x}$

Adding and Subtracting Fractions with Different Denominators

Finding the sums and differences of algebraic fractions with different denominators is similar to finding the sums and differences of arithmetic fractions with different denominators.

> **RULE** **To add or subtract fractions with different denominators**
>
> 1. Find the LCD of the denominators.
> 2. Express each fraction as an equivalent fraction having the same LCD as the denominator.
> 3. Add or subtract the newly formed fractions.

EXAMPLE 5 $\dfrac{x}{2} + \dfrac{x}{5}$ LCD (2,5) = 10

$$= \frac{x}{2} \cdot \frac{5}{5} + \frac{x}{5} \cdot \frac{2}{2}$$

$$= \frac{5x}{10} + \frac{2x}{10}$$ Express each fraction as an equivalent fraction having 10 as the denominator.

$$= \frac{5x + 2x}{10}$$ Add the numerators.

$$= \frac{7x}{10}$$

EXAMPLE 6 $\dfrac{4}{x} + \dfrac{2}{y}$ LCD $(x, y) = xy$

$$= \frac{4}{x} \cdot \frac{y}{y} + \frac{2}{y} \cdot \frac{x}{x}$$

$$= \frac{4y}{xy} + \frac{2x}{xy}$$

$$= \frac{4y + 2x}{xy}$$

EXAMPLE 7 $$\frac{23x}{y} - \frac{13y}{x}$$

$$= \frac{23x^2}{xy} - \frac{13y^2}{xy}$$

$$= \frac{23x^2 - 13y^2}{xy}$$

Expressions containing algebraic fractions can include both the operations of addition and subtraction.

EXAMPLE 8 $$\frac{3}{ab} + \frac{5}{ab} - \frac{7}{ab} = \frac{3 + 5 - 7}{ab}$$

$$\frac{8 - 7}{ab} = \frac{1}{ab}$$

Practice Problems

Add or subtract as indicated.

4. $\dfrac{y^2}{3} - \dfrac{y}{2}$

5. $\dfrac{5}{p} + \dfrac{3}{p^2}$

6. $\dfrac{1}{x} + \dfrac{3}{y}$

7. $\dfrac{1}{x} + \dfrac{3}{x} - \dfrac{2}{x}$

8. $\dfrac{5}{a} - \dfrac{6}{b} + \dfrac{4b}{a^2}$

9. $\dfrac{5x}{y} + \dfrac{3y}{x} + \dfrac{6}{xy}$

ANSWERS TO PRACTICE PROBLEMS

1. $\dfrac{8}{t}$ 2. $\dfrac{7x}{3a}$ 3. $\dfrac{3y - 6}{5x}$
4. $\dfrac{2y^2 - 3y}{6}$ 5. $\dfrac{5p + 3}{p^2}$
6. $\dfrac{y + 3x}{xy}$ 7. $\dfrac{2}{x}$
8. $\dfrac{5ab - 6a^2 + 4b^2}{a^2b}$
9. $\dfrac{5x^2 + 3y^2 + 6}{xy}$

PROBLEM SET 6.4

Add or subtract as indicated. Reduce results.

1. $\dfrac{4c}{3} + \dfrac{c}{3}$

2. $\dfrac{3q}{4} + \dfrac{q}{4}$

3. $\dfrac{5f}{7} - \dfrac{3f}{7}$

4. $\dfrac{8q}{5} - \dfrac{2q}{5}$

5. $\dfrac{-3x}{14} + \dfrac{9x}{14}$

6. $\dfrac{-7y}{10} + \dfrac{17y}{10}$

7. $\dfrac{5}{a} + \dfrac{2}{a}$

8. $\dfrac{6}{b} + \dfrac{4}{b}$

9. $\dfrac{3x}{10} - \dfrac{x}{10}$

10. $\dfrac{4y}{7} - \dfrac{y}{7}$

11. $\dfrac{x+3}{4} + \dfrac{x+2}{4}$

12. $\dfrac{y+1}{3} + \dfrac{2y+3}{3}$

13. $\dfrac{2t+7}{5} - \dfrac{t+8}{5}$

14. $\dfrac{6s+3}{8} - \dfrac{s+5}{8}$

15. $\dfrac{x}{8} + \dfrac{3}{4}$

16. $\dfrac{4}{5} + \dfrac{2x}{3}$

17. $\dfrac{2}{x} - \dfrac{1}{3}$

18. $\dfrac{7}{8} - \dfrac{3}{y}$

19. $\dfrac{1}{8x} + \dfrac{1}{12x}$

20. $\dfrac{4}{3x} - \dfrac{3}{2x}$

21. $\dfrac{x+4}{3} + \dfrac{x+3}{2}$

22. $\dfrac{2y+1}{4} + \dfrac{y+6}{5}$

23. $\dfrac{3}{b} - \dfrac{2}{c}$

24. $\dfrac{12}{x} - \dfrac{13}{y}$

25. $\dfrac{2x}{y} + \dfrac{3y}{x}$

26. $\dfrac{5v}{w} + \dfrac{7w}{v}$

27. $\dfrac{1}{2xy} + \dfrac{3y}{x^2} + \dfrac{5x}{y^2}$

28. $\dfrac{2a}{b^2} + \dfrac{3b}{a^2} + \dfrac{1}{3ab}$

29. $\dfrac{10s}{11t} - \dfrac{1}{11} + \dfrac{3t}{s}$

30. $\dfrac{1}{3p} - \dfrac{2p}{q} + \dfrac{q}{3}$

6.5 Evaluating Algebraic Expressions

OBJECTIVE

- Evaluate algebraic expressions.

In everyday life, we are frequently called on to evaluate algebraic expressions. Often we need to "use a formula" to obtain the information we're seeking. The phrase "use a formula" is just another way of saying "evaluate an algebraic expression." We use a formula when we want to fence in the garden or lay carpet in a room. In this section, we'll continue the study of this topic, which we began in Section 3.3.

> **RULE** To evaluate an algebraic expression
>
> **1.** Replace each variable in the expression with its indicated value.
> **2.** Perform the indicated operations. (Simplify.)

EXAMPLE 1   For $y = 2$, evaluate $y + 3$

$$2 + 3 = 5$$

EXAMPLE 2   For $y = -4$, evaluate $y + 3$

$$-4 + 3 = -1$$

A handy way to evaluate an expression is to set up a table.

y	$y + 3$	*Answer*
2	$2 + 3$	5
-4	$-4 + 3$	-1

Practice Problems

Evaluate

1. $3x^2 - 2x + 5$, for $x = 4$

2. $xy + x^2$, for $x = 10$, $y = -8$

3. $\frac{1}{3}x^2 - \frac{1}{6}x + \frac{1}{12}$, for $x = \frac{1}{2}$

EXAMPLE 3 Evaluate $3x + 2y$, for $x = 7$ and $y = -1$.

x	y	$3x + 2y$	*Answer*
7	-1	$3 \cdot 7 + 2(-1) = 21 + (-2)$	19

EXAMPLE 4 Evaluate $-x^2 + 1$, for $x = 4$.

x	$-x^2 + 1$	*Answer*
4	$-(4^2) + 1$	-15

EXAMPLE 5 Evaluate $2x^2 - x + 7$, for $x = 3$.

x	$2x^2 - x + 7$	*Answer*
3	$2 \cdot 3^2 - 3 + 7$	22

USING THE CALCULATOR

Let's look at this example again, this time using a calculator to evaluate it. Recall that a calculator with an algebraic operating system has the rules for order of operations built in.

Example A: Evaluate $2x^2 - x + 7$, when $x = 3$.

Enter 2 × 3 x^2 − 3 + 7 = **22** ← Display

Example B: Use a "fraction" calculator to evaluate

$$\frac{2}{3}x^2 + \frac{3}{2}x, \text{ when } x = \frac{1}{4}$$

Enter 2 $a^{b/c}$ 3 × 1 $a^{b/c}$ 4 x^2 + 3 $a^{b/c}$ 2 × 1 $a^{b/c}$ 4 = **5⌋12** ← Display

APPLICATION Suppose you're in Toronto for the weekend, and you're trying to decide if you need to take a jacket to go on your day-long tour of the city. The TV channel predicts the high will be 30° C, but you only understand the Fahrenheit scale. Should you take a jacket? The manual for your pocket calculator (You carry it everywhere!) lists this formula: $F = \frac{9}{5}C + 32$.

▦ **USING THE CALCULATOR**

For C = 30°

Enter 9 ÷ 5 × 30 + 32 = **86** ← Display

Answer: 86° F. You leave without a jacket. ◆

ANSWERS TO PRACTICE
PROBLEMS

1. 45   2. 20   3. $\frac{1}{12}$

● PROBLEM SET 6.5

Evaluate these expressions for the given values of the variables. Use a table.

1. $3 + t$, for $t = 2$

2. $4 + s$, for $s = 3$

3. $4 + s$, for $s = -3$

4. $3 + t$, for $t = -2$

5. $13r - 1$, for $r = 5$

6. $12q - 3$, for $q = 7$

7. $3z + 7$, for $z = -22$

8. $2w + 8$, for $w = -18$

9. $10x + 5y$, for $x = 2, y = -1$

10. $5x + 10y$, for $x = 2, y = -1$

11. $2x^2 + 6x + 5$, for $x = 3$

12. $3x^2 + 5x + 6$, for $x = 2$

13. $3x^2 + 5x + 6$, for $x = -1$

14. $2x^2 + 6x + 5$, for $x = -1$

15. $\frac{a + b}{2}$, for $a = 14, b = 12$

16. $\frac{c + d}{3}$, for $c = 13, d = 14$

17. $3(x + 7) - 5$, for $x = 13$

18. $4(y + 6) - 8$, for $y = 14$

19. $\frac{3c - 4f}{2e}$, for $c = -1, f = 3, e = 4$

20. $\frac{5q - 3s}{4t}$, for $q = 2, s = -1, t = 3$

Use a calculator to evaluate these expressions. Use a "fraction" calculator to evaluate the expressions containing fractions.

21. $\frac{1}{2}x^3 + \frac{2}{3}x^2 - 15x + 6$, for $x = \frac{1}{3}$

22. $\frac{1}{3}y^3 + \frac{3}{4}y^2 - 12y + 7$, for $y = \frac{1}{2}$

23. $\dfrac{5}{6}a^2 + \dfrac{7}{8}a + 11$, for $a = -\dfrac{2}{3}$

24. $\dfrac{2}{3}b^2 + \dfrac{8}{9}b + 15$, for $b = \dfrac{1}{4}$

25. $1\dfrac{1}{10}t^2 - 3\dfrac{2}{3}t + 5\dfrac{3}{4}$, for $t = 2\dfrac{1}{3}$

26. $7\dfrac{1}{5}w^2 - 1\dfrac{1}{2}w + 3\dfrac{9}{10}$, for $w = -4\dfrac{1}{4}$

27. $\dfrac{a - b}{c}$, for $a = \dfrac{1}{2}$, $b = \dfrac{1}{3}$, $c = \dfrac{1}{4}$

28. $\dfrac{w + y}{z}$, for $w = \dfrac{1}{5}$, $y = -\dfrac{1}{6}$, $z = \dfrac{1}{2}$

29. $\dfrac{jk - k^2}{3j}$, for $j = 27$, $k = 31$

30. $\dfrac{xy - x^2}{4y}$, for $x = 13$, $y = 25$

Evaluate the following formulas for the given values of the variables.

31. $P = 2L + 2W$, find P, when $L = 18$ and $W = 6$

32. $A = L \cdot W$, find A, when $L = 20$ and $W = 11$

33. $A = \pi r^2$, find A, when $r = 3$ ($\pi \approx 3.14$)

34. $C = d\pi$, find C, when $d = 5$ ($\pi \approx 3.14$)

35. $A = \dfrac{1}{2}bh$, find A, when $b = 4$, $h = 1$

36. $V = \dfrac{4}{3}\pi r^3$, find V, when $r = 6$

Applications:

37. Your friends from Canada are visiting. They plan to meet you for lunch after class. They want to know how to dress for the weather. They turn on The Climate Channel (TCC) and learn that the high today will be 54° F, but they're Celsius people. How can you use the formula stated below to convince them to wear sweaters or jackets?

Formula: $C = \dfrac{5}{9}(F - 32)$

38. Refer to Problem 37, how would you convince your friends *not* to wear sweaters if TCC reported that the high today will be 88° F?

SMALL GROUP ACTIVITY   *Pascal's Triangle*

Pascal's Triangle is a triangular array of numbers named after the seventeenth-century mathematician, Blaise Pascal. Look for a pattern in Pascal's Triangle, then use it to complete Rows 5 and 6.

Row 0						1						
Row 1					1		1					
Row 2				1		2		1				
Row 3			1		3		3		1			
Row 4		1		4		6		4		1		
Row 5	1		—		—		—		—		1	
Row 6	1	—		—		—		—		—		1

Consider the binomial $(a + b)$ raised to these powers.

$$(a + b)^0 = \quad 1 \qquad\qquad 1$$
$$(a + b)^1 = a + b \qquad 1a + 1b$$

In the right column, we have rewritten the results with the coefficients emphasized. Compare the coefficients with those in Rows 0 and 1 of Pascal's Triangle. Use the definition of exponents and FOIL to expand.

$$(a + b)^2 = \text{\underline{\hspace{4cm}}}$$

Compare the coefficients to the numbers in Row 2 of Pascal's Triangle. Use the definition of exponents and the generalized Distributive Property from Section 6.2 to expand.

$$(a + b)^3 = \text{\underline{\hspace{4cm}}}$$

Compare the coefficients to the numbers in Row 3. Try expanding $(a + b)^4 + (a + b)^5$, $(a + b)^6$ using Pascal's Triangle to find the coefficients. What is the pattern of the exponents?

$$(a + b)^4 = \text{\underline{\hspace{4cm}}}$$
$$(a + b)^5 = \text{\underline{\hspace{4cm}}}$$
$$(a + b)^6 = \text{\underline{\hspace{4cm}}}$$

● CHAPTER 6 REVIEW

The terms listed below are among those used in this chapter. Write a brief definition for each term. (Refer to the appropriate section in the chapter as needed.)

1. Polynomial _____

2. Binomial _____

3. Trinomial _____

4. Degree of a polynomial in one variable _____

5. Algebraic fraction _____

In this chapter, you used several rules and methods. State the following ones. (Refer to the appropriate section as needed.)

6. To add and subtract polynomials _____

7. To multiply a polynomial by a monomial _____

8. To multiply two binomials _____

9. To find the common monomial factors of a polynomial _____

10. To reduce an algebraic fraction to lowest terms _____

11. To multiply algebraic fractions _____

12. To divide algebraic fractions _____

13. To add and subtract algebraic fractions with the same denominators _____

14. To add and subtract algebraic fractions with different denominators _____

15. To evaluate an algebraic expression _____

After completing Chapter 6, you should have mastered the objectives listed at the beginning of each section. Test yourself by working the following problems. (Refer to the sections indicated as needed.)

• *Apply the laws of exponents (6.1).*

16. Express in exponential form: $\dfrac{125}{32}$

17. Express in exponential form: $\dfrac{x}{3} \cdot \dfrac{x}{3} \cdot \dfrac{x}{3}$

18. Evaluate: $\left(-\dfrac{5}{6}\right)^3$

• *Recognize polynomials (6.2).*

Identify the type and degree of each of the following.

19. $3x^2 + 7x + 1$

20. $\dfrac{7}{8}$

21. $2x - 3$

22. $\dfrac{x^3}{3}$

• *Add and subtract polynomials (6.2).*

Perform the indicated operations.

23. $(17y + 5) + (2y - 18)$

24. $(3x + 11) - (5x - 7)$

25. Add:

$4x^2 + 12x - 1$
$\underline{3x^2 - 27x + 5}$

26. Subtract:

$16y^2 - 12y + 23$
$\underline{11y^2 - 5y + 38}$

• *Multiply polynomials (6.2).*

27. $-3y(5y^2 - 7y + 13)$

28. $(3z^2 + z + 2)(5z)$

29. $(-x)(2x^2 + x - 1)$

• *Multiply binomials using the FOIL method (6.2).*

Multiply. Simplify results.

30. $(z + 5)(z + 6)$

31. $(3x + 1)(9x - 2)$

32. $(q - 2)(q + 2)$

33. $(2x - 3y)(3x - 2y)$

• *Find the common monomial factors of a polynomial (6.2).*

Factor out the common monomial factors.

34. $6w + 12x - 4y$

35. $12y + 3y^2$

36. $-15x^2y + 25xy^2 - 5xy$

• *Reduce algebraic fractions to lowest terms (6.3).*

Reduce to lowest terms.

37. $\dfrac{-24x}{30x^2y}$

38. $\dfrac{520\ gh}{-5g^2}$

• *Multiply algebraic fractions (6.3).*

Find the products.

39. $\dfrac{2}{x} \cdot \dfrac{y}{x}$

40. $\dfrac{3h^2}{4j} \cdot \dfrac{12j^2}{6h}$

41. $\dfrac{6s}{3r} \cdot \dfrac{9r^2}{3s^2}$

• *Divide algebraic fractions (6.3).*

Divide.

42. $\dfrac{6}{r^2} \div \dfrac{12}{r}$

43. $\dfrac{35c^2d}{5c} \div \dfrac{7cd}{10d^2}$

• *Add and subtract algebraic fractions (6.4).*

Add or subtract as indicated.

44. $\dfrac{3}{a} + \dfrac{2}{a}$

45. $\dfrac{6}{xy} - \dfrac{4}{xy}$

46. $\dfrac{6x}{2c} + \dfrac{10x}{2c}$

47. $\dfrac{11y}{5x} - \dfrac{y}{5x}$

48. $\dfrac{7a + 14}{5b} - \dfrac{2a + 4}{5b}$

49. $\dfrac{a}{3} + \dfrac{a}{2}$

50. $\dfrac{3b}{10} - \dfrac{b}{5}$

51. $\dfrac{x}{4} + \dfrac{x^2}{6}$

52. $\dfrac{13r}{2s} - \dfrac{r}{3s}$

53. $\dfrac{2}{y} + \dfrac{1 - y}{y}$

54. $\dfrac{3}{w} + \dfrac{5}{w} - \dfrac{2}{w}$

55. $\dfrac{17}{x^2} + \dfrac{8}{y} + \dfrac{10}{x}$

• *Evaluate algebraic expressions (6.5).*

Evaluate these expressions. Use a calculator when appropriate.

56. $2x + 1$, for $x = 3$ and for $x = -1$

57. $3a + 5b$, for $a = 2$ and $b = 7$ and for $a = -3$ and $b = 5$

58. $6x^4 + 5x^2 - 3x + 11$, for $x = 1$ and for $x = -4$

59. $\dfrac{1}{2}y^3 + \dfrac{2}{3}y^2 - \dfrac{1}{5}y + \dfrac{3}{8}$, for $y = 30$

60. $\dfrac{3w + 4x - 5y}{2z}$, for $w = 5$, $x = -4$, $y = 3$, $z = -2$

CHAPTER 6 PRACTICE TEST

Simplify by performing the indicated operations. Express results with positive exponents and in lowest terms.

1. $(10x + 2y) + (7x - 3y)$

2. $(a + 5b)(2a - b)$

3. $(2x + 3y)(3x + 2y)$

4. $(c + 4d)(c - 4d)$

5. $x^2(5x^2 + 21x - 11)$

6. $(-2)(y^2 - 10y + 5)$

7. $(ab)(a^2 + 3ab + b^2)$

8. $(17v - 2w) - (23v + 5w)$

9. $(3a + 54b - 6c) - (31a - 42b + 19c)$

10. $(5x^2 + 2x + 6) - (x^2 - 12x + 18)$

11. $\dfrac{15}{x} + \dfrac{12}{y}$

12. $\dfrac{3z}{2xy} - \dfrac{1}{xy}$

13. $\dfrac{2a+b}{5} - \dfrac{a-b}{2}$

14. $-2(25x^2 - 6x + 33)$

15. $\dfrac{x}{y} \div \dfrac{y}{x}$

16. $\dfrac{3a^2}{b} \div \dfrac{b^2}{8a}$

17. $\dfrac{(-6k)}{m^3} \div \dfrac{(-2k)}{m}$

18. $\dfrac{-32r^2s}{18r^3s^5}$

19. $\dfrac{3a}{2b} + \dfrac{5a}{2b}$

20. $(4q^3)^2$

21. $(r^4)^{-1}$

22. $\left(\dfrac{2g}{3h}\right)^3$

23. $x^2(x^{15})$

24. $-(5x)^2$

25. $(72y)^0$

26. $(4ab^3)^{-2}$

27. $(-3q)^2$

28. $5x^{-7}$

Evaluate.

29. $16 - 5y$, for $y = 5$

30. $7b + 9c$, for $b = 8$ and $c = -6$

31. $4x^2 - 10x + 8$, for $x = 3$

32. $3y^2 - 7y + 21$, $y = -5$

33. $2a^2 + 3ab + b^2$, for $a = -1$ and $b = 4$

34. 10^{-1}

35. $\left(\dfrac{3}{4}\right)^3$

36. $2^{-2} \cdot 3^{-3}$

Identify the type of each polynomial, and state its degree.

37. $-\dfrac{5}{6}$

38. $2t + 3$

39. $17x^4 - 5x^2 + 81$

40. $-5x^2 + 27$

		THE MATHEMATICS JOURNAL						
		Revision of Time Management Plan						
	Task:	Monitor your time-management plan closely for a week. Write a one-page (150-word) entry discussing it.						
	Starters:	Is your plan working? Why or why not? What aspects (if any) of it are working well? What aspects are not? Do you need to make any changes in it? What are they? Use the accompanying schedule to revise your plan.						
		Sun.	Mon.	Tue.	Wed.	Thurs.	Fri.	Sat.
7:00								
8:00								
9:00								
10:00								
11:00								
12:00								
1:00								
2:00								
3:00								
4:00								
5:00								
6:00								
7:00								
8:00								
9:00								
10:00								
11:00								
12:00								

CUMULATIVE REVIEW: CHAPTERS 4, 5, and 6

DIRECTIONS *Test your mastery of the topics covered in Chapters 4, 5, and 6 by taking this practice test. All items are worth 2 points unless indicated otherwise. The total number of points possible is 100. Remember, this is a review, so check your answers and then refer to the appropriate chapter and section to get assistance in correcting any mistakes.*

COMPLETION *Choose the appropriate word, phrase, or property from the list below.*

Distributive Property	Identity Property for Addition
$a^0 = 1$	$b^{-n} = \dfrac{1}{b^n}$
Constant	Binomials
Trinomial	Polynomial
Multiplication Property of Equality	Addition Property of Equality
Variable	Identity Property of Multiplication
Property of Reciprocals	Property of Additive Inverses

1. The letter x when used in mathematics is an example of a _____

2. The _____ states that, for all real numbers a, $a \cdot 1 = a$.

3. A polynomial with three terms is called a _____.

4. Rewriting the equation $6x + 9 = 7$ as $6x + 9 + (-9) = 7 + (-9)$ is an example of applying the _____

   _____.

5. $14x^4 - 3x^2 + 7x + 18$ is an example of a(n) _____.

6. The _____ states that, for all $a \in$ R, $a + (-a) = 0$.

7. $(14b + 7)$ and $(-8b + 3)$ are examples of _____.

8. The _____ states that for all $a, b, c \in$ R, $a(b + c) = ab + ac$ and $ab + ac = a(b + c)$.

9. Rewriting the equation $7x = 14$ as $\dfrac{1}{7} \cdot 7x = 14 \cdot \dfrac{1}{7}$ is an example of applying the _____

   _____.

10. The _____ states that for all real numbers $a(a \neq 0)$, $a \cdot \dfrac{1}{a} = 1$.

Follow the directions given for each item.

11. Evaluate: $2y^2 - y + 3$, when $y = -3$ _____

12. Evaluate: $2u - v^{-3}$, when $u = 33$ and $v = 2$ _____

13. Express $\dfrac{25}{7}$ as a mixed number. _____

14. $-(-7) =$ _____

15. $\dfrac{0}{4} =$ _____

16. $\dfrac{4}{0}$ = _____

17. $(-6)(-4)$ = _____

18. $7(-3)$ = _____

19. The reciprocal of $-6\dfrac{2}{7}$ is _____ .

20. $6 - (-5)$ = _____

21. $-22 + (-11)$ = _____

22. Reduce to lowest terms: $\dfrac{17}{34}$ _____

23. Write in exponential form: $\dfrac{9^8}{9^3}$ _____

Perform the indicated operations and simplify. Reduce answers to lowest terms. Express all answers with positive exponents.

24. $-\dfrac{3}{7} + \dfrac{5}{4}$ = _____

25. $-2\dfrac{3}{4} + 5\dfrac{4}{5}$ = _____

26. $\dfrac{\dfrac{1}{3} + \dfrac{1}{4}}{\dfrac{5}{6} + \dfrac{1}{12}}$ = _____ (3 pts.)

27. Simplify: $28 + 7 - 6 - (-8)$ = _____

28. Simplify: $\{21 - [5 - (7 - 2 \div 2)]\} \div (3 + 16 \div 2)$ = _____

29. $(-2m^5 - 7m^3 - 5) + (5m^5 - 2m^3)$ = _____

30. $(8x^3)(4xy^2)$ = _____

31. $(-3x^3y)^2$ = _____

32. $(x + 2)(x + 1)$ = _____

33. $(4a - 2)(a + 1)$ = _____

34. Express as a product of two factors: $2x^3 + 3x$ = _____ (3 pts.)

35. $(2x)^0$ = _____

36. $\dfrac{x^6}{x^5}$ = _____

37. $\dfrac{k^{10}m^{14}}{k^8}$ = _____

38. $\dfrac{3x^2}{18} \cdot \dfrac{6}{x^3} =$ _____

39. $\dfrac{3}{4} \div \dfrac{9}{x} =$ _____

40. $\dfrac{7}{12n^2} \div \dfrac{35}{n^3} =$ _____

41. $\dfrac{181y}{11} + \dfrac{27y}{11} =$ _____

42. $\dfrac{3a}{5} + \dfrac{a}{7} =$ _____

43. $\dfrac{4}{x^2} + \dfrac{6}{x^3} =$ _____ (3 pts.)

44. $\dfrac{1}{a} + \dfrac{1}{b} =$ _____ (3 pts.)

Solve for the variable.

45. $23 + a = \dfrac{36}{6}$

46. $x + 8 - 3 = 5 + x$

47. $-\dfrac{1}{5}y = 3$

48. $7x = 56$

7

Decimals

7.1 Place Value and the Comparison of Decimals
7.2 Addition and Subtraction of Decimals
7.3 Multiplication of Decimals
7.4 Division of Decimals
7.5 Conversions Between Fractions and Decimals

Introduction to Decimals

Islamic mathematicians of the Middle Ages knew about decimal fractions, but decimal fractions were first used systematically by the Flemish mathematician Simon Stevin (1548–1620). The use of the decimal point to separate whole numbers from decimal fractions was introduced in the late sixteenth century by John Napier (1550–1617). Thomas Jefferson devised a system of decimal coinage and was instrumental in passing the Mint Act of 1792, which introduced decimal coinage to the United States.

The metric system of weights, measures, and coinage (a system based on decimals) was introduced in France during the French Revolution in the 1790s. Today, the metric system of weights, measures, and coinage is used by most countries in the world. Then President Gerald Ford signed the Metric Conversion Act in 1975. We use metric measures in many areas of our lives. We buy pop in 2-liter bottles. We drive cars that also use metric measure. Initially, there was enthusiasm to change all weights and measures in the United States to the metric system, but our road signs are still in miles and we still buy milk by the gallon.

7.1 Place Value and the Comparison of Decimals

OBJECTIVES

- Recognize the position of the decimal point in the place value system.
- Order decimals.
- Name decimals.
- Write decimals in words.
- Express numbers in scientific notation.
- Round decimals.

You learned in Section 2.1 about place value and whole numbers. In this section, we will apply these concepts to decimals.

DEFINITION **Decimals:** A way of expressing rational numbers. All rational numbers can be expressed as fractions or decimals.

EXAMPLE 1
$$\frac{1}{2} = 0.5$$

$$\frac{3}{4} = 0.75$$

213

DEFINITION **Irrational numbers:** Decimals that do not terminate or repeat and cannot be expressed as fractions.

EXAMPLE 2 The following are some examples of irrational numbers:

$$\pi = 3.14159265358979323846264338327 9 \ldots ;$$
$$\sqrt{2} = 1.414213562 \ldots ; \sqrt{3} = 1.73205080 \ldots ;$$
$$\sqrt{43} = 6.557438524 \ldots .$$

Decimals expand the place value system we considered in Section 2.1. In that section, we considered the whole numbers.

1 is read one.
10 is read ten.
100 is read one hundred.
1000 is read one thousand.
10,000 is read ten thousand.
100,000 is read one hundred thousand.
1,000,000 is read one million.

Our number system is based on powers of 10. Each place value moving to the left is multiplied by 10. The reverse also holds true. Moving to the right, each place value is divided by 10. This system continues to the right of the decimal point.

EXAMPLE 3 Divide 1 by 10, the result is $\frac{1}{10}$; the place value to the right of the decimal point is the tenths place.

$$0.1 = \frac{1}{10}$$

EXAMPLE 4 Divide $\frac{1}{10}$ by 10, the result is $\frac{1}{100}$; the place value to the right of the $\frac{1}{10}$ is the $\frac{1}{100}$ ths place.

$$0.01 = \frac{1}{100}$$

EXAMPLE 5 Divide $\frac{1}{100}$ by 10, the result is $\frac{1}{1000}$; the place value to the right of the $\frac{1}{100}$ is the $\frac{1}{1000}$ ths place.

$$0.001 = \frac{1}{1000}$$

Reading decimals:

0.1 is read one-tenth.
0.01 is read one-hundredth.
0.001 is read one-thousandth.
0.0001 is read one ten-thousandth.
0.00001 is read one hundred-thousandth.
0.000001 is read one-millionth.

Note Every whole number has a decimal point (implied if not written) to the right of the units place.

EXAMPLE 6 We usually write 54, but if we remember the decimal point, we would write "54." instead.

$$123 = 123.$$
$$5 = 5.$$

RULE **To read or write a decimal number**

1. Read or write the whole number, which is to the left of the decimal point, followed by *and* to show where the decimal point appears.
2. Read or write the number to the right of the decimal point as a whole number, followed by the place value of the digit farthest to the right.

Write in words.

1. 46.4

2. 79.12

3. 345.678

4. 3008.08

5. 3.009

EXAMPLE 7   25.3 is written or read:

Twenty-five and three-tenths.

EXAMPLE 8   36.59 is written or read:

Thirty-six and fifty-nine hundredths.

EXAMPLE 9   0.05 is written or read:

Five-hundredths.

EXAMPLE 10   106.05 is written or read:

One hundred six and five-hundredths.

EXAMPLE 11   0.7 is written or read:

Seven-tenths.

EXAMPLE 12   2004.7 is written or read:

Two thousand four and seven-tenths.

EXAMPLE 13   200.002 is written or read:

Two hundred and two-thousandths.

The Principle of Trichotomy

As is true for all real numbers, decimals can be ordered or compared by using the symbols for "greater than," "less than," or "equal to": $>$, $<$, or $=$. This is the Principle of Trichotomy; see Section 4.2. Remember that the smaller part of the inequality symbol always points to the smaller number.

To compare decimals, each number may be expressed as a fraction; to compare them as fractions, they must have the same denominator.

EXAMPLE 14   Compare 0.2 and 0.26.

$$0.2 = \frac{2}{10} \qquad 0.26 = \frac{26}{100}$$

$$\frac{2}{10} = \frac{20}{100}$$

Which is less: $\frac{20}{100}$ or $\frac{26}{100}$?

$$\frac{20}{100} < \frac{26}{100} \text{ or } 0.2 < 0.26$$

Note   To help compare decimals, add zeros as needed to give them the same number of places after the decimal point.

EXAMPLE 15 Use Example 14 and compare 0.2 and 0.26. Add a zero to 0.2 and compare 0.20 to 0.26.

$$0.20 < 0.26$$

EXAMPLE 16 Compare -0.32 and -0.4. Which is greater: -0.32 or -0.40? Which is to the right on the number line?

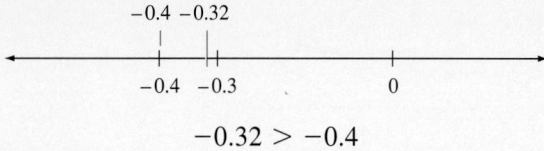

$$-0.32 > -0.4$$

Scientific Notation

Scientists often use extremely large or small numbers. Scientific notation is used to make these numbers more manageable. The notation also makes it possible to work with much larger and smaller numbers on a calculator than the display generally allows—only eight or ten spaces.

DEFINITION   A number is in **scientific notation** if it is written as a product of a number greater than or equal to 1 and less than 10 and a power of 10.
   If the exponent is positive, multiply by a power of 10. This can be accomplished by moving the decimal point to the right the number of places shown in the exponent. This will change the number from scientific notation into standard notation.

DEFINITION   **Standard notation:** The way you are accustomed to writing numbers.

EXAMPLE 17 $3.4 \times 10^2 = 340. = 340$

EXAMPLE 18 $7.46 \times 10^8 = 746000000. = 746,000,000$

USING THE CALCULATOR

$$7.46 \times 10^8$$

Enter 7.46 ⎡EE⎤ 8 ⎡=⎤ **746000000** ← Display

EXAMPLE 19 $8.384 \times 10^6 = 8384000. = 8,384,000$

To express a number in scientific notation, it must be written as a decimal greater than or equal to 1 and less than 10 multiplied by a power of 10.

EXAMPLE 20 Express 94,000 in scientific notation.

$9.4 \times 10^?$ To find the correct power of 10, count the number of places the decimal point was moved, in this case, four places.

94000.

9.4×10^4

EXAMPLE 21 Express 87,460,000 in scientific notation.

$$87\,460\,000. = 8.746 \times 10^7$$

EXAMPLE 22 Express 389,000 in scientific notation.

$$389\,000. = 3.89 \times 10^5$$

 USING THE CALCULATOR

Express 389,000 in scientific notation.

Enter 2nd Sci 389000 = **3.89** **⁰⁵** ← Display

If the exponent is negative, divide by a power of 10. This can be accomplished by moving the decimal point to the left the number of places shown in the exponent.

EXAMPLE 23 $3.4 \times 10^{-2} = .034 = 0.034$

EXAMPLE 24 $7.46 \times 10^{-8} = .0000000746 = 0.0000000746$

EXAMPLE 25 $8.3684 \times 10^{-6} = .0000083684 = 0.0000083684$

 USING THE CALCULATOR

Enter 8.3684 EE 6 +/− = **0.0000083684** ← Display

EXAMPLE 26 Express 0.0032 in scientific notation.

$$.0032 = 3.2 \times 10^{-3}$$

EXAMPLE 27 Express 0.00000231 in scientific notation.

$$.00000231 = 2.31 \times 10^{-6}$$

EXAMPLE 28 Express 0.00082 in scientific notation.

$$.00082 = 8.2 \times 10^{-4}$$

USING THE CALCULATOR

Enter 2nd Sci .00082 = **8.2** **⁻⁰⁴** ← Display

Rounding Decimals

We will use the same rules for rounding decimals as we used in Section 2.2 to round whole numbers.

RULE To round whole numbers

To round a number to a given place value

1. Look at the digit to the immediate right of that place.
2. If the digit is less than 5, replace it with 0 and replace all the digits to the right of it with 0s.
3. If the digit in that position is 5 or greater, add 1 to the digit to the left of it, and then replace it with a 0.
4. Replace all digits to the right of it with 0s.

Note   We will round decimals in the same way, except that we omit "trailing zeros" in the last positions to the right of the decimal point.

$$6.4 = 6.40 = 6.400 = 6.4000 = 6.40000$$

The zeros after the last digit in a decimal do not change the decimal and are deleted.

EXAMPLE 29 Round 569.2438 to the nearest hundred. The hundreds place is 5; the digit to the right is greater than 5, so add 1 to 5.

569.2438 ≈ 600.0000.   We do not need the zeros after the decimal point.

569.2438 ≈   Rounded to the nearest hundred is

600.

EXAMPLE 30 Using the same number 569.2438, round it to the nearest hundredth. In the hundredths place is the 4. The digit to its right is 3, which is less than 5, so 569.2438 rounded to the nearest hundredth is 569.24.

Numbers Rounded to the Nearest Units, Tenths, or Hundredths			
Number	**Units**	**Tenths**	**Hundredths**
231.346	231	231.3	231.35
30.02	30	30.0	30.02
800.004	800	800.0	800.00
999.999	1000	1000.0	1000.00

Practice Problems

Round the following numbers.

19. Round 45.5498 to the nearest hundredth.

20. Round 234.364 to the nearest tenth.

21. Round 239.56 to the nearest whole number.

22. Round 76.987 to the nearest tenth.

ANSWERS TO PRACTICE PROBLEMS
1. Forty-six and four-tenths.
2. Seventy-nine and twelve-hundredths.   **3.** Three hundred forty-five and six hundred seventy-eight thousandths.
4. Three thousand eight and eight-hundredths.   **5.** Three and nine-thousandths.   **6.** <   **7.** =
8. <   **9.** >   **10.** >   **11.** 5.6×10^7
12. 7.834×10^5 **13.** 932,000,000
14. 73,000   **15.** 0.0000342
16. 0.0097   **17.** 5.46×10^{-5}
18. 8.3692×10^{-2} **19.** 45.55
20. 234.4   **21.** 240   **22.** 77.0

PROBLEM SET 7.1

What is the place value of the 4 in each of the following numbers?

1. 405

2. 1.04

3. 12.4653

4. −43.98

5. 76.9804

6. −0.35204

7. 74,900

8. 0.000564

9. 7.3946

Write each number in words.

10. 0.5

11. 0.05

12. 0.005

13. −50.23

14. 0.023

15. 0.0023

16. 60.00023

17. 4.6

18. −29.02

19. 7.875

20. 56.124

21. −5300.0035

22. 4.3005

23. 463.0463

Write each of the following as a decimal.

24. Four-tenths

25. Four-hundredths

26. Four-thousandths

27. Four ten-thousandths

28. Sixty-seven hundredths

29. Negative sixty-seven thousandths

30. Negative sixty-seven ten-thousandths

31. Negative eight thousand four and two-hundredths

32. Sixty-seven and seventeen-thousandths

33. Negative four hundred and forty-two hundredths

34. Two hundred and two-hundredths

35. Sixteen thousand and sixteen-thousandths

36. Three hundred seventy-eight thousandths

37. Ninety-nine ten thousandths

38. Nine thousand ninety-nine and nine hundred ninety-nine thousandths

Fill in a <, >, *or* = *symbol between the pairs of numbers.*

39. 1.4   1.04   **40.** 0.3   0.32   **41.** 4.03   4

42. 0.666   0.6666   **43.** 0.58   0.580   **44.** 1.98   2.06

45. 0.96   0.149   **46.** 47.9   48.9   **47.** 0.33   0.3

48. 0.0006   0.006   **49.** 0.3   0.3   **50.** −4.8   −4.08

51. −0.38   −0.381   **52.** 0.495   0.4995   **53.** −0.495   −0.4995

Complete the table, rounding as requested.

	Number	*Hundreds*	*Tens*	*Units*	*Tenths*	*Hundredths*
54.	17.0015					
55.	6.529					
56.	99.999					
57.	0.04					
58.	123.456					
59.	56.192					
60.	86.939					
61.	76.001					
62.	4.9309					

Express in standard notation.

63. 7.8×10^4 **64.** 9.3×10^{-3}

65. 8.46×10^{-5} **66.** 3.5×10^8

67. 2.1×10^{-4} **68.** 7.8×10^{-6}

69. 3.91×10^7

70. 1.4×10^3

Express in scientific notation.

71. 0.0000863

72. 97,800,000

73. 2000

74. 0.0004

75. 6800

76. 54,900

77. 0.00386

78. 0.00005732

Applications:

79. People in the United States throw away 207,000,000 tons of trash each year. How much trash is that in scientific notation?

80. Mercury is 36,000,000 miles from the sun. Write this distance in scientific notation.

81. Mercury is 1.36×10^8 miles from the earth. How many miles is this in standard notation?

82. The nucleus of an atom is 6.66×10^{-5} smaller than an atom. Write that number in standard notation.

83. The diameter of an atom is 0.00000001 of an inch. How would a scientist write this number in scientific notation?

84. In 1992, the population of Australia was about 1.75×10^7. What was the population of Australia in standard notation?

85. In 1992, the United States sent $2,240,000 in aid to countries in the Western Hemisphere. How is that number written in scientific notation?

86. Lake Mead, a reservoir for the Hoover Dam in Nevada, is 28,250,000 acres. How many acres is that in scientific notation?

87. The population of the United States is 2.56×10^8. Write that number in standard notation.

88. The population of China is 1,169,000,000. Write this number in scientific notation.

89. Jupiter is 6.1×10^8 miles from the sun. How many miles is this in standard notation?

90. A virus is approximately 1.67×10^{-8} of an inch long. How long is that in standard notation?

91. A micron is a millionth of a meter and is approximately 0.0000004 of an inch. Write that number in scientific notation.

92. The mutation rates in humans for muscular dystrophy is 1×10^{-4}. What is this in standard notation?

93. Alpha particles are 6.65×10^{-24}. Beta particles are smaller, 9.19×10^{-28}. Find out how much bigger the alpha particles are than the beta particles. Use standard notation.

94. The mutation rate for kidney disease in humans is 0.00095. Write this number in scientific notation.

95. The energy of an alpha particle is 0.000013 erg (a measure of energy). Write that number in scientific notation.

96. The weight of the DNA of a common spotted newt is 0.00000000000758 grams. Write that number in scientific notation.

97. The weight of the DNA of a dusty salamander is 0.0000000000035 grams. Write that number in scientific notation.

7.2 **Addition and Subtraction of Decimals**

OBJECTIVES
● Add decimals.
● Subtract decimals.

How much do you pay for a calculator that costs $19.95 and $1.25 in tax?

$$\begin{array}{r} \$19.95 \\ + \ \$1.25 \\ \hline \$21.20 \end{array}$$ is the total cost of the calculator.

Adding money is a practical use for decimals, since most monetary systems are based on decimals.

Note Decimals are added in the same way as whole numbers. Line up the decimal point so that all the place values are aligned.

EXAMPLE 1 $5.37 + 6.4$

$$\begin{array}{r} 5.37 \\ + \ 6.40 \\ \hline 11.77 \end{array}$$ Stack the numbers vertically with the decimal points aligned. Add zeros to hold empty places. Add as for whole numbers starting at the right.

Subtraction of decimals is performed in the same way. Line up the decimal points (which automatically aligns all the place values), and then subtract.

EXAMPLE 2 $4.38 - 3.4$

$$\begin{array}{r} 4.38 \\ - \ 3.40 \\ \hline 0.98 \end{array}$$ Stack the numbers vertically with the decimal points aligned. Add zeros to hold empty places. Subtract as for whole numbers, borrowing as needed.

Practice Problems

Add or subtract as indicated.

1. $5.37 + 15.6$

2. $67.339 - 3.46$

3. $45 + 4.9 + 0.87$

4. $70.03 - 1.9$

5. $2.3 - 1.9$

The calculator is useful when working with decimals. Do not enter the extra zeros; the calculator will do that.

USING THE CALCULATOR

Add 2.58 + 9.4 + 3.685.

       Enter 2.58 `+` 9.4 `+` 3.685 `=` **15.665** ← Display

Subtract 5.253 − 2.97.

       Enter 5.253 `−` 2.97 `=` **2.283** ← Display

Use estimation to be sure the answer is reasonable. If necessary, reread Section 2.7 on estimating. Return to the examples just worked on the calculator.

EXAMPLE 3 Estimate the sum of 2.58, 9.4, and 3.685 to the nearest whole number.

$$
\begin{array}{rcl}
2.58 & \approx & 3 \\
9.4 & \approx & 9 \\
+ \ 3.685 & \approx & 4 \\
\hline
\end{array}
$$

             16 is the estimate of the sum.

The answer using the calculator was 15.665, which seems reasonable.

EXAMPLE 4 Estimate the difference of 5.253 and 2.97 to the nearest whole number.

$$
\begin{array}{rcl}
5.253 & \approx & 5 \\
- \ 2.97 & \approx & 3 \\
\hline
\end{array}
$$

             2 is the estimate of the difference.

The answer using the calculator was 2.283, which seems reasonable.

Note When adding or subtracting positive and negative decimals, follow the rules for addition and subtraction of integers found in Sections 4.3 and 4.4.

EXAMPLE 5 Calculate 3.24 − 8.4

    3.24 − 8.4   Use the rule for subtraction of integers. Change from subtraction of a positive to addition of a negative by rewriting the problem.

    3.24 + (−8.4)   The easiest way to determine which number has the greater absolute value is to estimate to the nearest whole number. In this case, which is greater: 3 or 8? Because 8 is greater, 8.4 has a greater absolute value than 3.24. Therefore, we subtract 3.24 from 8.4. The answer to the original problem will be negative because the number with the greater absolute value is negative.

$$
\begin{array}{r}
-8.40 \\
+ \ 3.24 \\
\hline
-5.16
\end{array}
$$

Thus, 3.24 − 8.4 = −5.16

EXAMPLE 6 To determine the absolute value in a problem, such as 3.07 + (−3.2), where the whole numbers are the same, estimate to the tenths place. In this example, −3.2 has the larger absolute value.

$$\begin{array}{r} -3.20 \\ +\ 3.07 \\ \hline -0.13 \end{array}$$

Thus, 3.07 + (−3.2) = −0.13

EXAMPLE 7 6.23 − (−2.1)

6.23 + 2.1   Rewrite using the rule for subtraction of integers.

$$\begin{array}{r} 6.23 \\ +\ 2.10 \\ \hline 8.33 \end{array}$$

Estimate the answer 6.23 ≈ 6
2.1 ≈ 2

6 + 2 = 8

Thus, 6.23 + 2.1 ≈ 8

USING THE CALCULATOR

Enter 6.23 − 2.1 +/− = **8.33** ← Display

APPLICATION 1

Melanie went to the grocery store with a $20.00 bill and purchased a half gallon of milk for $2.09, a loaf of bread for $1.78, a jar of jam for $2.39, and orange juice for $2.99. How much change did she have after paying for her purchases?

First consider how much Melanie spent.

$$\begin{array}{r} \$2.09 \\ \$1.78 \\ \$2.39 \\ +\ \$2.99 \\ \hline \end{array}$$

$9.25 is what Melanie spent.
$20.00 − $9.25 = $10.75

Answer: $10.75 is the money she had left. ◆

APPLICATION 2

Jay runs a mile in 5.386 minutes. Sam runs a mile in 6.78 minutes. How much faster than Sam does Jay run the mile?

Subtract Jay's time from Sam's time to find how much faster Jay runs.

$$\begin{array}{r} 6.780 \\ -\ 5.386 \\ \hline 1.394 \end{array}$$ minutes faster.

Answer: Jay runs 1.394 minutes faster. ◆

Practice Problems

Add or subtract as indicated.

6. 8.09 − 15.4

7. (−3.7) + (−35.09)

8. 6.98 − (−9.1)

9. −8.07 − 2.315

10. −2.38 + 19.4

APPLICATION 3

It will take 3.65 meters of fabric to make curtains for Janell's dorm-room windows. She also wants a matching comforter, which will require 7.5 meters, and a matching floor pillow, which will require 1.85 meters. If a bolt of fabric contains 15 meters, how much fabric will be left on the bolt?

$$
\begin{array}{r}
3.65 \text{ meters for curtains} \\
7.50 \text{ meters for a comforter} \\
+ \quad 1.85 \text{ meters for a pillow} \\
\hline
13.00 \text{ meters total required}
\end{array}
$$

$$
\begin{array}{r}
15.00 \text{ meters on the bolt} \\
- \quad 13.00 \text{ meters used} \\
\hline
2.00 \text{ meters left on the bolt.}
\end{array}
$$

Answer: 2 meters left on the bolt. ◆

ANSWERS TO PRACTICE PROBLEMS
1. 20.97 **2.** 63.879 **3.** 50.77
4. 68.13 **5.** 0.4 **6.** −7.31
7. −38.79 **8.** 16.08 **9.** −10.385
10. 17.02

PROBLEM SET 7.2

Add. Work these problems with paper and pencil, then check with a calculator.

1. 3.24 + 2.45

2. 3.89 + 5.57

3. 7.57 + 4.59

4. 13.6 + 8.45

5. 123.56 + 8.912

6. 0.9 + 23.5 + 47

7. 4.532 + 1.71 + 2.5

8. 3.4 + 9.006 + 7.77

9. 34.9845 + 2.11 + 56.091

10. 76.34 + (−19.6)

11. 9.09 + 0.95

12. 3.403 + 340.3

13. 82.44 + (−81.96)

14. 13.51 + (−5.6) + 7.295

15. 7.91 + (−7) + 8.32

16. 9.88 + (−6.999) + (−9.001)

17. 1754 + 3.0001 + 23.9

18. (−2.3) + (−5.006) + 7.306

19. (−4.53) + (−8.6651) + (−0.009)

20. 5.44763 + 0.9 + 6.999 + 267

21. 8.9 + 9.56 + 45 + 13.8476

22. 76.8 + 7.68 + 0.768 + 768

23. $23.4 + 46.39 + 7.869$

24. $154.6 + 78.93$

25. $-43.9 + (-8.64)$

26. $-0.64 + 0.364 + (-0.21)$

27. $4.3 + (-532) + (-2.004)$

28. $-5.03 + 82.3 + (-0.36)$

Subtract. Work these problems with paper and pencil, then check with a calculator.

29. $85.46 - 71.21$

30. $984.87 - 513.43$

31. $76.36 - 48.45$

32. $823.96 - 598.48$

33. $7.53 - 9.24$

34. $(-3.65) - 2.14$

35. $9.8 - 8.957$

36. $7.363 - 9.8$

37. $24.45 - (-4.768)$

38. $-10 - 8.76$

39. $-78.4 - 2.34$

40. $40.04 - 4.4$

41. $70.005 - 2.35$

42. $100 - 3.08$

43. $75 - (-68.89)$

44. $3.473 - (-24.47)$

45. $-7.54 - (-7.54)$

46. $657.874 - 472.097$

47. $905.008 - 799.999$

48. $460.7 - 4.607$

49. $46.36 - 782.3$

50. $-76.3 - (-89)$

51. $-3.86 - 93.4$

52. $782.34 - 764.8$

53. $100 - 0.863$

54. $-100 - 0.863$

55. $9.7 - 99.34$

56. $-6.4 - 63.54$

Simplify. Use a calculator.

57. 4.2 + (6.8 − 2.4)

58. (9.3 − 8.5) + 4.7

59. 6.4 − (3.2 + 5.6)

60. (6.77 + 3.53) − 11

61. 3.45 − (5.43 − 4.3)

62. (9.75 − 3.5) − 1.2

63. (36.4 − 76.4) + 2.6

64. 9.6 + (86.3 − 99.7)

65. (4.8 − 9.46) + (3.8 − 1.29)

66. (32.1 + 4.82) − (4.61 + 28.3)

67. 100 − [76.8 − (−8.63)]

68. (546.3 − 2.96) − 836.54

69. (1.864 − 32.7) + 3.86

70. 0.346 − (0.46 − 0.973)

Use a calculator as you work these problems.

71. Find the sum of 9.46 and 5.39.

72. Subtract 4.56 from 8.9.

73. What number must be added to 7.332 to get 10.4?

74. Subtract 3.98 from the sum of 5.2 and 4.876.

75. 9.86 increased by 32.7.

76. Add 932.64 and 72.63.

77. 8.93 decreased by 47.3.

78. Find the difference between 3.86 and 0.049.

79. Find the sum of 2.8, 7.49, 0.36, and 296.

80. What is 9.63 less than 12.3?

81. What is 6.004 reduced by 2.38?

82. What is 8.46 subtracted from 2.83?

Applications:

83. Your math textbook cost $35.59; your psychology textbook, $47.98; your history textbook, $54.45; and the three books required in your language class, $7.99, $4.68, and $10.98. How much did you spend on books this semester?

84. Peter spent $23.59 on his mother's holiday gift, $19.97 on his father's, $20.07 on his sister's, and $45.50 on a special gift for his girlfriend. How much did Peter spend for his holiday shopping?

85. The school record for running the mile is 4.3897 minutes. If Barry can run the mile in 5.48819 minutes, how much faster does he need to run to match the school record?

86. Laura had $306.94 in her checking account at the beginning of the month. She paid her $175.75 rent, her $54.67 grocery bill, and her $32.65 utility bill each by check. How much is left in her checking account, assuming she wrote no other checks?

87. Amanda is having new carpet laid in her house. The carpet salesperson takes the measurements and tells her she will need 14.5 square yards in the hall, 17.86 square yards in the dining room, and 22.5 square yards in the living room. How much carpet will she need to order? If a bolt of carpet contains 50 square yards, how much more carpet than 1 bolt will be needed?

88. Find the distance around this triangle (the perimeter) by finding the sum of the lengths of the sides.

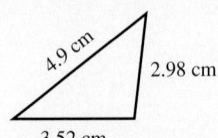

4.9 cm   2.98 cm

3.52 cm

89. What is the distance around this rectangle (the perimeter)? Find the perimeter by adding the lengths of all the sides.

8.64 in

3.96 in   3.96 in

8.64 in

90. The length of each side of this square is $2\frac{1}{2}$ ft. What is the distance around the square (the perimeter)?

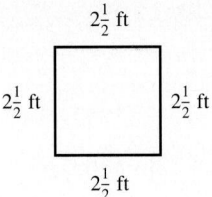

$2\frac{1}{2}$ ft

$2\frac{1}{2}$ ft $2\frac{1}{2}$ ft

$2\frac{1}{2}$ ft

7.3 Multiplication of Decimals

OBJECTIVE

● Multiply decimals.

We begin this section by changing decimals to fractions. We will multiply the fractions using a procedure you understand, and examples will then be used to demonstrate how we multiply decimals.

EXAMPLE 1   To multiply $(0.3)(0.4)$, consider these decimals as fractions.

$$0.3 = \frac{3}{10}$$

$$0.4 = \frac{4}{10}$$

$$\frac{3}{10} \cdot \frac{4}{10} = \frac{12}{100} \qquad \text{As a decimal, this is 0.12. Therefore,}$$

$$(0.3)(0.4) = 0.12$$

EXAMPLE 2   To multiply $(0.03)(0.4)$, consider these decimals as fractions.

$$0.03 = \frac{3}{100}$$

$$0.4 = \frac{4}{10}$$

$$\frac{3}{100} \cdot \frac{4}{10} = \frac{12}{1000} \qquad \text{As a decimal, this is 0.012. Therefore,}$$

$$(0.03)(0.4) = 0.012$$

We do not need to change all decimals to fractions to multiply them; these two examples do show a pattern for multiplying decimals.

RULE **Multiplying decimals**

1. Multiply the two numbers ignoring the decimal points.
2. Determine the location of the decimal point in the answer by counting the number of places after the decimal point in each of the numbers multiplied.
3. This will be the number of decimal places in the answer. Count them from the right.

EXAMPLE 3   Return to our first example, $(0.3)(0.4) = 0.12$.

$$(3)(4) = 12 \qquad \text{Multiply the numbers, and ignore the decimal points.}$$
$$(0.3)(0.4) = 0.12$$

There is one decimal place in each of the numbers to be multiplied and, therefore, two decimal places in the answer.

Note   When multiplying positive and negative decimals, use the rules for multiplying integers found in Section 4.5.

EXAMPLE 4 $(-1.2)(0.6)$ A negative number multiplied by a positive number is a negative number.

$(12)(6) = 72$ The answer will be a negative number.

$(-1.2)(0.6) = -0.72$ There are two decimal places in the numbers being multiplied, so there will be two decimal places in the answer.

USING THE CALCULATOR

$$-(1.2)(0.6)$$

Enter 1.2 $\boxed{+/-}$ $\boxed{\times}$ .6 $\boxed{=}$ **−0.72** ← Display

---

Practice Problems

How many decimal places are there in the answers to these problems?

1. $(0.7)(0.9)$

2. $(1.02)(3)$

3. $(2.5001)(0.4)$

4. $9(5.6)$

Multiplying by Powers of 10

Multiplying by powers of 10 is multiplying by

$$10^1 = 10$$
$$10^2 = 100$$
$$10^3 = 1000$$
$$10^4 = 10,000 \text{ and on to all the powers of 10.}$$

Multiply.

5. $(-0.5)(-0.6)$

6. $4(-3.23)$

7. $(-8)(1.9)$

8. $(7.654)10$

9. $(93.2)100$

10. $(24)1000$

11. $(0.0005)1000$

EXAMPLE 5 $(7.42)10$ Ignore the decimal point.

$742 \times 10 = 7420$ There are 2 decimal places in the answer.

74.20 We may omit the trailing zero.

74.2 We could arrive at the same answer by moving the decimal point one place to the right.

$$(7.42)10 = 74.2 = 74.2$$

EXAMPLE 6 $(8.362)100$ Ignore the decimal point.

$8362 \times 100 = 836200$ There are 3 decimal places.

$= 836.200 = 836.2$ We could have arrived at the same answer by moving the decimal point 2 places to the right.

$$(8.362)100 = 836.2 = 836.2$$

EXAMPLE 7 $(32.64)1000$ To multiply by 1000, move the decimal point 3 places to the right.

$$(32.64)1000 = 32640. = 32640$$

RULE **Multiplying by powers of 10**
To multiply any number by a power of 10

1. Count the zeros in the power of 10.
2. Move the decimal point of the other number to the right that many places.

APPLICATION 1

Caroline could buy 12 school notebooks for $1.29 each, or she could buy a package of 12 for $14.50. Which is the better buy? How much could she save?

At $1.29 each, 12 would cost

$$\$1.29 \times 12 = \$15.48$$

The package of 12 is cheaper.

$$\$15.48$$
$$- \$14.50$$

$0.98 is the savings.

Answer: The package of 12 is cheaper by 98 cents.   ◆

APPLICATION 2

During a two-hour TV movie, there were nineteen 1.58-minute commercials. How many minutes did the movie actually run?

The commercials ran $1.58 \times 19 = 30.02$ minutes. Two hours is 120 minutes.

$$\begin{array}{r} 120.00 \\ -\ \underline{30.02} \\ 89.98 \text{ minutes} \end{array}$$

Answer: The movie is 89.98 minutes long. ◆

APPLICATION 3

A trans-Atlantic telephone call to England costs 85 cents for the first minute and 79 cents for each additional minute. How much must Paul pay for a trans-Atlantic conversation lasting 10 minutes?

79¢ is $0.79.

Paul paid 85 cents for the first minute and 79 cents for each additional minute. He paid $0.79(9) for the remainder of the conversation.

$$.85 + [0.79(9)]$$
$$= .85 + [7.11]$$
$$= 7.96$$

Answer: Paul paid $7.96 for the phone call. ◆

ANSWERS TO PRACTICE PROBLEMS
1. 2   2. 2   3. 5   4. 1   5. 0.3
6. −12.92   7. −15.2   8. 76.54
9. 9320   10. 24,000   11. 0.5

PROBLEM SET 7.3

Multiply. Work these problems with paper and pencil, then check with a calculator.

1. 0.9(0.7)

2. (0.6)0.5

3. 1.1(0.9)

4. (0.4)0.8

5. (0.5)(−0.4)

6. (0.3)(0.8)

7. (0.03)(0.8)

8. (0.003)(0.8)

9. (0.3)(0.08)

10. (0.3)(0.008)

11. (0.03)(0.08)

12. (0.003)(0.08)

13. (0.03)(0.008)

14. (−0.6)(0.9)

15. (−0.6)(−0.7)

16. 0.0008(0.008)

17. (−0.0009)(0.00001)

18. (−0.007)(−0.003)

19. 231.352(1.5)

20. 78.59(36.1)

21. 92.7(−0.007)

22. 67.8(0.7)

23. 67.8(0.07)

24. 67.8(0.007)

25. 67.8(0.0007)

26. (−43.7)100

27. 75.43(−100)

28. 5.003(10)

29. (−98.999)(−1000)

30. 6.7(1000)

31. 5.42(−10)

32. (0.00005)10

33. (0.00005)100

34. (0.00005)1000

35. (0.00005)10,000

Perform the indicated operations. Use a calculator. Use the order of operations.
See Section 5.6.

36. 7.2 + (0.2)(0.4)

37. 5.34(0.01) + 21.87

38. [2.1 + 3.7][3.3 − 5.7]

39. 0.9(2.5 + 5.2)

40. −0.3(−7.6 − 2.4)

41. [2.5 − 9.6][−7.6 + 3.9]

42. 9.5008 + (0.3)(−0.7)

43. (7.5 + 3.1) + (8.6 − 4.2)

44. (7.003 − 9.1)(3.001)

45. (9.2 − 0.003) − (8.6 − 9.2)

46. 7.41(8.29 − 12.006)

47. (3.86 + 2.93)(2.1 − 6.89)

48. 7.2(8.23) − 2.6(3.92)

Use a calculator to work these problems.

49. What is the product of 6.23 and 0.6?

50. Multiply 0.004 by 0.0029.

51. What is 7 times 0.0086?

52. What is twice 8.375?

53. Triple the sum of 7.89 and 3.2.

54. Find the product of 3.89 and 6.2.

55. Multiply the product of 3 and 0.002 by 7.8.

56. Multiply the sum of 5.32 and 3.64 by 4.

57. Find the sum of 2.5 and the product of 5.8 and 2.1.

Applications:

58. A candy bar can be purchased for 35 cents each or in a box of 15 for $4.99. Which is the better buy? How much can be saved?

59. A university ordered new curtains made for a whole dorm, which has 82 windows. If each window requires 7.83 meters of fabric, how much will be needed? If each meter of fabric costs $4.99, how much will the fabric for all the dorm windows cost?

60. A phone call to California costs 30 cents for the first minute and 26 cents for each addition minute. How much will a 13-minute phone call cost?

61. During a two-hour TV movie, there were 21 commercials, each lasting 1.42 minutes. How long did the movie actually run?

62. Bath soap can be purchased for 42 cents a bar or for $2.99 for a box of 8. Which is the better buy? How much can be saved?

63. A home has 17 windows, all the same size. If each window needs 14.85 yards of fabric, how much fabric will be needed for all 17 windows? If each yard of fabric costs $8.95, how much will the fabric for the windows cost?

64. A trans-Atlantic phone call to Oslo costs $2.50 for the first 3 minutes and 85 cents for each additional minute. How much will a 19-minute call cost?

65. A phone call to Australia costs $2.95 for the first 3 minutes and 87 cents for each additional minute. How much will a 23-minute call cost?

66. Socks cost $2.03 for a pair or $11.99 for a package of 6 pairs. Which is the better buy?; How much can be saved?

67. A taxi company charges $2 to pick up a passenger plus 75 cents a mile. How much would a 4-mile ride cost?

68. A taxi company charges $2 to pick up a passenger plus 70 cents a half mile. How much would a 9-mile ride cost?

69. Find the distance around this square by multiplying the length of one side by 4.

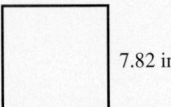
7.82 in

7.4 Division of Decimals

OBJECTIVES

● Divide decimals.

The dividend and divisor of any division problem can both be multiplied by the same number without changing the quotient. Multiplying the dividend and divisor by the same number results in an equivalent fraction.

EXAMPLE 1 $\dfrac{6}{2} = 3$

$\dfrac{6}{2} \cdot \dfrac{2}{2} = \dfrac{12}{4} = 3$ Multiply the original dividend and divisor by 2.

$\dfrac{6}{2} \cdot \dfrac{10}{10} = \dfrac{60}{20} = 3$ Multiply the original dividend and divisor by 10.

$\dfrac{6}{2} \cdot \dfrac{5}{5} = \dfrac{30}{10} = 3$ Multiply the original dividend and divisor by 5.

$\dfrac{6}{2} = \dfrac{12}{4} = \dfrac{60}{20} = \dfrac{30}{10}$ These fractions are equivalent because in each case the quotient is 3.

To use the division algorithm for decimals, the divisor needs to be an integer. The dividend can be an integer or a decimal. If a problem has a decimal divisor, multiply both the dividend and divisor by the same power of 10. To accomplish this, move the decimal point in the divisor as far to the right as possible, then move the decimal point in the dividend the same number of places to the right. If there are not enough places to move the decimal in the dividend, add zeros.

EXAMPLE 2 $(0.12) \div (0.3)$

$$0.3\overline{)0.12}^{\;.4}$$

Move the decimal point one place to make the divisor an integer. Move the decimal in the dividend the same number of places.

🖩 **USING THE CALCULATOR**

$(0.12) \div (0.3)$

Enter .12 ÷ .3 = **0.4** ← Display

EXAMPLE 3   75 ÷ 2.5

$$\begin{array}{r} 3\,0. \\ 2.5\overline{)75.0} \end{array}$$

Move the decimal point one place to make the divisor an integer. Move the decimal in the dividend the same number of places. Remember to divide into the zero.

USING THE CALCULATOR

75 ÷ 2.5

Enter 75 ÷ 2.5 = **30** ← Display

In some cases, divisors do not divide evenly into the dividends. Remember that every whole number has a decimal point after the units place; also remember that zeros can be added after the decimal point without changing the number. Therefore, if the division problem does not divide evenly, add zeros after the decimal point in the dividend and continue dividing.

EXAMPLE 4   29.8 ÷ 0.4

$$\begin{array}{r} 7\,4\,.\,5 \\ 0.4\overline{)29.8\ 0} \\ 28 \\ \hline 1\,8 \\ 1\,6 \\ \hline 2\ \ 0 \\ 2\ \ 0 \\ \hline \end{array}$$

Move the decimal points in both the dividend and the divisor. Add a zero to the dividend, and continue dividing.

USING THE CALCULATOR

29.8 ÷ 0.4

Enter 29.8 ÷ .4 = **74.5** ← Display

EXAMPLE 5   This division problem will never divide evenly.

$$20 \div 3$$

$$\begin{array}{r} 6.666 \\ 3\overline{)20.000} \\ 18 \\ \hline 2\ 0 \\ 1\ 8 \\ \hline 20 \\ 18 \\ \hline 20 \\ 18 \\ \hline 2 \end{array}$$

This is a repeating decimal and is written $6.\overline{6}$. The bar above the 6 in the tenths place indicates that the 6 will repeat forever.

Practice Problems

Divide.

1. 0.497 ÷ 0.7

2. 73.8 ÷ 0.9

3. 1.482 ÷ .06

4. 27.81 ÷ 0.9

5. 85.52 ÷ 1.7

6. 9.323 ÷ .03

EXAMPLE 6   This division will never divide evenly.

$$5 \div 11$$

```
       .4545
  11)5.0000
      4 4
      ‾‾‾
       60
       55
       ‾‾
       50
       44
       ‾‾
       60
       55
       ‾‾
        5
```

As you can see, this decimal will repeat .4545 and is written $.\overline{45}$ with the repeating bar over both the 4 and 5 because both numbers repeat.

EXAMPLE 7 $$5 \div 12$$

```
        .4166
  12)5.0000
      4 8
      ‾‾‾
       20
       12
       ‾‾
       80
       72
       ‾‾
       80
       72
       ‾‾
        8
```

In this example, only the 6 repeats, so the answer is written $.41\overline{6}$ with the repeating bar over only the 6.

If the quotient does not terminate or show a repeating pattern, round to the nearest hundredths place. To do this, divide to the thousandths place (three decimal places), and round to the hundredths place. Review rounding in Section 7.1.

Dividing by Powers of 10

Dividing by powers of 10 is dividing by 10 or multiples of 10.

EXAMPLE 8 $$743.8 \div 10$$

```
        74.38
  10)743.80
      70
      ‾‾
      43
      40
      ‾‾
       3 8
       3 0
       ‾‾‾
        80
        80
        ‾‾
```

We could have arrived at the same answer by counting the number of zeros in the divisor and moving the decimal point that number of places to the left in the dividend.

$$743.8 \div 10 = 7 \ 4 \ 3 \smile 8 = 74.38$$

EXAMPLE 9 $9367.84 \div 100 = 9\,3\underbrace{6\,7}.8\,4 = 93.6784$

EXAMPLE 10 $2.64 \div 1000 = \underbrace{0\,0\,2}.6\,4 = 0.00264$

APPLICATION 1

Four students living together in a house decide to buy a washing machine. The machine they buy costs $287.99, and the tax is $17.97. If they divide the total cost equally, how much will each student pay?

$287.99 Cost of the washer
+ $ 17.97 Tax
$305.96 Total cost
$305.96 \div 4 = 76.49$

Answer: Each student pays $76.49 for the washing machine. ◆

APPLICATION 2

The rent on a student house is $850 a month. Each month, the gas bill is $79.97, the electric bill is $93.75, and the water bill is $40.03. If five students share the house, how much does it cost each student to live in the house?

$850.00	Rent
$79.97	Gas
$93.75	Electric
$40.03	Water
$1063.75	Total cost

$1063.75 \div 5 = 212.75$

Answer: It costs each student $212.75 a month. ◆

APPLICATION 3

Maria, Clarissa, and Ann go to the College Restaurant for lunch. They forget to ask for separate checks and decide to split the bill three ways. If the bill was $28.14, how much did they each pay?

$28.14 \div 3 = 9.38$

Answer: They each paid $9.38 ◆

APPLICATION 4

Pop costs $3.98 for a six-pack of 20-ounce bottles. The same pop costs $1.99 for a six-pack of 12-ounce cans. Neither the bottles nor the cans are returnable. Which package is the better buy? How much per ounce would you save?

6 bottles of 20 oz = 120 oz
6 cans of 12 oz = 72 oz
$3.98 \div 120 \approx 3.32$ cents per oz
$1.99 \div 72 \approx 2.76$ cents per oz

Answer: Canned pop is a better buy. Bottled pop costs $\approx$ .56 of a cent (about $\frac{1}{2}$ cent) per ounce more than canned pop. ◆

PROBLEM SET 7.4

Divide. Work these problems with paper and pencil, then check with a calculator.

1. $0.534 \div 0.3$

2. $19.888 \div 0.8$

3. $408.03 \div 7$

4. $938.05 \div 5$

5. $15.0004 \div 0.05$

6. $1.8123 \div 0.06$

7. $8 \div 0.2$

8. $0.8 \div 2$

9. $0.8 \div 0.2$

10. $0.08 \div 0.2$

11. $0.008 \div 0.2$

12. $0.0008 \div 0.2$

13. $0.00008 \div 0.2$

14. $0.8 \div 0.02$

15. $0.8 \div 0.002$

16. $0.8 \div 0.0002$

17. $0.8 \div 0.00002$

18. $-55.368 \div 2.4$

19. $1595.139 \div (-3.9)$

20. $-71.064 \div 7.2$

21. $206.55 \div 4.5$

22. $-0.7921 \div (-0.89)$

23. $145.8 \div 0.36$

24. $101.455 \div 1.03$

25. $2372.49 \div (-2.61)$

26. $-13{,}208.08 \div 21.4$

27. $46.83 \div 10$

28. $0.864 \div 10$

29. $8536 \div 100$

30. $793.8 \div 100$

31. $53{,}864.6 \div 1000$

Divide. Round the answer to the nearest hundredth. Use a calculator.

32. $2.587 \div 0.7$

33. $33.39 \div 0.4$

34. $-128.9 \div 0.6$

35. $0.065 \div -0.8$

36. $9.8 \div 0.13$

37. $54.7 \div -2.3$

38. $97.6 \div 17$

39. $72.7 \div 1.9$

40. $0.633 \div 5.7$

41. $749 \div 29$

42. $0.0035 \div 2.56$

43. $4.23 \div 3.8$

44. $672 \div 0.82$

45. $7.638 \div 7$

46. $0.049 \div 7$

Use a calculator to work these application problems.

47. 2.38 divided by 4.

48. 2.9 divided into 22.7737.

49. Find the quotient of 54.33953 and 7.83.

Applications:

50. Brand A chocolate chip cookies cost $3.22 for a 14-ounce box. Brand B chocolate chip cookies cost $4.18 for a 19-ounce box. Which box of cookies is the better buy? How much per ounce would you save?

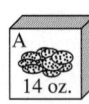 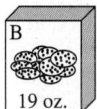

51. A box of 32 candy bars cost $10.88. How much does each candy bar cost?

52. Pencils are sold for $2.52 a dozen. How much do they cost each?

53. Jane, Liz, and Beth are all taking the same classes. The math book costs $54.99; the history book, $67.42; and the art history book, $59.00. If the students share their textbooks, how much will each student pay?

54. A trans-Atlantic phone call to Paris cost $11.12. The phone company charges 85 cents for the first minute and 79 cents for each additional minute. How long was the conversation?

55. A block of 15 theater tickets costs $126.45. How much must each theater patron pay for a seat?

56. Cleano soap powder costs $2.36 for a 29.5 oz. box. Betto soap costs $1.89 for a 31.5 oz. box. Which box of soap is the better buy? How much per ounce would you save?

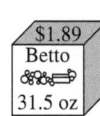 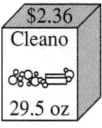

Use a calculator to perform the indicated operations. Round answers to hundredths.

57. 9.679 + 3.49 + 0.6325

58. 5.432 + 4.32 + 3.2 + 2 + 0.23

59. 0.396 ÷ (7 + 3.96)

60. 796.45 − 32.682

61. $5.97(-2.4 - 0.631)$

62. $(9.37 - 12.5)(-8)$

63. $3.3 - 4(0.22)$

64. $54.987 - 2(3.05 + 0.151)$

65. $-54.987 - 2(-3.05 + 0.151)$

66. $36.01 \div (4.101 - 1.01)$

67. $(2.3 + 1.2) \div 4$

68. $(7.6 \div 0.2)3.2$

69. $(3.04 + 2.66) \div (0.34 - 2.34)$

70. $9.8(12.63 - 128.9)$

71. $(7.01)(9.63) \div 2.8$

Application:

72. The distance around the square (the perimeter) is 8.6 inches. Find the length of one side by dividing the perimeter by 4.

7.5 Conversions Between Fractions and Decimals

OBJECTIVES

● Convert terminating decimals to fractions in lowest terms.

● Convert fractions to decimals.

● Simplify expressions containing both fractions and decimals.

When working with examples containing both decimals and fractions, it is necessary to work the example either as a fraction or as a decimal. In this section, we will learn to convert fractions to decimals and decimals to fractions.

A fraction is one of the notations used for division. The fraction $\frac{3}{2}$ is actually 3 divided by 2, and $\frac{3}{4}$ is 3 divided by 4. We will change the form $\frac{3}{2}$ by dividing 2 into 3. Remember that all whole numbers have a decimal point after the units place.

EXAMPLE 1

$$
\begin{array}{r}
1.5 \\
2{\overline{\smash{\big)}\,3.0}} \\
\underline{2} \\
1\,0 \\
\underline{1\,0} \\
\end{array}
$$

The fraction $\frac{3}{2}$ is equal to the decimal 1.5.

$$\frac{3}{2} = 1.5$$

EXAMPLE 2 $\frac{3}{4}$ is 3 divided by 4.

$$
\begin{array}{r}
0.75 \\
4\overline{)3.00} \\
2\,8 \\
\hline
20 \\
20 \\
\hline
\end{array}
$$

The fraction $\frac{3}{4}$ is equal to the decimal 0.75.

EXAMPLE 3 $\frac{5}{12}$ is 5 divided by 12.

$$
\begin{array}{r}
.4166 \qquad \text{6 will continue to repeat.} \\
12\overline{)5.0000} \\
4\,8 \\
\hline
20 \\
12 \\
\hline
80 \\
72 \\
\hline
80 \\
72 \\
\hline
\end{array}
$$

$\frac{5}{12}$ is equal to $0.41\overline{6}$ with a repeating bar over the 6.

$$\frac{5}{12} = 0.41\overline{6}$$

EXAMPLE 4 $\frac{5}{11}$ is 5 divided by 11.

$$
\begin{array}{r}
0.4545 \\
11\overline{)5.0000} \\
4\,4 \\
\hline
60 \\
55 \\
\hline
50 \\
44 \\
\hline
60 \\
55 \\
\hline
\end{array}
$$

The fraction $\frac{5}{11}$ is equal to the decimal $0.\overline{45}$ repeating. The repeating bar is over both the 4 and the 5.

$$\frac{5}{11} = 0.\overline{45}$$

RULE To change fractions to decimals
Divide the numerator by the denominator.

Note   When changing mixed numbers to decimal equivalents, the whole number stays the same. The proper fraction will change form into a decimal.

EXAMPLE 5

$$2\frac{1}{2} = 2.?$$

$$\begin{array}{r} 0.5 \\ 2\overline{)1.0} \\ \underline{1\ 0} \end{array}$$

$$2\frac{1}{2} = 2.5$$

Terminating decimals can be converted to fractions in lowest terms.

RULE **To change decimals to fractions**

1. Read the number.
2. Write it as a fraction.
3. Reduce it to lowest terms.

EXAMPLE 6 Read 0.75 as seventy-five hundredths. Write it as $\frac{75}{100}$. Reduce $\frac{75}{100}$ to lowest terms by dividing both the numerator and denominator by 25.

$$0.75 = \frac{3}{4}$$

USING THE CALCULATOR

Check the answer. Does $\frac{3}{4} = 0.75$?

Enter 3 $a^{b/c}$ 4 ▢=▢ 3 ⌋ 4 ▢2nd▢ → **0.75** ← Display

▢2nd▢ → **3 ⌋ 4** ← Display

EXAMPLE 7 Write 0.05 as a fraction in lowest terms.

$$0.05 = \frac{5}{100} = \frac{1}{20}$$

USING THE CALCULATOR

Enter 1 $a^{b/c}$ 20 ▢=▢ 1 ⌋ 20 ▢2nd▢ → **0.05** ← Display

▢2nd▢ → **1 ⌋ 20** ← Display

EXAMPLE 8 Write 0.6 as a fraction in lowest terms.

$$0.6 = \frac{6}{10} = \frac{3}{5}$$

EXAMPLE 9 Write 0.625 as a fraction in lowest terms.

$$0.625 = \frac{625}{1000} = \frac{25}{40} = \frac{5}{8}$$

EXAMPLE 10 Write 2.5 as a mixed number in lowest terms.

$$2.5 = 2 \text{ and } 5 \text{ tenths}$$

$$2\frac{5}{10} \text{ reduced to lowest terms} = 2\frac{1}{2}$$

Whole numbers remain the same in the fraction or decimal form of the number.

 USING THE CALCULATOR

What is $2\frac{1}{2}$ as a decimal?

Enter 2 $a^{b}/_{c}$ 1 $a^{b}/_{c}$ 2 = 2nd → 2.5 ← Display

2nd → **2 _ 1 ⌐ 2** ← Display

Problems containing both fractions and decimals can be solved using either fractions or decimals.

EXAMPLE 11 $\frac{1}{2}$ + 0.25

Method 1

$$0.25 = \frac{1}{4}$$

$$\frac{1}{2} + \frac{1}{4}$$

$$= \frac{2}{4} + \frac{1}{4}$$

$$= \frac{3}{4}$$

Method 2 $\frac{1}{2}$ + 0.25

$$\frac{1}{2} = 0.5$$

$$\begin{array}{r} 0.5 \\ + 0.25 \\ \hline 0.75 \end{array}$$

$\frac{3}{4}$ = 0.75   The answers will be equivalent if you use fractions or decimals.

 USING THE CALCULATOR

$$\frac{1}{2} + 0.25$$

Enter 1 $a^{b}/_{c}$ 2 + .25 = **0.75** ← Display

EXAMPLE 12 $\dfrac{1}{2} + 0.5\left(\dfrac{3}{5}\right)$ As this problems has two fractions and one decimal, it is easier to work it as a fraction.

$$0.5 = \dfrac{1}{2}$$

$$\dfrac{1}{2} + \left(\dfrac{1}{2}\right)\left(\dfrac{3}{5}\right)$$ Multiply.

$$\dfrac{1}{2} + \dfrac{3}{10}$$ Find the LCD.

$$\dfrac{5}{10} + \dfrac{3}{10}$$ Add.

$$\dfrac{8}{10} = \dfrac{4}{5}$$

USING THE CALCULATOR

$$\dfrac{1}{2} + 0.5\left(\dfrac{3}{5}\right)$$

Enter 1 a^{b}/c 2 + (.5 × 3 a^{b}/c 5) = **0.8** ← Display

Note   If the problem contains fractions that would become repeating decimals, work the problem using fractions. The answer will be more accurate.

EXAMPLE 13 $$3\left(\dfrac{1}{3}\right) = 1$$
$$3(0.3) = 0.9$$
$$3(0.33) = 0.99$$
$$3(0.333) = 0.999$$

EXAMPLE 14 $\dfrac{2}{3}(4.6 + 4.4)$ Even though there are two decimals and one fraction, work this
$\dfrac{2}{3}(9)$ problem as a fraction because $\dfrac{2}{3}$ will convert to a repeating
$= 6$ decimal. Inside the parentheses the decimals can be added to equal 9.

USING THE CALCULATOR

$$\dfrac{2}{3}(4.6 + 4.4)$$

Enter 2 a^{b}/c 3 × (4.6 + 4.4) = **6** ← Display

EXAMPLE 15

Method 1
$$\frac{1}{4}(2.4) + \frac{1}{5}(5.5) \quad \text{Multiply.}$$

$$\left(\frac{1}{4} \cdot \frac{24}{10}\right) + \left(\frac{1}{5} \cdot \frac{55}{10}\right)$$

$$\frac{3}{5} + \frac{11}{10} \qquad \text{Add.}$$

$$\frac{6}{10} + \frac{11}{10}$$

$$\frac{17}{10} = \frac{17}{10}$$

Method 2
$$\frac{1}{4}(2.4) + \frac{1}{5}(5.5)$$

$$0.25(2.4) + 0.2(5.5) \qquad \text{Multiply.}$$

$$0.6 + 1.1 \qquad \text{Add.}$$

$$1.7$$

USING THE CALCULATOR

$$\frac{1}{4}(2.4) + \frac{1}{5}(5.5)$$

Enter 1 $a^{b}/_{c}$ 4 $\times$ 2.4 $=$ **0.6** $\leftarrow$ Display $+$ 1 $a^{b}/_{c}$ 5 $\times$ 5.5

$=$ **1.7** $\leftarrow$ Display

PROBLEM SET 7.5

Change these fractions and mixed numbers to decimals (calculate until the number terminates or repeats). Check answers with a calculator.

1. $\frac{1}{2}$

2. $\frac{1}{3}$

3. $\frac{1}{4}$

4. $\frac{2}{5}$

5. $1\frac{1}{2}$

6. $\frac{7}{8}$

7. $\frac{2}{3}$

8. $\frac{3}{11}$

9. $3\frac{5}{6}$

10. $\frac{7}{5}$

11. $\frac{3}{7}$

12. $\frac{4}{9}$

13. $2\frac{1}{12}$

14. $\frac{3}{8}$

15. $2\frac{1}{8}$

16. $4\frac{3}{4}$

17. $\frac{1}{20}$

18. $\frac{7}{100}$

19. $\frac{3}{20}$

20. $\frac{61}{1000}$

21. $\frac{4}{5}$

22. $\frac{9}{8}$

23. $\frac{3}{4}$

24. $\frac{3}{5}$

25. $\frac{1}{6}$

26. $\frac{5}{9}$

27. $\frac{5}{6}$

28. $\frac{15}{16}$

29. $\frac{2}{17}$

30. $\frac{3}{40}$

Change these decimals to fractions or mixed numbers in lowest terms. Check answers with a calculator.

31. 0.7

32. 0.35

33. 0.8

34. 0.5

35. 0.05

36. 0.005

37. 0.0005

38. 0.25

39. 2.75

40. 3.2

41. 0.015

42. 0.625

43. 0.07

44. 0.065

45. 0.012

46. 0.17

47. 0.34

48. 0.875

49. 3.125

50. 0.0002

51. 3.12

52. 1.45

53. 236

54. 4.5

55. 7.004

56. 0.2

57. 0.02

58. 0.002

59. 0.0002

60. 0.00002

Use a calculator to simplify the following expressions.

61. $\dfrac{3}{4} + 0.5$

62. $\dfrac{3}{4} - 0.5$

63. $\dfrac{1}{3} + 0.25$

64. $\dfrac{1}{3} - 0.25$

65. $0.8 - \dfrac{4}{5}$

66. $3\dfrac{1}{2} \cdot 0.25$

67. $\left(\dfrac{7}{8}\right)2.4$

68. $(0.04)\left(\dfrac{3}{4} + 0.25\right)$

69. $\left(\dfrac{3}{8} - 0.125\right)\dfrac{2}{3}$

70. $1.7\left(2 - \dfrac{3}{17}\right)$

71. $\dfrac{1}{3} + 0.6\left(\dfrac{1}{3}\right)$

72. $\dfrac{3}{5} + 0.7\left(\dfrac{4}{7}\right)$

73. $\dfrac{1}{3}(1.2) + 0.5\left(\dfrac{2}{3}\right)$

74. $0.75\left(1\dfrac{1}{3}\right) + \dfrac{5}{8}(0.25)$

75. $\dfrac{2}{5}(0.86 + 2.83)$

76. $\left(\dfrac{3}{4} - 0.125\right)\left(0.5 - \dfrac{3}{8}\right)$

77. $-1 - 4.25 + \dfrac{34}{8}$

78. $-2 - 3.75 + \dfrac{15}{4}$

79. $0.9 - \dfrac{9}{25} + 0.18$

80. $5 - 0.5 + \dfrac{5}{8}$

81. $0.875 + \dfrac{5}{6} + \dfrac{1}{12}$

82. $0.75 - \dfrac{1}{2} + 0.25 + 5\dfrac{1}{2}$

83. $-\dfrac{1}{2} + 1.25 - \dfrac{7}{8} - 0.5$

84. $0.75 - \dfrac{1}{5} + 0.1$

85. $\frac{1}{3} + 5 - 0.2$

86. $0.9 - \frac{1}{2} + 0.25$

87. $2.5 + 7 + 0.4 - 20$

Applications:

88. Find the distance around this rectangle (the perimeter) by finding the sum of the sides.

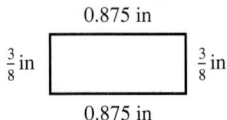

0.875 in

$\frac{3}{8}$ in $\frac{3}{8}$ in

0.875 in

89. Find the distance around this triangle (the perimeter) by finding the sum of the three sides.

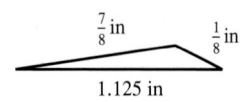

$\frac{7}{8}$ in $\frac{1}{8}$ in

1.125 in

90. This is a square. All four of its sides are the same length, 2.75 inches. Find the distance around this square (the perimeter) in both decimals and fractions.

2.75 in

SMALL GROUP ACTIVITY   *Estimating Grocery Shopping*

Materials: Grocery tapes with about ten items listed, but with the totals removed. Calculator.

One member of the group acts as a collector. Each member of the group estimates, within 30 seconds, the total of the grocery tape. The group member who estimates closest without going under the total calculated by the collector on a calculator is the winner.

CHAPTER 7 REVIEW

The terms listed below are among those used in this chapter. Write a brief definition for each term. (Refer to the appropriate section in the chapter as needed.)

1. Decimals _____

2. Irrational numbers _____

3. Scientific notation _____

4. Standard notation _____

After completing Chapter 7, you should be familiar with the following concepts and be able to apply them.

• *Recognize the position of the decimal point in the place value system (7.1).*
  Name the place value of the 5s in the following numbers.

5. 4.653 _____

6. 76.5 _____

7. 0.0005 _____

8. 5.36 _____

9. 946.753 _____

10. 23.24509 _____

11. 0.0000005 _____

12. 53 _____

• *Order decimals (7.1).*

Fill in a <, >, *or* = *symbol between the following pairs of numbers.*

13. 0.06   0.6 _____

14. 0.3 $0.\overline{3}$ _____

15. −0.6   −0.65 _____

16. 1.7   1.17 _____

17. 0.0004   0.004 _____

18. −0.72   −0.73 _____

19. 9.01   9.1 _____

20. 0.0006   0.00006 _____

• *Name decimals (7.1).*

• *Write decimals in words (7.1).*

Write the following in numbers.

21. Twenty-seven and twenty-seven thousandths _____

22. Four hundred and seven-hundredths _____

23. Five thousand and five-thousandths _____

24. Three hundred two and two-hundredths _____

25. Seven hundred sixty-three thousandths _____

Write the following numbers in words.

26. 75.75 _____

27. 930.003 _____

28. 4.5002 _____

29. 0.00678 _____

30. 12.4004 _____

• *Express numbers in scientific notation (7.1).*
Express in standard notation.

31. 6.73×10^6 _____

32. 2.4×10^{-5} _____

33. 6.08×10^{-3} _____

34. 7×10^9 _____

Express in scientific notation.

35. 7,860,000 _____

36. 3000 _____

37. 0.00000098 _____

38. 0.009684 _____

• *Round decimals (7.1).*

39. Round 36.279 to the nearest tenth. _____

40. Round 793.83 to the nearest ten. _____

41. Round 0.83709 to the nearest thousandth. _____

42. Round 99.999 to the nearest hundredth. _____

43. Round 99.999 to the nearest hundred. _____

44. Round 700.007 to the nearest tenth. _____

• *Add decimals (7.2).*
Add the following.

45. 2.83 + 17.4 + 362 _____

46. 0.3 + 3 + 3.03 _____

47. 3.648 + (−2.08) + 0.9376 _____

48. 0.003 + 300 + (−3.03) _____

49. −45.2 + (−3.5) + (−628) _____

50. 342.09 + (−71.309) + 0.261 _____

51. $-456 + 23.33 + 86.9$ _____

52. $0.009 + 0.981 + 972$ _____

• *Subtract decimals (7.2).*

Subtract the following.

53. $9.63 - 6.897$ _____

54. $26.3 - 9.37$ _____

55. $-7.84 - 83.7$ _____

56. $-15.2 - (-2.38)$ _____

57. $236.72 - 23.672$ _____

58. $7.86 - 19.4$ _____

59. $-3.28 - (-9.6)$ _____

60. $-6.73 - 6.73$ _____

• *Multiply decimals (7.3).*

Find the products.

61. $(0.3)1.2$ _____

62. $(56.1)(0.003)$ _____

63. $(0.0005)(0.0004)$ _____

64. $(-2.7)(4.03)$ _____

65. $(-9.32)(-0.11)$ _____

66. $(39.87)1000$ _____

67. $0.0098(100)$ _____

68. $79.00987(1000)$ _____

• *Divide decimals (7.4).*

Find the quotients. Round to the nearest hundredth.

69. $78.6 \div 3$ _____

70. $3.79 \div 0.05$ _____

71. $1.88 \div 8$ _____

72. $94 \div 5$ _____

73. 16.64 ÷ 3.2 _____

74. 7.26 ÷ 1.8 _____

75. 74.83 ÷ 1000 _____

76. 238.976 ÷ 100 _____

77. 0.009 ÷ 1000 _____

• *Convert fractions to decimals (7.5).*

Change each fraction to a decimal.

78. $\dfrac{4}{5}$ _____

79. $\dfrac{2}{3}$ _____

80. $\dfrac{2}{11}$ _____

81. $\dfrac{7}{8}$ _____

82. $\dfrac{5}{2}$ _____

83. $\dfrac{43}{100}$ _____

84. $\dfrac{7}{100}$ _____

85. $\dfrac{3}{8}$ _____

• *Convert terminating decimals to fractions (7.5).*

Change each decimal to a fraction in lowest terms.

86. 0.25 _____

87. 2.125 _____

88. 0.005 _____

89. 0.61 _____

90. 0.375 _____

91. 1.75 _____

92. 0.875 _____

93. 3.5 _____

• *Simplify expressions containing both fractions and decimals (7.5).*

Simplify.

94. $\dfrac{4}{3} - \dfrac{1}{6} + 0.75$ _____

95. $\dfrac{7}{8} + 1\dfrac{1}{2}\left(0.125 + \dfrac{3}{4}\right)$ _____

96. $0.5 - 2\dfrac{2}{3}\left(0.25 + \dfrac{3}{4}\right)$ _____

97. $\left(\dfrac{1}{6} + 0.5\right)(4 + 0.375)$ _____

98. $1 - 2.75 + \dfrac{11}{4}$ _____

99. $0.6 + \dfrac{7}{10} + 6.1$ _____

100. $0.3 - 8 - 9 + 0.2$ _____

101. $18 - 0.4 + 2.6 - 3$ _____

CHAPTER 7 PRACTICE TEST

1. Circle the tenths place and underline the hundredths place. 3923.049

2. Which decimal is larger: 0.6 or $0.\overline{6}$?

3. Change 0.005 to a fraction in lowest terms.

4. Which decimal is smaller -0.05 or -0.005?

5. Write in numbers: Four hundred seventy and seventy-seven thousandths.

6. Change $\frac{4}{9}$ to a decimal.

7. Write in words: 9004.0403

8. Round 708.056 to the nearest hundred.

9. Change 0.875 to a fraction in lowest terms.

10. Round 939.909 to the nearest tenth.

11. Change $\frac{3}{7}$ to a decimal rounded to the nearest thousandth.

12. Add: 49.93 + 604 + 0.793

13. Multiply: (0.35)(76.6)

14. Add: 349.004 + (−26.6) + (−45.3)

15. Divide: 1106.46 ÷ (−2.7)

16. Multiply: (−535.5)(9.2)

17. Subtract: 763.87 − 84.392

18. Divide and round to the nearest hundredth: 264.8 ÷ 1.7.

19. Subtract: −36.08 − 21.39

20. Subtract: −4.38 − (−7.89)

21. Simplify: 8 − 3.3 − 0.5 − 0.45

22. Simplify: 4.8 − 0.83 − 9.95

23. Simplify: $0.2 + \dfrac{4}{5} + \dfrac{7}{5}$

24. Simplify: $0.7\left(\dfrac{3}{8} + 0.125\right)$

25. Simplify: $\left(\dfrac{1}{3}\right)^2(5.4) + \left(\dfrac{1}{2}\right)^3(3.2)$

26. Express in scientific notation: 0.00000765

27. Express in standard notation: 2.48×10^6

28. Pens are sold for $23.64 a dozen. How much does each pen cost?

29. A phone call to Florida costs 22 cents for the first minute and 19 cents for each additional minute. How much will a 17-minute call cost?

30. Your math textbook cost $42.97; your English textbook, $23.45; and the three books required for your history class, $2.99, $13.54, and $10.02. How much did your textbooks cost this semester?

	THE MATHEMATICS JOURNAL	
	Midterm Progress Report	
Task:	As you reach the midpoint in the semester, take a few minutes to	
	reflect on how the semester is progressing for you.	
Starters:	What do you want to get out of the course? Are your plans helping	
	you accomplish your goals?	

8

Equations (Revisited) and Inequalities

8.1 Solving Equations Using Both the Addition and Multiplication Properties of Equality

8.2 Solving Equations Containing Fractions

8.3 Solving Equations Containing Decimals

8.4 Solving Linear Inequalities in One Variable

Introduction to Equations

We can trace the history of algebra back to about 1700 B.C. Originally, algebra, as developed by the Egyptians, Babylonians, and Arabs, was limited to the study of equations. It was not, however, until the seventeenth and eighteenth centuries in Europe that the symbols we use in equations including, for example, the equal sign as = originated. This development of notation as a way to express mathematical thought made possible the mathematics as we know it today.

8.1 Solving Equations Using Both the Addition and Multiplication Properties of Equality

OBJECTIVES

- Solve first-degree equations in one variable using the Addition Property of Equality

- Solve first-degree equations in one variable using the Multiplication Property of Equality.

- Solve first-degree equations in one variable using both the Addition and Multiplication Properties of Equalities.

In this section, we will continue our study of solving first-degree equations in one variable.

Solving Equations Using the Addition Property of Equality (Revisited)

In Section 4.8, we solved equations by applying the Addition Property of Equality. It states that if two real numbers (or algebraic expressions) are equal, and if the same real number (or algebraic expression) is added to both of them, then the sums formed will also be equal.

> **Addition Property of Equality**
> For all a, b, and $c \in \mathbb{R}$,
>
> $$\text{If } a = b, \text{ then } a + c = b + c$$

The equations we will discuss in this section may vary in appearance, but the procedure is the same for solving them.

> **RULE** To solve an equation using the addition property of equality
>
> 1. Simplify each side.
> 2. Apply the Addition Property of Equality as many times as needed to isolate the variable on one side of the equation.

EXAMPLE 1

$$0 = 15x + 32 - 14x$$

$$0 = x + 32 \qquad \text{Combine like terms.}$$

$$0 + (-32) = x + 32 + (-32) \qquad \text{Apply the Addition Property of Equality.}$$

$$-32 = x$$

EXAMPLE 2

$$\frac{30}{3} + x = 25 - 5$$

$$10 + x = 20 \qquad \text{Simplify each side.}$$

$$10 + x + (-10) = 20 + (-10) \qquad \text{Apply the property.}$$

$$x = 10$$

Note Some equations may have variables on both sides. To solve an equation of this type, begin by using the Addition Property of Equality to get all terms containing the variable on one side of the equation.

EXAMPLE 3

$$
\begin{array}{rl}
17x = & 16x + 25 \\
-16x & -16x \\
\hline
1x = & 25 \\
x = & 25
\end{array}
$$

EXAMPLE 4

$$
\begin{array}{rl}
5x + 17 = & 4x - 22 \\
-4x & -4x \qquad \text{Apply the Addition Property.} \\
\hline
1x + 17 = & -22 \\
-17 & -17 \qquad \text{Apply it again.} \\
\hline
1x = & -39 \\
x = & -39
\end{array}
$$

Solving Equations Using the Multiplication Property of Equality (Revisited)

In Section 5.7, we solved equations by applying the Multiplication Property of Equality. It states that if two real numbers (or algebraic expressions) are equal and if they are multiplied by the same real number (or algebraic expression), the products formed are also equal.

> **Multiplication Property of Equality** For all a, b, and $c \in R$,
>
> If $a = b$, then $a \cdot c = b \cdot c$.

Practice Problems

Solve for the variable.

1. $3a - 2a = 17$

2. $17b + 18 = 16b - 2$

3. $23 = 5x + (-4x)$

4. $0 = 30 - 3 \cdot 2 + y$

5. $x + (-5) = x$

6. $2y + 15 = 5 + 10 + y$

7. $5f - 3f = f + 41$

EXAMPLE 5

$$-a = 87$$
$$(-1) \cdot (-a) = 87 \cdot (-1) \qquad \text{Apply the Multiplication Property.}$$
$$1a = -87$$
$$a = -87$$

EXAMPLE 6

$$\frac{7}{8}y = 56$$

$$\frac{8}{7} \cdot \frac{7}{8}y = 56 \cdot \frac{8}{7} \qquad \text{Apply the Multiplication Property.}$$

$$\frac{8}{7} \cdot \frac{7}{8}y = \frac{\overset{1}{\cancel{7}} \cdot 8}{1} \cdot \frac{8}{\cancel{7}} \qquad \text{Simplify.}$$

$$y = 64$$

Solving Equations Using Both Properties

Often we need both the Addition and Multiplication Properties of Equality to find the solution sets to equations. In most of these cases, *first* we would use the Addition Property of Equality to isolate the term containing the variable on one side of the equation; *second,* we would use the Multiplication Property of Equality to get the coefficient of the variable to be 1.

EXAMPLE 7

$$2x + 7 = 15$$
$$\underline{\quad -7 \quad -7 \quad} \qquad \text{Apply the Addition Property.}$$
$$2x \quad = \quad 8$$
$$\frac{1}{2} \cdot 2x = 8 \cdot \frac{1}{2} \qquad \text{Apply the Multiplication Property.}$$

$$1x = \frac{\overset{4}{\cancel{8}}}{1} \cdot \frac{1}{\cancel{2}} \qquad \text{Simplify.}$$

$$x = 4$$

Note Many times we need to solve an equation with variables on both sides of it. In solving this type of equation, begin by performing the operations needed to get all terms containing the variable on one side of the equation.

EXAMPLE 8

$$3x - 4 = x + 6$$
$$\underline{-x \qquad\quad -x \quad} \qquad \text{Add } -x \text{ to both sides.}$$
$$2x - 4 = \quad 6$$
$$\underline{\qquad 4 \qquad\quad 4 \quad} \qquad \text{Add 4 to both sides.}$$
$$2x \quad = \quad 10$$
$$\frac{1}{2} \cdot 2x = 10 \cdot \frac{1}{2} \qquad \text{Multiply both sides by } \frac{1}{2}.$$

$$1x = \frac{\overset{5}{\cancel{10}}}{1} \cdot \frac{1}{\cancel{2}}$$

$$x = 5$$

Practice Problems

Solve for the variable.

8. $\frac{3}{4}p = 21$

9. $-\frac{2}{3}r = -28$

10. $\frac{6z}{7} = -\frac{24}{5}$

Note   The following example illustrates a type of equation in which the variable has a coefficient of -1 rather than a coefficient of $+1$. This type of equation may be solved several ways. We'll show two methods.

EXAMPLE 9

$$7 - x = 4 \qquad \text{Method 1}$$
$$\underline{-7 -7} \qquad \text{Isolate the term containing the variable.}$$
$$-x = -3$$
$$(-1)(-x) = (-1)(-3) \qquad \text{Multiply both sides by } -1.$$
$$x = 3$$

$$7 - x = 4 \qquad \text{Method 2}$$
$$\underline{+x +x} \qquad \text{Add the term containing the variable to both sides.}$$
$$ \text{Proceed as usual.}$$
$$7 = x + 4$$
$$\underline{-4 -4}$$
$$3 = x \quad \text{or} \quad x = 3$$

Note   No matter how complicated a linear equation looks, we can find the solution set by using the following procedure.

RULE **To solve an equation by using both properties**

1. Simplify each side. (Use the Distributive Property and the order of operations.)
2. Use the Addition and Multiplication Properties of Equality as many times as needed to isolate the variable. Begin with the Addition Property. Follow it with the Multiplication Property. (Remember to simplify after each step.)

EXAMPLE 10

$$11(x + 3) - 5 = -7 + 4x$$
$$11x + 11 \cdot 3 - 5 = -7 + 4x \qquad \text{Apply the Distributive Property.}$$
$$11x + 33 + (-5) = -7 + 4x \qquad \text{Simplify.}$$
$$11x + 28 = -7 + 4x$$
$$\underline{-4x -4x} \qquad \text{Use the Addition Property.}$$
$$7x + 28 = -7$$
$$\underline{ -28 -28} \qquad \text{Use the Addition Property.}$$
$$7x = -35$$
$$\frac{7x}{7} = \frac{-35}{7} \qquad \text{Use the Multiplication Property.}$$
$$x = -5$$

Practice Problems

Solve for the variable.

11. $2x + 7 = 3$

12. $3x - 8 = -5$

13. $7 = -2x + 3$

14. $12 + 5x = x + 7$

15. $5y - 10 = -y + 5$

EXAMPLE 11

$$13(2y - 1) + 19 = 5(2y - 8) + 10$$

$$26y - 13 + 19 = 10y - 40 + 10 \qquad \text{Use the Distributive Property.}$$

$$26y + 6 = 10y - 30$$

$$\underline{-10y \qquad\quad -10y} \qquad\qquad \text{Use the Addition Property.}$$

$$16y + 6 = \qquad -30$$

$$\underline{- 6 \qquad\qquad -6} \qquad\qquad \text{Use the Addition Property.}$$

$$16y \quad = \qquad -36$$

$$\frac{16y}{16} = -\frac{\overset{9}{\cancel{36}}}{\underset{4}{\cancel{16}}} \qquad\qquad \text{Use the Multiplication Property.}$$

$$y = -\frac{9}{4} \quad \text{or} \quad -2\frac{1}{4}$$

PROBLEM SET 8.1

Solve for the variable. Check your solution.

1. $21 + a - 2 = \dfrac{36}{6}$

2. $16 + b - 3 = \dfrac{24}{8}$

3. $30 + c - 15 = 10 \cdot 2$

4. $14 + f - 3 = 11 \cdot 3$

5. $11z + 2 - 10z = 41$

6. $7q + 8 - 6q = 29 + 13$

7. $20x + 17 = 19x - 5$

8. $8b - 10 = 6 + 7b$

9. $-5y = 60$

10. $-4z = 108$

11. $8x = -24$

12. $12w = -120$

13. $2x + 5 = 17$

14. $2x + 3 = 15$

15. $2y - 7 = 9$

16. $3y - 8 = 7$

17. $10x + 7 = -3$

18. $9x + 8 = -10$

19. $15 = 4x + 3$

20. $17 = 5x + 2$

21. $-12 = -6x + 6$

22. $-14 = -8x + 2$

23. $7y - 8 = 5y - 16$

24. $5x + 7 = 3x - 5$

25. $15z + 3 = 5z - 8$

26. $20s + 5 = 13s - 6$

27. $-7t + 2 = 4t - 1$

28. $-8t + 3 = 3t - 19$

29. $3(2y + 1) = 9$

30. $4(3y + 2) = 20$

31. $-2(10x + 7) = 26$

32. $-5(3x + 4) = 5$

33. $2(3y - 4) = 4y + 10$

34. $5(y + 2) = 3y + 12$

35. $2(3x + 2) - 12 = 3x - 11$

36. $3(2y + 10) - 8 = 3y + 7$

37. $3(x + 7) = 2(2x + 1)$

38. $2(x + 5) = 3(2x + 1)$

39. $-3(2q - 4) = 8q - 18$

40. $-2(4r + 7) = 2r - 20$

41. $2(3y + 2) - 12 = 3y - 11$

42. $4(2y - 1) + 11 = y - 7$

43. $3(2 + 5x) - (1 + 14x) = 6 - 12$

44. $2(3 + 2x) - (7 + 10x) = 5 - 9$

45. $4s + 5s - 3s + 8 + (-5s) = 12 - (-8)$

46. $10q + 7q - 3q + 5 - 12q = 16 - (-4)$

8.2 Solving Equations Containing Fractions

OBJECTIVE

● Solve first-degree equations in one variable containing fractions and/or mixed numbers.

In Sections 5.7 and 8.1, we solved some equations containing fractions. In most cases, these were equations in which the coefficient of the variable was a fraction. In a very real sense, you already know how to solve equations containing fractions. *No new properties are needed.* The advice for working with mixed numbers in equations remains the same: *Convert all mixed numbers to improper fractions.*

In this section, we will describe a new procedure to facilitate solving an equation containing two or more fractions. It consists of three steps.

RULE **To solve equations containing two or more fractions**

1. Find the LCD of the denominators of all fractions (also called LCM) in the equation.
2. Multiply both sides of the equation by the LCM found in Step 1.
3. Solve the resulting equation.

Practice Problems

Solve for the variable.

1. $\dfrac{1}{7}x + \dfrac{3}{7}x = \dfrac{4}{7}$

Note To review the procedure for finding the LCM/LCD, consult Section 5.5.

2. $x - \dfrac{1}{3} = \dfrac{1}{6}$

EXAMPLE 1 $\qquad \dfrac{x}{5} + \dfrac{2x}{5} = \dfrac{4}{5}$

3. $\dfrac{y}{3} - \dfrac{1}{4} = \dfrac{2}{5}$

$\qquad\qquad$ Find LCM. LCM (5, 5, 5) = 5

$$5\left(\frac{x}{5} + \frac{2x}{5}\right) = \frac{4}{5} \cdot 5 \qquad \text{Multiply both sides by LCM.}$$

4. $\dfrac{1}{4}(x - 6) = \dfrac{2}{3}$

$$\left(5 \cdot \frac{x}{5}\right) + \left(5 \cdot \frac{2x}{5}\right) = \frac{4}{5} \cdot 5 \qquad \text{Apply the Distributive Property.}$$

$$\left(\frac{\overset{1}{\cancel{5}}}{1} \cdot \frac{x}{\cancel{5}}\right) + \left(\frac{\overset{1}{\cancel{5}}}{1} \cdot \frac{2x}{\cancel{5}}\right) = \frac{4}{\cancel{5}} \cdot \frac{\overset{1}{\cancel{5}}}{1} \qquad \text{Simplify.}$$

$$1x + 2x = 4$$

$$3x = 4 \qquad \text{Combine similar terms.}$$

$$\frac{1}{3} \cdot 3x = 4 \cdot \frac{1}{3} \qquad \text{Apply the Multiplication Property.}$$

$$x = \frac{4}{3} \quad \text{or} \quad 1\frac{1}{3}$$

EXAMPLE 2 $\qquad \dfrac{y}{2} + \dfrac{y}{4} = \dfrac{30}{8}$

$\qquad\qquad$ Find the LCM. LCM (2, 4, 8) = 8.

$$8\left(\frac{y}{2} + \frac{y}{4}\right) = \frac{30}{8} \cdot 8 \qquad \text{Multiply both sides by LCM.}$$

$$8\left(\frac{y}{2}\right) + 8\left(\frac{y}{4}\right) = \frac{30}{8} \cdot \frac{8}{1} \qquad \text{Apply the Distributive Property.}$$

$$\left(\frac{\overset{4}{\cancel{8}}}{1} \cdot \frac{y}{\cancel{2}}\right) + \frac{\overset{2}{\cancel{8}}}{1} \cdot \frac{y}{\cancel{4}} = \frac{30}{\cancel{8}} \cdot \frac{\overset{1}{\cancel{8}}}{1} \qquad \text{Simplify.}$$

$$4y + 2y = 30$$

$$6y = 30 \qquad \text{Collect similar terms.}$$

$$\frac{6y}{6} = \frac{\overset{5}{\cancel{30}}}{\cancel{6}} \qquad \text{Apply the Multiplication Property.}$$

$$y = 5$$

EXAMPLE 3 $\quad 2\frac{1}{3}x - 6 = 1\frac{1}{4}x + 2\frac{2}{3}$

$$\frac{7}{3}x - 6 = \frac{5}{4}x + \frac{8}{3} \qquad \text{Convert mixed numbers to fractions.}$$

$$12\left(\frac{7}{3}x - 6\right) = 12\left(\frac{5}{4}x + \frac{8}{3}\right) \qquad \text{Multiply both sides by 12, the LCM.}$$

$$12\left(\frac{7}{3}x\right) - 12(6) = 12\left(\frac{5}{4}x\right) + 12\left(\frac{8}{3}\right) \qquad \text{Apply the Distributive Property.}$$

$$\frac{\overset{4}{\cancel{12}}}{1} \cdot \frac{7}{\underset{1}{\cancel{3}}}x - 72 = \frac{\overset{3}{\cancel{12}}}{1} \cdot \frac{5}{\underset{1}{\cancel{4}}}x + \frac{\overset{32}{\cancel{96}}}{\underset{1}{\cancel{3}}} \qquad \text{Simplify.}$$

$$\begin{array}{rcl} 28x - 72 &=& 15x + 32 \\ \underline{-15x \qquad\qquad} & & \underline{-15x \qquad\qquad} \qquad \text{Apply the Addition Property.} \\ 13x - 72 &=& \qquad 32 \\ \underline{+ 72 \qquad\qquad} & & \underline{+ 72 \qquad\qquad} \qquad \text{Apply the Addition Property again.} \\ 13x &=& \qquad 104 \end{array}$$

$$\frac{1}{13} \cdot 13x = 104 \cdot \frac{1}{13} \qquad \text{Apply the Multiplication Property.}$$

$$x = 8$$

Note   Try using a calculator to check the solutions to equations that look complicated.

⌨ **USING THE CALCULATOR**

Check the solution to this equation.

Does $2\frac{1}{3}x - 6 = 1\frac{1}{4}x + 2\frac{2}{3}$, for $x = 8$?

Enter 2 $\boxed{a^{b/c}}$ 1 $\boxed{a^{b/c}}$ 3 $\boxed{\times}$ 8 $\boxed{-}$ 6 $\boxed{=}$ **$12\frac{2}{3}$** ← Display

Enter 1 $\boxed{a^{b/c}}$ 1 $\boxed{a^{b/c}}$ 4 $\boxed{\times}$ 8 $\boxed{+}$ 2 $\boxed{a^{b/c}}$ 2 $\boxed{a^{b/c}}$ 3 $\boxed{=}$ **$12\frac{2}{3}$** ← Display

$12\frac{2}{3} = 12\frac{2}{3}$, so the solution is correct.

EXAMPLE 4 $\quad \frac{1}{4}(3n - 2) - \frac{1}{5}(2n + 1) = 0$

$$\frac{(3n - 2)}{4} - \frac{(2n + 1)}{5} = 0 \qquad \begin{array}{l}\text{Use the definition of division to} \\ \text{make the denominators stand} \\ \text{out. Find LCM. LCM(4, 5) = 20.}\end{array}$$

$$20\left[\frac{(3n - 2)}{4} - \frac{(2n + 1)}{5}\right] = 0 \cdot 20 \qquad \text{Multiply both sides by 20.}$$

$$20 \cdot \frac{(3n - 2)}{4} - 20 \cdot \frac{(2n + 1)}{5} = 0 \cdot 20 \qquad \text{Apply the Distributive Property.}$$

$$\frac{\overset{5}{\cancel{20}}}{1} \cdot \frac{(3n-2)}{\underset{1}{\cancel{4}}} - \frac{\overset{4}{\cancel{20}}}{1} \cdot \frac{(2n+1)}{\underset{1}{\cancel{5}}} = 0$$ Simplify.

$$5(3n-2) - 4(2n+1) = 0$$ Don't forget the subtraction sign.

$$15n - 10 - 8n - 4 = 0$$

$$7n - 14 = 0$$

$$7n - 14 + 14 = 0 + 14$$ Apply the Addition Property.

$$7n = 14$$

$$\frac{7n}{7} = \frac{14}{7}$$ Apply the Multiplication Property.

$$n = 2$$

● PROBLEM SET 8.2

In your own words, explain the procedure for solving equations containing fractions and/or mixed numbers.

Solve for the variable. Show all work. Check the solution set.

1. $\dfrac{3}{8}x + 1 = \dfrac{1}{8}$

2. $\dfrac{2}{5}x + 3 = \dfrac{1}{5}$

3. $\dfrac{3}{4}y - 2 = \dfrac{3}{4}$

4. $\dfrac{5}{6}y - 1 = \dfrac{1}{6}$

5. $\dfrac{1}{3}x + \dfrac{2}{3} = 5$

6. $\dfrac{1}{6}y + \dfrac{5}{6} = 2$

7. $\dfrac{2}{9}y - \dfrac{1}{9} = 3$

8. $\dfrac{3}{10}x - \dfrac{1}{10} = 4$

9. $4x + \dfrac{1}{2} = \dfrac{1}{4}$

10. $5y + \dfrac{1}{5} = \dfrac{1}{10}$

11. $3x - \dfrac{1}{3} = \dfrac{1}{6}$

12. $6x - \dfrac{1}{8} = \dfrac{1}{16}$

13. $\dfrac{y}{3} + \dfrac{1}{2} = -\dfrac{1}{2}$

14. $\dfrac{x}{2} + \dfrac{4}{3} = -\dfrac{2}{3}$

15. $\dfrac{x}{5} + x = 4$

16. $\dfrac{x}{3} + x = 8$

17. $\dfrac{x}{3} + \dfrac{x}{6} = 4$

18. $\dfrac{x}{2} + \dfrac{x}{5} = 7$

19. $\dfrac{w}{7} - \dfrac{w}{2} = 1$

20. $\dfrac{v}{4} - \dfrac{v}{6} = 2$

21. $\dfrac{3x}{5} + \dfrac{7x}{6} = \dfrac{2}{3}$

22. $\dfrac{2x}{3} + \dfrac{3x}{4} = \dfrac{6}{5}$

23. $\dfrac{2y}{7} - \dfrac{3y}{5} = 1$

24. $\dfrac{5y}{8} - \dfrac{2y}{3} = 4$

25. $\dfrac{3y}{5} - \dfrac{1}{2} = \dfrac{y}{6}$

26. $\dfrac{4y}{3} - \dfrac{1}{4} = \dfrac{y}{6}$

27. $\dfrac{x}{2} - 3 = 7 - \dfrac{x}{3}$

28. $\dfrac{x}{4} - 2 = 5 - \dfrac{x}{5}$

29. $\dfrac{1}{2}(y - 5) = 3$

30. $\dfrac{1}{3}(y - 2) = 4$

31. $\dfrac{1}{3}(v + 8) - \dfrac{1}{4}(3 - 2v) = \dfrac{1}{6}$

32. $\dfrac{1}{3}(6w - 9) = \dfrac{1}{2}(8w - 4)$

33. $1\dfrac{1}{2}x + 2 = 3$

34. $2\dfrac{1}{3}x + 1 = 4$

35. $2\dfrac{1}{4}y - 6 = 8$

36. $3\dfrac{1}{5}y - 2 = 3$

37. $\dfrac{7}{6}x + \dfrac{5}{2} = \dfrac{10}{3}$

38. $\dfrac{9}{8}x + \dfrac{9}{4} = \dfrac{7}{2}$

39. $\dfrac{17}{8}y - \dfrac{7}{6} = \dfrac{9}{2}$

40. $\dfrac{9}{4}y - \dfrac{3}{2} = \dfrac{25}{8}$

41. $\dfrac{25}{12}x + \dfrac{43}{6} = \dfrac{11}{6}x$

42. $\dfrac{6}{5}y + \dfrac{43}{8} = \dfrac{13}{5}y$

43. $\dfrac{29}{6}x - \dfrac{13}{2} = \dfrac{10}{3}x$

44. $\dfrac{21}{8}x - \dfrac{49}{12} = \dfrac{5}{4}x$

45. $x + \dfrac{5}{3} = 3x + \dfrac{49}{4}$

46. $y + \dfrac{9}{4} = 2y + \dfrac{51}{5}$

47. $q - \dfrac{21}{10} = 5q + \dfrac{16}{5}$

48. $z - \dfrac{28}{9} = 4z + \dfrac{37}{6}$

49. $2\left(\dfrac{2}{3}x + \dfrac{5}{4}\right) = \dfrac{37}{12}$

50. $3\left(\dfrac{3}{4}x + \dfrac{7}{3}\right) = \dfrac{61}{12}$

8.3 Solving Equations Containing Decimals

OBJECTIVES

● Solve first-degree equations in one variable containing decimals.

● Solve first-degree equations in one variable containing both fractions and decimals.

Equations containing decimals can be solved by working with the decimals or by using the Multiplication Property of Equality to clear the equation of decimals. We will describe both methods in this section. In solving equations with decimals, choose the method that is more comfortable for you.

EXAMPLE 1   Method 1

$$0.35 + x = 2.31$$ Clear the decimals.

$$100(0.35 + x) = (2.31)100$$ Multiply both sides by 100.

$$100(0.35) + 100x = 231$$ Apply the Distributive Property.

$$35 + 100x = 231$$ Solve the resulting equation.

$$\underline{-35 \qquad\qquad -35}$$

$$100x = 196$$

$$\dfrac{100}{100}x = \dfrac{196}{100}$$

$$x = 1.96$$

Method 2

$$0.35 + x = 2.31$$ Keep the decimals, and apply the

$$\underline{-0.35 \qquad -0.35}$$ Addition Property.

$$x = 1.96$$

USING THE CALCULATOR

Check the solution to this equation.

$$0.35 + x = 2.31$$

Does $0.35 + \mathbf{1.96} = 2.31?$

Enter 0.35 $\boxed{+}$ 1.96 $\boxed{=}$ **2.31** ← Display

The solution is correct.

Practice Problems

Solve for the variable.

1. $0.24 + x = 3.64$

2. $3.6 + y = 1.38$

3. $4.62 = 0.3x - 1.23$

4. $7 - 0.25z = -10$

EXAMPLE 2   Method 1

$4.5 - 0.2x = 2.34$	Clear the decimals.
$4.50 - 0.20x = 2.34$	Express all decimals as hundredths.
$100(4.50 - 0.20x) = (2.34)100$	Multiply both sides by 100.
$100(4.50) - 100(0.20x) = 234$	Simplify.
$450 - 20x = 234$	Solve this equation.

$$\begin{array}{rr} -450 & -450 \\ \hline -20x = & -216 \end{array}$$

$$\frac{-20}{-20}x = \frac{-216}{-20}$$

$$x = 10.8$$

Method 2

$4.5 - 0.2x = 2.34$	Keep the decimals.

$$\begin{array}{rr} -4.5 & -4.50 \\ \hline -0.2x = & -2.16 \end{array}$$

$$\frac{-0.2}{-0.2}x = \frac{-2.16}{-0.2}$$

$$x = 10.8$$

USING THE CALCULATOR

Check the solution.

$$4.5 - 0.2x = 2.34$$

Does $4.5 - 0.2(\mathbf{10.8}) = 2.34?$

Enter 4.5 $\boxed{-}$ 0.2 $\boxed{\times}$ 10.8 $\boxed{=}$ **2.34** ← Display

The solution is correct.

RULE **To clear an equation of decimals**

1. Express all decimals to the same number of decimal places.
2. Multiply both members by 10 raised to the power needed to remove the decimals, i.e., if the decimals are expressed in tenths, multiply by 10; if the decimals are expressed in hundredths, multiply by 100.

Solving Equations Containing Both Decimals and Fractions

Two approaches can be used to solve equations that contain both decimals and fractions. The methods follow:

1. Convert the decimals to fractions, and solve the resulting equation, using the procedure for solving equations containing fractions; or
2. Convert the fractions to decimals, and solve the resulting equation, using one of the methods described for solving equations containing decimals.

EXAMPLE 3   Method 1

$$0.4x - 3 = \frac{1}{5}$$ Convert the decimal to a fraction.

$$\frac{4}{10}x - 3 = \frac{1}{5}$$ Solve this equation.

$$\frac{2}{5}x - 3 = \frac{1}{5}$$

$$5\left(\frac{2}{5}x - 3\right) = \left(\frac{1}{5}\right)5$$

$$5 \cdot \frac{2}{5}x - 5 \cdot 3 = 1$$

$$2x - 15 = 1$$

$$2x - 15 + 15 = 1 + 15$$

$$2x = 16$$

$$\frac{2}{2}x = \frac{16}{2}$$

$$x = 8$$

Method 2

$$0.4x - 3 = \frac{1}{5}$$ Convert the fraction to a decimal.

$$0.4x - 3 = 0.2$$ Solve this decimal equation.

$$10(0.4x - 3) = (0.2)10$$

$$10(0.4x) - 10 \cdot 3 = 2$$

$$4x - 30 = 2$$

$$4x - 30 + 30 = 2 + 30$$

$$4x = 32$$

$$\frac{4}{4}x = \frac{32}{4}$$

$$x = 8$$

Note   *Do not* convert the fractions to decimals if any of the fractions would convert to repeating decimals, that is, fractions with denominators that have factors other than powers of 2 and 5. This would introduce approximations, and the resulting equation would not be equivalent to the original one.

Practice Problems

Solve for the variable.

5. $x + 0.8 = \dfrac{3}{4}$

6. $0.5y - \dfrac{1}{2} = 21$

7. $0.16 - \dfrac{2}{3}t = 4$

8. $1.2r + 1.5 = \dfrac{2}{5}$

9. $0.5q + 2.7 = \dfrac{1}{6}$

EXAMPLE 4 $0.2x + \dfrac{1}{3} = 1.4$ Convert the decimals to fractions.

$$\dfrac{2}{10}x + \dfrac{1}{3} = \dfrac{14}{10}$$ Using decimals would result in an approximate solution, since $\dfrac{1}{3} = .\overline{3}$

$$30\left(\dfrac{2}{10}x + \dfrac{1}{3}\right) = \left(\dfrac{14}{10}\right)30$$

$$30 \cdot \dfrac{2}{10}x + 30 \cdot \dfrac{1}{3} = 14 \cdot 3$$

$$2 \cdot 3x + 10 = 42$$

$$6x + 10 - 10 = 42 - 10$$

$$6x = 32$$

$$\dfrac{1}{6} \cdot 6x = 32 \cdot \dfrac{1}{6}$$

$$x = \dfrac{16}{3} \quad \text{or} \quad 5\dfrac{1}{3}$$

ANSWERS TO PRACTICE PROBLEMS
1. $x = 3.4$ 2. $y = -2.22$
3. $x = 19.5$ 4. $z = 68$
5. $x = -0.05$ or $-\dfrac{1}{20}$
6. $y = 43$ 7. $t = -5.76$
8. $r = -\dfrac{11}{12}$
9. $q = -\dfrac{76}{15}$ or $-5\dfrac{1}{15}$

PROBLEM SET 8.3

Solve for the variable. Use a calculator to check each solution.

1. $6.054 = 0.43 + x$

2. $0.079 = 0.308 - x$

3. $4.83 = y + 2.107$

4. $-4.83 = y - 2.107$

5. $2z + 0.5 = 0.75$

6. $0.05z + 17 = 35$

7. $0.5q + 17 = 14$

8. $8.2q - 14 = 17.8$

9. $0.09 = 0.27 + 3s$

10. $0.9s + 1.1 = 0.2$

11. $-7.063 = 9.3 - x$

12. $-3.4x + 9.2 = -7.1$

13. $3 + 0.25x = 0.5x$

14. $12 + 0.62x = -0.7x$

15. $1.6h - 15.4 = 2.8h + 22.4$

16. $7g + 23 = 8g - 1.7$

17. $-1.5c - 1 = -0.5c + 9$

18. $0.5c + 0.13 = 0.25c + 1$

19. $0.45x + 0.35x - 17 = -0.15x$

20. $4.1x + 0.7x - 2.3 = -8.9x$

21. $0.35y + 0.65(70 - y) = 70(0.55)$

22. $5x + 5(x - 0.04) = 0.45$

23. $-x - 4.25 = \dfrac{34}{8}$

24. $-2x - 3.75 = \dfrac{15}{4}$

25. $0.9y - \dfrac{9}{25} = 0.18y$

26. $5y - 0.5 = \dfrac{5}{8}$

27. $0.875z + \dfrac{5}{6} = \dfrac{25}{12}$

28. $0.75z - \dfrac{1}{2} = 0.25z + \dfrac{11}{2}$

29. $-\dfrac{1}{2} + 1.25q = \dfrac{7}{8} - 0.5q$

30. $-\dfrac{3}{4} + 0.24q = \dfrac{3}{20} - 0.6q$

31. $\dfrac{1}{3}s + 5 = 0.2$

32. $\dfrac{1}{6}s + 8 = 0.3$

33. $0.9 - \dfrac{1}{2}r = 0.25r$

34. $2.5a + 0.7 = \dfrac{3}{4}a$

8.4 **Solving Linear Inequalities in One Variable**

OBJECTIVES

● Solve first-degree inequalities in one variable by using the Addition Property of Inequalities.

● Solve first-degree inequalities in one variable by using the Multiplication Properties of Inequalities.

● Solve first-degree inequalities in one variable by using a combination of the Addition and Multiplication Properties of Inequalities.

In Sections 8.1, 8.2, and 8.3 of this chapter, we focused on solving equations. In our daily lives, however, we must often deal with quantities that are not equal. For example, we may be aware that we need at least $600 each month to pay our bills or that we are considered college freshmen if we have accumulated less than 33 semester credit hours. In this section, we will consider linear inequalities in one variable by continuing the discussion on order on the number line that we began in Section 4.1.

Recall the Principle of Trichotomy, also known as the Axiom of Comparison, from Section 4.1. It states that given ay two real numbers, the first is either greater than the second, equal to the second, or less than the second.

Principle of Trichotomy
For all a and $b \in R$, exactly one of these statements is true.

$$a < b, \text{ or } a = b, \text{ or } a > b$$

From Section 4.1, we learned that less than (<) can be interpreted as to the left of on the number line and that greater than (>) can be interpreted as to the right of on the number line. Previously, we restricted our use of inequalities to compare numbers whose values we knew. For example, we would state that $3 < 5$ and that $-22 > -84$.

In this section, we will extend our use of inequalities to ranges of values we will indicate by variables. Unless indicated otherwise, the set of real numbers is the replacement set for the variables used, that is, for all $x \in R$. First, we will extend the concept of an inequality, and then we will introduce a new notation to indicate the solution sets of inequalities.

DEFINITION   The symbol $\geq$ is read greater than or equal to.

EXAMPLE 1 $x \geq 1$ is read x greater than or equal to 1. It refers to all real numbers, x, which are either equal to 1 or greater than 1.

DEFINITION   The symbol $\leq$ is read less than or equal to.

EXAMPLE 2 $x \leq -5$ is read x less than or equal to -5. It refers to all real numbers, x, which are either equal to (-5) or less than (-5).

Using Set Builder Notation

DEFINITION   **Set builder notation:** Notation that represents the solution of an equation or inequality in terms of sets.

EXAMPLE 3   The statement $x = 5$, the solution to the equation $2x = 10$, is written $\{x \mid x = 5\}$ in set builder notation and is read, the set of all x, such that $x = 5$.

EXAMPLE 4   The statement $x > 3$ is written $\{x \mid x > 3\}$.

EXAMPLE 5   The statement $x \leq -7$ is written $\{x \mid x \leq -7\}$.

In first-degree equations, the solutions were limited to one real number, all real numbers, or no real numbers (see Sections 4.8, 5.7, 8.1, 8.2, and 8.3). In first degree *inequalities*, the solutions often include more than one real number—a range of numbers that fulfills the given conditions. Set builder notation is a way to emphasize that the solution is a range of numbers rather than a unique number. In the previous example, writing the solution as $\{x \mid x > 3\}$ emphasizes that the solution set is all real numbers greater than 3.

Graphing Inequalities on the Number Line

Recall from Section 1.4 that we graph a specific real number by drawing a filled-in circle to indicate the corresponding point on the number line.

EXAMPLE 6   Graph $\{x \mid x = 2\}$ on a number line.
Method

1. Choose a point on the number line. Name it 0.
2. Locate a point 2 units to the right of 0.
3. Draw a filled-in circle at this position.

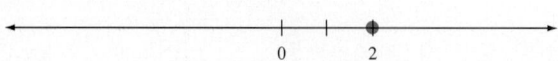

EXAMPLE 7 Graph $\{x \mid x = -5\}$ on a number line.
Method

1. Choose a point on the number line. Name it 0.
2. Locate a point 5 units to the left of 0.
3. Draw a filled-in circle at this position.

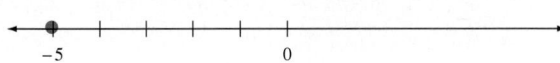

EXAMPLE 8 Graph $\{x \mid x \geq 3\}$ on a number line.
    Since $x \geq 3$ means $x = 3$ or $x > 3$, follow this method:

1. Choose a point on the number line. Name it 0.
2. Locate a point 3 units to the right of 0.
3. Draw a filled-in circle at this position.
4. Draw a horizontal arrow beginning at the circle and continuing to the right on the number line.

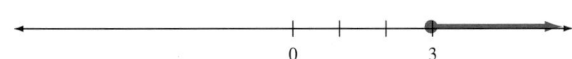

RULES To graph $\{x \mid x \leq a\}$ on a number line

1. Choose a point on the number line. Name it 0.
2. Locate a point a units from 0.
3. Draw a filled-in circle at this position.
4. Draw a horizontal arrow beginning at the circle and continuing to the left on the number line.

To Graph $\{x \mid x \geq a\}$ on a Number Line

1. Choose a point on the number line. Name it 0.
2. Locate a point a units from 0.
3. Draw a filled-in circle at this position.
4. Draw a horizontal arrow beginning at the circle and continuing to the right on the number line.

Note   A filled-in circle is also referred to as a closed circle.

EXAMPLE 9 Graph $x \leq -5$ on the number line.
    Since $x \leq -5$ means $x = -5$ or $x < -5$, follow this method:

1. Choose a point on the number line. Name it 0.
2. Locate a point 5 units to the left of 0.
3. Draw a filled-in circle at this position.
4. Draw a horizontal arrow beginning at the circle and continuing to the left of the number line.

EXAMPLE 10 Graph $x < 10$ on the number line.
  Since $x = 10$ is not included, follow this method.

1. Choose a point on the number line. Name it 0.
2. Locate a point 10 units to the right of 0.
3. Draw an open circle at this position.
4. Draw a horizontal arrow beginning at the circle and continuing to the left on the number line.

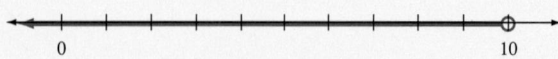

RULES To graph $\{x \,|\, x < a\}$ on a number line

1. Choose a point on the number line. Name it 0.
2. Locate a point a units from 0.
3. Draw a open circle at this position.
4. Draw a horizontal arrow beginning at the circle and continuing to the left on the number line.

To graph $\{x \,|\, x > a\}$ on a number line

1. Choose a point on the number line. Name it 0.
2. Locate a point a units from 0.
3. Draw a open circle at this position.
4. Draw a horizontal arrow beginning at the circle and continuing to the right on the number line.

EXAMPLE 11 Graph $x > -6$ on the number line.
  Since $x = -6$ is not included, follow this method:

1. Choose a point on the number line. Name it 0.
2. Locate a point 6 units to the left of 0.
3. Draw an open circle at this position.
4. Draw a horizontal arrow beginning at the circle and continuing to the right on the number line.

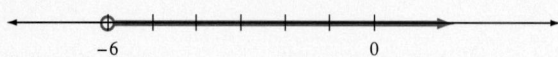

An English phrase for $\leq$ is *at most* and an English phrase for $\geq$ is *at least*.

Write an algebraic statement for each sentence in Applications 1 and 2.

APPLICATION 1

A child must be at least 40 inches tall to ride the Yellow Streak at the local amusement park.

  Let h = height of a child in inches.

Answer: $h \geq 40$ is the algebraic statement. ◆

APPLICATION 2
The FDIC insures at most $100,000 per bank account.

Let d = the amount in dollars in a bank account insured by FDIC.

Answer: $d \leq 100,000$ ◆

Properties of Inequalities

A few fundamental properties govern the operations we can perform on inequalities. We'll discuss two of them—the Addition and the Multiplication Properties of Inequalities.

Practice Problems

Solve and graph.

4. $x + 7 > 5$

5. $x - 25 \leq -27$

6. $x + 80 \geq 91$

Addition Property of Inequalities   For a, b, and $c \in R$,

$$\text{If } a < b,$$
$$\text{then } a + c < b + c.$$
$$\text{and}$$
$$\text{If } a > b,$$
$$\text{then } a + c > b + c.$$

When any real number is added to both sides of an inequality, the sums are unequal in the same order.

EXAMPLE 12 Solve and graph $\{x \mid x + 7 < 15\}$.

$$x + 7 < 15$$
$$x + 7 + (-7) < 15 + (-7)$$
$$x + 0 < 8$$
$$x < 8$$
$$\{x \mid x < 8\}$$

EXAMPLE 13 Solve and graph $\{x \mid x - 30 > -27\}$.

$$x - 30 > -27$$
$$\underline{+30 \qquad +30}$$
$$x \quad > \quad 3$$
$$\{x \mid x > 3\}$$

Using a combination of the Addition property of Equality and the Addition Property of Inequalities, we can solve inequalities of the form $x + b \leq c$ and $x + b \geq c$.

EXAMPLE 14 Solve and graph $\{x \mid 15 + x \leq 19\}$.

$$15 + x \leq 19$$
$$15 + x + (-15) \leq 19 + (-15)$$
$$x \leq 4$$
$$\{x \mid x \leq 4\}$$

Multiplication Properties of Inequalities For a, b, and $c \in$ R,

$$\text{If } a < b \text{ and } c > 0,$$
$$\text{then } a \cdot c < b \cdot c.$$
$$\text{and}$$
$$\text{If } a < b, \text{ and } c < 0,$$
$$\text{then } a \cdot c > b \cdot c.$$

Notice that two properties are stated here. The first property states that multiplying both sides of an inequality by a positive number does not affect the order of the inequality.

The second property states that multiplying both sides of an inequality by a negative number reverses the order of the inequality. To see the reasonableness of this property, consider this example. We know that $2 < 3$. Now multiply both sides by (-1). How do the products compare? How does (-2) compare to (-3)?

$$2 < 3$$
$$(-1) \cdot 2 \ ? \ (-1) \cdot 3$$
$$-2 \ ? \ -3$$

As we know (see Section 4.1), $-2 > -3$.

Because $b < a$ implies that $a > b$ (see Section 4.1), the properties may also be stated in this manner:

Multiplication Properties of Inequalities For a, b, and $c \in$ R,

$$\text{If } a > b \text{ and } c > 0,$$
$$\text{then } a \cdot c > b \cdot c.$$
$$\text{and}$$
$$\text{If } a > b \text{ and } c < 0,$$
$$\text{then } a \cdot c < b \cdot c.$$

EXAMPLE 15 Solve and graph $\{x \mid 5x < 15\}$.

$$5x < 15$$
$$\frac{1}{5} \cdot 5x < 15 \cdot \frac{1}{5}$$
$$x < 3$$

Note Be sure to reverse the inequality symbol as soon as you multiply by a negative number.

EXAMPLE 16 Solve and graph: $\{x \mid -3x > -18\}$.

$$-3x > -18$$
$$\left(-\frac{1}{3}\right)(-3x) < (-18)\left(-\frac{1}{3}\right)$$
$$x < 6$$

Using one of the Multiplication Properties of Inequalities, we can solve inequalities of the form $ax \leq c$ and $ax \geq c$.

EXAMPLE 17 Solve and graph $\{x \mid 2x \leq 28\}$.

$$2x \leq 28$$

$$\frac{1}{2} \cdot 2x \leq \frac{1}{2} \cdot 28$$

$$x \leq 14$$

$$\{x \mid x \leq 14\}$$

EXAMPLE 18 Solve and graph $\left\{ x \mid \frac{1}{5}x \geq -3 \right\}$.

$$\frac{1}{5}x \geq -3$$

$$5 \cdot \frac{1}{5}x \geq 5 \cdot (-3)$$

$$x \geq -15$$

$$\{x \mid x \geq -15\}$$

Solving Inequalities Using Both Properties

The method we'll use to solve inequalities using both properties is similar to the one we followed for solving equations using both properties (see Section 8.1).

RULE **To solve inequalities using both properties**

1. Use the Addition Property of Inequalities to isolate the term containing the variable on one side of the inequality.
2. Use one of the Multiplication Properties of Inequalities to obtain a coefficient of 1 for the variable.

EXAMPLE 19 Solve and graph $2x + 7 < 3$.

$$2x + 7 - 7 < 3 - 7$$

$$2x < -4$$

$$\frac{1}{2} \cdot 2x < -4 \cdot \frac{1}{2}$$

$$x < -2$$

$$\{x \mid x < -2\}$$

EXAMPLE 20   Solve and graph $-6x + 6 \geq 12$.

$$-6x + 6 - 6 \geq 12 - 6$$
$$-6x \geq 6$$
$$\left(-\frac{1}{6}\right)(-6x) \leq 6\left(-\frac{1}{6}\right)$$
$$x \leq -1$$
$$\{x \mid x \leq -1\}$$

Note   The inequalities we have solved in this section have been written so that the variable is in the left hand member (on the left side) of the inequality. However, the inequalities may also be written so that the variable is in the right hand member. Here's one method for solving this type of problem: Use the fact that if $a < b$, then $b > a$ to rewrite the inequality so that the variable is in the left hand member before solving it.

EXAMPLE 21   Solve and graph $35 < 5x$.

$$5x > 35 \quad \text{Rewrite the inequality with } x \text{ on the left side.}$$
$$\frac{1}{5} \cdot 5x > \frac{1}{5} \cdot 35 \quad \text{Solve for } x.$$
$$x > 7$$

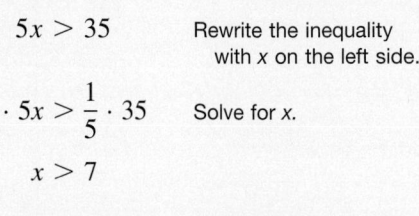

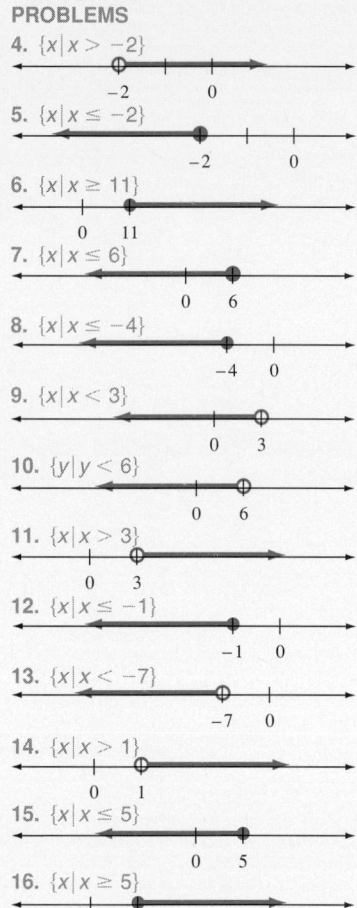
PROBLEM SET 8.4

Express using set builder notation.

1. $y = 56$ **2.** $z = 102$

3. $x = -17$ **4.** $g = -29$

5. $r \leq 18$ **6.** $s \geq 21$

7. $c > -51$ **8.** $d < -47$

9. $x \geq 0$ **10.** $y \leq 0$

Solve and graph.

11. x + 7 < 12

12. $x - 3 > 1$

13. $x - 8 \geq -7$

14. $x + 8 \leq -1$

15. $2 + x \leq 10$

16. $5 + x \geq 11$

17. $5x > 60$

18. $4x < 48$

19. $9x \geq -36$

20. $8x \leq -56$

21. $-2x < 16$

22. $-3x > 18$

23. $-14x \geq -28$

24. $-13x \leq -39$

25. $2x - 7 \leq 9$

26. $3x - 8 \geq 19$

27. $10x + 7 < -3$

28. $9x + 8 > -10$

29. $4x + 3 \geq 15$

30. $5x + 2 \leq 17$

31. $-6x + 6 \leq -12$

32. $-8x + 2 \geq -14$

33. $-14 - 20x < 26$

34. $-20 - 15x > 5$

35. $8 + 12x \geq 20$

36. $12 - 5y \leq -48$

37. $-3x - 6 < -6$

38. $-4x - 9 > -9$

39. $-x + 5 \geq 5$

40. $-x + 7 \leq 7$

41. $15 > 3x$

42. $21 < 7x$

43. $9 \leq -8x - 7$

44. $22 \geq -5x - 3$

SMALL GROUP ACTIVITY   *The Coin Toss*

Tossing a coin to decide which of two teams goes first, or which one gets the ball, or even which team gets to decide if it wants the ball is a common practice at some sporting events. Tossing a coin is viewed as a fair way to make decisions when two teams are involved because each team has an equal chance of guessing the result of the toss. The assumption is that the coin will come up heads in $\frac{1}{2}$ of the tosses and tails in $\frac{1}{2}$ of the tosses. Will this actually happen? Let's test the theory.

Activity 1:
Have each member of the group toss a coin 10 times and tally the results.

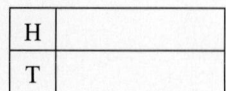

Next, record the results as fractions: $\dfrac{\#H}{10} = \dfrac{}{10}$ $\dfrac{\#T}{10} = \dfrac{}{10}$.

How close to $\frac{1}{2}$ is each of these results?

Activity 2:
Combine the 10 tosses of the first group member with those of the second member.

Record the results as: $\dfrac{\#H}{20} = \dfrac{}{20}$ $\dfrac{\#T}{20} = \dfrac{}{20}$.

Is each of these results closer to $\frac{1}{2}$?

Activity 3:
Continue combining the tosses until all the members' tosses have been used. Keep recording the results, and keep asking, "Is each of these results closer to $\frac{1}{2}$?"

What conclusions can you draw from this experiment? Do you still think tossing a coin is a good way to start a game between two teams? Why or why not?

CHAPTER 8 REVIEW

After completing Chapter 8, you should have mastered the objectives listed at the beginning of each section. Test yourself by completing this review. (Refer to the sections indicated as needed.)

1. State the Addition Property of Equality (8.1). _____

2. State the Multiplication Property of Equality (8.1). _____

3. State the Multiplication Properties of Inequalities (8.4) _____

4. Identify the property illustrated by each example.

  a. If $\frac{1}{2}x = 3$, then $x = 6$. _____

  b. If $y = 8$, then $y + 2 = 10$. _____

  c. If $x + 2 < 5$, then $x < 3$. _____

5. Write the method or rule you would follow in solving each of the following.

  a. $3z + 15 = 36$ _____

  b. $-3g + 15 = 2g + 10$ _____

  c. $\frac{3}{4}y = 27$ _____

  d. $0.7x + 27.6 = 0.2$ _____

  e. $\frac{15}{4} = -2s - 3.75$ _____

  f. $3(2 + 5h) - (1 + 14h) = -6$ _____

  g. $-15x \geq 45$ _____

  h. $2x - 5 \leq 7$ _____

• *Solve first-degree equations in one variable using the Addition Property of Equality (8.1). Solve for the variable.*

6. $3k - 6 = 18 + 2k$ _____

7. $2x + 6 - x = 10$ _____

8. $7y + (-6y) - 4 = 4$ _____

9. $-4a + 3 + 5a = 6$ _____

10. $10q = 10q + 3$ _____

11. $7 + t = 3 + t + 4$ _____

12. $12 = r + 8$ _____

- *Solve first-degree equations in one variable using the Multiplication Property of Equality (8.1).*

Solve for the variable.

13. $2y = 36$ _____

14. $-6x = 102$ _____

15. $7t = -28$ _____

16. $\frac{1}{4}z = 40$ _____

17. $-\frac{2}{3}w = 20$ _____

18. $\frac{2}{5}x = -\frac{3}{10}$ _____

19. $3y = 2\frac{1}{3}$ _____

- *Solve first-degree equations in one variable using both the Addition and Multiplication Properties of Equalities (8.1).*

Solve for the variable.

20. $3y + 8 = -16$ _____

21. $-3 = 2x + 7$ _____

22. $7 + 5y = 12 + y$ _____

23. $-q + 10 = 2q + 11$ _____

24. $3(5h + 2) = 13h + 4$ _____

25. $-2(7 + w) + 20 = -4w + 4$ _____

- *Solve first-degree equations in one variable containing fractions and/or mixed numbers (8.2).*

Solve for the variable.

26. $\frac{x}{5} - x = 4$ _____

27. $3y + \frac{1}{2} = \frac{1}{4}$ _____

28. $\frac{t}{2} + \frac{3}{4} = -\frac{2}{3}$ _____

29. $\frac{2w}{7} - \frac{w}{2} = \frac{15}{14}$ _____

30. $2\frac{1}{2}q + 5 = 7$ _____

31. $-1\frac{1}{3}z + \frac{1}{2} = 1$ _____

• *Solve first-degree equations in one variable containing decimals (8.3).*
  Solve for the variable.

32. $7x + 2.3 = 8x - 1.7$ _____

33. $4.2 - x = 8.9$ _____

34. $0.9x - 0.36 = 0.18x$ _____

35. $7.2x - 8.37 = 6.03$ _____

36. $1.6x - 15.4 = 2.8x + 22.4$ _____

37. $0.3y - 4.2 = (y - 3) + 1.7$ _____

• *Solve first-degree equations in one variable containing both fractions and decimals (8.3).*
  Solve for the variable.

38. $2x - 3.75 = \dfrac{15}{4}$ _____

39. $-x + 4.5 = \dfrac{21}{3}$ _____

40. $0.52 + \dfrac{2}{5} = 0.24y$ _____

41. $0.4(0.2 - w) = \dfrac{3}{10}$ _____

42. $0.5n + 0.75n = 3\dfrac{1}{8}$ _____

43. $\dfrac{2}{3}y - 2.5 = \dfrac{5}{6}$ _____

• *Solve first-degree inequalities in one variable by using the Addition Property of Inequalities (8.4).*
  Solve and graph.

44. $x + 12 > 8$

45. $7 + x < 3$

46. $x - 15 \geq -20$

47. $-3 + x \leq -3$

• *Solve first-degree inequalities in one variable by using the Multiplication Properties of Inequalities (8.4).*
Solve and graph.

48. $4x \leq 32$

49. $7x > -35$

50. $-11x < 44$

51. $-13x \leq -52$

• *Solve first-degree inequalities in one variable using a combination of the Addition and Multiplication Properties of Inequalities (8.4).*
Solve and graph.

52. $5x + 17 \geq 22$

53. $20 + 7x < -1$

54. $-4x + 7 \leq 15$

55. $-20 - 6x > 4$

56. $17 > 2x - 1$

CHAPTER 8 PRACTICE TEST

Match the statement with the property. Use the list provided (statements may be used more than once). Write the appropriate letter in the blank.

_____ **1.** If $q = 3$, then $q + 2 = 5$.

_____ **2.** If $2x < 18$, then $x < 9$.

_____ **3.** If $-z > -12$, then $z < 12$.

_____ **4.** If $x + 5 > 2$, then $x > -3$.

_____ **5.** If $-3p = 21$, then $p = -7$.

_____ **6.** If $15 = h$, then $30 = 2h$.

A. Multiplication Properties of Inequalities

B. Addition Property of Equality

C. Multiplication Property of Equality

D. Addition Property of Inequalities

Solve for the variable.

7. $5z = 25$

8. $y + 5 = 20$

9. $2x + 3 = 7$

10. $q + 4 = q$

11. $x + 5 > 3$

12. $-14x \leq -56$

13. $-4y - 10 = -2$

14. $8t = 2\dfrac{1}{4}$

15. $5a + 4 = 3a + 8$

16. $-2 + 2w + 5 = 3w + 3 - w$

17. $2b + 15 = -3b + 10$

18. $7 + \dfrac{c}{5} = 10$

19. $3(s - 5) = 2(s - 3)$

20. $-12 = -x + 2$

21. $0.4(x - 3) + 0.5x = 0.6x$

22. $\dfrac{3a}{4} + 15 = 8$

23. $\dfrac{x}{2} - \dfrac{5}{6} = 12 - \dfrac{3x}{2}$

24. $\dfrac{x + 2}{4} - \dfrac{2x + 3}{3} = 12$

25. $0.3(2 - x) = 0.4(0.7 - x) + 0.3$

26. $\dfrac{x}{8} - 1 = \dfrac{3}{4}$

27. $\dfrac{2x}{3} - \dfrac{3x}{4} = \dfrac{5}{6}$

28. $-2x - 3.75 = \dfrac{15}{4}$

29. $5x - 7 \geq -42$

30. $3 - 12x < 39$

31. $4\dfrac{1}{5} - x = 8.9$

32. $5 - 8v = 21$

33. $4c - 3 = 3c + 12$

34. $0.5n + 4 = 10$

35. $0.8q + 1.2q = -12$

36. $5[2y - (3y + 1) - y] = -5$

37. $15 = 3b - (b - 7)$

38. $5 = -15y$

39. $-1.3 - 0.7z = 0.42 - 1.1z$

40. $0.9t - 0.36 = 0.18t$

41. $\dfrac{1}{3}x + 0.5 = 1\dfrac{2}{3}$

42. $0.7y - 0.05 = \dfrac{1}{5}y + 0.35$

43. $5.4(h + 1) = \dfrac{1}{2}(1.5 + h)$

44. $\dfrac{1}{4}(1 - k) = 0.7(2 + 3k)$

Solve.

45. $3x \leq 12$

46. $-2x > 16$

47. $x + 7 < 8$

48. $3x - 1 \geq -13$

49. $5 - 2x > 7$

50. $11 \leq 2x + 1$

THE MATHEMATICS JOURNAL

Explaining A Procedure II

Task: Choose a method for solving an equation or an inequality you learned in this chapter. Make sure it's one you understand thoroughly. Work a sample problem using it. Imagine that you are explaining it to a new student who just entered the class. Provide a step-by-step solution to the problem, then write your step-by-step explanation next to the solution.

9

Problem Solving

9.1 Engaging in Problem Solving
9.2 Some Strategies for Solving Logic and Mathematics Problems
9.3 Solving Problems by Using Equations and Formulas
9.4 Using Equations in One Variable to Solve Coin and Ticket Problems (Optional)

Introduction to Problem Solving

Do you consider yourself a good problem solver? Before you answer that question, take an imaginary journey with us.

Suppose you are an 18-year-old college freshman, and a friend suggests that the two of you go to Florida for spring break.

Or suppose you are 30-something, married, with small children, and your spouse suggests a get-away weekend for just the two of you.

Or suppose you are a single parent, and a couple of friends suggest going to the World Series or going to New York to do some holiday shopping. Would you be interested? What if they admitted that they didn't have the trip planned yet and didn't even know whether it was possible? Would you still be interested? What if they invited you to join them in making it happen? Would you look on it as an adventure and say, "Why not?"

Improbable as any of these excursions may seem in the light of your present circumstances, we all know that thousands of college students spend spring break in Florida every year; the World Series is always sold out (a few people in attendance must be single parents); out-of-town shoppers crowd the New York stores each December; and hotels are often sold out on weekends because of their advertised get-away specials.

If you say, "OK, let's go!" what do you and your friends do? You call a few hotels, visit a travel agency, or contact AAA. In short, you begin gathering information—information you'll need to achieve your goal. Once you have the information, you'll start sorting it. You'll keep what will help you, and you'll discard the rest. Now you'll start planning the trip piece by piece. You will have to get tickets to the ball game. You will have to make living arrangements for the children. You will have to decide how you're going to finance the trip. Sometimes, your first ideas won't work. You may have to start over; you may have to devise a new plan. If that plan doesn't work, you may have to come up with another one, and another, and another. Each time, you'll refine the process; each time you'll get closer to Florida, the World Series, or New York.

You may, however, decide at any time to abort the trip. You may decide that you have other priorities, it's too expensive, you don't have the time, or you may get discouraged and quit. You may change your goals.

What have we been talking about during this exercise? Something we all do every day and several times a day. We've been talking about **problem solving.**

If we were to ask high school freshmen in an algebra class if they considered themselves good problem solvers, the majority would say, "No." If we were to ask them if they had attended a concert given by a major rock star in the past six months, the majority would say yes. Think about it—these students are 14-year-olds. Many of them have no money of their own; they don't have cars—they don't even drive. They're in school when the tickets go on sale, yet they manage to get to the rock concerts in their part of the country. Mathematically, however, they consider themselves to be "not very good at problem solving." What do you think?

9.1 Engaging in Problem Solving

OBJECTIVES
- Demonstrate a heightened awareness of problem solving.
- Develop a model to be used in problem solving.

Seeing Yourself as a Problem Solver

After reading the introduction, can you agree once and for all that we're all pretty good problem solvers? Some of us may be better than others, some of us may enjoy doing it more than others, but we can all solve problems. Our daily lives are filled with them. Before we continue, please complete Problems 1, 2, and 3 of Problem Set 9.1.

The Problem-Solving Process

When asked if solving mathematical problems is different from solving real-life problems, many people would respond with a definite yes. Is this really true? Let's examine the problem-solving models you constructed in Problem 3 of Problem Set 9.1 which you just completed. See if your models contain the following elements:

1. Have a clearly defined goal in mind.
2. Gather all the information you can about the goal and how to achieve it.
3. Sort the information. Keep what is helpful. Discard the rest.
4. Decide on a plan.
5. Carry out the plan.
6. Make sure you have achieved the goal.

It looks deceptively simple, doesn't it? In solving a complex problem, we know that we don't just go through these steps once, and as we go through the process, we have to keep asking ourselves, "Is this plan working? Is it leading me toward my goal?" If it is not, we have to go back and devise another plan. We also have to make sure we have enough information. If we don't, we have to go back to Step 2. We may even have to go back to Step 1 to make sure we have a clearly defined goal. Occasionally, we may have to reformulate our goal. During the whole process, we have to continually evaluate our progress keeping the goal in mind. A diagram called a **flow chart** (Figure 9.1.1 on next page) may help us understand the process.

There is another important aspect to problem solving that cannot be represented in a flow chart: time. Problem solving takes time. We need time to mull over the problem and look at it from different angles. We need time to let the problem sink into our unconscious, and then we need time to bring it out of our unconscious so we can take another look to see if it's ready to be solved. While we are not in totally passive states during this process, we can't be in frantic or agitated states either. We do our best problem solving when we're relaxed.

Historically, we can find many examples of scientists, poets, mathematicians, and artists who had been working on complex problems for a long time, only to

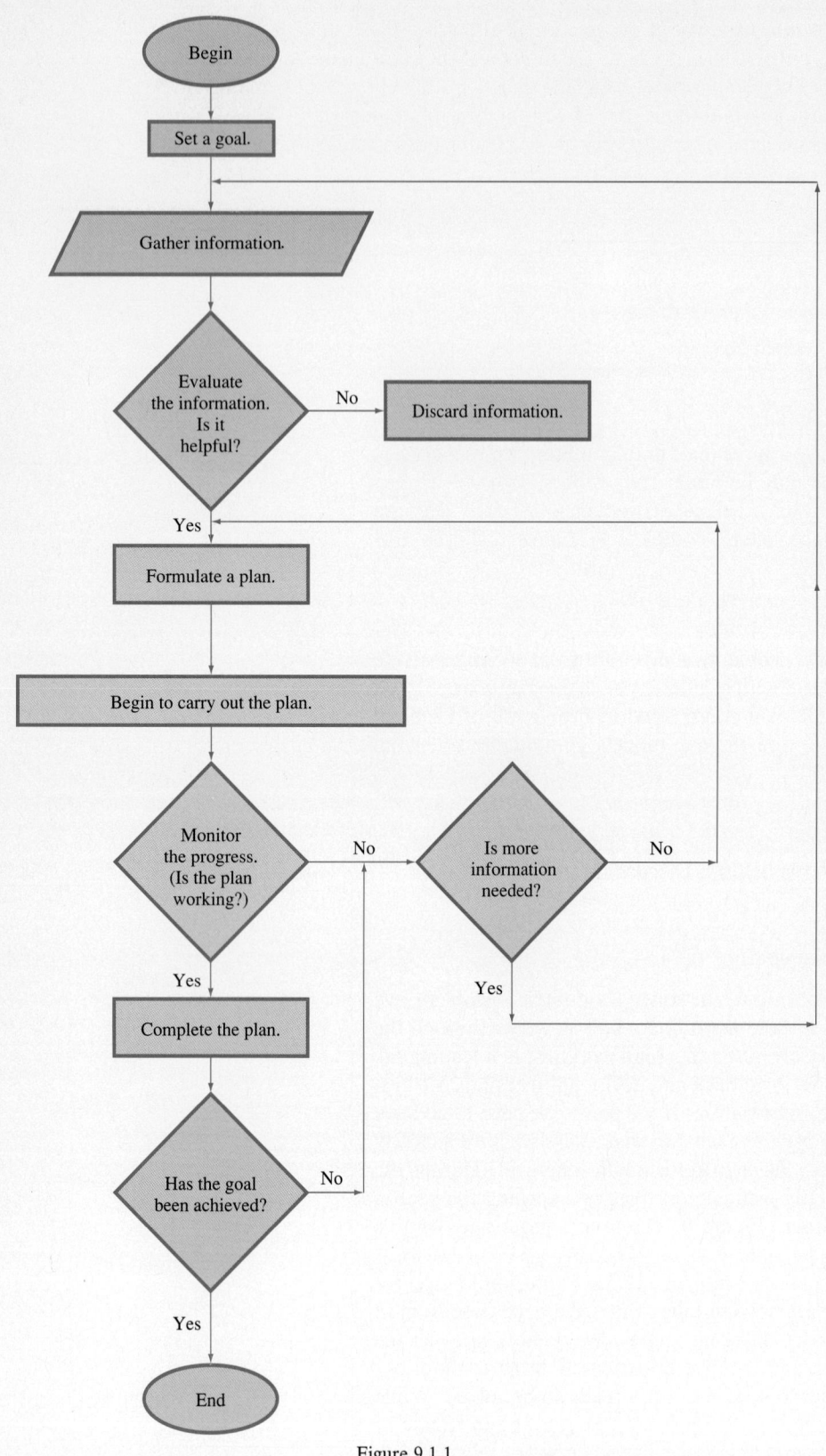

Figure 9.1.1

have the solutions "come to them" while they were just plain relaxing. We have all heard stories of poets and composers who woke up in the middle of the night and rushed to write down the poems and songs in their heads. Even though the account that an apple falling on Newton's head while he was sitting under a tree led to his discovery of the laws of gravity or that Archimedes' famous displacement principle came to him while he was taking a bath both may have become fictionalized over the centuries, the stories still testify to the relaxed states of these two great problem solvers.

What can we learn from these examples? First, we can learn to give ourselves time—incubation time—when solving complex problems. Second, we can learn to reflect on our behavior during problem solving. By doing this, we can discover our problem-solving modes and determine what works best for us. Once we know that, we can consciously reproduce those behaviors to improve our skills.

PROBLEM SET 9.1

1. Reflect for a few minutes on your problem-solving style. Write a 150-word essay on the problem solving you have done and/or are currently doing on one of these topics: deciding to go to college, choosing which college to attend, or choosing a major field of study.

2. Working in a group with two or three of your classmates, follow these steps to develop a model to be used in problem solving.
   a. Share the reports you wrote on the problem-solving involved in pursuing a college degree (see Problem 1).
   b. Discuss these reports, searching for common elements among them.
   c. At the conclusion of the session, write and submit a one-page report outlining the common steps group members followed when doing their individual problem-solving.

3. Consider a problem you've solved in the past—one whose solution gave you a lot of satisfaction. Reflect on the way you went about solving it. Write a one-page (150-word) essay on your behaviors during this time.

9.2 Some Strategies for Solving Logic and Mathematics Problems

OBJECTIVE
- Use appropriate strategies for solving mathematical problems.

Read to Comprehend

Read the statement of the problem slowly, carefully, and completely. Don't skim, but don't try to memorize the facts either. You can review the problem later for the facts. Don't stop to analyze—that will come later. After you've finished reading, test your comprehension of the problem by asking yourself a few general questions about it. What pertinent information has been given? What questions have been asked?

EXAMPLE 1   You and a friend are out for a walk. You pass a pizza shop, and you're both hungry. The sign in the window advertises a 12-inch pizza special for $8.40. You each reach into your pockets to see how much money you have. Counting your money, you find that you have $3.67, and your friend has $4.96. If you pool your money, can you buy the pizza?
Questions:
What information has been given?
What questions have been asked?
Answers:
Information given: price of the pizza and amout of money you each have.
Questions asked: Do you have enough money to pay for pizza?

EXAMPLE 2   Imagine you are a first-year college student with a part-time job on campus. You start off the month with $52.83 in your checking account. On the third of the month, you deposit your paycheck of $42.95. On the sixth of the month, you pay $61.00 to your roommate for your share of the rent. On the seventh, you write a check for $28.91 to VISA and one to the phone company for $11.70. Later that night, you balance your checkbook. Will you have any money left in the account, or will you be overdrawn? How much will you have left, or by how much will you be overdrawn?
Questions:
What information has been given?
What questions have been asked?
Answers:
Information given: balance in checkbook at beginning of the month, amount deposited, number of checks written, and amount of each.
Questions asked: Is there money left in the account, or is it overdrawn? How much is left, or by how much is the account overdrawn?

EXAMPLE 3   You fill the gas tank of your car to take a friend to the airport 78 miles away. You plan to drop your friend off and return home immediately. If your car has a 12-gallon tank and gets 18 miles per gallon (mpg), will you have to stop for gas before you get home?
Questions:
What information has been given?
What questions have been asked?
Answers:
Information given: distance from home to airport, capacity of gas tank, mpg ratio of car.
Question asked: Will you have to stop for gas on the way home?

Analyze the Problem

Any problem boils down to this: How can I use the information given to answer the question or questions asked? This is the most challenging part of the problem-solving process. There is no one way to proceed at this point. This is where the incubation time we discussed earlier may need to occur, so plan for it. We'll suggest some strategies, but don't be limited by them.

1. *Make a chart or table of the information given.* This strategy can be especially helpful if several pieces of information are given.
2. *Draw a diagram.* This works well for geometric problems and for those that can be visualized.

3. *Think backwards from the question asked to the information given.* Use this strategy if you find yourself saying, "I could answer the question if I knew a certain piece of information." Then ask, "Is there any way I can use what I know to get the information I need?"

4. *Look for relationships; look for patterns.* Look for relationships among the pieces of information given, as well as between each piece of information given and the question(s) asked. This strategy employs **inductive reasoning**—reasoning from the particular to the general.

5. *List all the possible outcomes, then eliminate those that don't conform to the information given in the problem.*

6. *Separate the problem into smaller parts.* Solve each part, then put the solutions together to solve the original problem.

Sometime the strategy itself may be the plan you use. Other times, the strategy may only suggest a plan. For example, the strategy may help you realize that you need to do one or more arithmetic operations to get the information that will lead to solving the problem.

Now we are ready to apply these strategies to a few of the earlier examples used in the "Read To Comprehend" section.

EXAMPLE 4   You and a friend are out for a walk. You pass a pizza shop and you're both hungry. The sign in the window advertises a 12-inch pizza special for $8.40. You each reach into your pockets to see how much money you have. Counting your money, you find that you have $3.67, and your friend has $4.96. If you pool your money, can you buy the pizza?

Use Strategy 3 and/or 4. Think backwards from the question asked to the information given, and/or look for relationships.
Analysis
You need $8.40. If you pool your money, will you have at least $8.40?

$$\$3.67 + \$4.96 = \$8.63$$
$$\$8.63 > \$8.40$$

Conclusion
Yes, you can buy the pizza.

EXAMPLE 5   You fill the gas tank of your car to take a friend to the airport 78 miles away. You plan to drop your friend off and return home immediately. If your car has a 12-gallon tank and gets 18 miles per gallon, will you have to stop for gas before you get home?

Use Strategies 3 and 4 again, i.e., work backwards and look for relationships. Then, use Strategy 6, and break the problem into smaller parts.

Practice Problems

For each of the following problems, list the information given and the questions asked.

1. You arrive at the laundromat with three loads of laundry and discover that you left your wallet in your room. You do, however, have $2.35 in change in your pockets. The prices posted are washer 50 cents per load, dryer 20 cents for the first 15 minutes and 10 cents for each additional 10 minutes. Since one of your loads is towels, you estimate you'll need to leave it in the dryer for 25 minutes. Will you have enough money to do your laundry?

2. You stop at the grocery store on your way home from class. You have $4.32. Do you have enough money to buy a head of lettuce at 59 cents each, a pound of hot-dogs at $2.39 per pound, a package of buns at $1.09 per package, and a can of baked beans at 79 cents each?

Analysis:

To answer the question, I need to know three things:

1. How far I can get on one tank of gas?
2. How many miles is a round-trip to the airport?
3. Which distance is greater?

Part 1. How far can I get on one tank of gas?

$$12 \text{ gallons} \times 18 \text{ mpg} = 216 \text{ miles}$$

Part 2. How many miles is a round-trip to the airport?

$$78 \text{ miles} \times 2 = 156 \text{ miles}$$

Part 3. Which distance is greater?

$$216 \text{ miles} > 156 \text{ miles}$$

Conclusion:

I will have enough gas for a round-trip to the airport. I won't have to stop for gas.

EXAMPLE 6   You and your three suite mates are playing partners in a new card game. In this game, only queens, kings, and aces are used. Each player is dealt 4 cards. For scoring purposes, a queen is 12 points, a king is 13 points, and an ace is 1 point. Your partner tells you that her hand is worth 38 points and the cards are all red. How many queens, kings, and aces could she be holding?

Use Strategies 5 and 1. List all the possible red hands, and eliminate those that don't add up to 38 points. Use the following chart to record these hands.

K_d	K_h	Q_d	Q_h	A_d	A_h	*Total points*
13	13	12	12			50
13	13	12		1		39
13	13	12			1	39
13	13		12	1		39
13	13		12		1	39
13		12	12	1		**38**
13		12	12		1	**38**
	13	12	12	1		**38**
	13	12	12		1	**38**
13		12		1	1	27
13			12	1	1	27
	13	12		1	1	27
	13		12	1	1	27
13	13			1	1	28
		12	12	1	1	26

Conclusion: Your partner is holding 1 king, 2 queens, and 1 ace.

Solve each problem. List the strategies used, analyze the solution, and answer the question or questions.

3. You arrive at the laundromat with three loads of laundry and discover that you left your wallet in your room. You do, however, have $2.35 in change in your pockets. The prices posted are washer 50 cents per load, dryer 20 cents for the first 15 minutes and 10 cents for each additional 10 minutes. Since one of your loads is towels, you estimate you'll need to leave it in the dryer for 25 minutes. Will you have enough money to do your laundry?

4. You stop at the grocery store on your way home from class. You have $4.32. Do you have enough money to buy a head of lettuce at 59 cents each, a pound of hotdogs at $2.39 per pound, a package of buns at $1.09 per package, and a can of baked beans at 79 cents each?

EXAMPLE 7   Write the next two numbers.

$$1, 3, 7, 15, 31, \underline{\quad}, \underline{\quad}$$

Analysis: Use Strategy 4. Look for a pattern.

1 + 2 = 3	or	**1 + 2^1 = 3**
3 + 4 = 7	or	**3 + 2^2 = 7**
7 + 8 = 15	or	**7 + 2^3 = 15**
15 + 16 = 31	or	**15 + 2^4 = 31**
31 + ? = ?		
? + ? = ?		

Following the pattern above, we should add 2^5 to 31 to get the next number, and then add 2^6 to it to get the second number.

$$31 + 2^5 = 31 + 32 = 63$$
$$63 + 2^6 = 63 + 64 = 127$$

Conclusion: Following this pattern, the next two numbers are 63 and 127. Can you find other patterns?

EXAMPLE 8   In this cryptogram, a sentence in which one set of letters is substituted for another, z is substituted for m, and e is substituted for u.

Qezzi   fez   biera

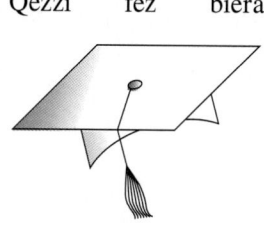

Decode the sentence. Hint: It is a Latin phrase associated with graduation.
Analysis
Use Strategies 2 and 4. Make a diagram and look for relationships.

<pre>
Q e z z i   f e z   b i e r a .
_ u m m _   _ u m   _ _ u _ _ .
</pre>

Now use the definition. Did you think of *summa cum laude?*

Try it.

<pre>
Q e z z i   f e z   b i e r a .
S u m m a   c u m   l a u d e .
</pre>

Notice that the i stands for a in both the words *summa* and *laude,* so the solution works.

EXAMPLE 9   All 36 members of the Alpha Alpha Pi sorority are taking at least one of the following courses: English 099, History 101, and Math 202.

Twenty are taking English 099.
Sixteen are taking History 101.
Twelve are taking Math 202.
Ten are taking both English 099 and History 101.
Eight are taking both History 101 and Math 202.
Seven are taking both English 099 and Math 202.
Three are taking all three courses.

How many members of Alpha Alpha Pi are taking only English 099? How many take only History 101? How many take only Math 202?

Analysis:

Use Strategy 4. Make a diagram. The Venn Diagrams we used in Chapter 1 will work well here. Next, use the concepts of intersection and union of sets to determine how many members are taking only one of the three courses.

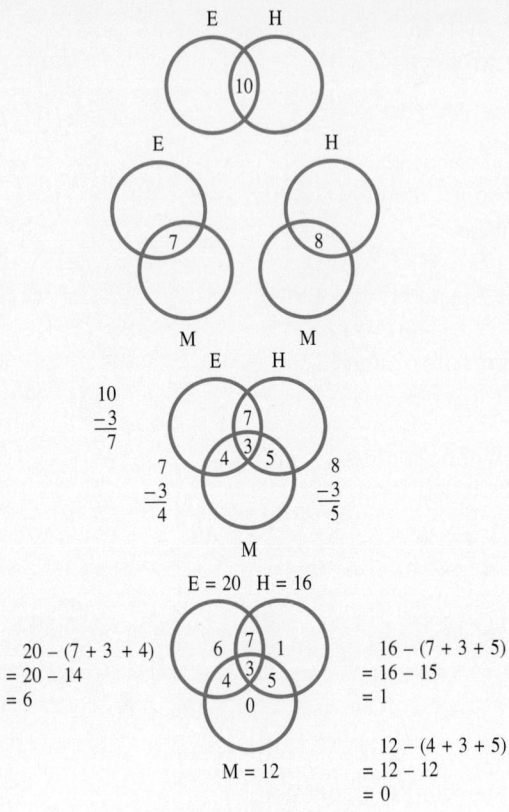

$$20 - (7 + 3 + 4)$$
$$= 20 - 14$$
$$= 6$$

$$16 - (7 + 3 + 5)$$
$$= 16 - 15$$
$$= 1$$

$$M = 12$$

$$12 - (4 + 3 + 5)$$
$$= 12 - 12$$
$$= 0$$

Six take only English 099.
One takes only History 101.
None take only Math 202.

PROBLEM SET 9.2

Solve these problems, making notes of the strategies you use. (Suggestion: Work with one or two other students in the class.)

1. A weekly literary magazine you need for your semester (15-week) English class sells for $2.50 an issue at the bookstore. Your English instructor announces that the class can place a group order at a special rate of $22.50 per year (52 issues). What is the better buy?

2. Decode this cryptogram to reveal a famous literary quotation. "Sq fi qz rqs sq fi, shus mo shi vwiosmqr."—Shakespeare

3. Use the numbers in the accompanying square with the operations of addition, subtraction, multiplication, and division for whole numbers along with the order of operations to obtain 32.

6	2
3	8

4. There are two recycling bins. Each contains two smaller recycling bins. Both of them contain three smaller recycling bins each. How many recycling bins are there?

5. Write the next three letters.
A B A C B C B _ _ _

6. Fill in the missing number.
5, 2, 4, 1, 2, -1, -2, __

7. Three students—Jaron, Rebeca, and Cliff—are raising 30 kittens. Eighteen of the kittens are males. Jaron has 6 females, and Rebeca has the same number of males. Rebeca has 2 more kittens than Jaron, who has 8 kittens. Cliff has 8 more males than females and has the same number of females as Jaron has males. How many male kittens does Jaron have? How many does Cliff have?

8. Replace x in each number with the same digit so that each number will be divisible by 6.

x2,741,008   1,705,03x

9. Rearrange the digits of each addend in the following addition problem so that the sum is correct.

$$479 + 625 = 1203$$

10. Gamma Pi Alpha has developed a secret counting system based on 13 digits instead of the 10 used in our base 10 system. Their numerals are:

0, 1, 2, 3, 4, 5, 6, 7, 8, 9,

γ (our 10), π (our 11), α (our 12), and 10 (our 13).

Give the base 10 equivalents of these Gamma Pi Alpha numbers.

a. 75   b. 3γ4   c. 4010   d. 2$\alpha\pi$

11. Remember the man who was going to St. Ives? He met a man with 7 wives. Each wife had 7 cats, each cat had 7 kits, and each kit had 7 mice. "Kits, cats, mice, and wives, how many were going to St. Ives?" the story asks. The answer, of course, is one—the man. Suppose all the people and animals in the story really were going to St. Ives. How many would be making the trip?

12. To provide better service, the cafeteria manager asks the student cashier to count everyone who purchases food on Tuesday at the dinner hour. Tuesday evening she checks the register tape and finds that 96 spaghetti dinners were sold, 154 pizza dinners, and 28 chicken dinners and that no other dinners were sold. She also recalls that 14 of the customers who had spaghetti dinners also had pizza dinners and that none of the customers who had chicken dinners had spaghetti or pizza. What number should the cashier turn in?

13. Maria, LaTisha, and Evelyn each attend a different school. One of them is 19, one is 21, and one is 35 years old. They each take a different course. One takes Math 501, another takes Engl 700, and the third takes Psych 205. Use the clues listed to determine each student's name, age, school, and course.

1. La Tisha does not attend Red College.
2. Evelyn attends Olive Community College.

3. The student who attends Green State University does not take Math 501.
4. The 19-year-old takes Psych 205.
5. Evelyn is not the youngest.
6. Neither LaTisha nor Evelyn is the oldest.
7. The 35-year-old does not take Engl 700.

*Use the accompanying chart to record the information. Use an • to indicate a match and an **X** otherwise. Note: Once a match has been found, the other two entries in that category should be marked with an **X**.*

	E	*L*	*M*	*19*	*21*	*35*	*GSU*	*OCC*	*RC*	*Engl*	*Math*	*Psych*
E												
L												
M												
19												
21												
35												
GSU												
OCC												
RC												
Engl												
Math												
Psych												

9.3 Solving Problems by Using Equations and Formulas

OBJECTIVES

- Solve mathematical problems by using equations.
- Solve mathematical problems by using formulas.

Some mathematical problems lend themselves to finding solution by equations or by using formulas. This is true when a quantity or number is represented two ways in the problem or when we know a formula that applies to the problem. In the latter case, we can solve the problem by evaluating a formula. In this section, we will take a close look at these types of problems and develop specific procedures for solving them.

Solving Problems by Using Equations

As we mentioned earlier, start the problem solving process by reading the problem:

1. Read it once to comprehend the problem and to get an overall view of the situation being described.
2. Read it the second time to set up the assignment statements.
3. Read it the third time to set up the equation.
4. Solve the equation.
5. Check the solution for accuracy.
6. Refer back to the problem statement and, by using the information gained by solving the equation, answer the question or questions asked.
7. Reread the problem one last time substituting your answer for the question asked. Ask yourself this question: "Is the answer I'm proposing reasonable?" If the answer is no, reject your proposed answer and begin again.

1. *Read to comprehend.* In Section 9.2, we emphasized the importance of reading to comprehend the problem, and you practiced this skill by working through some examples. Review that section now if you feel you need to.
2. *Reread to set up the assignment statements.*
   a. Assign a variable to the quantity or number you have the least information about. Many times this quantity or number will become the answer to the question posed in the problem.
   b. Represent other quantities or numbers in terms of this variable.
3. *Read the problem statement once more to set up the equation.*
   a. If a formula exists for the type of problem being considered, state the formula. Then, substitute the variables and the information given into this formula.
   b. If a formula does not exist, translate the language used in the problem from English into algebra. Look for a sentence in the problem that states that two or more quantities or numbers are equal.
4. *Solve the equation.* Refer to Chapter 8 to review the procedures for solving equations.
5. *Check the solution for accuracy.* Use a calculator, if appropriate.
6. *Referring back to the problem statement, answer the questions asked.* If you had more than one assignment statement, use the solution to the equation to evaluate the others now. Keep in mind that the solution to the equation may not be the answer to the question asked, but may only give you new information you can use in answering the question. Write down the answer or answers.
7. *Reread the problem one last time substituting your answer for the question asked.* Ask yourself this question: "Is the answer I'm proposing reasonable?" If the answer is no, reject your proposed answer and begin again.

Now that we have an overview of the steps to follow, let's look at Steps 2 through 7 more closely and practice using them.

Step 2. Reread the problem statement to set up the assignment statements. As we pointed out in Section 3.4, knowing how to translate phrases and sentences from English into algebra is essential to solving problems that can be set up as equations. You may find Tables 9.1 and 9.2 helpful as a quick review of the material covered in Section 3.4. If you need more practice in developing this skill, review Section 3.4 on your own. In Tables 9.1 and 9.2, let $n = $ a number. Any number can replace the number 2. Notice that because there are no commutative properties for subtraction and division, the order in which quantities are named is important, e.g., $4 \div 2 \neq 2 \div 4$ and $n \div 2 \neq 2 \div n$.

English phrase	*Algebraic expression*
A number plus two	$n + 2$
Two added to a number	$n + 2$
Two more than a number	$n + 2$
A number increased by two	$n + 2$
The sum of two and a number	$2 + n$
A number minus two	$n - 2$
Two minus a number	$2 - n$
A number subtracted from two	$2 - n$
Two subtracted from a number	$n - 2$
Two less than a number	$n - 2$
A number decreased by two	$n - 2$
The difference of a number and two	$n - 2$
A number multiplied by two	$n \cdot 2$
Two times a number	$2 \cdot n$
The product of two and a number	$2 \cdot n$
A number divided by two	$n \div 2$
Two divided by a number	$2 \div n$
The quotient of a number and two	$n \div 2$
The quotient of two and a number	$2 \div n$
Twice a number	$2 \cdot n$
A number squared	n^2

Table 9.1

Some English phrases may include more than one operation. When translating them into algebra, be careful to keep the original intent of the English phrase. Consider the examples in Table 9.2.

English phrase	*Algebraic expression*
Four times *the sum of* a number and six	$4(n + 6)$
One-half *the difference of* three and a number	$\frac{1}{2}(3 - n)$
The sum of seven and twice a number	$7 + 2n$
The quotient of five and eight times a number	$\dfrac{5}{8n}$
The sum of four times a number and six	$4n + 6$
The difference of three and one-half a number	$3 - \frac{1}{2}n$
Twice *the sum of* seven and a number	$2(7 + n)$

Table 9.2

Note Rather than presenting entire problem statements as examples for practice, we will present phrases from problems.

Set up assignment statements for the following examples.

EXAMPLE 1 At City College, the average grade point average (GPA) of students with junior or senior standing is 1.06 points higher than the average GPA of students with freshman or sophomore standing. Set up the assignment statements.
Answer
Let g = average GPA of freshman and sophomore students
$g + 1.06$ = average GPA of junior and senior students

EXAMPLE 2 The minimum wage has gone up $1.27 in the past two years.
Answer
Let m = minimum wage in dollars two years ago
$m + 1.27$ = current minimum wage in dollars

EXAMPLE 3 Stacy's roommate studies twice as long as she does.
Answer
Let t = time in hours Stacy studies
$2t$ = time in hours her roommate studies

Step 3. Read the problem statement once more to set up the equation. After rereading the problem, look for a sentence with some form of the verb "to be" in it. This verb may take many forms: is, are, was, were, will be, has been, etc. It will indicate the equality of two or more quantities. Algebraically, it will be replaced by an equal sign, =. Other phrases that indicate equality are *results in* and *is the same as.*

In each of the following examples, we will first assign the variables and then set up the equation.

EXAMPLE 4 Find a number that is two less than 17.
Answer
Let n = the number
Equation: $n = 17 - 2$

EXAMPLE 5 Riordan, a student at CKU's regional campus, has a part-time job, with net earnings, or take-home pay, last year of $3182.57. The payroll deductions amounted to $563.12. What was his gross income for the year?
Answer:
Let g = gross income in dollars
Equation: $g - 563.12 = 3182.57$

EXAMPLE 6 A cheese souffle recipe calls for 6 ounces of cheese and $1\frac{1}{2}$ cups of milk. How much cheese should be used if the cook uses a quart of milk? (Note: 4 cups = 1 quart.)
Answer:
Let c = number of ounces of cheese to be used for 1 quart milk
Equation: $\dfrac{c}{6} = \dfrac{4}{1\frac{1}{2}}$.

At this point, let's take another look at the last three examples discussed and complete them by applying Steps 4 through 6. We're asking you to apply Step 7 for each example.

Translate into algebraic expressions.

1. The difference of six and a number

2. The quotient of a number and four

3. The sum of twice a number and eight

4. Three times the sum of six and a nuimber

5. One-third the difference of two and a number

6. The product of five and a number squared

Assign the variables, and set up the equation.

7. Your roommate studies 1 hour more than you do each day. If the sum of the hours you both study daily is 7, how many hours do you study?

8. Think of a number that is 3 more than 27. What is the number?

EXAMPLE 7   Find a number that is two less than 17.

Let n = the number   2. Assign the variable.

Equation: $n = 17 - 2$ 3. Set up the equation.

$n = 15$ 4. Solve for the variable.

Check: $15 ? 17 - 2$ 5. Check the solution.

The number is 15.   6. Answer the question.

7. Evaluate the answer.

EXAMPLE 8   Riordan, a student at CKU's regional campus, has a part-time job, with net earnings, or take-home pay, last year of $3182.57. The payroll deductions amounted to $563.12. What was his gross income for the year?

Let g = gross income in dollars   2. Assign the variable.

$g - 563.12 = 3182.57$ 3. Set up the equation.

$\underline{+ \ 563.12 \qquad + 563.12}$ 4. Solve the equation.

$g \qquad\quad = 3745.69$

$3745.69 - 563.12 ? 3182.57$ 5. Check the solution.

$3182.57 = 3182.57$

His gross income was $3745.69   6. Answer the question.

Is it a reasonable answer? Why?   7. Evaluate the answer.

EXAMPLE 9   A cheese souffle recipe calls for 6 ounces of cheese and $1\frac{1}{2}$ cups of milk. How much cheese should be used if a cook plans to use a quart of milk? (Note: 4 cups = 1 quart.)

Let c = number of ounces of cheese to be used for 1 quart milk

$$\frac{c}{6} = \frac{4}{1\frac{1}{2}}$$ 3. Set up the equation.

$$6 \cdot \frac{c}{6} = 6 \cdot \frac{4}{\frac{3}{2}}$$ 4. Solve the equation.

$$c = 6 \cdot 4 \cdot \frac{2}{3}$$

$$c = 24 \cdot \frac{2}{3}$$

$$c = 16$$

Check: $$\frac{16}{6} ? \frac{4}{1\frac{1}{2}}$$ 5. Check the solution.

$$\frac{8}{3} ? 4 \div \frac{3}{2}$$

$$\frac{8}{3} ? 4 \cdot \frac{2}{3}$$

$$\frac{8}{3} = \frac{8}{3}$$

The cook should use 16 ounces of cheese.   6. Answer the question.

Is 16 ounces reasonable? Why?   7. Evaluate the answer.

Assign the variables, set up the equations, solve them, and then answer the questions.

9. Juanita's net pay last month was $302.50. The total deductions were $76.25. What was her gross pay?

10. Fifty more women than men are enrolled in the freshman class. The total number of freshmen is 212. How many men are freshmen?

Solving Problems by Using Formulas

Over the centuries, scientists and mathematicians have discovered general formulas that show the relationship between certain quantities—for example, that distance is the product of rate and time ($r \cdot t = d$). When we confront a problem in which the facts are related by a known formula, we need only evaluate the formula to solve the problem. Applying a formula to a particular problem employs **deductive reasoning**—reasoning from the general to the specific.

Decide whether a given application problem can be solved by formula by asking yourself if you know a formula that applies in this case. Do this immediately after reading for comprehension and before assigning the variables. If the answer is yes, write the formula down. Now assign the variables. (We suggest using the letters that occur in the formula as your assignment variables. The following examples illustrate this model.)

EXAMPLE 10 Elita is planning a trip to the state capital, which is approximately 150 miles from her house. Since she'll be driving on interstate highways most of the way, she can average at least 50 miles per hour. If she needs to be there for a 9:00 A.M. meeting, what time should she leave her house?

Use the formula: Distance = Rate × Time
$$d = r \cdot t$$

Let t = time in hours.   Assignment Statement
$r \cdot t = d$ Formula
$50 \cdot t = 150$ Equation

Note Some of you may prefer to use a table like one of the following to set up both the assignment statements and the equation. All these methods are acceptable.

r	·	t	=	d
50	·	t	=	150

Equation: $50t = 150$ or
$$r \cdot t = d$$

r	t	d
50	t	150

$50t = 150$ 4. Solve the equation.
$$\frac{50t}{50} = \frac{150}{50}$$
$t = 3$
$50 \cdot 3 = 150$ 5. Check the solution.
The trip will take her about 3 hours.
She needs to be there by 9:00 A.M. so,
Elita needs to leave before 6:00 A.M.   6. Answer the question.
Is the answer reasonable? Why?   7. Evaluate the answer.

EXAMPLE 11   Find the length of the rectangle in feet. The perimeter is 20 feet and the width is 3 feet.

$$(\text{Perimeter of a rectangle} = 2 \cdot \text{length} + 2 \cdot \text{width})$$

Let L = length of the rectangle in feet   Assignment statement

$2 \cdot L + 2 \cdot W = P$ Formula.

$2 \cdot L + 2 \cdot 3 = 20$

$2L + 6 = 20$

$$\underline{\quad -6 \quad -6 \quad}$$

$2L \qquad = 14$

$$\dfrac{2L}{2} = \dfrac{14}{2}$$

$$L = 7$$

The length of the rectangle is 7 feet. Is this answer reasonable?

PROBLEM SET 9.3

Read each of the following problem statements once, then answer the questions: What information has been given? What questions have been asked?

1. In the United States, 4.2 million children were expected to be born this year; 1.7 million children had been born so far. How many more births can be expected?

2. A mathematics student receives the following scores on tests: 94, 82, 89, 75, and 79. What is that student's current test average?

3. Find a number that when multiplied by 2 and added to 6 results in 24.

4. The bookstore is having a half-price sale on pens that usually sell for $1.19 each. If Angelo buys 3, how much will he pay, excluding tax?

5. The accompanying chart shows the tuition costs at public universities from 1970 to 1990. Based on this information, how much has tuition increased from the year 1970 to the year 1990?

Year	In-state Tuition* (per semester)
1970	$323
1980	$583
1990	$1356

*U.S. Dept. of Education 1991–92

Set up assignment statements for the following phrases and statements.

6. Zak's brother has three times as much money as Zak.

7. In the United States, the life expectancy for women is 6.8 years longer than it is for men.

8. Chelsey's car gets 5.1 mpg fewer than your car does.

9. According to U.S. Department of Commerce statistics, in 1970 single parents had an average household income of $17,500. By 1990, this income had increased by $3900. What was the average household income of single parents in 1990?

10. If it took Connor's parents $2\frac{1}{2}$ hours to get to school to pick him up for semester break and their house is approximately 172 miles from school, what was their average driving rate?

Use an equation or a formula to solve the following problems. Attend to each of the seven steps.

11. In the United States, 4.2 million children were expected to be born this year; 1.7 million children had been born so far. How many more births can be expected?

12. A mathematics student receives the following scores on tests: 94, 82, 89, 75, and 79. What is that student's current test average?

13. Find a number that when multiplied by 2 and added to 6 results in 24.

14. Find a number that when multiplied by 7 and then subtracted from 98 results in the number 21.

15. Christian's wife makes twice as much money as he does. Their combined annual income is $53,135. How much money does Christian earn?

16. In the United States, the life expectancy for women is 6.8 years longer than it is for men. In 1991, the life expectancy for women was 78.8. What was the life expectancy for men?

17. According to U.S. Department of Commerce Statistics, in 1970, single parents had an average household income of $17,500. By 1990, this income had increased by $3900. What was the average household income of single parents in 1990?

18. If it took Kevin's parents $2\frac{1}{2}$ hours to get to school to pick him up for semester break and their house is approximately 172 miles from school, what was their average driving rate? $(d = r \cdot t)$

19. If one dozen long-stem roses cost $49.50, how much will 5 roses cost?

20. If it's been snowing at the rate of 1.2 inches per hour for the last 7 hours, how much snow has fallen?

21. One night at 10:10 o'clock when you were driving back to school on the interstate highway, you set the cruise control at 65 mph. At 12:20 A.M. you realize that the cruise control is still on. How many miles have you traveled?

22. The perimeter of a rectangle is 56 meters. The length is three times the width. Find the measure of both the length and the width. (Use the formula: $P = 2L + 2W$)

23. A 12-inch pizza costs $8.40. An 8-inch pizza costs half as much, and a 16-inch pizza costs twice as much. If you buy two 12-inch pizzas, one 8-inch pizza, and three 16-inch pizzas, how much will you pay?

24. The student who sits next to you in class won't tell you his grade point average (GPA), but he does tell you that it is 0.25 points higher than his twin brother's and that their combined GPAs are 6.71. What is his GPA?

25. Stu estimates that it will take him 3 hours to drive to his parent's house for semester break provided that he drives at an average rate of 60 mph. How far from his school do his parents live?

26. Sylvia took the bus home for the winter break. She left school at 3:00 P.M. and arrived home at 11:00 P.M. The average speed the bus was going was 55 mph. How many miles did she travel?

27. The perimeter of a square is 64 feet. Find the length of one side. $(P = 4s)$

28. Find the circumference of a circle whose radius is 5 inches. $(C = 2\pi r)$

29. Find the area of a circle whose radius is 4 inches. $(A = \pi r^2)$

30. A triangle has a 4-foot base and is 3 feet high. Find the area. $\left(A = \frac{1}{2} b \cdot h\right)$

9.4 **Using Equations in One Variable to Solve Coin and Ticket Problems (Optional)**

OBJECTIVE

● Solve coin and ticket problems using equations in one variable.

Certain types of problems—including coin, stamp, ticket, and motion problems, which take into account the rate of the wind or the current—are commonly solved by using two equations in two variables (a topic beyond the scope of this text.) In this section, we solve them by using only one variable.

EXAMPLE 1   Jack has 11 coins in his pocket. The total value of the coins is $0.85. If he has only nickels and dimes, how many of each type does he have?

   Strategy: Set up a table. Fill in the given information.

Type of coin	Unit value of each type	Number of each type	Total value
nickels	0.05		
dimes	0.10		
Totals		11	0.85

Assign a variable to the number of coins of one type.

Type of coin	Unit value of each type	Number of each type	Total value
nickels	0.05	n	
dimes	0.10		
Totals		11	0.85

Express the number of the other types of coins in terms of this variable. Since the sum of the number of nickels and dimes is 11, the number of dimes must be 11 minus the number of nickels. We've assigned the variable n to the number of nickels, so $11 - n$ = number of dimes.

Type of coin	Unit value of each type	Number of each type	Total value
nickels	0.05	n	
dimes	0.10	$11 - n$	
Totals		11	0.85

Complete the table.

Type of coin	Unit value of each type	Number of each type	Total value
nickels	0.05	n	$0.05n$
dimes	0.10	$11 - n$	$0.10(11 - n)$
Totals		11	0.85

Set up the equation. Use the information in the last column.

Formula: Total value of nickels + Total value of dimes = Total amount of money

$$0.05n + 0.10(11 - n) = 0.85$$
$$5n + 10(11 - n) = 85$$
$$5n + 110 - 10n = 85$$
$$-5n + 110 = 85$$
$$-5n + 110 - 110 = 85 - 110$$
$$-5n = -25$$
$$\frac{-5n}{-5} = \frac{-25}{-5}$$
$$n = 5$$
$$11 - n = 6$$

Answer: He has 5 nickels and 6 dimes. Justify this answer.

PROBLEM SET 9.4 (Optional)

Directions: Solve these problems. Use the grids provided.

1. Suda has $1.70 in dimes and quarters. She has a total of 8 coins. How many dimes and how many quarters does she have?

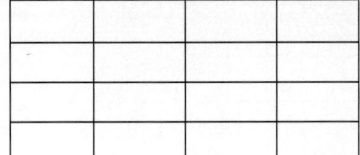

2. Paco and Elise plan to take family and friends to the XSU playoff game. They purchase 7 tickets for $60. Adult tickets cost $12 each, and children's tickets cost $8 each. How many adults and children are in their party?

3. Mandy and Cindy have $1 in nickels and dimes. They have twice as many dimes as nickels. How many dimes and how many nickels do they have?

4. On Family Airlines, children fly for half-fare. On a recent flight from New York to Florida, 130 paying passengers were aboard. The regular one-way fare for this flight was $110, and the total revenue from tickets was $11,825. How many adults and how many children were on the flight?

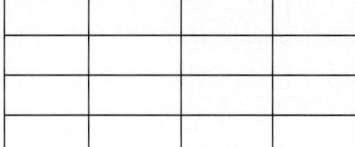

5. The Robinsons' 8 children ask their parents for $10.60 to pay for their tickets to the matinee at the Discount Cinema. If adult tickets (age 12 and over) cost $1.95 each and children's tickets cost $0.95 each, how many of the Robinson's children are 12 or older? How many are under 12 years of age?

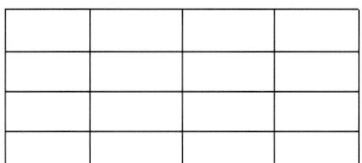

6. Dina goes to the post office to buy Christmas stamps for the cards she plans to send to family and friends. She has some leftover 29-cent stamps, so she buys some 3-cent stamps to add to them. She also buys some 32-cent stamps. She pays $5.13 for the stamps she purchases and buys 4 more 32-cent stamps than 3-cent stamps. How many of each does she buy?

7. Construct and solve a coin or ticket problem.

SMALL GROUP ACTIVITY *Strategies for Problem Solving*

Activity: Write and solve six mathematical problems, each using a different strategy from Section 9.2.

CHAPTER 9 REVIEW

After completing Chapter 9, you should have mastered the objectives listed at the beginning of each section. Test yourself by working the following problems. Refer to the sections indicated as needed.

• *Demonstrate a heightened awareness of problem solving (9.1).*

1. Complete The Mathematics Journal entry on "Thinking about Problem Solving." _____

• *Develop a model to be used in problem solving (9.1).*

2. Write a brief essay discussing and evaluating the model your group developed. _____

• *Use appropriate strategies for solving mathematical problems (9.2).*

3. List the six strategies developed in this section. Add one of your own. _____

• *Solve mathematical problems by using formulas (9.3).*

4. A commuter carpools to the station and then takes the train into the city every day. The entire trip is 50 miles and takes 1 hour. The car averages 30 mph and the train 60 mph. How much time is spent on the train? _____

5. Over the summer break, Rick took a plane trip from Cleveland to a city in California 3247 miles from the Cleveland airport. If the flight took him 4 hours and 36 minutes, what was the plane's speed in miles per hour? _____

6. Write and solve a problem using the distance formula. _____

7. A rectangular athletic field has a width of 60 feet and a length of 100 feet. What is the area in square feet? ($A = L \cdot W$) _____

8. Write and solve a problem using a formula you are familiar with. _____

• *Solve mathematical problems by using equations (9.3).*

9. The state of Ohio has one-third the population of California and twice the population of Indiana. If the total population of the three states is 45,000,000, find the population of each state. ___

10. Write a problem that can be solved by using an equation.

11. Write the step-by-step explanation you would give to a fellow student when telling him or her how to solve the problem from Problem 10.

• *Solve coin and ticket problems using equations in one variable (9.4) (Optional).*

12. The theater department raised $361 by selling 215 tickets for the fall play. If it charged $1 for each student tricket and $3 for each general admission ticket, how many of each type of ticket did the department sell?

_____

13. Write and solve a coin or ticket problem. _____

CHAPTER 9 PRACTICE TEST

Solve the following problems. Show all work. Label all answers.

1. Ken and his roommate, Jake, have 150 CDs between them. Ken has twice as many CDs as Jake. How many CDs does Ken have?

2. Your car has $\frac{1}{2}$ tank of gas at the beginning of a trip. You stop for gas once during the trip and put 11 gallons into the tank. At the end of the trip, you have $\frac{1}{3}$ tank of gas left. Your car has a 15-gallon tank and gets about 21 mpg.
   a. Approximately how many gallons of gas did you use during the trip?
   b. How many miles did you drive while on the trip?

3. On Friday, you and a friend travel from Cleveland to Pittsburgh, a distance of 125 miles, to go skiing for the weekend. Going to the ski resort you average 60 mph. On the trip back to Cleveland Sunday night, you get caught in a snowstorm and average only 45 mph. How much time did you spend driving to and from Pittsburgh?

4. There are four boxes. Inside each box are three smaller boxes. Inside each of these smaller boxes are two boxes. What is the total number of boxes?

5. The Smiths' living room is 19 feet by 13 feet. How many square feet of carpeting do they need to cover the floor? (Use the formula: $A = L \cdot W$.)

6. Write the next number. 5, 7, 4, 6, 3, 5, __

7. What is the missing letter? B Y __ X D W

8. Decode this cryptogram. Hint: *o* stands for *e* and *e* stands for *o*.

"quwo ro xutosbz es quwo ro koibp."
—Patrick Henry

9. As campaign manager for your best friend who is running for Student Senate, you want to station workers outside the left and right exit doors of the cafeteria to pass out flyers during lunch time next Tuesday. A recent campus survey shows that 35 of every 100 students exit by the left door. The cafeteria manager tells you that approximately 950 students each lunch there every day. How many flyers should you provide to the worker at the right door?

10. Your social club is planning a fund raiser for a local charity. The plan is to sell raffle tickets. The group has $350 to give away in prize money and would like to have 3 cash prizes. The second prize will be twice the third prize, and the first prize will be twice the second prize. How much money will each prize be worth?

11. Choose one.
A. The mathematician Henri Poincare (1854–1912) is credited with this quotation: "Thought is only a flash between two long nights, but this flash is everything." Write a paragraph interpreting this statement with respect to the problem-solving process.
B. In one of A. A. Milne's famous Winnie the Pooh children's books, Rabbit reminds Pooh and Piglet that problem solving ". . . always takes longer than you think." Write a two-paragraph discussion of this statement in light of this chapter and your own personal experiences.

Optional:

12. Ben has $1.70 in nickels and quarters. If he has a total of 10 coins, how many nickels and how many quarters does he have?

13. The perimeter of a rectangle is 600 feet. The length is twice the width. Find the width. $(P = 2L + 2W)$

	THE MATHEMATICS JOURNAL
	Thinking about Problem Solving
	Task: Reflect for a few minutes on any changes that have taken place in
	your thinking about problem solving as you studied this chapter.
	Write a one-page (150-word) entry on this topic.

 ## CUMULATIVE REVIEW: CHAPTERS 7, 8, and 9

DIRECTIONS *Test your mastery of the topics covered in Chapters 7, 8, and 9 by taking this practice test. All items are worth 2 points unless indicated otherwise. The total number of points possible is 100. Remember, this is a review, so check your answers and then refer to the appropriate chapter and section to get assistance in correcting any mistakes.*

COMPLETION *Choose the appropriate word or phrase from the list below.*

Distributive Property
Greater than or equal to
Less than or equal to
Multiplication Properties of Inequalities
Multiplication Property of Equality
Repeating decimal
Irrational number

Property of Reciprocals
Less than
Greater than
Addition Property of Inequalities
Addition Property of Equality
Identity Property of Multiplication
Scientific notation

1. The symbol $\geq$ is used in mathematics for the phrase _____.

2. The symbol $<$ is used to represent the phrase _____.

3. If $6x + 9 = 7$, then the statement that $6x + 9 + (-9) = 7 + (-9)$ is an example of the _____

_____.

4. The number π is a/an _____.

5. The _____ states that, for all a, b, $c \in$ Reals, $a(b + c) = ab + ac$ and $ab + ac = a(b + c)$.

6. If $7x = 14$, then the statement that $\frac{1}{7} \cdot 7x = 14 \cdot \frac{1}{7}$ is an example of the _____

_____.

7. If $-x < 6$, then the statement that $(-1)(-x) > (-1)(6)$ is an example of one of the _____

_____.

8. Writing 0.00345 as 3.45×10^{-3} is an example of expressing a number in _____.

9. $0.23\overline{5}$ is an example of a/an _____.

10. The _____ states that, for all real numbers $a(a \neq 0)$, $a \cdot \dfrac{1}{a} = 1$.

Follow the directions given for each item.

11. Write 3.153×10^5 in standard form. _____

12. Write 0.0000832 in scientific notation. _____

13. Express 0.015 as a fraction. _____

14. Express $0.\overline{3}$ as a fraction. _____

15. Which is greater 0.6 or $\dfrac{2}{3}$? _____

16. Express $\dfrac{2}{9}$ as a decimal. _____

17. Express $\dfrac{3}{20}$ as a decimal. _____

18. Express $5\dfrac{1}{4}$ as a decimal. _____

19. Round 0.461 to the nearest tenth. _____

20. Round 1.007 to the nearest hundredth. _____

Perform the indicated operations and simplify.

21. $0.402 \div 0.6 =$ _____

22. $(2.95)(1.232) =$ _____

23. $159.73 + 5.71 =$ _____

24. $27.83 - 165.894 =$ _____

25. $0.9 + \dfrac{1}{3} - 2.5 =$ _____

26. $\dfrac{3}{5}(0.68 + 2.75) =$ _____

*Solve for the variable. * Indicates 3 points each.*

27. $23 + a = \dfrac{36}{6}$

28. $x + 8 - 3 = 5 + x$

29. $-\dfrac{1}{5}y = 3$

30. $7x = 56$

31. $5x - 3 = 10$

32. $2n + 10 = 3n - 9$

33. $2 - b = -8 - 3b$

34. $-16 + 3a - 4 - 5a = 20 - 12a$

35. $-0.44y = 13.2$

36. $-2.3x - 2 = 1.7x$

37. $\dfrac{y}{4} + 1 = \dfrac{3}{8}$

38. $\dfrac{1}{5}x - \dfrac{2}{3} = 7$

Solve and graph.

39. $y + 17 \leq 20$

40. $-x + 5 > 3*$

41. $2x \geq -12$

42. $7 + 8y < -9$

43. $15 > 3y$

Assign variables. Use equations, formulas, or diagrams when appropriate. Label answers. (4 points each)

44. A motorist travels at a rate of 65 mph. Find the distance she has traveled in 3.5 hours.

45. Mary Ruth and Meg are 6600 feet apart and start running toward each other. When they meet 3 minutes later, Mary Ruth calculates that she has run 3600 feet. At what rate is Meg running?

46. Andy wants to join Alpha Pi Beta, the mathematics honorary society. To be eligible for membership, a student must have an interest in mathematics, have completed 32 semester hours, and have a cumulative GPA of 2.5 on a 4.0 scale. Prior to the current semester, Andy had taken 25 semester hours and had a cumulative GPA of 2.9. This semester he is taking 15 semester hours. What GPA must he get for this current semester to be eligible for membership at the end of the term?

47. Choose A or B.
A. Find a pattern. Use it to write the missing terms: 1, 3, 7, 15, __, __, __, 255

B. Jack has 11 coins in his pocket. The total value of the coins is $0.85. If he only has nickels and dimes, how many nickels and how many dimes does he have? (optional Section 9.4)

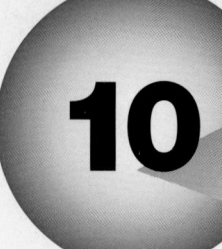

10 Ratio and Proportion

10.1 Ratios
10.2 The Fundamental Property of Proportion
10.3 Applications of Proportion

Introduction to Ratio and Proportion

The Golden Rectangle describes a rectangle in which the ratio of the longer side to the shorter side is the golden ratio, approximately 1.6 to 1; 1.6 is the approximation of the irrational number $\frac{1 + \sqrt{5}}{2}$. A 3×5-inch index card is a rectangle with the proportions of the golden ratio. Those proportions are aesthetically pleasing to the eye and were used by the Greeks when they built the Parthenon and by the Egyptians when they built the pyramids. The golden ratio appears frequently in the paintings of Leonardo DaVinci and Georges Seurat. This same golden ratio occurs in the quotients of the terms of the Fibonacci series 1, 1, 2, 3, 5, 8, 13, 21 . . . . Many natural objects have links with the Fibonacci series: sunflower petals, honey combs, and the spirals on a pine cone. The ratio of the circumference of a circle (the distance around a circle) to the circle's diameter is always π, an irrational number with a value of approximately 3.14. A construction of a golden rectangle follows.

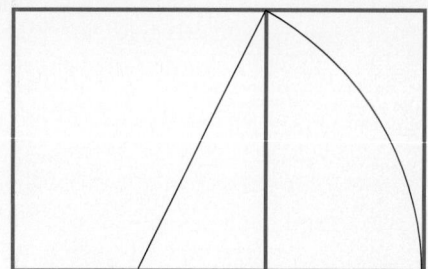

10.1 Ratios

OBJECTIVES

● Express ratios in lowest terms.

● Express units of length, weight, volume, and time as ratios.

● Express the relationship between two quantities as a rate.

We use ratios and rates to compare the exchange rates between different currencies, such as the dollar and the Japanese yen. Universities also use ratios when they boast about their student-instructor ratio to show that they have a large teaching faculty and small classes.

DEFINITION **Ratio:** A comparison of two numbers by division.

There are four ways to write a ratio: a is to b, $a:b$, $\frac{a}{b}$, and $a \div b$. The ratio of two numbers can be expressed as a fraction, with the first number being the numerator and the second number the denominator. The numerator and denominator of a ratio must be in the same units, unless they are labeled. For a ratio to be in lowest terms, the numerator and denominator will be relatively prime.

EXAMPLE 1   The ratio 2 to 3 or $2:3$ is written as a fraction $\frac{2}{3}$.

EXAMPLE 2   The ratio 15 to 10 or $15:10$ is written as a fraction $\frac{15}{10}$, reduced to lowest terms is $\frac{3}{2}$.

$$15:10 = 3:2$$

EXAMPLE 3   A ratio of 50 to 25 written as a fraction is $\frac{50}{25}$. Reduced to lowest terms it is $\frac{2}{1}$.

$$50:25 = 2:1$$

 USING THE CALCULATOR

$50:25$ Enter 50 $\div$ 25 $=$ **2** $\leftarrow$ Display

EXAMPLE 4   Simplify the ratio $\frac{5}{8} : \frac{3}{8}$.

$$\dfrac{\frac{5}{8}}{\frac{3}{8}}$$ Rewrite as a complex fraction.

$$= \frac{5}{8} \div \frac{3}{8}$$ Division is the same as multiplication by a reciprocal.

$$= \frac{5}{8} \cdot \frac{8}{3}$$

$$= \frac{5}{3} \text{ or } 5:3$$

EXAMPLE 5   Simplify the ratio $1\frac{1}{4}$ to $1\frac{3}{4}$.

$$= 1\frac{1}{4} : 1\frac{3}{4}$$

$$= \dfrac{1\frac{1}{4}}{1\frac{3}{4}}$$ Rewrite as a complex fraction.

$$= \frac{5}{4} \div \frac{7}{4}$$ Change the mixed numbers to fractions.

$$= \frac{5}{7} \text{ or } 5:7$$

EXAMPLE 6   Simplify the ratio 0.05 to 0.15.

$$0.05 : 0.15$$

$$= \frac{0.05}{0.15}$$ Rewrite as a fraction.

$$= \frac{0.05}{0.15} \cdot \frac{100}{100} = \frac{5}{15}$$ Since there are 2 decimal places, multiply by $\frac{100}{100}$.

$$= \frac{1}{3} \text{ or } 1 : 3$$

EXAMPLE 7   Simplify the ratio 0.3 to 3.

$$0.3 : 3$$

$$= \frac{0.3}{3}$$ Rewrite as a fraction.

$$= \frac{0.3}{3} \cdot \frac{10}{10} = \frac{3}{30}$$ Since there is 1 decimal place, multiply by $\frac{10}{10}$.

$$= \frac{1}{10} \text{ or } 1 : 10$$

Note   When writing ratios that involve length, time, weight, or other measurements, the quantities must be in the same units, unless they are labeled. We cannot compare inches to feet or minutes to hours. The quantities need to be changed to the same units. To avoid using fractions, it is easiest to change to the smaller unit or to change both quantities to an equivalent unit. For example, when working a problem with feet and inches, convert the feet to inches. When working a problem with dimes and quarters, convert to nickles or pennies.

Units of Linear Measure, Capacity, Weight, and Time

The following conversions will help you find equivalent measures, which will enable you to compare quantities that need to be changed to the same unit.

Linear Measure
12 inches = 1 foot
3 feet = 1 yard
5280 feet = 1 mile

Capacity
8 ounces = 1 cup
2 cups = 1 pint
2 pints = 1 quart
4 quarts = 1 gallon

Weight
16 ounces = 1 pound
2000 pounds = 1 ton

Time
60 seconds = 1 minute
60 minutes = 1 hour

A complete listing of linear measure can be found in Appendix B.

Practice Problems

Express as ratios.

10. 6.4 to 0.8

11. 8.1 : 0.3

12. 0.5 to 0.25

13. 5 : 0.005

EXAMPLE 8 Express 9 inches to 3 feet as a ratio. If 1 foot is equal to 12 inches, 3 feet are 12(3) inches = 36 inches.

3 feet = 36 inches.

Write the ratio of 9 inches to 36 inches.

$$\frac{9}{36} = \frac{1}{4} \text{ or } 1:4$$

EXAMPLE 9 Express 20 minutes to 1 hour as a ratio.

1 hour = 60 minutes. Write the ratio of 20 minutes to 60 minutes.

$$\frac{20}{60} = \frac{1}{3} \text{ or } 1:3$$

EXAMPLE 10 Write the ratio of 4 nickels to 4 dimes.

Method 1

1 nickel = 5 cents, so 4 nickels = 20 cents.
1 dime = 10 cents, so 4 dimes = 40 cents.
Write the ratio of 20 cents to 40 cents.

$$\frac{20}{40} = \frac{1}{2} \text{ or } 1:2$$

Method 2
Write the ratio of 4 nickels to 4 dimes.
Rewrite as nickels.
4 nickels = 4 nickels
1 dime = 2 nickels, so 4 dimes = 8 nickels.
A ratio of 4 nickels to 8 nickels

$$\frac{4}{8} = \frac{1}{2} \text{ or } 1:2$$

DEFINITION **Rate:** A ratio that compares the relationship between two kinds of quantities. It may be written as a quotient of two numbers and can be divided to result in a unit rate. When you write rate you mean unit rate.

Note Rates need to labeled. Ratios do not need to be labeled. Examples of rates are miles per hour, miles per gallon, and ounces per cubic foot.

EXAMPLE 11 What is the rate in miles per hour of a car traveling 100 miles in 5 hours?

$$\frac{100 \text{ miles}}{5 \text{ hours}}$$ Divide the denominator into the numerator to find miles per hour

$$\frac{20 \text{ miles}}{1 \text{ hour}}$$

= 20 miles per hour or 20 mph

USING THE CALCULATOR

100 miles per 5 hours.

Enter 100 ÷ 5 = **20** ← Display

A rate of 20 miles per hour or 20 mph.

Practice Problems

Write the following ratios as fractions in lowest terms.

14. 6 inches to 2 feet.

15. 5 minutes to 50 seconds.

16. 2 quarters to 8 nickels.

17. 2 feet to 2 yards.

EXAMPLE 12 Mark bought 3 pounds of fish for $13.50. What was the price per pound?

$$\frac{13.50}{3} \quad \frac{\text{price}}{\text{pounds}}$$

Divide the denominator into the numerator to find price per pound.

$$\frac{13.50}{3} = \$4.50 \text{ per pound}$$

$$= \$4.50 \text{ per pound. (unit rate)}$$

USING THE CALCULATOR

13.50 per 3 pounds.

Enter 13.50 ÷ 3 = **4.5** ← Display

A rate of $4.50 per pound.

EXAMPLE 13 In terms of gas consumption, what is the rate in miles per gallon of a car that can be driven 150 miles on 5 gallons of gas?

$$\frac{150}{5} \quad \frac{\text{miles}}{\text{gallons}}$$

Divide the denominator into the numerator to find miles per gallon.

$$= 30 \text{ miles to 1 gallon}$$

$$= 30 \text{ miles per gallon or 30 mpg.}$$

USING THE CALCULATOR

150 miles per 5 gallons.

Enter 150 ÷ 5 = **30** ← Display

A rate of 30 miles per gallon or 30 mpg.

EXAMPLE 14 What is the rate per pound of 50 pounds of potatoes costing $11.00?

$$\frac{\$11.00}{50} \quad \text{pounds}$$

$$\$0.22 \text{ to 1 pound}$$

$$= \$0.22 \text{ (22 cents) per pound.}$$

USING THE CALCULATOR

Enter 11.00 ÷ 50 = **0.22** ← Display

A rate of $0.22 (22 cents) per pound.

EXAMPLE 15 What is the rate in ounces per cubic foot of 15 ounces per 5 cubic feet?

$$\frac{15}{5} \quad \frac{\text{ounces}}{\text{cubic feet}}$$

$$= 3 \text{ ounces per cubic foot}$$

Practice Problems

Find the rates.

18. A car traveling 75 miles in 5 hours, in miles per hour.

19. A runner running 100 yards in 25 seconds, in yards per second.

20. What is the rate in miles per gallon of a car that can travel 450 miles on 12 gallons of gas?

21. What is the rate per ounce of a 64 ounce jar of juice costing $1.60?

22. What is the rate per ounce of a 12-ounce package of frozen peas costing $2.46?

USING THE CALCULATOR

15 ounces per 5 cubic feet.

Enter 15 ÷ 5 = **3** ← Display

A rate of 3 ounces per cubic foot.

Note Remember, rates must be labeled. Ratios do not need to be labeled.

EXAMPLE 16 Which is a better buy: a box of CoCo cookies that costs $3.68 and weighs 16 ounces or a box of Mave's cookies that weighs 13 ounces and costs $3.38?

What is the unit price of a 16-ounce box at $3.68?

$$\frac{3.68}{16 \text{ ounce}} = \$0.23 \text{ (23 cents) per ounce}$$

What is the unit price of a 13-ounce box at $3.38?

$$\frac{3.38}{13 \text{ ounce}} = \$0.26 \text{ (26 cents) per ounce}$$

The box of CoCo cookies is the better buy.

ANSWERS TO PRACTICE
PROBLEMS
1. 1:3 2. 6:7 3. 2:3 4. 1:3
5. 5:4 6. 2:3 7. 3:5 8. 1:2
9. 1:3 10. 8:1 11. 27:1
12. 2:1 13. 1000:1 14. 1:4
15. 6:1 16. 5:4 17. 1:3
18. 15 mph 19. 4 yards per sec
20. 37.5 mpg 21. 2.5 cents per oz
22. 20.5 cents per oz

PROBLEM SET 10.1

Simplify the following ratios. Express each answer in lowest terms.

1. 19 to 38

2. 42 to 72

3. 75 to 50

4. 34 to 51

5. 30 to 35

6. 7 to 18

7. 92 to 42

8. 40 to 60

9. 88 to 55

10. 250 to 100

11. 24 : 60

12. 92 : 27

13. 54 : 45

14. 84 : 63

15. $72:90$

16. $\dfrac{2}{3}$ to $\dfrac{5}{6}$

17. $\dfrac{3}{5}$ to $\dfrac{9}{10}$

18. $2\dfrac{1}{2}$ to $1\dfrac{3}{4}$

19. $\dfrac{1}{2}$ to $3\dfrac{1}{2}$

20. $\dfrac{5}{9}$ to $\dfrac{9}{10}$

21. $\dfrac{3}{4}$ to $\dfrac{3}{8}$

22. $4\dfrac{1}{8}$ to $4\dfrac{1}{2}$

23. $\dfrac{7}{10}$ to $3\dfrac{1}{2}$

24. $2\dfrac{1}{3}$ to $1\dfrac{2}{3}$

25. $\dfrac{2}{7}$ to $\dfrac{4}{7}$

26. $\dfrac{2}{3}:\dfrac{7}{8}$

27. $\dfrac{3}{4}:\dfrac{1}{8}$

28. $3\dfrac{1}{2}:\dfrac{7}{8}$

29. $\dfrac{4}{5}:\dfrac{7}{10}$

30. $\dfrac{3}{19}:\dfrac{7}{38}$

31. 0.8 to 6.4

32. 4 to 3.2

33. 0.7 to 21

34. 7 to 2.1

35. 9.5 to 3.8

36. 0.002 to 2

37. $\frac{3}{4}$ to 0.75

38. 1.25 to $\frac{1}{2}$

39. 3.7 to $\frac{7}{10}$

40. 0.6 to 1.2

41. 4 : 0.04

42. 9.9 : 0.09

43. 0.75 : 2.5

44. 5.4 : 0.45

45. 0.49 : 7

46. 0.4 : $\frac{1}{4}$

47. $\frac{7}{8}$: 0.375

48. 0.5 : $\frac{2}{3}$

49. 3.2 : $\frac{7}{10}$

50. $\frac{3}{7}$: $\frac{5}{14}$

Simplify the following ratios. Answers should be in lowest terms.

51. 2 feet to 2 inches

52. 15 minutes to 2 hours

53. 7 yards to 2 feet

54. 4 hours to 45 minutes

55. 10 seconds to 2 minutes

56. 32 ounces to 2 pounds

57. 5000 pounds to 4 tons

58. 6 inches to 2 yards

59. 5 minutes to 50 seconds

60. 10 feet to 80 inches

61. 3 weeks to 14 days

62. 11 nickels to 2 dimes

63. 6 pints to 5 quarts

64. 8 cents to 2 nickels

65. 5 gallons to 5 quarts

66. 10 ounces to 3 pints

67. 3 pounds to 14 ounces

68. 400 pounds to 2 tons

69. 2 hours to 19 minutes

70. 14 inches to 3 feet

71. 10 inches to 3 feet

72. 3 hours to 20 minutes

73. 4 feet to 3 yards

74. 50 minutes to 3 hours

75. 20 seconds to 3 minutes

76. 3 pounds to 14 ounces

77. 2 tons to 3000 pounds

78. 9 inches to 3 yards

79. 45 seconds to 4 minutes

80. 7 inches to 4 feet

81. 5 days to 2 weeks

82. 3 quarters to 6 nickels

83. 7 dimes to 4 nickels

84. 3 pints to 3 quarts

85. 2 quarts to 3 gallons

86. 15 ounces to 3 pounds

87. 8 ounces to 2 pints

88. 250 pounds to 3 tons

89. 5 hours to 25 minutes

90. 7 inches to 5 feet

91. 5 months to 3 years

Applications:
Simplify the following unit rates. Write answers as decimals. Label all answers.

92. What is the rate in miles per hour of a train traveling 300 miles in 5 hours?

93. A man drives 495 miles in 9 hours. What is his rate in miles per hour?

94. Carla worked for 7 months and earned $9086. What is her rate of pay in dollars per month?

95. Thomas earned $173.70 for 30 hours work at the pizza shop. What is his rate of pay in dollars per hour?

96. A bathtub holds 51 gallons of water and can be filled in 17 minutes. At what rate is the bathtub filled in gallons per minute?

97. There are 299 students in 13 sections of math. What is the rate of students per section?

98. A car's gas tank is filled and holds 11 gallons of gas. If the gas tank is empty after 297 miles, what rate in miles per gallon does the car get?

99. A feather comforter weighs 8 ounces and has a volume of 2 cubic feet. What is the rate in ounces per cubic foot?

100. Compare the rates in cents per ounce of a 64-ounce bottle of juice that costs $2.56 and a 48-ounce can of juice that costs $1.44. Which is the better buy?

101. Which is the better buy: a jar of peanut butter that costs $2.98 and weighs 28 ounces or a jar that weighs 20 ounces and costs $2.15?

102. A car travels 260 miles in 4 hours. Express the rate in miles per hour.

103. At a small college, there are 2691 students and 207 professors. What is the rate of students per professor?

104. Find the rate in dollars per cubic yard of garden mulch that costs $141 for 6 cubic yards.

105. What is the rate in miles per hour of a train traveling 518 miles in 7 hours?

106. A man drives 708 miles in 12 hours. What is his rate in miles per hour?

107. Mary Ann worked for a year and earned $40,536. What is her rate of pay in dollars per month?

108. Kevin earned $142.56 for 27 hours as a waiter. What is his rate of pay in dollars per hour?

109. A bathtub holds 65 gallons of water and can be filled in 26 minutes. At what rate is the bathtub filled in gallons per minute?

110. A heavy wooden walking stick weighs 2 pounds and has a volume of 36 cubic inches. What is the rate in ounces per cubic inch?

111. Find the rates in cents per ounce to determine which is the better buy: an 18-ounce box of cereals that costs $2.34 or a 13-ounce box that costs $2.21.

112. Which is the better buy: a jar of strawberry jam that costs $3.68 and weighs 16 ounces or a jar that weighs 13 ounces and costs $3.38?

113. At a large state college, there are 33,912 students and 1256 professors. What is the rate of students per professor?

10.2 The Fundamental Property of Proportion

OBJECTIVE
- Use the Fundamental Property of Proportion, which states that the product of the extremes equals the product of the means.

In Sections 4.8, 5.7, and in Chapter 8, we studied how to solve equations. In this section, we will study how to solve another type of equation, a proportion.

DEFINITION   **Proportion:** An equation in which both sides are fractions or ratios. The proportion 5:10 = 1:2 is read 5 is to 10 as 1 is to 2, and written as fractions is $\frac{5}{10} = \frac{1}{2}$. In symbols, a proportion is written

$$\frac{a}{b} = \frac{c}{d}$$

and is read a is to b as c is to d.

RULE **The fundamental property of proportion**
For all real numbers a, b, c, d $b, d \neq 0$.

$$\text{If } \frac{a}{b} = \frac{c}{d}$$

then $ad = bc$

$$\frac{a}{b} = \frac{c}{d} \quad \text{LCD } (b, d) = bd$$

$$bd\left(\frac{a}{b}\right) = \left(\frac{c}{d}\right)bd$$

$$\frac{a\cancel{b}d}{\cancel{b}} = \frac{bc\cancel{d}}{\cancel{d}}$$

$$ad = bc \quad \text{See Chapter 8.}$$

Note   This property is stated in English as the product of the means is equal to the product of the extremes.

The word *means* comes from the Latin word meaning *in the middle,* and the word *extremes* comes from the Latin word meaning *at the ends.* When we write the proportion this way, $a : b = c : d$, we can readily see that a and d are at the ends (extremes), and b and c are in the middle (means).

EXAMPLE 1   Check this with the original example.

$$\frac{5}{10} = \frac{1}{2}$$
$$5(2) = 10(1)$$
$$10 = 10$$

The Fundamental Property of Proportion can be used to find the value of a missing extreme or mean to make the proportion true. When we find that missing term, we are solving the proportion.

EXAMPLE 2 $\dfrac{2}{3} = \dfrac{x}{6}$

$3x = 12$ For what value of x will the proportion be true?
$\dfrac{3x}{3} = \dfrac{12}{3}$ Put the product of the means equal to the product of the extremes. Divide both sides by the coefficient to the variable.
$x = 4$

The term missing from the proportion is 4.

USING THE CALCULATOR

Enter 2 $\times$ 6 $\div$ 3 $=$ **4** ← Display

Practice Problems

Are these true proportions?

1. $\dfrac{9}{7} = \dfrac{27}{21}$

2. $\dfrac{10}{9} = \dfrac{40}{36}$

Write the proportion.

3. 6 is to 3 as 18 is to 9

4. 4 is to 12 as 1 is to 3

Solve these proportions.

5. $\dfrac{2}{5} = \dfrac{4}{x}$

6. $\dfrac{4}{0.2} = \dfrac{20}{x}$

7. $\dfrac{\frac{1}{3}}{x} = \dfrac{3}{1}$

8. $\dfrac{7}{21} = \dfrac{x}{105}$

9. $\dfrac{\frac{1}{3}}{\frac{4}{9}} = \dfrac{\frac{1}{4}}{x}$

EXAMPLE 3

$$\frac{5}{0.5} = \frac{20}{x}$$

$$5x = 20(0.5) \quad \text{Solve the proportion.}$$

$$5x = 10$$

$$\frac{5x}{5} = \frac{10}{5}$$

$$x = 2$$

The missing term from the proportion is 2.

USING THE CALCULATOR

Enter 20 ☐× .5 ☐÷ 5 ☐= **2** ← Display

EXAMPLE 4

$$\frac{n}{12} = \frac{\frac{1}{4}}{\frac{1}{2}}$$

$$\frac{1}{2}(n) = \frac{1}{4}(12) \quad \text{Solve the proportion.}$$

$$\frac{1n}{2} = 3$$

$$\frac{2}{1} \cdot \frac{1n}{2} = \frac{3}{1} \cdot \frac{2}{1}$$

$$n = 6$$

EXAMPLE 5 $\frac{1}{2}$ is to $\frac{3}{4}$ as $\frac{5}{6}$ is to what number?

$$\frac{\frac{1}{2}}{\frac{3}{4}} = \frac{\frac{5}{6}}{x} \quad \text{Set up as a proportion.}$$

$$\frac{1}{2}(x) = \left(\frac{3}{4}\right)\left(\frac{5}{6}\right) \quad \text{Solve the proportion.}$$

$$x = \frac{5}{4}$$

ANSWERS TO PRACTICE
PROBLEMS
1. $189 = 189$ **2.** $360 = 360$
3. $\frac{6}{3} = \frac{18}{9}$ **4.** $\frac{4}{12} = \frac{1}{3}$
5. $x = 10$ **6.** $x = 1$
7. $x = \frac{1}{9}$ **8.** $x = 35$ **9.** $x = \frac{1}{3}$

PROBLEM SET 10.2

Determine if each proportion is true or false.

1. $\frac{1}{2} = \frac{5}{10}$

2. $\frac{5}{8} = \frac{10}{16}$

3. $\dfrac{27}{81} = \dfrac{33}{99}$

4. $\dfrac{0.25}{\frac{3}{5}} = \dfrac{1}{2}$

5. $\dfrac{0.04}{1.2} = \dfrac{1}{3}$

6. $\dfrac{100}{\frac{1}{2}} = \dfrac{50}{\frac{1}{4}}$

Solve these proportions.

7. $\dfrac{x}{1} = \dfrac{12}{6}$

8. $\dfrac{2}{5} = \dfrac{4}{x}$

9. $\dfrac{3}{8} = \dfrac{9}{y}$

10. $\dfrac{5}{a} = \dfrac{4}{8}$

11. $\dfrac{18}{4} = \dfrac{y}{10}$

12. $\dfrac{16}{12} = \dfrac{24}{n}$

13. $\dfrac{30}{k} = \dfrac{12}{20}$

14. $\dfrac{9}{z} = \dfrac{3}{11}$

15. $\dfrac{5}{7} = \dfrac{5}{x}$

16. $\dfrac{x}{10} = \dfrac{10}{4}$

17. $\dfrac{a}{25} = \dfrac{18}{30}$

18. $\dfrac{8}{9} = \dfrac{32}{m}$

19. $\dfrac{35}{21} = \dfrac{x}{36}$

20. $\dfrac{17}{51} = \dfrac{100}{c}$

21. $\dfrac{x}{12} = \dfrac{10}{14}$

22. $\dfrac{a}{3} = \dfrac{15}{16}$

23. $\dfrac{40}{9} = \dfrac{y}{5}$

24. $\dfrac{25}{36} = \dfrac{x}{20}$

25. $\dfrac{18}{p} = \dfrac{4}{7}$

26. $\dfrac{0.01}{0.1} = \dfrac{x}{100}$

27. $\dfrac{\frac{1}{2}}{\frac{1}{4}} = \dfrac{y}{7}$

28. $\dfrac{\frac{2}{3}}{6} = \dfrac{x}{18}$

29. $\dfrac{7}{23} = \dfrac{\frac{4}{5}}{x}$

30. $\dfrac{5}{0.5} = \dfrac{50}{x}$

31. $\dfrac{0.05}{1.2} = \dfrac{x}{6}$

32. $\dfrac{x}{6} = \dfrac{3\frac{1}{2}}{5}$

33. $\dfrac{0.64}{0.4} = \dfrac{0.96}{y}$

34. $\dfrac{0.5}{\frac{1}{4}} = \dfrac{n}{6}$

35. $\dfrac{2\frac{1}{4}}{4} = \dfrac{x}{10}$

36. $\dfrac{0.55}{0.72} = \dfrac{1.65}{m}$

37. $\dfrac{0.6}{x} = \dfrac{1.2}{0.84}$

38. $\dfrac{1.2}{2.7} = \dfrac{3.4}{x}$

39. $\dfrac{1.8}{6.4} = \dfrac{x}{0.16}$

40. $\dfrac{x}{\frac{3}{4}} = \dfrac{10}{4}$

41. $\dfrac{9}{x} = \dfrac{7}{9}$

42. $\dfrac{x}{13} = \dfrac{5}{9}$

43. $\dfrac{3}{11} = \dfrac{x}{7}$

44. $\dfrac{7}{9} = \dfrac{18}{x}$

45. $\dfrac{x}{5} = \dfrac{2\frac{1}{2}}{3}$

46. $\dfrac{2.3}{6\frac{9}{10}} = \dfrac{6}{y}$

47. $\dfrac{9\frac{1}{2}}{x} = \dfrac{2.6}{8}$

48. $\dfrac{0.3}{y} = \dfrac{4.3}{12.9}$

49. $\dfrac{x}{2\frac{1}{3}} = \dfrac{1.9}{\frac{4}{5}}$

50. $\dfrac{\frac{1}{6}}{\frac{1}{3}} = \dfrac{x}{18}$

Set up the proportions and solve.

51. 5 is to 6 as 15 is to x.

52. 3 is to 4 as 12 is to y

53. 2 is to x as 6 is to 10

54. a is to 25 as 2 is to 5

55. 2 is to 7 as x is to 7

56. y is to 9 as 24 is to 12

57. 17 is to 51 as a is to 12

58. 15 is to x as 7 is to 14

59. 3 is to 12 as 5 is to x

60. 15 is to 45 as 7 is to x

10.3 **Applications of Proportion**

OBJECTIVE

● Solve application problems using proportions.

It is important that proportions be set up accurately. The easiest way to do this is to determine the two quantities being compared.

EXAMPLE 1   A train can travel 80 miles in 60 minutes. At this rate, how far can the train travel in 36 minutes?

The two quantities are miles and minutes.

$$\frac{\text{Miles}}{\text{Minutes}} \quad \frac{}{} = \frac{}{}$$

Before you try to translate a problem into a proportion, determine the two quantities and write them in. The left side of the proportion will be the original rate, in this case the original rate of the train. Fill in the number of miles lined up with the word *miles* and the same with the number of minutes.

$$\frac{\text{Miles}}{\text{Minutes}} \quad \frac{80}{60} = \frac{}{}$$

The question asks how far the train travels; therefore, the miles $= x$ and the minutes are given in the problem as 36.

Note   A proportion is a statement that two ratios or rates are equal.

In this problem, 80 miles is to 60 minutes as x miles is to 36 minutes.

$$\frac{\text{Miles}}{\text{Minutes}} \quad \frac{80}{60} = \frac{x}{36} \qquad \text{Solve the proportion.}$$

$$60x = 80(36) \qquad \text{Apply the Fundamental Property of Proportion.}$$

$$\frac{60x}{60} = \frac{80(36)}{60} \qquad \text{Reduce to lowest terms and multiply.}$$

$$x = 48$$

USING THE CALCULATOR

Enter 80 ⊠ 36 ÷ 60 ⊟ **48** ← Display

Note   As you translate problems into proportions, always line up the numbers with like quantities, so you have two rates that compare the same quantities, in this case miles to minutes.

EXAMPLE 2   A baseball player gets 9 hits in 21 games. If he continues to hit at this rate, how many hits will he get in 56 games?

The number of hits is in the numerator on both sides of the equation and the number of games is in the denominator.

$$\frac{\text{Hits}}{\text{Games}} \quad \frac{9}{21} = \frac{x}{56} \qquad \text{Solve the proportion.}$$

$$21x = 9(56)$$

$$\frac{21x}{21} = \frac{9(56)}{21} \qquad \text{Reduce to lowest terms and multiply.}$$

$$x = 24$$

USING THE CALCULATOR

Enter 9 ⊠ 56 ÷ 21 ⊟ **24** ← Display

Practice Problems

Solve these problems.

1. A basketball player scores 162 points in 8 games. At this rate, how many points will he score in 24 games?

2. In the rectangles below, the ratio of length to width is the same. Find the width of the larger rectangle.

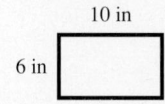

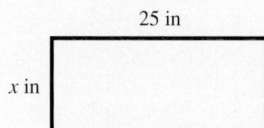

EXAMPLE 3   In the rectangles below, the ratio of length to width is the same. Find the length of the larger rectangle.

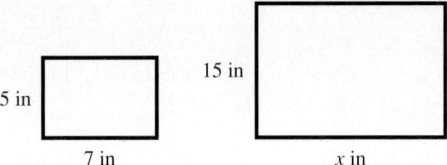

5 in

15 in

7 in

x in

$$\frac{\text{Length}}{\text{Width}} \quad \frac{7}{5} = \frac{x}{15} \qquad \text{Solve the proportion}$$

$$5x = 15(7)$$

$$\frac{5x}{5} = \frac{15(7)}{5} \qquad \text{Reduce to lowest terms and multiply.}$$

$$x = 21$$

The length of the larger rectangle is 21 inches.

USING THE CALCULATOR

Enter 15 $\boxed{\times}$ 7 $\boxed{\div}$ 5 $\boxed{=}$ **21** ← Display

EXAMPLE 4   On a map of Ohio, the scale indicates that one-half inch corresponds to an actual distance of 40 miles. If two locations are 3.5 inches apart on the map, what is the actual distance between them?

$$\frac{\text{Inches}}{\text{Miles}} = \frac{\frac{1}{2}}{40} = \frac{3.5}{x}$$

$$\left(\frac{1}{2}\right)x = 40(3.5) \qquad \text{Solve the proportion.}$$

$$\left(\frac{1}{2}\right)x = 140 \qquad \begin{array}{l}\text{Use the Multiplication Property of}\\ \text{Equality to isolate the variable.}\end{array}$$

$$(2)\left(\frac{1}{2}\right)x = 140(2)$$

$$x = 280$$

The actual distance is 280 miles.

EXAMPLE 5   If bread is on sale at 5 loaves for $8.75, how much will 11 loaves cost?

$$\frac{\text{Loaves}}{\text{Cost}} \quad \frac{5}{8.75} = \frac{11}{x} \qquad \text{Solve the proportion.}$$

$$5x = 11(8.75)$$

$$\frac{5x}{5} = \frac{11(8.75)}{5}$$

$$x = 19.25$$

11 loaves will cost $19.25.

USING THE CALCULATOR

Enter 11 ☒ 8.75 ÷ 5 ═ **19.25** ← Display

EXAMPLE 6   An airplane flies 3300 miles in 6 hours. At this rate, how far will it fly in 120 minutes?

In this problem, we are comparing flying time in hours in one sentence and in minutes in another sentence. We need to choose either minutes or hours. Convert 6 hours to 360 minutes, or 120 minutes to 2 hours.

Using Hours

$$\frac{\text{Miles}}{\text{Hours}} \quad \frac{3300}{6} = \frac{x}{2}$$
$$6x = 3300(2)$$
$$\frac{6x}{6} = \frac{3300(2)}{6}$$
$$x = 1100$$

Using Minutes

$$\frac{\text{Miles}}{\text{Minutes}} \quad \frac{3300}{360} = \frac{x}{120}$$
$$360x = 3300(120)$$
$$\frac{360x}{360} = \frac{3300(120)}{360}$$
$$x = 1100$$

Calculating the problem either way will result in the same answer. The plane can fly 1100 miles in 2 hours or 120 minutes.

EXAMPLE 7   A solution contains 6 milliliters (ml) of alcohol and 18 ml of water. If another solution is to have the same ratio of alcohol to water and must contain 27 ml of water, how much alcohol should it contain?

$$\frac{\text{Alcohol}}{\text{Water}} \quad \frac{6}{18} = \frac{x}{27}$$
$$18x = 27(6)$$
$$\frac{18x}{18} = \frac{27(6)}{18} \quad \text{Reduce to lowest terms and multiply.}$$
$$x = 9$$

A solution of 9 ml of alcohol to 27 ml water would be the same ratio.

USING THE CALCULATOR

Enter 27 ☒ 6 ÷ 18 ═ **9** ← Display

Practice Problems

Solve these problems.

5. An airplane flies 256 miles in 20 minutes. At this rate, how far will it fly in 6 hours?

6. A punch recipe contains 8 cups of juice to 5 cups of ginger ale. How many cups of ginger ale must be added to 24 cups of juice to make the same punch?

ANSWERS TO PRACTICE PROBLEMS
1. 486 points 2. 15 in
3. 11.6 mi or $11\frac{2}{3}$ mi 4. $2.45
5. 4608 mi 6. 15 cups of ginger ale

PROBLEM SET 10.3

Write a proportion and solve.

1. A train can travel 252 miles in 4 hours. At this rate, how far can the same train travel in 7 hours?

2. A baseball player makes 22 hits in 16 games. At this rate, how many hits would he make in 56 games?

3. In the rectangles below, the ratio of length to width is the same. Find the length of the larger rectangle.

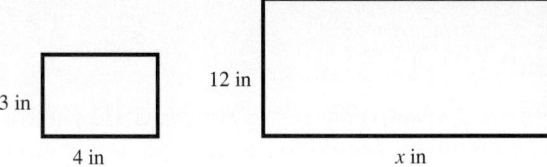

4. The scale on a city map indicates that $2\frac{1}{2}$ inches represents 1 mile. If two locations are $5\frac{3}{4}$ miles apart, how far apart are they on the map?

5. If an airplane can travel 120 miles in 30 minutes, how far can it travel in 3 hours?

6. A bottle contains a mixture of 14 fluid ounces of water and 22 fluid ounces of antifreeze. If you have 35 fluid ounces of antifreeze, how much water should be added to make the same mixture?

7. The scale of the plan of John's new house is $\frac{1}{2}$ inch to 2 feet. How long is the living room if it is represented by $6\frac{1}{2}$ inches on the plan? How long is the kitchen if it is represented by $2\frac{1}{2}$ inches? How wide is the garage if it is represented by 6 inches? How long is the hallway if it is represented by $3\frac{3}{4}$ inches?

8. Mary was planning a trip of 2619 miles. Her car gets 27 miles per gallon. How many gallons of gas should she expect to use on her trip?

9. Before the election, a poll was taken of 1280 people. Three hundred forty said they were planning to vote yes on the levy. If people vote at the same rate, how many yes votes could you expect from a voter population of 21,760? Would the levy pass or fail?

10. A 50-pound bag of fertilizer will cover a 5000-square-yard lawn. If you have a 25,000-square-yard area of lawn, how much fertilizer will you need? How many 50-pound bags will you need?

11. If an owner of a $132,500 house pays $875 in property taxes, how much would an owner of a $53,000 house pay in property taxes in the same town?

12. If Joan can read 14 pages of a novel in 12 minutes, how many minutes will it take her to read 203 pages? How long will it take her in hours?

13. A manufacturer knows that 8 of every 250 parts will be defective. If he makes 1875 parts, how many can he expect to be defective?

14. One roll of carpet will cover $3\frac{1}{2}$ stairways in a library. There are 14 stairways in the library to be carpeted. How many rolls will be needed? Will there be carpet left over?

15. A car uses a gallon of gas every 22 miles. If the car has 12 gallons of gas in the tank at the beginning of a 187-mile journey, how much gas is left at the end of the journey?

16. If 3 movie tickets cost $23.25, how much must be paid for 7 tickets?

17. It takes 102 bricks to edge a walk $76\frac{1}{2}$ feet long. How many bricks will be needed to edge a path 191.25 feet long? If bricks came in lots of 100, how many lots would need to be purchased, and how many bricks would be left over?

18. A 3-pound ham serves 8 people. How much ham would be needed to serve 13 people? Express the answer in pounds and ounces.

19. Six skirts can be made from 8 yards of fabric. How many yards of fabric will be needed to make 21 skirts?

20. It takes 478 pounds of gravel to make a path 8 feet long and 2 inches deep. How many pounds of gravel will it take to make a path 20 feet long and 2 inches deep?

21. A baseball team wins 26 of its first 44 games. At this rate, how many games can the team expect to win during the season of 154 games? How many games can the team expect to lose during the season?

22. It takes $\frac{3}{5}$ of a skein of yarn to make 1 square for an afghan. How many skeins will be needed to make a 35-square afghan?

23. Jane can swim 9 lengths of the 25-yard pool in 5 minutes. How long would it take her to swim 450 yards?

24. A baseball player gets 7 hits in 12 games. At this rate, how many hits will she get in 60 games?

25. Two cities are 28 miles apart. If the scale on a map is 3 inches to 7 miles, how far apart are the cities on the map?

26. A car can travel 112 miles in 4 hours. At this rate, how far will it travel in 15 minutes?

27. The chocolate chip cookie recipe uses 2 cups of chocolate chips for every 3 cups of flour. If you have only $1\frac{1}{2}$ cups of chocolate chips, how much flour should you use?

28. A train can travel 285 miles in 5 hours. At this rate, how far can the same train travel in 9 hours?

29. A baseball player makes 19 hits in 17 games. At this rate, how many hits would she make in 60 games?

30. A photograph that is 5 inches by 7 inches is to be enlarged to be 17.5 inches long. What is the width of the enlarged photograph?

31. If an airplane can travel 775 miles in 5 hours, how far can it travel in 36 minutes?

32. Elizabeth was planning a trip of 2635 miles. Her car gets 31 miles per gallon. How many gallons of gas should she expect to use on her trip?

33. Before the election, a poll was taken of 798 people, of which 546 said they were planning to vote yes on the levy. If people vote at that same rate, how many yes votes could you expect from a voter population of 75,810? Would the levy pass or fail?

34. A 50-pound bag of fertilizer will cover a 500-square-yard lawn. If you have a 22,000-square-yard area lawn, how much fertilizer will you need? How many 50-pound bags will you need?

35. If the owner of a $48,000 house pays $298 in property taxes, how much would the owner of a $168,000 house pay in property taxes in the same town?

36. If Jacob can read 15 pages of a novel in 13.5 minutes, how long in minutes will it take him to read 315 pages? How long will it take him in hours?

37. A manufacturer knows that 12 of every 174 parts will be defective. If she makes 609 parts, how many can she expect to be defective?

38. Seven boxes of tiles will cover the floors of $1\frac{1}{2}$ classrooms. Fifteen classrooms are to be tiled. How many boxes will be needed? Will there be tiles left over?

39. A car uses a gallon of gas every 19 miles. If the car has 21 gallons of gas in the tank at the beginning of a 361-mile journey, how much gas is left at the end of the journey?

40. If 4 theatre tickets cost $35.16, how much must be paid for 13 tickets?

41. It takes 79 bricks to edge a walk $59\frac{1}{4}$ feet long.

How many bricks will be needed to edge a walk 231.75 feet long? If bricks come in lots of 100, how many lots would need to be purchased, and how many bricks would be left over?

42. A 5-pound turkey breast can serve 12 people. How much turkey breast would be needed to serve 57 people? Express the answer in pounds and ounces.

43. Three skirts can be made from 5 yards of fabric. How many yards will be needed to make 36 skirts?

44. It takes 950 pounds of gravel to make a path 12 feet long and 2 inches deep. How many pounds of gravel will it take to make a path 30 feet long and 2 inches deep?

45. A baseball team wins 42 of its first 50 games. At this rate, how many games can the team expect to win during the season of 225 games? How many games can the team expect to lose during the season?

46. It takes $\frac{3}{4}$ of a skein of yarn to make 1 square for an afghan. How many skeins will be needed to make a 40-square afghan?

47. Ryan can swim 18 lengths of the 25-yard pool in 12 minutes. How long would it take him to swim 875 yards?

48. It takes 5 minutes to hard boil 1 egg. How long will it take to boil 6 eggs?

49. The triangles below are similar triangles; the ratio of the sides $a:d$, the sides $b:f$, and the sides $c:e$ are the same. Using proportion find the lengths of e and f.

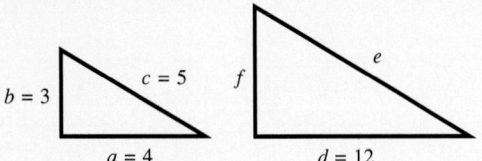

50. In the rectangles below, the ratio of length to width is the same. Find the width of the smaller rectangle.

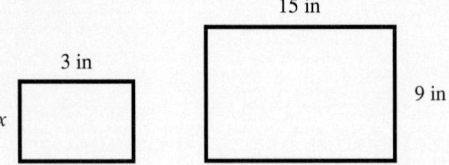

SMALL GROUP ACTIVITY   *Ratios and Fitness* The American Heart Association has determined a way to measure body fat using the waist-to-hip ratio. To find the ratio of the measurement of your waist to the measurement of your hip, measure your waist and divide that number by the measure of your hip. The result is your waist-to-hip ratio. When you have your waist-to-hip ratio, check your risk level for heart attack on the following waist-to-hip ratio chart.

Risk level	Men	Women
Low	< 0.85	< 0.75
Moderate	0.85–0.95	0.75–0.80
High	> 0.95	> 0.80

CHAPTER 10 REVIEW

The terms listed below are among those used in this chapter. Write a brief definition for each term. (Refer to the appropriate section in the chapters as needed.)

1. Ratio _____

2. Rate _____

3. The Fundamental Property of Proportion _____

After completing Chapter 10, you should be familiar with the following concepts and be able to apply them. Test yourself by working the following problems. (Refer to the indicated section as needed.)

- *Express ratios in lowest terms (10.1).*

  Simplify the following ratios. Give each answer as a reduced fraction.

4. 17 to 34 _____

5. 75 to 25 _____

6. $\dfrac{3}{4}$ to $\dfrac{8}{3}$ _____

7. $\dfrac{14}{5}$ to $\dfrac{7}{10}$ _____

8. 0.02 to 2 _____

9. 0.003 to 30 _____

10. $0.9 : \dfrac{4}{5}$ _____

11. $\dfrac{2}{3} : 0.4$ _____

- *Express units of length, weight, volume, and time as ratios (10.1).*

  Express as ratios in lowest terms.

12. 10 inches to 1 foot _____

13. 2 yards to 2 feet _____

14. 16 ounces to 4 pounds _____

15. 20 seconds to 2 minutes _____

16. 3 hours to 30 minutes _____

17. 5 pints to 2 gallons _____

18. 3 months : 7 years _____

19. 2 quarts : 3 gallons _____

• *Express the relationship between two quantities as a unit rate (10.1).*

  Find the following unit rates and label them.

20. What is the rate in miles per hour of a train traveling 550 miles in 11 hours? _____

21. John earned $11,250 in 9 months. What is his rate of pay in dollars per month? _____

22. A car can travel $195\frac{1}{2}$ miles on 8.5 gallons of gas. What rate is that in miles per gallon? _____

23. What is the rate in cents per pound of a 5-pound bag of apples costing $1.85? _____

24. There are 209 students in 11 sections of developmental math. What is the rate of students per section?

_____

25. A car's gas tank is full and holds 17 gallons of gas. If the gas tank is empty after 469.2 miles, what rate in

  miles per gallon does the car get? _____

26. A car travels 325 miles in 5 hours. Give the rate in miles per hour. _____

27. Find the rate in dollars per cubic yard of garden mulch costing $196 for 8 cubic yards. _____

• *Use the Fundamental Property of Proportion, which states that the product of the extremes
equals the product of the means (10.2).*

  Solve these proportions.

28. $\dfrac{7}{14} = \dfrac{3}{x}$ _____

29. $\dfrac{17}{51} = \dfrac{20}{y}$ _____

30. $\dfrac{0.4}{1.2} = \dfrac{7}{x}$ _____

31. $\dfrac{\frac{3}{8}}{\frac{2}{3}} = \dfrac{a}{\frac{5}{3}}$ _____

32. $\dfrac{4}{9} = \dfrac{8}{x}$ _____

33. $\dfrac{5}{2} = \dfrac{y}{9}$ _____

34. $\dfrac{35}{42} = \dfrac{7}{x}$ _____

35. $\dfrac{0.3}{0.4} = \dfrac{12}{x}$ _____

• *Solve application problems using proportion (10.3).*
Write a proportion and solve.

36. Three pounds of apples cost 36 cents. How much will you pay for 11 pounds?

37. An airplane can fly 150 miles in 30 minutes. At this rate, how far will the plane fly in 5 hours?

38. A photograph 3 inches by 4 inches is to be enlarged to 10 inches long. What is the width of the enlarged photograph?

39. During the first 432 miles of their trip, Peter and Phil use 16 gallons of gas. At this rate, how much gas will they need for the other 513 miles of their trip?

40. The scale on a city map indicates that 4 inches represents 1 mile. If two locations are $13\frac{1}{2}$ miles apart, how far apart are they on the map?

41. A bottle contains a mixture of 20 fluid ounces of water and 28 fluid ounces of antifreeze. If you have 70 fluid ounces of antifreeze, how much water should be added to make the same mixture?

42. A 50-pound bag of fertilizer will cover a 5000-square-yard lawn. If you have a 32,000-square-yard area of lawn, how much fertilizer will you need? How many 50-pound bags will you need?

43. If the owner of a $78,000 house pays $438 in property taxes, how much would the owner of a $195,000 house pay in property taxes in the same town?

CHAPTER 10 PRACTICE TEST

1. Write this ratio as a reduced fraction: $42:72$. _____

2. Write this ratio as a reduced fraction: $\frac{3}{4}$ to $\frac{7}{8}$ _____

3. Write this ratio as a decimal: 40 to 60. _____

4. Write this ratio as a reduced fraction: $4\frac{1}{8}$ to $4\frac{1}{2}$. _____

5. Write this ratio as a decimal: $\frac{3}{5}:\frac{9}{10}$. _____

6. Write this ratio as a decimal: $2\frac{1}{3}:1\frac{2}{3}$. _____

7. Write this ratio as a reduced fraction: 2 yards to 2 feet. _____

8. What is the ratio as a decimal of 2 minutes to 20 seconds? _____

9. Write the ratio as a decimal of 5 cents to 2 dimes? _____

10. What is the ratio as a reduced fraction of 20 ounces to 2 pounds? _____

11. Write the ratio as a decimal of 2 gallons to 3 quarts. _____

12. What is the ratio as a reduced fraction of 3 hours to 30 minutes? _____

13. What is the rate in miles per hour of a car traveling 160 miles in 4 hours? _____

14. Kevin earns $8584 in 8 months. What is his rate of pay in dollars per month? _____

15. A car can travel 164 miles on 8 gallons of gas. What is the rate in miles per gallon? _____

16. Which is the better buy: a package of writing paper containing 120 sheets for $2.40 or a package of 90 sheets for $2.25? _____

17. Solve: $\dfrac{2}{5} = \dfrac{4}{x}$ _____

18. Solve: $\dfrac{18}{4} = \dfrac{y}{10}$ _____

19. Solve: $\dfrac{\frac{1}{2}}{\frac{1}{4}} = \dfrac{x}{7}$ _____

20. Solve: $\dfrac{0.05}{1.2} = \dfrac{x}{6}$ _____

21. Solve: $\dfrac{2\frac{1}{4}}{4} = \dfrac{x}{10}$ _____

22. A baseball playuer makes 25 hits in 18 games. At this rate, how many hits would he make in 54 games?

_____

23. If 3 pots of pansies sell for 57 cents, how much will I pay in dollars for 11 pots? _____

24. A swimmer can swim 16 lengths in 20 minutes. How many lengths can she swim in 45 minutes? _____

25. If Mary can read 15 pages of her history textbook in 28 minutes, how many minutes will it take her to read the 252-page assignment? _____

		THE MATHEMATICS JOURNAL
		Present Experiences in Mathematics
	Task:	Reflect on your experiences and progress in learning mathematics in this course. Write a one-page (150-word) entry on your reflections.

11 Percents

11.1 Conversions: Fractions, Decimals, and Percents
11.2 Basic Percent Problems or Cases of Percent
11.3 Percent Application Problems
11.4 Percent Increase and Decrease
11.5 Simple Interest

Introduction to Percent

Percent means per hundred. Ten percent, which is written 10%, is 10 per hundred or $\frac{10}{100}$. The term 35% is 35 per hundred or $\frac{35}{100}$. Percent has been used since Roman times when it was used to collect taxes; for every 100 sheep a person had, taxes would take 7. Seven per hundred is 7%. The percent symbol % is said to be a modification of p/c written quickly by scribes in the fifteenth century.

Today, percents are used by businesses to compute profits, costs, and losses. It is used by scientists to show the results of their experiments; by sports teams to compare players' and teams' standings; and by all of us to compute property taxes, income taxes, sales taxes, and tips.

11.1 Conversions: Fractions, Decimals, and Percents

OBJECTIVES

- Convert percents to reduced fractions.
- Convert fractions to percents.
- Convert percents to decimals.
- Convert decimals to percents.

We will begin our study of percents by looking at the relationships between fractions, decimals, and percents.

DEFINITION **Percent:** A ratio with a denominator of 100. Percent means per hundred. A percent can be expressed as a fraction or as a decimal. To work with percents, convert them to decimals or equivalent fractions.

RULE Converting percents to fractions

Percent means to divide by 100. Use the percent number as the numerator and 100 as the denominator. Reduce to lowest terms.

EXAMPLE 1

$$93\% = \frac{93}{100}$$

$$3\% = \frac{3}{100}$$

$$25\% = \frac{25}{100} = \frac{1}{4}$$

$$50\% = \frac{50}{100} = \frac{1}{2}$$

$$37.5\% = \frac{37.5}{100} = \frac{375}{1000} = \frac{3}{8}$$

$$0.5\% = \frac{0.5}{100} = \frac{5}{1000} = \frac{1}{200}$$

$$125\% = \frac{125}{100} = \frac{5}{4} \text{ or } 1\frac{1}{4}$$

$$6\frac{1}{2}\% = \frac{6\frac{1}{2}}{100} = \frac{\frac{13}{2}}{100} = \frac{13}{2} \div 100 = \frac{13}{2} \cdot \frac{1}{100} = \frac{13}{200}$$

> **RULE** Converting fractions to percents
> To convert a fraction to a percent, multiply the fraction by 100%.

EXAMPLE 2

$$\frac{93}{100} = \frac{93}{100}(100\%) = 93\%$$

$$\frac{3}{100} = \frac{3}{100}(100\%) = 3\%$$

$$\frac{1}{4} = \frac{1}{4}(100\%) = 25\%$$

$$\frac{1}{2} = \frac{1}{2}(100\%) = 50\%$$

$$\frac{1}{8} = \frac{1}{8}(100\%) = 12.5\%$$

$$\frac{1}{200} = \frac{1}{200}(100\%) = \frac{1}{2}\% \text{ or } 0.5\%$$

$$\frac{5}{4} = \frac{5}{4}(100\%) = 125\%$$

$$\frac{13}{200} = \frac{13}{200}(100\%) = \frac{13}{2}\% = 6\frac{1}{2}\%$$

> **RULE** Converting percents to decimals
> To convert a percent to a decimal, drop the percent symbol and divide by 100. The easy way to divide by 100 is to move the decimal point two places to the left.

EXAMPLE 3

$93.\% = 0.93$

$3.\% = 0.03$

$25.\% = 0.25$

$50.\% = 0.50 = 0.5$

$37.5\% = 0.375$

$0.5\% = 0.005$

$125.\% = 1.25$

$6\frac{1}{2}\% = 6.5\% = 0.065$

RULE Converting decimals to percent

To convert a decimal to a percent, multiply by 100 and add the percent symbol. The easy way to multiply by 100 is to move the decimal point two places to the right.

EXAMPLE 4

$0.93 = 93\%$

$0.03 = 3\%$

$0.25 = 25\%$

$0.5 = 50\%$

$0.375 = 37.5\%$

$0.005 = 0.5\%$

$1.25 = 125\%$

$0.065 = 6.5\% = 6\frac{1}{2}\%$

RULE Converting fractions to decimals

To convert a fraction to a decimal, divide the denominator into the numerator. See Section 7.5.

EXAMPLE 5

$\dfrac{93}{100} = 0.93$

$\dfrac{3}{100} = 0.03$

$\dfrac{1}{4} = 0.25$

$\dfrac{1}{2} = 0.5$

$\dfrac{1}{8} = 0.125$

$\dfrac{1}{200} = 0.005$

$\dfrac{5}{4} = 1.25$

Practice Problems

Express as decimals.

7. 8%

8. 175%

9. 0.3%

Express as percents.

10. 0.5

11. 2.25

12. 0.006

Express as decimals.

13. $\dfrac{3}{4}$

14. $\dfrac{7}{8}$

15. $\dfrac{3}{200}$

RULE Converting decimals to reduced fractions
To convert a decimal to a reduced fraction, read the decimal, write it as a fraction, and reduce it to lowest terms. See Section 7.5.

Express as reduced fractions.

16. 0.6

EXAMPLE 6

$$0.93 = \frac{93}{100}$$

$$0.03 = \frac{3}{100}$$

$$0.25 = \frac{25}{100} = \frac{1}{4}$$

$$0.5 = \frac{5}{10} = \frac{1}{2}$$

$$0.375 = \frac{375}{1000} = \frac{3}{8}$$

$$0.005 = \frac{5}{1000} = \frac{1}{200}$$

$$1.25 = 1\frac{25}{100} = 1\frac{1}{4}$$

17. 0.625

18. 2.03

Fill in the missing numbers in the following chart. This chart should be completed in class, either as a class project or as a group activity.

Fraction	Decimal	Percent
		50%
	0.25	
$\frac{3}{4}$		
$\frac{1}{3}$		
$\frac{2}{3}$		
		175%
		$\frac{1}{2}$%
$\frac{1}{8}$		
$\frac{3}{8}$		
$\frac{5}{8}$		
$\frac{7}{8}$		

See next page for completed chart.

Fraction	Decimal	Percent
$\frac{1}{2}$	0.5	50%
$\frac{1}{4}$	0.25	25%
$\frac{3}{4}$	0.75	75%
$\frac{1}{3}$	$0.\overline{3}$	$33\frac{1}{3}\%$
$\frac{2}{3}$	$0.\overline{6}$	$66\frac{2}{3}\%$
$1\frac{3}{4}$	1.75	175%
$\frac{1}{200}$	0.005	$\frac{1}{2}\%$
$\frac{1}{8}$	0.125	12.5%
$\frac{3}{8}$	0.375	37.5%
$\frac{5}{8}$	0.625	62.5%
$\frac{7}{8}$	0.875	87.5%

PROBLEM SET 11.1

Write each number as a percent.

1. $\frac{2}{3}$

2. $\frac{7}{10}$

3. $1\frac{1}{4}$

4. $2\frac{1}{8}$

5. $\frac{3}{8}$

6. 4

7. $1\frac{5}{6}$

8. $\frac{3}{16}$

9. 0.37

10. 0.62

11. 0.14

12. 3.1

13. 0.07

14. 0.002

15. 1.06

Write each percent as a decimal.

16. 4%

17. 3.5%

18. $5\frac{1}{4}\%$

19. $11\frac{1}{2}\%$

20. $17\frac{3}{4}\%$

21. $\frac{1}{2}\%$

22. $1\frac{1}{4}\%$

23. 31%

24. 49%   **25.** $62\frac{1}{2}\%$ **26.** 112%   **27.** $124\frac{1}{2}\%$

28. 0.3%   **29.** 0.06%   **30.** 8.1%

Write each percent as a fraction in lowest terms.

31. 8%   **32.** 6%   **33.** 15%   **34.** 24%

35. 30%   **36.** 42%   **37.** 112%   **38.** $37\frac{1}{2}\%$

39. 7.5%   **40.** 140%   **41.** 265%   **42.** 0.06%

43. $82\frac{1}{2}\%$ **44.** $21\frac{1}{4}\%$ **45.** $6\frac{1}{4}\%$

Fill in the missing numbers in the following problems.

	Percent	*Decimal*	*Fraction in lowest terms*
46.	10%		
47.	25%		
48.	61%		
49.	3%		
50.		0.12	
51.		1.2	
52.		0.004	
53.		0.35	
54.			$\frac{1}{2}$
55.			$\frac{3}{4}$
56.			$\frac{1}{200}$

	Percent	Decimal	Fraction in lowest terms
57.	80%		
58.			$\frac{4}{5}$
59.		0.065	
60.			$\frac{5}{4}$
61.	5%		
62.			$\frac{1}{8}$
63.	145%		
64.	7%		
65.		0.25	
66.	20%		
67.	67.5%		
68.	$\frac{1}{2}$%		
69.			$\frac{1}{3}$
70.	87.5%		
71.	3.5%		
72.		0.15	
73.			$\frac{3}{20}$
74.	0.02%		
75.	50%		
76.		0.14	
77.			$\frac{2}{5}$
78.	60%		
79.	3.7%		
80.		0.012	
81.	0.12%		

11.2 Basic Percent Problems or Cases of Percent

OBJECTIVES

● Use the formula Base · Rate = Amount to solve basic percent problems.

● Use proportions to solve basic percent problems.

In this section, we will consider two approaches to solving percent problems. We will work each problem, using both the equation method and the proportion method. After working all the problems, find the method you feel most comfortable with and use it. Percent problems contain three quantities: base, rate, and amount.

DEFINITION **Base:** The "original amount." It follows *of* in the problem.

EXAMPLE **1**   35 is 10% of 350. 350 is the base.

DEFINITION **Rate:** The percent in the problem.

EXAMPLE **2**   25% of 400 is 100. 25% is the rate.

DEFINITION **Amount:** Part of the base, but it can exceed the base when the percent exceeds 100. The amount is linked with *IS* in the problem.

EXAMPLE **3**   75 is 15% of 500. 75 is the amount.

The Equation Method

The formula used to solve a percent problem is, using the terms defined, *Base* multiplied by *Rate* equals *Amount*.

The rate or percent must be changed into a decimal or fraction for use in the formula. Use the formula $B \cdot R = A$

EXAMPLE **4**   What number is 25% of 80?

The base follows *of* and is 80. The rate is 25%, written as a decimal 0.25. The amount is the quantity we are looking for.

$$B \cdot R = A$$
$$80 \cdot 0.25 = A \quad \text{Solve for } A.$$
$$20 = A$$

Therefore, 20 is 25% of 80. Is this answer reasonable? 25% is equivalent to $\frac{1}{4}$ and $\frac{1}{4}$ of 80 is 20. This answer is reasonable.

USING THE CALCULATOR

Enter 80 × 25 % = **20** ← Display

EXAMPLE 5 What percent of 62 is 31? The base follows *of* and is 62.

The rate is the quantity we are looking for. The amount is linked with *is* in the problem and is 31.

$$B \cdot R = A$$
$$62 \cdot R = 31 \qquad \text{Solve for } R.$$
$$62 \cdot \frac{R}{62} = \frac{31}{62}$$
$$R = \frac{31}{62}$$
$$R = \frac{1}{2} \text{ or } 50\%$$

Note Remember that in the formula the rate is expressed as either a fraction or a decimal. The answer to the equation will be a fraction or decimal that must be changed to a percent.

$$50\% \text{ of } 62 \text{ is } 31.$$

Is this answer reasonable? Is 31 half of 62? Yes, the answer is reasonable.

 USING THE CALCULATOR

Once you have the equation with the R on one side,

Enter 31 ÷ 62 = **0.5** ← Display

50%, 0.5, and $\frac{1}{2}$ are all equivalent. $0.5 = 50\%$

EXAMPLE 6 18 is 36% of what number?

The base follows *of* and is the quantity we are looking for. The rate is 36%, written as a decimal 0.36. The amount is linked with *is* in the problem and is 18.

$$B \cdot R = A$$
$$B \cdot 0.36 = 18 \qquad \text{Solve for B.}$$
$$\frac{0.36B}{0.36} = \frac{18}{0.36}$$
$$B = \frac{18}{0.36}$$
$$B = 50$$

18 is 36% of 50.

Is this answer reasonable? Since $36\% \approx 40\% = \frac{2}{5}$ and $\frac{2}{5}$ of 50 is 20, the answer is reasonable.

 USING THE CALCULATOR

Once you have the equation with the B on one side,

Enter 18 ÷ 0.36 = **50** ← Display

Practice Problems

Solve these percent problems.

1. What number is 10% of 30?

2. What percent of 56 is 7?

3. 20 is 25% of what number?

Note The amount can be larger than the base when the percent exeeds 100.

EXAMPLE 7 What percent of 22 is 55?

The base follows *of* and is 22. The rate is the quantity we are looking for. The amount is linked with *is* in the problem and is 55.

$$B \cdot R = A$$
$$22 \cdot R = 55$$
$$22R = 55$$
$$\frac{22R}{22} = \frac{55}{22}$$
$$R = \frac{55}{22}$$
$$R = 2.5 = 250\%$$

250% of 22 is 55. Is this answer reasonable? Since $22 \approx 20$, $250\% = 2.5$ and 2.5×20 is 50, the answer is reasonable.

USING THE CALCULATOR

Once you have the equation with the *R* on one side,

Enter 55 ÷ 22 = **2.5** ← Display

To change the decimal to a percent,

Enter × 100 = 250%.

EXAMPLE 8 18 is 180% of what number?

The base follows *of* and is the quantity we are looking for. The rate is 180%, written as a decimal 1.8. The amount is the number linked with *is* in the problem and is 18.

$$B \cdot R = A$$
$$B \cdot 1.8 = 18 \qquad \text{Solve for B.}$$
$$1.8B = 18$$
$$\frac{1.8B}{1.8} = \frac{18}{1.8}$$
$$B = \frac{18}{1.8}$$
$$B = 10$$

18 is 180% of 10.

USING THE CALCULATOR

Once you have the equation with the *B* on one side,

Enter 18 ÷ 1.8 = **10** ← Display

EXAMPLE 9 $\frac{1}{12}$ is what percent of $\frac{1}{3}$? The base is $\frac{1}{3}$.

It follows *of* in the problem. The rate is what we are looking for. The amount is $\frac{1}{12}$. It is linked with *is* in the problem.

$$B \cdot R = A$$

$$\frac{1}{3} \cdot R = \frac{1}{12}$$

$$\frac{1}{3} R = \frac{1}{12}$$

$$(3)\frac{1}{3} R = (3)\frac{1}{12}$$

$$R = \frac{1}{4} = 25\%$$

$\frac{1}{12}$ is 25% of $\frac{1}{3}$

USING THE CALCULATOR

Once you have the R on one side,

Enter 3 ☐×☐ 1 ☐$a^{b/c}$☐ 12 ☐=☐ **1⌐4** ← Display

EXAMPLE 10 $\frac{1}{4}\%$ of 10,000 is what number?

The base is 10,000. It follows *of* in the problem. The rate is $\frac{1}{4}\%$, written 0.0025 as a decimal. The amount is what we are looking for.

$$B \cdot R = A$$

$$10,000 \cdot \frac{1}{4}\% = A$$

$$10,000(0.0025) = A$$

$$25 = A$$

$\frac{1}{4}\%$ of 10,000 is 25.

USING THE CALCULATOR

Enter 10000 ☐×☐ 1 ☐$a^{b/c}$☐ 4 ☐%☐ ☐=☐ **25** ← Display

The Proportion Method

For this method we will be using the terms as defined at the beginning of the section. The rate is one side of the proportion and is written as a fraction with a denominator of 100.

$$R = \frac{A}{B}; \quad \frac{x}{100} = \frac{A}{B}$$

To set up a percent problem as a proportion, one side of the proportion will always be the *percent* side. If the problem asks for the percent, that side will be $\frac{x}{100}$ because percent is a special ratio meaning per 100 or out of 100. If the problem states the percent, write that percent as the first ratio of the proportion. Check Section 10.2 to review solving proportions.

EXAMPLE 11 What number is 25% of 80?

$$\frac{25}{100} = \frac{x}{80}$$

Note   The number that follows *of* in the problem is the base and is the denominator of the second ratio. The base and 100 are always the denominators of the ratios.

$$\frac{25}{100} = \frac{A}{80} \qquad \text{Solve for } A.$$

$$100A = 25(80)$$

$$\frac{100A}{100} = \frac{25(80)}{100}$$

$$A = 20$$

20 is 25% of 80.

USING THE CALCULATOR
Enter 25 $\times$ 80 $\div$ 100 = **20** ← Display

EXAMPLE 12 What percent of 62 is 31?

What percent means the percent ratio is $\frac{x}{100}$. 62 follows *of*, so it is the base and thus the denominator of the second ratio. 31 is the amount and the numerator of the second ratio.

$$\frac{x}{100} = \frac{31}{62} \qquad \text{Solve for } x.$$

$$62x = 31(100)$$

$$\frac{62x}{62} = \frac{31(100)}{62}$$

$$x = 50$$

50% of 62 is 31.

USING THE CALCULATOR
Enter 31 $\times$ 100 $\div$ 62 = **50** ← Display

EXAMPLE 13 18 is 36% of what number?

36% becomes $\frac{36}{100}$, the first ratio of the proportion. The number we are to find, the base (which we will call B) follows *of* and, therefore, becomes the denominator of the second ratio.

$$\frac{36}{100} = \frac{18}{B} \qquad \text{Solve for B.}$$

$$36B = 100(18)$$

$$\frac{36B}{36} = \frac{100(18)}{36}$$

$$B = 50$$

18 is 36% of 50.

USING THE CALCULATOR
Enter 100 $\times$ 18 $\div$ 36 = **50** ← Display

Practice Problems

Solve these percent problems.

7. What number is 12% of 200?

8. What percent of 84 is 21?

9. 11 is 25% of what number?

Note The amount can exceed the base when the percent exceeds 100.

EXAMPLE 14 What percent of 22 is 55?

We need to find the percent so the first ratio of the proportio will be $\frac{x}{100}$. The denominator of the second ratio will be 22 because it is the base and it follows *of* in the problem.

$$\frac{x}{100} = \frac{55}{22} \qquad \text{Solve for } x.$$

$$22x = 100(55)$$

$$\frac{22x}{22} = \frac{100(55)}{22}$$

$$x = 250$$

250% of 22 is 55.

USING THE CALCULATOR

Enter 100 $\times$ 55 $\div$ 22 $=$ **250** $\leftarrow$ Display

EXAMPLE 15 18 is 180% of what number?

The percent ratio will be $\frac{180}{100}$. The number we need to find, the base, which follows the *of*, will be the denominator of the second ratio.

$$\frac{180}{100} = \frac{18}{B} \qquad \text{Solve for } B.$$

$$180B = 18(100)$$

$$B = 10$$

18 is 180% of 10.

USING THE CALCULATOR

Enter 18 $\times$ 100 $\div$ 180 $=$ **10** $\leftarrow$ Display

EXAMPLE 16 $\frac{1}{12}$ is what percent of $\frac{1}{3}$?

The percent ratio of the proportion will be $\frac{x}{100}$. The denominator of the other ratio will be $\frac{1}{3}$, the base, which follows *of* in the problem.

$$\frac{x}{100} = \frac{\frac{1}{12}}{\frac{1}{3}}$$

$$\frac{1}{3}x = (100)\frac{1}{12}$$

$$(3)\frac{1}{3}x = (100)\left(\frac{1}{12}\right)(3)$$

$$R = 25$$

$\frac{1}{12}$ is 25% of $\frac{1}{3}$.

Enter 100 $\times$ 1 $a^{b/c}$ 12 $\div$ 1 $a^{b/c}$ 3 $=$ **25** $\leftarrow$ Display

Note Multiplying by 3 is the same as dividing by $\frac{1}{3}$.

EXAMPLE 17 $\frac{1}{4}$% of 10,000 is what number?

The base follows *of* and is 10,000. The rate is $\frac{1}{4}$%. The amount is what we are looking for.

$$\frac{\frac{1}{4}}{100} = \frac{A}{10,000}$$

$$100A = \frac{1}{4}(10,000)$$

$$\frac{100A}{100} = \frac{\frac{1}{4}(10,000)}{100}$$

$$A = 25$$

$\frac{1}{4}$% of 10,000 is 25.

USING THE CALCULATOR

Enter 1 $a^{b/c}$ 4 $\times$ 10000 $\div$ 100 $=$ **25** $\leftarrow$ Display

ANSWERS TO PRACTICE PROBLEMS
1. 3 **2.** 12.5% **3.** 80 **4.** 20
5. 8.8 **6.** $66\frac{2}{3}$% **7.** 24 **8.** 25%
9. 44 **10.** 22.5 **11.** 2 **12.** 75%

PROBLEM SET 11.2

In the following percent problems, list the base, the rate, and the amount.

1. 50% of 90 is 45.

2. $33\frac{1}{3}$% of 60 is 20.

3. 300% of 7 is 21.

4. 90% of 200 is 180.

5. 175% of 88 is 154.

6. 12 is 75% of 16.

7. 4.5 is 5% of 90.

8. 25 is 40% of 62.5.

9. 150% of 30 is 45.

10. $\frac{1}{3}$ is 20% of $\frac{5}{3}$.

Find the missing quantity in the following basic percent problems. Use either the equation or proportion method.

11. 50% of 200 is what number?

12. 25% of 372 is what number?

13. 5% of 20 is what number?

14. What percent of 38 is 19?

15. What percent of 284 is 71?

16. What percent of 115 is 92?

17. 18 is 75% of what number?

18. 21 is 60% of what number?

19. 23 is $33\frac{1}{3}$ of what number?

20. What percent of 400 is 76?

21. 4 is what percent of 15?

22. 7 is what percent of 25?

23. 8% of 30 is what number?

24. 6 is what percent of 27?

25. 52 is what percent of 39?

26. 11% of 6 is what number?

27. 30% of 20 is what number?

28. 42% of 15 is what number?

29. 110% of 16 is what number?

30. 120% of 55 is what number?

31. 4 is what percent of $\frac{1}{2}$?

32. 15 is what percent of 12?

33. 120 is what percent of 90?

34. 140 is what percent of 112?

35. $6\frac{1}{2}\%$ of 36 is what number?

36. $7\frac{1}{2}\%$ of 46 is what number?

37. $\frac{1}{4}\%$ of 10 is what number?

38. $\frac{1}{5}\%$ of 40 is what number?

39. 40% of what number is 6?

40. 64% of what number is 12?

41. 33% of what number is 55?

42. 30% of what number is 60?

43. 120% of what number is 16?

44. 130% of what number is 65?

45. $\frac{1}{2}$ is what percent of $\frac{3}{4}$?

46. $\frac{7}{8}$ is what percent of $\frac{3}{4}$?

47. 0.2% of what number is 10?

48. 0.4% of 25 is what number?

49. $5\frac{1}{4}\%$ of what number is 14?

50. $7\frac{1}{2}\%$ of 40 is what number?

11.3 **Percent Application Problems**

OBJECTIVE

● Solve application problems using either proportions or the formula: $B \cdot R = A$.

The percent application problems contain the same three quantities as the basic percent problems. (See Section 11.2 for definitions of base, rate, and amount.) Because an application problem describes a real-life situation, it is usually fairly easy to decide which is the base, the rate, and the amount.

EXAMPLE 1 A baseball player hit 63 out of 75 pitches in practice games. What percent did he hit?

The base is 75, the total number of pitches. The rate is what we are looking for. The amount is 63, the number of pitches actually hit.

$$B \cdot R = A \qquad\qquad R = \frac{A}{B}$$

$$75 \cdot R = 63 \qquad\qquad \frac{x}{100} = \frac{63}{75}$$

$$\frac{75R}{75} = \frac{63}{75} \qquad\qquad 75x = 63(100)$$

$$R = \frac{63}{75} \qquad\qquad \frac{75x}{75} = \frac{63(100)}{75}$$

$$R = 0.84 = 84\% \qquad x = 84;\ R = 84\%$$

USING THE CALCULATOR

Using the proportion method,

Enter 63 $\times$ 100 $\div$ 75 $=$ **84** $\leftarrow$ Display

EXAMPLE 2   The sales tax rate in Ohio is $6\frac{1}{2}\%$.

How much tax will Joe pay on a VCR that costs $539? The base is the cost of the VCR, which is $539. The rate is $6\frac{1}{2}\%$ or 6.5% or, written as a decimal, 0.065. The amount is the tax.

$$B \cdot R = A \qquad\qquad R = \frac{A}{B}$$

$$539 \cdot 0.065 = A \qquad\qquad \frac{6.5}{100} = \frac{A}{539}$$

$$35.035 = A \qquad\qquad 6.5(539) = 100A$$

$$3503.5 = 100A$$

$$35.035 = A$$

Is this answer reasonable? $6\frac{1}{2}\% \approx \$7$ on each $100, $7 \times 5 = 35$. The answer is reasonable.

Note   Money must be rounded to two decimal places because you cannot pay someone part of a penny! Rounding $35.035 to the nearest penny is $35.04.

USING THE CALCULATOR

Using the equation method,

Enter 539 $\times$ 6.5 2nd % $=$ **35.035** $\leftarrow$ Display

EXAMPLE 3   Melanie sold a house. Her real estate commission was $6860 and was 7% of the selling price of the house. What was the selling price of the house? The base is the selling price. The rate is 7%, written as a decimal 0.07. The amount is her commission, which is $6860.

$$B \cdot R = A \qquad\qquad R = \frac{A}{B}$$

$$B \cdot 0.07 = 6860 \qquad\qquad \frac{7}{100} = \frac{6860}{B}$$

$$0.07B = 6860 \qquad\qquad 7B = 6860(100)$$

$$\frac{0.07B}{0.07} = \frac{6860}{0.07} \qquad\qquad \frac{7B}{7} = \frac{6860(100)}{7}$$

$$B = 98,000 \qquad\qquad B = 98,000$$

The selling price of the house was $98,000.

USING THE CALCULATOR

Using the proportion method,

Enter 6860 $\times$ 100 $\div$ 7 $=$ **98000** $\leftarrow$ Display

---

Practice Problems

Solve these percent problems.

1. A baseball player hit 81 out of 108 pitches in practice games. What percent did he hit?

2. The sales tax rate is $4\frac{1}{4}\%$. How much tax will be paid on a bike which cost $200?

3. The car salesperson's commission is 12%. If Tina made $1800 commission on a sale, what was the price of the car?

EXAMPLE 4   A lawyer will pursue a claim and will take as his fee 31% of the award. If Yohanna is awarded $75,000 for her fall on the ice, how much does she pay her lawyer?

The base is Yohanna's award, which is $75,000. The rate is 31%, written as a decimal 0.31. The amount is what Yohanna pays her lawyer.

Practice Problems

Solve these problems.

4. If Yohanna had been awarded $53,000 for her fall, how much does she pay her lawyer if his fee is 31% of the award?

5. Another family in the same neighborhood as the Nadirs' purchased a home for $103,000. How much has the cost of this family's home increased in value in the last year?

$$B \cdot R = A \qquad R = \frac{A}{B}$$

$$75{,}000 \cdot 0.31 = A \qquad \frac{31}{100} = \frac{A}{75{,}000}$$

$$23{,}250 = A \qquad 100A = 31(75{,}000)$$

$$\frac{100A}{100} = \frac{31(75{,}000)}{100}$$

$$A = 23{,}250$$

Yohanna pays her lawyer $23,250. Is this answer reasonable? $31\% \approx \frac{1}{3}$, $\frac{1}{3}$ of 75,000 is 25,0000. The answer is reasonable.

> **USING THE CALCULATOR**
>
> Using the equation method,
>
> Enter 75000 $\times$ 31 2nd % = **23250** ← Display

EXAMPLE 5   The Nadirs purchased their home a year ago. Prices of homes in their neighborhood rose $8\frac{1}{2}\%$ during the year. If they purchased their home for $75,900, how much has their home increased in value?

The base is the cost of their home a year ago, which is $75,900. The rate is $8\frac{1}{2}\%$, written as a decimal 0.085. The amount is the increase in the home's value.

$$B \cdot R = A \qquad R = \frac{A}{B}$$

$$75{,}900 \cdot 0.085 = A \qquad \frac{8.5}{100} = \frac{A}{75{,}900}$$

$$6451.50 = A \qquad \frac{100A}{100} = \frac{8.5(75{,}900)}{100}$$

$$A = 6451.50$$

The value of the Nadirs' home has risen $6451.50 during the year they have lived there. Is this answer reasonable? $8\frac{1}{2}\% \approx 10\%$, 10% of 75,900 is 7590. The answer is reasonable.

> **USING THE CALCULATOR**
>
> Using the equation method,
>
> Enter 75900 $\times$ 8.5 2nd % = **6451.5** ← Display

EXAMPLE 6 Pure gold is 24 karat. Jill's bracelet is 9 karat gold. What percent of Jill's bracelet is gold?

The base is 24, which is the whole. The rate is what we are looking for. The amount is 9, the part of the bracelet that is gold.

$$B \cdot R = A \qquad\qquad R = \frac{A}{B}$$

$$24 \cdot R = 9 \qquad\qquad \frac{x}{100} = \frac{9}{24}$$

$$24R = 9 \qquad\qquad 24x = 9(100)$$

$$\frac{24R}{24} = \frac{9}{24} \qquad\qquad \frac{24x}{24} = \frac{9(100)}{24}$$

$$R = 0.375 \qquad\qquad x = 37.5\%$$

$$= 37.5\%$$

Jill's bracelet is 37.5% pure gold.

USING THE CALCULATOR

Using the proportion method,

Enter 9 × 100 ÷ 24 = **37.5** ← Display

EXAMPLE 7 What percent of an hour is 36 minutes?

The base is 1 hour, which is 60 minutes. The rate is what we are looking for. The amount is 36 minutes.

$$B \cdot R = A \qquad\qquad R = \frac{A}{B}$$

$$60 \cdot R = 36 \qquad\qquad \frac{x}{100} = \frac{36}{60}$$

$$60R = 36 \qquad\qquad 60x = 36(100)$$

$$\frac{60R}{60} = \frac{36}{60} \qquad\qquad \frac{60x}{60} = \frac{36(100)}{60}$$

$$R = 0.6 = 60\% \qquad\qquad x = 60; R = 60\%$$

36 minutes is 60% of an hour. Is this answer reasonable?

36 minutes $\approx \frac{1}{2}$ an hour, $\frac{1}{2}$ would be 50%. The answer is reasonable.

USING THE CALCULATOR

Using the proportion method,

Enter 36 × 100 ÷ 60 = **60** ← Display

PROBLEM SET 11.3

Solve the following problems.

1. The ticket price on a sweater is $45. If the sales tax rate is 6%, how much tax is paid on the sweater?

2. In a shipment of airplane parts, 2% are defective. If there are 8 defective parts, how many parts were in the shipment?

3. The sales tax on a new microwave oven is $15. If the sales tax rate is 5%, what is the selling price?

4. If the commission on an $800 washer is $168, what is the commission rate?

5. Sally earns $1000 a month; she gets a raise of $60 a month. What is the percent increase in her salary?

6. The state sales tax rate is 4%. How much sales tax is charged on a new washing machine if the selling price is $350?

7. The sales tax on a new car is $384. If the sales tax rate is 6%, what is the selling price?

8. Suppose the price of a stereo system is $550. If the sales tax is $27.50, what is the sales tax rate?

9. A real estate agent charges a commission rate of 7%. If she sells an $85,000 house, what is her commission?

10. A sales rep has a commission rate of 10% on the electronics he sells. If the commission on a telephone answering machine is $32.50, how much did the machine cost?

11. If the commission on a $700 mountain bike is $56, what is the commission rate?

12. If a teacher earning $1200 per month gets a raise of $96 a month, what is the percent increase in her salary per month?

13. The price of a new car is increased by 12.5%. If the amount of increase is $1562.50, what was the original price of the car?

14. A $250 suit is on sale for 15% off. What is the discount?

15. A department store is having a "30% Off" sale. A woman buying a dress saves $36. What was the original price?

16. Paint is advertised for sale at $39.50 a gallon plus tax. If the tax is 20%, how much tax is paid per gallon of paint?

The Paint Company			
Paint	$	39	50
Tax at 20%	$		
Total	$		
	Received by		
Please come again.			

17. A discount of 12% is offered to all college students buying at the bookstore. What do you save on a $28.45 purchase?

18. If you must put down 15% to purchase a $73,000 house, how much money do you need?

19. A real estate investment of $6000 earns you a $720 profit in one year. What rate of interest is this?

20. Social Security taxes amount to 8.67%. How much will be deducted from your $1254 monthly paycheck?

21. If the rate of commission is 12%, find the commission on a sale of $1200.

22. How much tax must you pay on an article priced at $17.50 if the sales tax rate is 8%?

23. If a car costs $7500 and the price is decreased 5%, how much less does it cost now?

24. A baseball player hit 48 out of 56 pitches in the game. What percent did she hit?

25. A basketball player made 36 baskets out of 48 tries. What percent did she make?

26. A lawyer will pursue a claim and take as his fee 28% of the award. If Maisie is awarded $60,000 for her back injury, how much does she pay her lawyer?

27. In another case, the lawyer agreed to pursue the case for 26% of the award. If José is awarded $20,000 for the wrongful death of his cat, how much does he pay his lawyer?

28. The price of homes in the heights rose 11% last year. If a home was worth $225,000 last year, how much has the house increased in value?

29. The price of homes in the lowlands rose only $2\frac{1}{2}$% this year. If a home was worth $29,000 last year, how much has the home increased in value?

30. If you work 8 hours a day, what percent of the day do you work?

31. Watching a movie takes $12\frac{1}{2}$% of your day. How long is the movie?

32. Twelve minutes is what percent of an hour?

33. How many minutes is $33\frac{1}{3}$% of an hour?

34. What percent of a year is 3 months?

35. How many months is 75% of a year?

36. What percent of a year is 13 weeks?

37. How many weeks is 75% of a year?

38. Kent has 18,000 registered voters. A total of 6660 of them voted in the election. What percent of the registered voters voted?

39. A charity raised $9850. If 6% is overhead (money used to run the charity), how much of the money raised is overhead?

40. The math class had 7 absent students on Friday. If there are 28 students enrolled in the class, what percent were absent on Friday?

41. If 20% of the students in a math class are absent on Monday and 9 students are absent, what is the total number of students in the class?

42. William paid $720 in sales tax on his new car. If the tax rate in his county is 6%, what was the list price of William's new car?

43. The Bellini family's take-home pay is $3950 a month. They spend $1185 on rent, $790 on food, $592.50 on clothing, $395 on incidentals, $276.50 on entertainment, and they save $711. What percent of their take-home pay do they spend on rent?

What percent of their take-home pay do they spend on food?

What percent of their take-home pay do they spend on clothing?

What percent of their take-home pay do they spend on incidentals?

What percent of their take-home pay do they save?

What percent of their take-home pay is spent on entertainment?

44. To curb violence on the streets and raise money for health care, it has been suggested that a 10,000% tax be levied on bullets. If a bullet costs 10 cents, how much would each bullet be taxed?

11.4 Percent Increase and Decrease

OBJECTIVE

● Solve percent increase and decrease problems using either proportions or the formula: $B \cdot R = A$.

Most real-life problems using percents involve finding more than one of the three quantities of the basic percent problem. In this section, we will explore finding percent increase and decrease, for example, finding the new price of discounted merchandise. We will also explore finding the markup required by stores to make a profit, or knowing the markup, determining how much a car dealer paid for a car.

   In calculating increase or decrease problems, consider the same three quantities as in the basic percent problems. In increase and decrease problems, the base is always the *original* number.

EXAMPLE 1   Crystal earns $10,000 a year. She gets a 5% increase in pay. What is her new salary?

   Her old salary is 100%; if she gets a 5% increase, her new salary will be $100\% + 5\% = 105\%$. To find her new salary, we find 105% of $10,000.

$$B \cdot R = A$$
$$10,000 \cdot 1.05 = A$$
$$10,500 = A$$
$$105\% \text{ of } \$10,000 \text{ is } \$10,500.$$

Her new salary is $10,500.

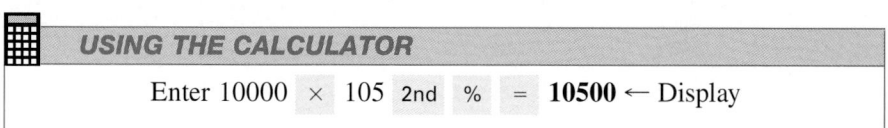

USING THE CALCULATOR

Enter 10000 ✕ 105 2nd % = **10500** ← Display

EXAMPLE 2   A coat is marked 30% off the selling price of $360. What is the sale price?

   100% is the original cost of the coat. 30% is the discount, so $100\% - 30\% = 70\%$. The base is the original price $360. The rate is 70%, written 0.7 as a decimal.

30% Off

$$B \cdot R = A$$
$$360 \cdot 0.7 = A$$
$$252 = A$$

The sale price of the coat is $252.

EXAMPLE 3 The tax rate in the state where the coat was sold is 6%. How much would you pay for a $252 coat?

100% for the coat, 6% for the tax. 100% + 6% = 106%, written 1.06 as a decimal.

$$B \cdot R = A$$
$$252 \cdot 1.06 = A$$
$$267.12 = A$$

The coat plus the tax would cost $267.12.

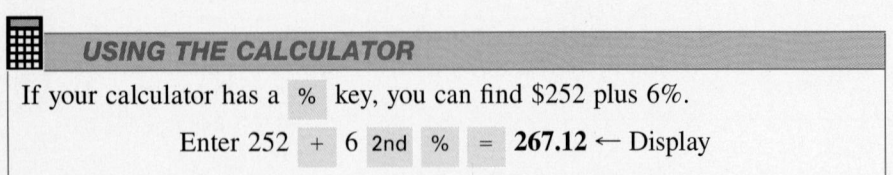

USING THE CALCULATOR

If your calculator has a % key, you can find $252 plus 6%.

Enter 252 + 6 2nd % = **267.12** ← Display

EXAMPLE 4 A pair of boots were originally selling for $60. The boots are on sale for $51. What percent is the discount on the boots?

The original price is the base and is $60. The amount (the actual discount) is $60 − $51 = $9.

$$R = \frac{A}{B}$$

$$\frac{x}{100} = \frac{9}{60} \qquad \text{Discount rate and discount price}$$
$$\phantom{\frac{x}{100} = \frac{9}{60}} \qquad \text{Total percent and original price}$$

$$60x = 100(9)$$
$$x = \frac{100(9)}{60}$$
$$x = 15$$

The boots were discounted 15%; there was a 15% decrease in price.

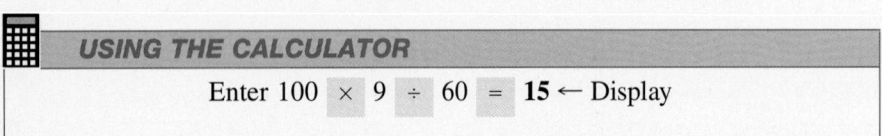

USING THE CALCULATOR

Enter 100 × 9 ÷ 60 = **15** ← Display

EXAMPLE 5 A car dealer sells cars at 20% over invoice (the price paid for the car). If the dealer is selling a car for $15,000, what did she pay for the car?

What the dealer paid for the car is 100%; her markup is 20%, so the selling price of the car is 100% + 20% = 120% or $15,000. The amount is the selling price, which is $15,000. The rate is 120%, written 1.2 as a decimal.

$$B \cdot R = A$$
$$B \cdot 1.2 = 15{,}000$$
$$B = \frac{15{,}000}{1.2}$$
$$B = 12{,}500$$

The dealer paid $12,500 for the car.

USING THE CALCULATOR

Enter 15000 ÷ 1.2 = **12500** ← Display

Practice Problems

Solve these problems.

3. A shirt that originally sold for $35 is on sale for $28. What percent is the discount on the shirt?

4. A car dealer sells cars at 15% over the invoice (the price he paid for the car). If he is selling the car for $26,450, what did he pay for the car?

EXAMPLE 6   If Jay weighs 200 pounds and gains 50 pounds, what percent of his weight has he gained?

The base is his original weight, which is 200 pounds. The amount is the weight he gained or 50 pounds.

$$B \cdot R = A$$
$$200 \cdot R = 50$$
$$R = \frac{50}{200}$$
$$R = 0.25 = 25\%$$

Jay gained 25% of his weight.

EXAMPLE 7   Later Jay, who now weighs 250 pounds, loses 50 pounds. What percent of his weight did he lose?

$$B \cdot R = A$$
$$250 \cdot R = 50$$
$$R = \frac{50}{250}$$
$$R = 0.2 = 20\%$$

Note   When the original number changes, as it did in these examples, the percent changes.

EXAMPLE 8   The government of Nigeria increased the price of gas 600% in the fall of 1993. If the original price of the gas was 50 cents a gallon, what is the new price of gas?

The base is the original price 50 cents or .50 of a dollar. The rate was originally 100%, the increase is 600%, which gives a rate of 100% + 600% = 700%.

$$R = \frac{A}{B}$$
$$\frac{700}{100} = \frac{A}{0.50}$$
$$100A = 700(0.50)$$
$$A = \frac{700(0.50)}{100}$$
$$A = 3.50$$

The new price of gas is $3.50 a gallon. Is this answer reasonable?

$700\% \approx 7.$ 0.50 is $\frac{1}{2}$. $7 \times \frac{1}{2} = 3.5$. The answer is reasonable.

ANSWERS TO PRACTICE PROBLEMS
1. 1284   **2.** $3300   **3.** 20%
4. $23,000

PROBLEM SET 11.4

Solve the following problems. Read the problem carefully to be sure you are answering the question being asked.

1. The ticket price of a sweater is $45. If the sales tax rate is 6%, what is the *total* price of the sweater?

2. Seven out of 25 students did not attend the last session of class. What percent of the class was present?

3. A gallon of paint was advertised for $39.50 plus tax. If the tax is 20%, what does the paint cost?

4. A table sells for $60 before a 20% discount. What was its price after the discount?

5. A $19.95 vest is on sale at a 20% discount. What is the sale price? What must you pay for the vest if there is a 6% sales tax?

6. The Klobber Kloset is having a sale and all clothes are reduced 30%. What is the sale price of each of the following?
a. A shirt listed at $18.50.

b. A pair of shoes listed at $124.75

c. A necktie listed at $14

d. A pair of socks for $2

7. The population of Ilkley is 2000. If it increases to 2500, what is the percent of increase?

8. A china platter costs $45 before it is reduced by 20%. What is the new price of the platter?

9. How much must you pay for a calculator priced at $17.50 if the sales tax is 8%?

10. If a car costs $9500 and the price is increased 6%, what is the new price of the car?

11. A meat market increased by 4% the price of steak at $3.25 per pound. What is the new price of the steak?

12. The price of a new car is increased 12.5%. If the amount of increase is $1562.50, what is the new selling price of the car?

13. A microwave oven that usually sells for $125 is on sale for $100. What is the percent decrease in price?

14. The sales tax rate in Ohio is 6%. If a computer costs $3500, what is the total price of the computer?

15. Ted was recently promoted and will be earning $25,000 a year. If Ted has been making $20,000 a year, what was his percent of increase in pay?

16. Lucy receives 15% discount on purchases at the clothing store where she is employed. How much would she pay for a suit having a list price of $110?

17. Alex purchased a new pair of shoes marked $40. Alex must pay an additional $5\frac{1}{2}\%$ in sales tax. What will be the total cost of the shoes?

18. Of the 2400 students enrolled in developmental mathematics, only 12% are not freshmen. How many freshmen are enrolled in developmental mathematics?

19. Jodi purchased a new dress that had been marked down from $80 to $64. What discount rate did Jodi receive?

20. Gina is being promoted from a position where she is currently earning $1200 per month. She will receive a pay raise of 8% per year. What will be her new monthly salary?

21. Aaron purchased a new pocket calculator, paying a *total* of $12.98. The price included 98 cents in sales tax. What sales tax rate did Aaron get charged?

22. At a local school, 90 students have chicken pox. If this represents 15% of the students, how many students are enrolled in the school?

23. A teacher who earns $25,000 a year gets an 8% increase in pay. What is his new salary?

24. An accountant who earns $102,000 a year gets a 9% increase in pay. What is her new salary?

25. A recliner chair is on sale for 22% off the original price of $450. What is the sale price?

26. An electronic game is marked 35% off. The original price of the game was $99. What is the sale price?

27. An oil and lube usually costs $24. This week it is on sale for $18. What is the percent discount?

28. The Ewbanks had listed their home for $125,000. It sold for $115,000. What was the percent decrease between the asking price and the sale price?

29. Shirley sews stuffed bunnies. She sells them for 60% more than the materials cost her. If she sells a bunny for $15, how much did the materials cost?

30. A landscape gardener plants trees for 40% more than he pays for them. If he charges me $126 to plant a maple tree, how much did he pay for the tree?

31. The price of coffee has risen 550% since 1980. If coffee was 90 cents a pound in 1980, what does it cost today?

11.5 Simple Interest

OBJECTIVE
● Use the formula $P \cdot R \cdot T = I$.

In this section, we learn how to calculate simple interest. The ability to understand and calculate interest will be useful when you borrow or invest money. Interest is generally charged when money is borrowed to buy a house, or a car, or to attend college. Credit cards issued by banks, department stores, and oil companies add interest to unpaid accounts. The formula for calculating *simple interest* follows:

$$\text{Principal} \cdot \text{Rate} \cdot \text{Time} = \text{Interest}$$
$$P \cdot R \cdot T = I$$

DEFINITION   **Time:** The length of time of a loan or investment.

DEFINITION   **Principal:** The amount of money borrowed or invested.

DEFINITION   **Rate:** The percent of the principal that will be earned or paid over the specified period of time.

DEFINITION   **Interest:** The money paid for the use of the principal. Interest is the money earned by the investor of money invested or paid by the borrower on money borrowed.

EXAMPLE 1   How much interest will be earned on $500 invested at 6% for a year?

$$P \cdot R \cdot T = I$$
$$500 \cdot 0.06 \cdot 1 = 30$$

$500 invested at 6% for 1 year will earn $30 interest. How much would the investor have in the bank at the end of the year? $500 + $30 = $530.

USING THE CALCULATOR

Enter 500 × 6 2nd % × 1 = **30** ← Display

Note   Example 1 calculates interest for a year. When the problem involves calculating interest for more than 1 year, multiply by the number of years.

EXAMPLE 2   How much interest will be earned on $8000 invested at 7% for 3 years?

$$P \cdot R \cdot T = I$$
$$8000 \cdot 0.07 \cdot 3 = I$$
$$1680 = I$$

$8000 invested at 7% for 3 years will earn $1680. How much will the investor have in the bank at the end of 3 years? $8000 + $1680 = $9680.

USING THE CALCULATOR

Enter 8000 × 7 2nd % × 3 = **1680** ← Display

Note   To calculate interest earned for less than a year, use a fraction with the number of months as the numerator and 12 as the denominator.

Practice Problems

Solve these problems.

1. How much interest will be paid on a loan of $700 at 11% loaned for a year?

2. How much interest will be paid on a loan of $2000 at 8% for 2 years?

EXAMPLE 3 How much interest will be earned on $600 invested at 4% for 3 months?

$$P \cdot R \cdot T = I$$

$$600 \cdot 0.04 \cdot \frac{3}{12} = I$$

$$6 = I$$

$600 invested at 4% will earn $6 in 3 months. At the end of 3 months, there will be $600 + $6 = $606 in the bank.

USING THE CALCULATOR

Enter 600 × 0.04 × 3 $a^{b/c}$ 12 = **6** ← Display

Not all problems will require finding the interest. To find another quantity in the formula, fill in the quantities given and solve the equation.

EXAMPLE 4 How much should be invested at 11% for 3 months to earn $11 interest?

$$P \cdot R \cdot T = I$$

$$P \cdot 0.11 \cdot \frac{3}{12} = 11$$

$$0.0275P = 11$$

$$P = \frac{11}{0.0275}$$

$$P = 400$$

$400 invested at 11% for 3 months will earn $11. At the end of 3 months, the investor could withdraw $411 from the bank.

EXAMPLE 5 At what rate of interest must $900 be invested for 6 months to earn $27 interest?

$$P \cdot R \cdot T = I$$

$$900 \cdot R \cdot \frac{6}{12} = 27$$

$$450R = 27$$

$$R = \frac{27}{450}$$

$$R = 0.06$$

$$R = 6\%$$ A fraction or decimal must be changed to a percent.

$900 must be invested at 6% to earn $27 in 6 months.

EXAMPLE 6 How long must $2000 be invested at 7% to earn $280 in simple interest?

$$P \cdot R \cdot T = I$$

$$2000 \cdot 0.07 \cdot T = 280$$

$$140T = 280$$

$$T = \frac{280}{140}$$

$$T = 2$$

$2000 must be invested for 2 years at 7% to earn $280.

PROBLEM SET 11.5

Use the formula $P \cdot R \cdot T = I$ to write an equation and solve the following problems. All interest is assumed to be simple.

1. How much interest will you earn on $900 invested at 12% for a year?

2. An investment of $14,000 invested at 7% for 3 years will earn how much interest?

3. How much money invested at 5% will earn $4650 interest in a year?

4. If $280 interest is earned in 6 months at 7%, how much was invested?

5. How much money must be invested at 11% to earn $33 in 3 months?

6. At what rate of interest must $5400 be invested to earn $648 interest in a year?

7. If $650 is invested for 6 months and earns $19.50 interest, what is the rate?

8. How long must $500 be invested at 8% to earn $40 interest?

9. How long must $7500 be invested at 9% to earn $168.75 interest?

10. If you deposit $8000 in a bank and receive $5\frac{1}{2}$% interest, how much interest would it earn in 18 months?

11. What is the interest on a $750 bank loan at $8\frac{1}{2}$% for 30 months?

12. At what rate of interest must $1500 be invested to earn $15 interest in 3 months?

13. If $102,000 is borrowed for 2 months and owes $935 interest, what is the rate?

14. At what rate of interest must $79,000 be invested for 3 years to earn $15,997.50 interest?

15. If you owe $138 interest on a bank loan at 8%, borrowed for 9 months, how much was borrowed?

16. How long do I have to repay a loan of $2300 borrowed at 4.5% so I pay $51.75 interest?

17. How long must $700 be invested to earn $36.75 at 7%?

18. How long must $77,000 be invested at $9\frac{1}{2}$% to earn $3657.50 interest?

19. How much money is invested at $6\frac{1}{2}\%$ for 2 years to earn $1235 interest?

20. Maida invests $10,500 in a special account that pays 5% interest. How much will she have at the end of one year?

21. Juan deposits $1800 in an account that pays 6% interest. If he withdraws all the money in the account after 3 months, how much does he withdraw?

22. What interest will be earned on an investment of $8000 at 4% for 9 months?

23. How much interest will be earned on $7500 at 6% for 3 months?

24. If you invest $2300 at $5\frac{1}{2}\%$ for 3 years, how much interest do you earn?

25. If you borrow $3300 at 9% for 3 months, how much interest do you owe?

26. The credit union will loan $2900 at 4.5% for 15 months. How much interest will be owed?

27. How much interest will be earned on $11,750 at $5\frac{1}{4}\%$ for 2 years?

28. How much interest will be earned on $390 invested for 12 months at 12%?

29. What amount of money invested at 6% for 18 months will yield $540 in interest?

30. Astra invested $8000 at $8\frac{1}{4}\%$ interest. How much did she have in the bank at the end of 6 months?

31. If $27,000 is invested at 7% interest for a year, how much can be withdrawn if the account is closed at the end of the year?

32. A student takes out a loan for $8000 to attend Kent State University this year. If the interest rate is 6%, how much does the student owe at the end of one year? Hint: He will owe the principle and interest!

33. Stacie deposits $500 into an account paying 6% a year. Steve deposits $450 into an account paying 12% a year. Who will have the larger total balance at the end of one year?

SMALL GROUP ACTIVITY *Estimating Tips*

Have one person in each group write the total of a restaurant bill, a beauty shop bill, a taxicab ride, or any other bill that requires a tip.

The other members of the group take 30 seconds (without using paper and pencil or a calculator) to estimate 15% of the bill and write it down. The person who originally wrote the bill should use a calculator to find the 15%. The person whose estimate is closest (either higher or lower) gets to present the next bill to be estimated.

CHAPTER 11 REVIEW

The terms listed below are among those used in this chapter. Write a brief definition for each term. (Refer to the appropriate section in the chapter as needed.)

1. Percent _____

2. Base _____

3. Rate _____

4. Amount _____

5. Simple interest _____

6. Principal _____

After completing Chapter 11, you should be familiar with the following concepts and be able to apply them.

• *Convert percents to reduced fractions (11.1).*

Change each percent to a reduced fraction.

7. 40%   **8.** 4%

9. $33\frac{1}{3}\%$ **10.** 150%

11. 0.8%

• *Convert fractions to percents (11.1).*

Change each fraction to a percent.

12. $\frac{1}{2}$ **13.** $\frac{7}{8}$

14. $\frac{2}{3}$ **15.** $\frac{3}{10}$

16. $\frac{2}{9}$

• *Convert percents to decimals (11.1).*
Change each percent to a decimal.

17. 27%   **18.** 55%

19. 250%   **20.** 0.4%

21. 3%

• *Convert decimals to percents (11.1).*
Change each decimal to a percent.

22. 0.8   **23.** 0.04

24. 3.42   **25.** 0.0346

26. 0.261   **27.** 0.24

28. 0.125   **29.** 9.75

30. 0.004

• *Solve basic percent problems using either proportion or the formula: $B \cdot R = A$ (1.12).*

31. 9 is 90% of what number? _____

32. What percent of 60 is 30? _____

33. 80% of what number is 28? _____

34. What percent of 56 is 14? _____

35. 38% of 70 is what number? _____

36. 0.5% of 200 is what number? _____

37. 175% of 300 is what number? _____

38. What percent is 16 of 48? _____

39. 37% of what number is 333? _____

40. 0.25% of 3800 is what number? _____

• *Solve percent application problems using either proportion or the formua B · R = A (11.3).*

41. The sales tax rate in Ohio is $6\frac{1}{2}\%$. How much tax must I pay on a tape deck that costs $300?

42. In a "30% Off" sale, how much would be saved on a coat originally priced at $275?

43. Marissa scored 20 of 25 problems on her test. What percent did she score?

44. Akita received a 25% discount on a TV. If she saved $210, what was the list price of the TV?

45. Agnus made 9 of 15 tries in the basketball game. What percent did he make?

46. In a mathematics class, there are 390 female students, which is 52% of the total number of students taking the class. How many students are taking this mathematics class?

• *Solve percent increase or decrease problems using either proportions or the formula:*
B · R = A (11.4).

47. Jayci missed 5 of 25 problems on a test. What percent did he get correct?

48. A shirt that usually sells for $65 has been marked down to $52. What is the rate of discount on this shirt?

49. Janet receives a 25% discount from the store where she works. If she buys a sweater that costs $84, how much will she pay for the sweater?

50. Ronald missed 14 problems on his test. He had 80% of the problems correct. How many problems were on the test?

51. In a math class, 40% of the students are freshmen. If there are 18 freshmen, how many students are in the class?

52. There are 35 students in a class. On Monday, 28 were present. What percent were absent?

• *Use the formula P · R · T = I (11.5).*

53. How much interest will be earned on $15,000 invested for 2 years at $6\frac{1}{2}\%$

54. Jerry borrowed $3200 to buy a used car. The bank charged him $9\frac{1}{2}\%$ interest. He paid the loan and the interest 9 months later. How much did he repay the bank?

55. How much must be invested at $7\frac{1}{4}$ to earn $1305 interest in 6 months?

56. How long must $103,000 be left in a savings account to earn $3347.50 interest at $6\frac{1}{2}\%$?

57. At what rate must $785 be invested to earn $15.70 in 4 months?

58. The bank charges 11% on a loan. Joe can only pay $33 for 3 months' interest. How much can Joe borrow?

59. How long must $1400 be left in a savings account at 9% to earn $189 interest?

60. If $7000 earns $236.75 in 9 months, what is the rate of interest?

61. If $12,500 is deposited in a savings account paying $8\frac{1}{2}\%$ interest, how much will be in the account in 30 months?

CHAPTER 11 PRACTICE TEST

Fill in the blank spaces with equivalent fractions, decimals and percents.

	Fraction	Decimal	Percent
1.	$\frac{1}{4}$		
2.			62.5%
3.		0.8	
4.	$\frac{1}{3}$		
5.			175%
6.			0.4%
7.			$\frac{1}{2}\%$
8.	$\frac{5}{8}$		
9.		0.005	

10. What is 150% of 62?

11. 32 is 25% of what number?

12. 76 is what percent of 400?

13. 135% of what number is 54?

14. 36.75% of 28 is what number?

15. 1 is what percent of 1000?

16. Barry bought a used car for $1200. He made a down payment of 15%. How much cash did he need for the down payment?

17. The baseball team hit 36 of the 90 pitches during the game. What percent did they hit?

18. Daika got 63 problems correct on her test. Her percent correct was 70%. How many problems were there on the test?

19. A store raised the price of its TVs 8%. If a TV now sells for $216, what was the original price?

20. Azil was being paid $5 an hour. His wages were decreased to $4.75 an hour. What percent was the decrease?

21. The population of Kent increased 9% at the last census. If the former population was 23,900, what is the present population?

22. Monique had invested $6800 in a savings account at $15\frac{1}{2}$% interest. How much was in the savings account at the end of 6 months?

23. How much must be invested at $8\frac{1}{2}$% to earn $110.50 interest in 3 months?

24. At what rate must $7500 be invested to earn $825 interest in 2 years?

25. How long must $13,000 be invested at 4% to earn $390?

STUDY SKILLS TIP

Preparation for the Final Exam

Preparation for the final exam really begins the first day of the semester and continues through every class, every homework assignment, and every tutoring session. It is impossible to pay little or no attention all semester and make a supreme effort for the final and then do well. Doing well in a course requires constant attention *all* semester. It is important to understand that in learning mathematics, every skill builds upon the last skill learned.

There are, however, ways you can improve your chances of doing well on the final. Take the time management plan on page 380 and fill it out for finals week. Remember, there may be no classes that week, and your finals could be at different times than your classes have been all semester. If you work, check well ahead of time at your place of employment to ensure that your employer can give you time off that week. You need to be totally involved in your school work during finals. Record on the time management plan your work hours and any other responsibilities you have. Schedule the hours you sleep. It is very important that you get plenty of rest—eight to nine hours a night. Do not cheat yourself on rest, and allow enough time on the time plan to prepare and eat wholesome meals. You will function more productively if you are well rested and well fed. Once you have a time management plan, follow it. You spent valuable time arranging your time so stay with the plan.

There will not be time before finals to review each section of the textbook. You need a practice final with answers. Your instructor will give you one, or you can make your own using the "Study Skills Tip: Test Preparation." It is important to *not* have the problems in the order you studied them. Problems on the final or on any test will not necessarily be in the order you learned them.

Work the problems on the practice test, then check your answers. Highlight the problems you found difficult. Make yourself a second problem sheet containing only the problems you had difficulties with. Try to rework each problem. If the procedure does not come quickly, find the appropriate section in your notebook or textbook. Reread the section to better understand the problem. If you are able to resolve your difficulty, write out the procedure in *detail* and move on. Do the same for all the other problems. If after studying your notes and the textbook, you are still unable to understand a problem, make another sheet of just those problems. Take these problems to a tutor or your instructor. They will be able to explain them for you and probably will be able to show you why these particular problems were giving you difficulties. Never give up on a problem. Be determined to understand each one—it will pay off during the final.

	Sun.	Mon.	Tue.	Wed.	Thurs.	Fri.	Sat.
7:00							
8:00							
9:00							
10:00							
11:00							
12:00							
1:00							
2:00							
3:00							
4:00							
5:00							
6:00							
7:00							
8:00							
9:00							
10:00							
11:00							
12:00							

THE MATHEMATICS JOURNAL

Mathematics in Everyday Life

Task: Reflect on the ways you use mathematics every day. Write a one-page (150-word) entry on your reflections.

Starters: You use money every day.

Do you shop at sales? Do you look for 25% reductions?

Are you buying a car with a loan? What is the rate of interest?

12

Geometry and Measurement

12.1 Linear Measure and Applications
12.2 Square Measure and Applications
12.3 Cubic Measure and Applications

Introduction to Geometry and Measurement

We do not speak of one geometry, but of several geometries because so many have developed through the centuries. The geometry that examines the physical characteristics of shapes or figures—as mathematicians like to call them—dates back to early civilization and has its origin in the shapes of nature: the roundness of stones, the cylindrical shapes of trees, the pyramidlike mountains. The geometry that deals with measurement also dates back to ancient times, when questions such as "How far?" and "How long?" and "How big?" were first being asked. We do know from a book called *The Elements,* written by Euclid, that by approximately 300 B.C. a new geometry based solely on deductive reasoning had developed. It was followed several centuries later by projective geometry as the Renaissance artists became interested in perspective. By the seventeenth century, mathematicians had created coordinate geometry. In modern times, new geometries have emerged—in particular, several non-Euclidean geometries in the nineteenth century and topology in the twentieth.

In this chapter, using Euclidean terms and formulas, we will explore the geometry of measurement.

12.1 Linear Measure and Applications

OBJECTIVES

- Perform conversions of linear measures within both the U.S. and metric systems.
- Apply the Pythagorean Theorem.
- Calculate the perimeters of polygons, including triangles, rectangles, and squares.
- Calculate the circumferences of circles.
- Find the perimeters of certain irregular plane figures.
- Solve application problems involving the concepts of perimeter and circumference.

Euclidean geometry is based on three terms considered undefined. They are point, line, and plane. While we will accept them as undefined, we can describe their properties.

A point has position only. It has no dimensions.
A line refers to what we commonly call a straight line. It extends infinitely far in two directions.

A **line segment** is a part of a line. It has one dimension—length. $\overline{AB}$ is a line segment. Its endpoints are A and B.

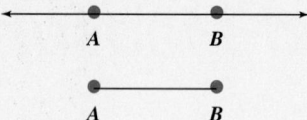

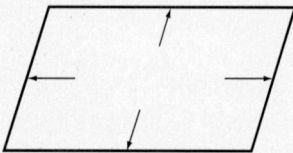

Plane. To think of a **plane,** think of a flat surface that extends infinitely far in four directions.

Linear Measure

Linear measure is used to determine the length of a line segment by using a measuring device such as ruler or a tape measure. We will use two common systems: the U.S. customary system and the metric system.

U.S. Customary System of Linear Measure

Inches, feet, yards, and miles are common units of measure in this system. The smallest of these units is **inches.** We would commonly measure the length of our fingers in inches, the distance from a stop sign to the corner in **feet,** the distance the football is from the goal line in **yards,** and the distance from one city to another in **miles.** Inches, feet, yards, and miles all measure distance or length. Which label or unit of measure we use depends on how long or short the segment we're measuring is. We can convert from one unit of measure to another by using the following chart.

U.S. Linear Measures
12 in = 1 ft
3 ft = 1 yd
5280 ft = 1 mi

Abbreviations
in = inches
ft = feet
yd = yards
mi = miles

EXAMPLE 1   How many inches are there in 2 feet?

Let x = number of inches in 2 feet.   Assign the variable.

12 in = 1 ft   Refer to the chart. Choose the equivalence that states
                          the relationship between the measures.

$\overline{12\text{ in}} = \overline{1\text{ ft}}$ Make these measures the denominators of the
                          ratios that form the proportion.

$$\frac{x \text{ in}}{12 \text{ in}} = \frac{2 \text{ ft}}{1 \text{ ft}}$$

Set up the proportion by filling in the numerators from the assignment statement. Remember to keep inches on one side of the equal sign and feet on the other.

$$\frac{x}{12} = \frac{2}{1}$$

$$1 \cdot x = 2 \cdot 12$$

Use the product of the means. . . to solve the proportion.

$$x = 24$$

There are 24 inches in 2 feet.   Answer the question.

EXAMPLE 2   Convert 6 miles to yards. Since the chart does not state the relationship between yards and miles, we must use two proportions to solve this problem. First, we'll use the equivalence that shows the relationship between miles and feet and then the one that shows the relationship between feet and yards.

Part I: Convert miles to feet.

Let f = number of feet in 6 miles.

$$5280 \text{ ft} = 1 \text{ mi}$$

$$\overline{5280 \text{ ft}} = \overline{1 \text{ mi}}$$

$$\frac{f \text{ ft}}{5280 \text{ ft}} = \frac{6 \text{ mi}}{1 \text{ mi}}$$

$$\frac{f}{5280} = \frac{6}{1}$$

$$1 \cdot f = 6 \cdot 5280$$

$$f = 31{,}680 \text{ so } 31{,}680 \text{ ft} = 6 \text{ mi}$$

Part II: Convert feet to yards.

Let y = number of yards in 31,680 ft.

$$3 \text{ ft} = 1 \text{ yd}$$

$$\overline{3 \text{ ft}} = \overline{1 \text{ yd}}$$

$$\frac{31680 \text{ ft}}{3 \text{ ft}} = \frac{y \text{ yd}}{1 \text{ yd}}$$

$$\frac{31680}{3} = \frac{y}{1}$$

$$10{,}560 = y$$

There are 10,560 yards in 6 miles.

Metric System of Linear Measure

Millimeters, centimeters, meters, and kilometers are commonly used units of measure in this system. The **meter,** which is about 3 inches longer than a yard, is the central unit of linear measure. The other measures are defined in terms of their relationship to the meter. A **millimeter** is one-thousandth of a meter; a **centimeter** is one-hundredth of a meter; and a **kilometer** is a thousand meters.

We would measure the width of a piece of lead in millimeters and the length of a pencil in centimeters. We would measure the distance from a stop sign to the intersection in meters and the distance between cities in kilometers. The metric system is based on powers of 10. To convert from one metric measure to another, we need only multiply or divide by 10 raised to the appropriate power.

Abbreviations
mm = millimeters
cm = centimeters
m = meters
km = kilometers

Metric Linear Measures

$$10 \text{ mm} = 1 \text{ cm}$$
$$100 \text{ cm} = 1 \text{ m}$$
$$1000 \text{ m} = 1 \text{ km}$$

EXAMPLE 3   Convert 5.6 meters to centimeters.

Method 1
Use the metric chart for reference and multiply by the appropriate power of 10.

$$100 \text{ cm} = 1 \text{ m}$$
$$100 \cdot 5.6 \text{ cm} = 1 \cdot 5.6 \text{ m}$$
$$560 \text{ cm} = 5.6 \text{ m}$$

Method 2
Use the same procedure we used in the two preceding examples. Use the metric chart for reference.

Let s = number of centimeters in 5.6 meters.
$$100 \text{ cm} = 1 \text{ m}$$

$$\overline{100 \text{ cm}} = \overline{1 \text{ m}}$$
$$\frac{s \text{ cm}}{100 \text{ cm}} = \frac{5.6 \text{ m}}{1 \text{ m}}$$
$$\frac{s}{100} = \frac{5.6}{1}$$
$$1 \cdot s = 100 \cdot 5.6$$
$$s = 560$$

There are 560 centimeters in 5.6 meters.

EXAMPLE 4   Convert 5727 meters to kilometers.

$$1 \text{ m} = \frac{1}{1000} \text{ km}$$
$$1 \cdot 5727 \text{ m} = \frac{1}{1000} \cdot 5727 \text{ km}$$
$$5727 \text{ m} = 5.727 \text{ km}$$

There are 5.727 kilometers in 5727 meters.

Practice Problems

Make these conversions.

1. 5 km = _____ m

2. 17 yd = _____ ft

3. 3.2 yd = _____ in

4. 20 mm = _____ cm

5. 1.6 mi = _____ ft

6. 108 in = _____ ft

7. 6.05 m = _____ cm

Plane Figures

Plane figures have only two dimensions—length and width.

DEFINITION **Polygon:** A plane figure bounded by line segments that form its sides. It is a closed figure, which means each of its sides shares each of its endpoints with another side. A polygon has two dimensions: length and width.

DEFINITION **Triangle:** A polygon that has three sides.

DEFINITION **Quadrilateral:** A polygon that has four sides.

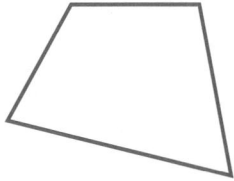

DEFINITION **Rectangle:** A quadrilateral that has four 90° angles.

DEFINITION **Square:** A rectangle with four equal sides.

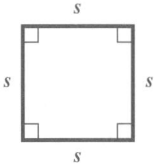

The Pythagorean Theorem

A **theorem** is a statement that can be proved. The Pythagorean Theorem gets its name from Pythagoras, a mathematician who lived in the sixth century B.C. It deals with triangles containing right, or 90° angles. It states that the square of the length of the side opposite the 90° angle is equal to the sum of the squares of the lengths of the other two sides. In a right triangle, the sides are given special names. The side opposite the right angle is called the **hypotenuse,** and the other two sides are called the **legs.**

Pythagorean Theorem
The square of the length of the hypotenuse is equal to the sum of the squares of the lengths of the legs.

In terms of the figure, it may be stated:

$$c^2 = a^2 + b^2$$

This theorem can be used to find the length of one side of a right triangle when the other two sides are known.

EXAMPLE 5 Find the length of the hypotenuse of a right triangle whose legs measure 3 inches and 4 inches.

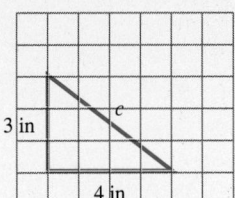

Draw a picture.

$$a^2 + b^2 = c^2$$ State the theorem.

$$3^2 + 4^2 = c^2$$ Substitute the known values.

$$9 + 16 = c^2$$ Simplify and solve for the variable.

$$25 = c^2$$

$$5^2 = c^2$$

$$5 = c$$

The hypotenuse is 5 inches long.

USING THE CALCULATOR

To solve this example on the calculator,

Enter 3 $\boxed{x^2}$ + 4 $\boxed{x^2}$ = $\boxed{\sqrt{x}}$ **5** ← Display.

EXAMPLE 6 Refer to the figure. Find the length of the third side.

Let x = the length in cm of the third side.

$$a^2 + b^2 = c^2$$

$$x^2 + 12^2 = 13^2$$

$$x^2 + 144 = 169$$

$$\underline{-144 = -144}$$

$$x^2 = 25$$

$$x^2 = 5^2$$

$$x = 5$$

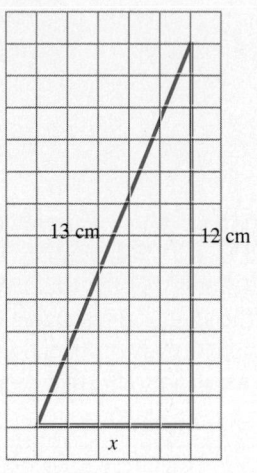

The length of the third side is 5 centimeters.

Practice Problems

Consider the right triangle drawn at the right. For each problem, find the length of the indicated side.

8. $a = 6$ m
$b = 8$ m
$c = ?$

9. $a = ?$,
$b = 24$ ft
$c = 26$ ft

10. $a = 12$ cm
$b = ?$,
$c = 20$ cm

11. $a = 8$ in
$b = 15$ in
$c = ?$

12. $a = 9$ yd
$b = ?$
$c = 41$ yd

▦ *USING THE CALCULATOR*

Enter 13 x^2 $-$ 12 x^2 $=$ $\sqrt{x}$ **5** ← Display

EXAMPLE 7   Find the length of the hypotenuse of the right triangle with legs of 1 inch and 2 inches in length.

$$c^2 = a^2 + b^2$$
$$c^2 = 1^2 + 2^2$$
$$c^2 = 1 + 4$$
$$c^2 = 5$$
$$c = \sqrt{5}$$
$$c = 2.236067977 \ldots$$
$$c \approx 2.2$$

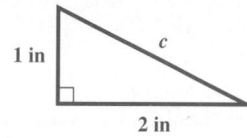

1 in

c

2 in

The hypotenuse is approximately 2.2 inches.

Note   Because 5 is not a perfect square, its square root is an irrational number. It can be expressed only as a nonterminating, nonrepeating decimal, so we only approximate its value to any given number of decimal places.

Perimeters of Polygons

DEFINITION   **Perimeter:** The distance around a plane figure.

RULE **Perimeter of a polygon**
The perimeter of a polygon is the sum of the lengths of its sides.

$P = a + b + \ldots + g$ where $a, b, \ldots, g$ are the lengths of its sides

---

EXAMPLE 7   Find the perimeter of the polygon. To find the perimeter, find the sum of the lengths of the sides.

$$P = a + b + c + d + e$$
$$P = 2 \text{ ft} + 5 \text{ ft} + 1 \text{ ft} + 6 \text{ ft} + 3 \text{ ft}$$
$$P = 2 + 5 + 1 + 6 + 3$$
$$P = 17$$

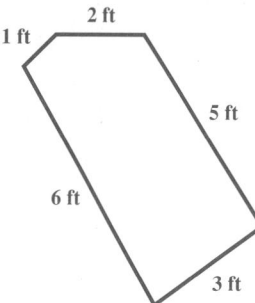

2 ft

1 ft

5 ft

6 ft

3 ft

The perimeter is 17 feet.

RULE **Perimeter of a triangle**

$P = a + b + c$ where $a, b,$ and c are the lengths of its sides.

EXAMPLE 8   Find the perimeter of the triangle drawn at the right.

$$P = a + b + c$$
$$P = 3 \text{ m} + 7 \text{ m} + 5.2 \text{ m}$$
$$P = 15.2 \text{ m}$$

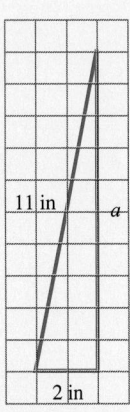

7 m

5.2 m

3 m

The perimeter is 15.2 meters.

EXAMPLE 9   The perimeter of the triangle is approximately 23.8 inches. Find the length of side a.

$$P = a + b + c$$
$$23.8 = a + 2 + 11$$
$$23.8 = a + 13$$
$$\underline{-13 \quad = \quad -13}$$
$$10.8 = a$$

11 in

a

2 in

Side a is approximately 10.8 inches long.

EXAMPLE 10 Find the perimeter of a triangle with sides of 5 ft, 12 ft, and 3 yd. Be careful! Before we can apply the formula for perimeter, we must first convert 3 yards to feet.

Part I: Let x = number of feet in 3 yards.

$$\frac{3 \text{ yd}}{1 \text{ yd}} = \frac{x \text{ ft}}{3 \text{ ft}}$$

$$\frac{3}{1} = \frac{x}{3}$$

$$1 \cdot x = 3 \cdot 3$$

$$x = 9$$

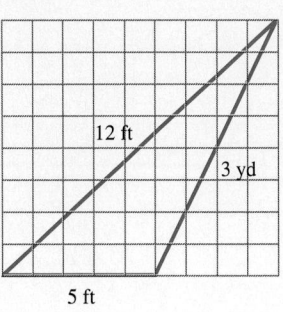

12 ft

3 yd

5 ft

There are 9 feet in 3 yards.

Part II: $$P = a + b + c$$

$$P = 5 \text{ ft} + 12 \text{ ft} + 9 \text{ ft}$$

$$P = 26 \text{ ft}$$

The perimeter is 26 feet.

APPLICATION   Hank Jones wants to fence off the triangular gardens at each corner of his back yard to keep out the rabbits. He knows that the two plots have the same measurements. Before he goes to the hardware store to buy fencing, he measures the distance around the edge of one plot. Once at the store, he realizes he left the paper with the measurements at home. He does, however, remember that the triangles the gardens form are right triangles and that one of the short sides measures 9 feet and the other measures 12 feet. Help him decide how much fencing he needs. If the fencing he likes is packaged by the yard and priced at $1.95 per yard, how many yards should he buy, and how much will it cost him including tax? The sales tax rate in his county is 6%.

Part I: Draw a diagram. Label the sides.

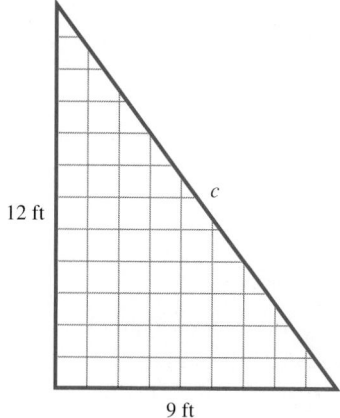

12 ft

c

9 ft

$a^2 + b^2 = c^2$ Since the triangles are right triangles,
$9^2 + 12^2 = c^2$ use the Pythagorean Theorem to
$81 + 144 = c^2$ find the length of the third side.
$225 = c^2$
$15 = c$ The third side is 15 feet.
$P = a + b + c$ Find the perimeter of one garden.
$P = 9 + 12 + 15$
$P = 36$

He needs 36 feet of fencing for each garden. He needs 72 feet of fencing for both gardens.

Part II: Let x = number of yards in 72 feet.

$$\frac{72 \text{ ft}}{3 \text{ ft}} = \frac{x \text{ yd}}{1 \text{ yd}}$$

$$\frac{72}{3} = \frac{x}{1}$$

$$24 = x$$

He needs to buy 24 yards of fencing.

$$\$1.95 \cdot 24 + (\$1.95 \cdot 24)(0.06) = \text{Total cost}$$
$$\$46.80 + (\$46.80)(0.06) = \text{Total cost}$$
$$\$46.80 + \$2.808 = \text{Total cost}$$
$$\$49.608 = \text{Total cost}$$
$$\$49.61 = \text{Total cost}$$

Answer: He needs 72 feet (24 yards) of fencing for both gardens. The total cost is $49.41.   ◆

Find the perimeters of polygons whose sides have these measurements.

13. 15 ft, 13 ft, 9 ft

14. 6 cm, 4 cm, 1 cm, 12 cm, 8 cm

15. 12 mm, 6 mm, 1 cm

16. The perimeter of a triangle is 16 in. One side measures 5 in, and another side measures 6 in. Find the length of the third side.

For Part II.

Enter 1.95 $\times$ 24 $+$ 6 % $=$ **49.608** $\leftarrow$ Display

Perimeters of Rectangles

Since a rectangle is a polygon with four sides, its perimeter is the sum of the lengths of those four sides. We can shorten this rule because the opposite sides of a rectangle are equal in length. If we refer to one of the two opposite sides of the rectangle as *l* for length and one of the other two opposite sides as *w* for width, then

$$P = l + w + l + w$$

so $P = 2l + 2w.$

RULE **Perimeter of a rectangle**

The perimeter of a rectangle is twice the length plus twice the width.

$$P = 2l + 2w$$

EXAMPLE 11   Find the perimeter of the rectangle.

$P = 2l + 2w$

$P = 10 + 34$

$P = 44$

5 cm

17 cm

The perimeter is 44 centimeters.

For this example,

Enter 2 $\times$ 5 $+$ 2 $\times$ 17 $=$ **44** $\leftarrow$ Display.

EXAMPLE 12   Find the perimeter of the rectangle whose length is 5 inches and whose width is 1 foot.

Draw a diagram.

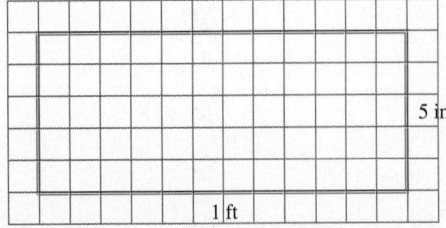

5 in

1 ft

1 ft = 12 in   Convert feet to inches

$P = 2 \cdot L + 2 \cdot W$ Apply the formula.

$P = 2 \cdot 5 + 2 \cdot 12$

$P = 10 + 24$

$P = 34$

The perimeter is 34 inches.

EXAMPLE 13 The perimeter of a rectangle is 16 meters. The width is 5 meters. Find the length.

$P = 2L + 2W$

$16 = 2L + 2 \cdot 5$

$16 = 2L + 10$

$\underline{-10 \qquad\quad -10}$

$6 = 2L$

$\dfrac{6}{2} = \dfrac{2L}{2}$

$3 = L$

5 m

L

The length is 3 meters.

APPLICATION Joe decides to put a chain-link fence around part of the back yard so that his dog will have a place to run and play. He measures the rectangular plot he wants to fence in and learns that it is 30 feet by 40 feet. How many feet of fencing does he need? (Don't forget to allow space for a 3-foot gate.)

$P = 2l + 2w$

$P = 2 \cdot 30 + 2 \cdot 40$

$P = 60 + 80$

$P = 140$

$P - 3 = 140 - 3$

$P - 3 = 137$

40 ft

30 ft

3 ft

Answer: Joe will need 137 feet of fence. ◆

Perimeters of Squares

A square is a rectangle with four equal sides. Since the length of a square equals the width, we can rewrite the formula for the perimeter of a rectangle as follows:

Let s = length of one side of a square.

Then $P = 2l + 2w$

becomes $P = 2s + 2s$

or $P = 4s$

s

s

RULE Perimeter of a square

The perimeter of a square is 4 times the length of one side.

$P = 4s$ where s = length of one side

Practice Problems

Find the perimeter of each rectangle.

17. l = 15 in, w = 12 in

18. l = 2.6 cm, w = 1.9 cm

19. l = 2 yd, w = 8 ft

Find the missing dimension of each rectangle.

20. P = 56 ft, w = 8 ft

21. P = 2 m, l = 30 cm

EXAMPLE 14 Find the perimeter of a square with a side 6 feet in length.

$$P = 4s$$
$$P = 4 \cdot 6$$
$$P = 24$$

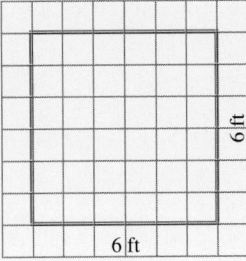

The perimeter is 24 feet.

EXAMPLE 15 The perimeter of a square is 100 meters. Find the length of one side. Use the formula for perimeter of a square to find the length of one side in meters.

Let s = length of one side in meters.

$$P = 4s$$
$$100 = 4 \cdot s$$
$$\frac{100}{4} = \frac{4s}{4}$$
$$25 = s$$

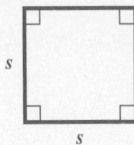

The length of one side is 25 meters.

APPLICATION Molly decides to put a wallpaper border around her child's 10×10 foot bedroom. How much wallpaper should she buy? When she gets to the store and chooses the pattern, she finds that it is sold in 15 foot rolls at $10 per roll. How many rolls does she need? How much will it cost? (Her sales tax rate is 5.5%.)

Part I

$$P = 4s$$
$$P = 4 \cdot 10$$
$$P = 40$$

Answer: Molly needs 40 feet of wallpaper.

Part II: Let x = number of rolls she needs.

$$\frac{x \text{ rolls}}{1 \text{ roll}} = \frac{40 \text{ ft}}{15 \text{ ft}}$$
$$\frac{x}{1} = \frac{40}{15}$$
$$x = 2\frac{2}{3}$$

Answer: She needs to buy 3 rolls.

Part III

$$3 \times \$10 = \$30$$
$$\$30 \times 0.055 = \$1.65$$
$$\$30 + \$1.65 = \$31.65$$

Answer: The wallpaper border will cost $31.65. ◆

Find the perimeter of each square.

22. s = 2.6 in

23. s = 3 km

Find the side of the square.

24. P = 18 cm

Circles

We can all recognize a circle when we see one, and we see circles every day: a penny, a ring, and a CD are but a few examples of circular shapes.

DEFINITION   **Circle:** A plane figure bounded by a closed curved line. All points on the curve are a fixed distance from a fixed point. The fixed point is called the **center** of the circle. The fixed distance is called the **radius.**

Let's draw a circle and refer to it as we analyze the definition. In this circle, we called the center, O, and the radius, r. Choose two points on the circle. Call them A and B. Draw the line segment $\overline{AO}$ and the line segment $\overline{BO}$. Notice that $\overline{AO}$ and $\overline{BO}$ are both radii (plural of radius) of the circle.

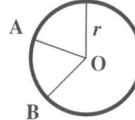

DEFINITION   **Diameter:** A line segment that contains the center of the circle and whose endpoints are on the circle.

DEFINITION   **Semicircle:** One-half of a circle.

DEFINITION   **Circumference:** The distance around a circle.

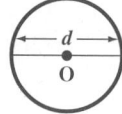

RULE Length of the diameter of a circle
The length of the diameter is equal to twice the length of the radius.

$$d = 2r$$

Since the third century B.C., mathematicians have known that the ratio of the circumference to its diameter, $\frac{C}{d}$, is a constant. Since the eighteenth century, they have called that constant "pi" and used the Greek letter π to represent it. Pi is an irrational number; that is, it cannot be expressed as the quotient of two integers. We can and do approximate its value and often say, $\pi \approx 3.14$. (Push the $\boxed{\pi}$ key on your calculator and see what answer you get.)

Since

$$\frac{C}{d} = \pi$$

$$\frac{C}{d} \cdot d = \pi \cdot d$$

or

$$C = \pi d$$

RULE Circumference of a circle
The circumference of a circle equals pi times the diameter of the circle.

$$C = \pi d$$

EXAMPLE 16 Find the circumference of a circle whose diameter is 5 centimeters.

$$C = \pi d$$
$$C = \pi \cdot 5$$
$$C \approx 3.14 \cdot 5$$
$$C \approx 15.7$$

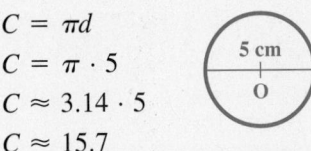

The circumference is approximately 15.7 centimeters.

 USING THE CALCULATOR

For this example,

Enter π × 5 = **15.707963** ← Display.

The number of decimal places displayed will vary depending on the calculator used. Round the displayed answer to one or two decimal places.

EXAMPLE 17 Find the circumference of a circle whose radius is 3.7 centimeters.

$$C = \pi d \quad \text{and} \quad d = 2r$$

so
$$C = \pi \cdot (2r)$$

or
$$C = 2\pi r$$
$$C = 2\pi(3.7)$$
$$C = 7.4\pi$$

or
$$C \approx (7.4)(3.14)$$
$$C \approx 23.24$$

The circumference is 7.4π centimeters or approximately 23.24 centimeters.

 USING THE CALCULATOR

Enter 2 × π × 3.7 = **23.247786** ← Display

Note   Often mathematicians leave π in the answer rather than using an approximation for it. A good rule of thumb is to leave π in if you have additional calculations to perform or if the question asked is purely theoretical. Approximate π, and perform the multiplication to get an approximate answer if the question is an application type, i.e., of a practical nature.

APPLICATION   Sophie wants to sew a ribbon border around a circular tablecloth as a holiday decoration. She asks for your help in deciding how much ribbon to buy. Advise her.

She needs to determine the circumference of the tablecloth. To find the circumference, she needs to find the diameter or the radius of the cloth. She can determine the diameter by folding the tablecloth in half and using a yardstick or

---

Practice Problems

Find the circumference of each circle.

25. $d = 6$ ft

26. $r = 10.2$ in

Find the value of the diameter or radius as indicated.

27. $C \approx 17.84$ cm, $d \approx$?

28. $C \approx 235.9$ m, $r \approx$?

29. $C = 15\pi$, $r =$?

tape measure to find the length of the folded edge. (See diagram.) Next, she needs to apply the formula $C = \pi d$ to find the circumference. That answer will tell her approximately how much ribbon to buy.

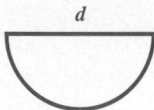

Perimeters of Other Plane Figures

Some plane figures are formed by combining two or more polygons or by combining circles or semicircles with polygons. The following examples illustrate how to find the perimeters of these figures.

EXAMPLE 18 Find the perimeter of the accompanying figure. To find the perimeter, we need to determine the lengths of the two sides whose lengths are not given, then use the definition of perimeter. Begin by labeling these sides a and b. Next, draw the dotted lines as shown in the diagram. By doing this, we have divided the figure into three rectangles. Using the properties of rectangles, we can now determine that a is 5 centimeters and b is 4 centimeters.

$$P = a + b + 4 + 7 + 9 + 3$$
$$P = 5 + 4 + 4 + 7 + 9 + 3$$
$$P = 32$$

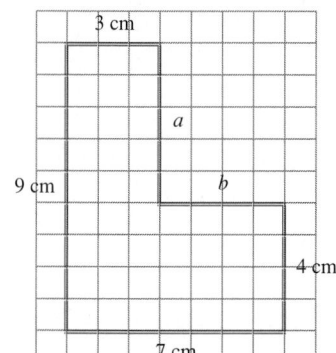

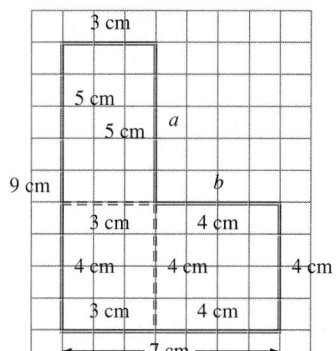

The perimeter is 32 centimeters.

● PROBLEM SET 12.1

Make the following conversions within the U.S. customary system.

1. 4 feet to inches

2. 9.5 feet to inches

3. 50 inches to feet

4. 103 inches to feet

5. 5 yards to feet

6. 2.5 yards to feet

7. 75 feet to yards

8. 23 feet to yards

9. 2.5 miles to yards

10. 3 yards to inches

11. 1.75 miles to feet

12. 95 inches to yards

Make the following conversions within the metric system.

13. 12 meters to centimeters

14. 382 centimeters to meters

15. 15 meters to millimeters

16. 3 meters to centimeters

17. 3 meters to millimeters

18. 5 kilometers to meters

19. 7.5 kilometers to meters

20. 853 meters to kilometers

21. 2000 meters to kilometers

22. 93 millimeters to centimeters

23. 83 centimeters to millimeters

24. 748 centimeters to meters

Applications: Make the following conversions.

25. How high in inches is a door 2.5 yards high?

26. How wide in meters is a carpet 256 centimeters wide?

27. A person can high jump 4 feet 5 inches. How many inches is the jump?

28. The desk is 1.27 meters wide. How many centimeters is that?

29. A computer disk is 13 centimeters wide. Express the width in millimeters.

30. A notebook is 9 inches wide. Express the width in feet.

Measure the following line segments in inches. State the results.

31. *A* •————————————————————• *B*

32. *C* •————————————————————————• *D*

33. E •————————————————• F

34. G •——————————————————————————• H

Measure the following line segments in centimeters. State the results.

35. J •——————————————————————• K

36. L •————————————• M

37. N •————————————————• O

38. P •————————————————————————• Q

39. R •————————————————• S

40. T •————————————————————————• U

Find the length of the missing side in each right triangle. Round the answer to the nearest hundredth.

41.

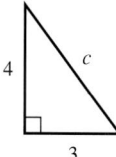

42.

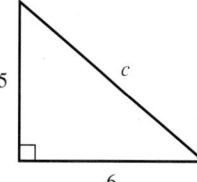

43.

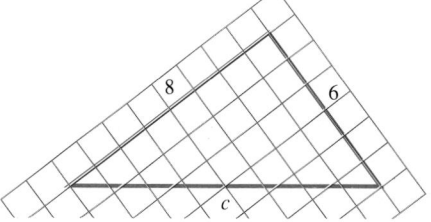

44.

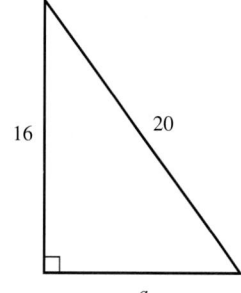

45.

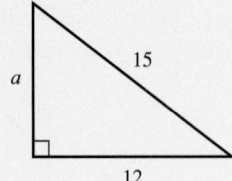

46.

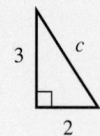

47.

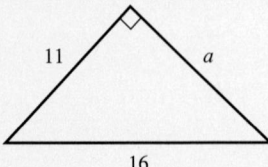

48.

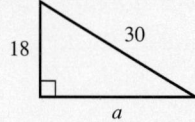

Use the Pythagorean Theorem to solve these problems.

49. How long must a wire be to reach from the top of a 24-foot flagpole to the ground 10 feet from the base of the pole?

50. A ladder is leaning against the top of a 13-meter wall. If the bottom of the ladder is 9 meters from the wall, how long is the ladder?

51. How long a ladder is needed to reach from the top of a building 15 feet high to the ground 8 feet from the building?

52. A baseball diamond is actually a square 90 feet on a side. How far is it from home plate to second base?

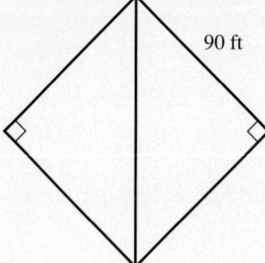

Find the perimeters of these polygons.

53.

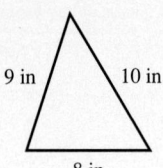

54.

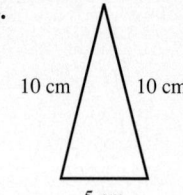

55.

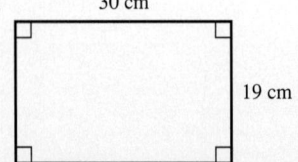

56.

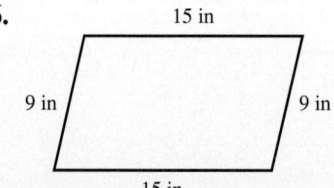

57.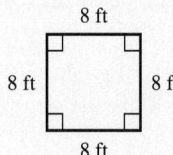

58.

59.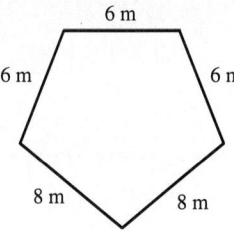

60.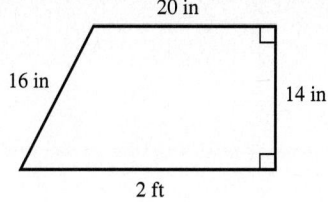

Find the perimeters of these squares.

61. 21 feet on a side

62. 2.3 kilometer on a side

63. 27 millimeter on a side

64. $3\frac{1}{2}$ inches on a side

Find the perimeters of these rectangles.

65. 8 feet by 7 feet

66. 90 millimeter by 3.5 millimeter

67. 4 inches by 3 feet

68. 3 kilometer by 4 kilometer

Find the circumferences of these circles.

69. Diameter of 3 inches

70. Radius of 8 feet

71. Diameter of $2\frac{1}{2}$ feet

72. Radius of 4.6 millimeters

73. Diameter of 7 centimeters

74. Radius of 5 meters

75. Radius of $6\frac{1}{3}$ yards

76. Diameter of 8.5 meters

77.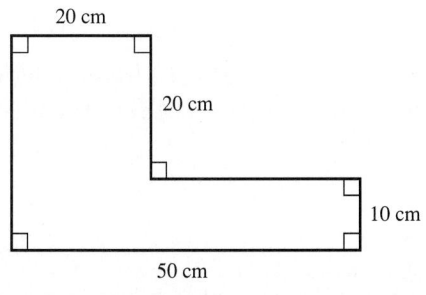

78.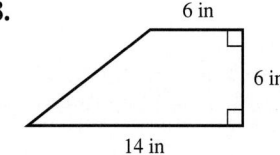

79.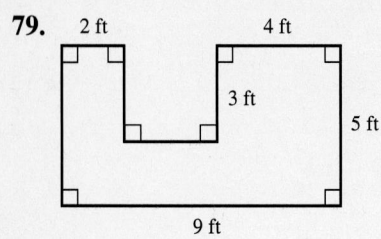

80.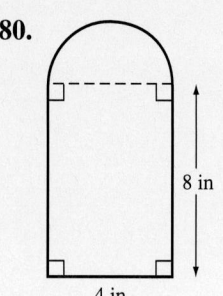

81.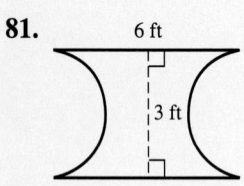

82.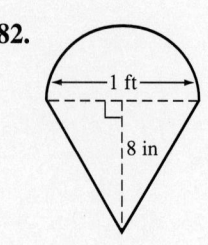

Applications:

83. A triangle has a perimeter of 23 centimeters. One side has length of 5 cm and the second side has a length of 10 cm. Find the length of the third side.

84. A square has a perimeter of 14 inches. What is the length of each side?

85. A rectangle has a perimeter of 92 feet. If the length of the rectangle is 25 feet, what is the width?

86. The circumference of a circle is 43.96 millimeters. Find the length of the diameter.

87. The outside measure of a frame is 21 inches long by 17 inches wide. How many feet of picture framing will be needed?

88. How many feet of fencing will be needed to build a square dog run 12 feet on each side (there will be a 3-foot gate in the run)?

89. How many feet of decorative fencing do you need to edge a circular flower garden 13 meters in diameter?

90. If Mary has $25\frac{1}{2}$ feet of ribbon, how large a circle can she make on her tablecloth?

91. How many yards of Christmas decoration will be needed to edge a tablecloth 9 feet long by 4 feet wide?

92. How many yards of Valentine decoration will be needed for a circular tablecloth 4 feet in diameter?

93. A swimming pool needs a new top row of tiles. The pool is 25 yards long and 15 yards wide. What is the perimeter of the pool in yards? What is the perimeter in feet? What is the perimeter in inches? If the new tiles are 6 inches square, how many will be needed for the new top row of tiles?

94. Jane wants to decorate her dorm room with a wallpaper border. Her dorm room is 10 feet long by 9 feet wide. (The border will go all the way around the room.) How many feet of border will she need? The borders are sold in rolls of 15 feet; how many rolls will she need?

12.2 Square Measure and Applications

OBJECTIVES

● Perform conversions of square measures within both the U.S. and the metric systems.

● Calculate the areas of selected plane figures.

● Calculate the areas of certain irregular plane figures.

● Solve application problems involving area.

Area and Square Measure

The concept of area is not a mysterious one. It deals with the question of how much room a plane figure takes up. We use **square measure** to calculate area. We literally announce how many squares we could fit within the boundaries of the plane figure.

DEFINITION **Area:** The amount of surface enclosed by a plane figure.

When we say that the area of a 9×12-foot rug is 108 square feet, we mean that we can fit 108 squares each 1×1 foot on the rug. If we wanted to tile a 3×5-foot counter with tiles that were each 1×1 foot, we would need to use 15 of them.

Count them yourself in the accompanying diagram. We'd have 5 rows of 3 tiles each, or 3 rows of 5 tiles each. Either way, that's 15 tiles, each a 1×1-foot square. Speaking mathematically, we say the area of the 3×5-foot counter is 15 square feet.

5 ft

3 ft

As we mentioned, square measure is used to calculate area. We will use the U.S. customary system as well as the metric system of square measure. We can convert from one unit to another within each system by using the following charts.

U.S. Square Measures
$144 \text{ in}^2 = 1 \text{ ft}^2$
$9 \text{ ft}^2 = 1 \text{ yd}^2$
$43{,}560 \text{ ft}^2 = 1 \text{ acre}$
$640 \text{ acres} = 1 \text{ mi}^2$

Abbreviations
in^2 = square inches
ft^2 = square feet
yd^2 = square yards
mi^2 = square miles

Note Square measures may also be abbreviated by writing sq in front of in, ft, yd, etc. There is no abbreviation for the term acre.

EXAMPLE 1 How many square inches are there in 2 square feet?
Method 1

$$144 \text{ in}^2 = 1 \text{ ft}^2$$
$$2 \cdot 144 \text{ in}^2 = 2 \cdot 1 \text{ ft}^2$$
$$288 \text{ in}^2 = 2 \text{ ft}^2$$

Method 2
Follow the same procedure we used in Section 12.1.

Let x = the number of in^2 in 2 ft^2

$$\frac{x \text{ in}^2}{144 \text{ in}^2} = \frac{2 \text{ ft}^2}{1 \text{ ft}^2}$$

$$\frac{x}{144} = \frac{2}{1}$$

$$1 \cdot x = 2 \cdot 144$$

$$x = 288$$

There are 288 in^2 in 2 ft^2.

EXAMPLE 2 Convert 2 acres to square yards.
Since the table does not state the relationship between yd^2 and acres, we must use two proportions to solve this problem.

Part I: Convert 2 acres to square feet.

Let f = number of ft^2 in 2 acres.

$$\frac{f \text{ ft}^2}{43{,}560 \text{ ft}^2} = \frac{2 \text{ acres}}{1 \text{ acre}}$$

$$\frac{f}{43{,}560} = \frac{2}{1}$$

$$1 \cdot f = 2 \cdot 43{,}560$$

$$f = 87{,}120$$

There are 87,120 ft^2 in 2 acres.

Part II: Convert 87,120 square feet to square yards.

Let y = number of yd^2 in 87,120 ft^2.

$$\frac{y \text{ yd}^2}{1 \text{ yd}^2} = \frac{87{,}120 \text{ ft}^2}{9 \text{ ft}^2}$$

$$\frac{y}{1} = \frac{87{,}120}{9}$$

$$y = 9680$$

There are 9680 square yards in 2 acres.

Note The term *acre* is a square measure term. Do *not* say square acre.

Practice Problems

Perform these conversions.

1. 15 yd^2 = _____ ft^2

2. 3 ft^2 = _____ in^2

3. 2.3 mi^2 = _____ acres

Metric Square Measures

$$100 \text{ mm}^2 = 1 \text{ cm}^2$$
$$100 \text{ cm}^2 = 1 \text{ dm}^2$$
$$100 \text{ dm}^2 = 1 \text{ m}^2$$
$$1{,}000{,}000 \text{ m}^2 = 1 \text{ km}^2$$

Abbreviations
mm^2 = square millimeters
cm^2 = square centimeters
dm^2 = square decimeters
m^2 = square meters
km^2 = square kilometers

EXAMPLE 3 Convert 2.3 square meters to square decimeters.

Method 1
Since $1\text{m}^2 = 100 \text{ dm}^2$, convert m^2 to dm^2 by moving the decimal point two places to the right.

$$1 \text{ m}^2 = 100 \text{ dm}^2$$
$$2.3 \cdot 1 \text{ m}^2 = 2.3 \cdot 100 \text{ dm}^2$$
$$2.3 \text{ m}^2 = 230 \text{ dm}^2$$

Method 2
Let d = number of dm^2 in 2.3 m^2.

$$\frac{d \text{ dm}^2}{100 \text{ dm}^2} = \frac{2.3 \text{ m}^2}{1 \text{ m}^2}$$
$$\frac{d}{100} = \frac{2.3}{1}$$
$$1 \cdot d = 2.3 \cdot 100$$
$$d = 230$$

There are 230 dm^2 in 2.3 m^2.

Area of a Rectangle

As we saw in the discussion about covering a counter top with 1 ft $\times$ 1 ft square tiles in the beginning of this section, if the length of the counter top is 3 feet and the width is 5 feet, the area is calculated by adding the 3 rows each having 5 tiles.

1st row
2nd row
3rd row

Since multiplication is a shortcut for addition, we can say $3 \times 5 = 15$. The area of a rectangle with length 3 feet and width 5 feet is 3 ft $\times$ 5 ft or 15 sq ft.

RULE **Area of a rectangle**
The area of a rectangle is length times width.

$$A = l \cdot w \quad \text{where } l = \text{length and } w = \text{width}$$

Practice Problems

Perform these conversions.

4. 6 dm^2 = _____ cm^2

5. 15.4 cm^2 = _____ mm^2

6. 2.7 m^2 = _____ dm^2

EXAMPLE 4 Find the area of a rectangle whose width is 2 cm and whose length is 7 cm.

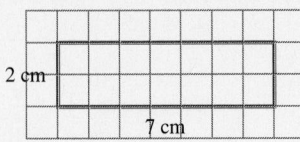

Draw and label a diagram.

2 cm

7 cm

$A = l \cdot w$ State the formula or rule.

$A = 2 \text{ cm} \cdot 7 \text{ cm}$ Substitute the values for l and w.

$A = 14 \text{ sq cm}$ Evaluate the formula.

The area is 14 cm².

Note   We often write sq cm as cm². Either way is acceptable.

EXAMPLE 5   Find the area of a rectangle whose dimensions are 14 in $\times$ $2\frac{1}{2}$ ft.

Part I: Convert $2\frac{1}{2}$ feet to inches

Let x = number of inches in $2\frac{1}{2}$ feet.

$$\frac{x \text{ in}}{12 \text{ in}} = \frac{2\frac{1}{2} \text{ ft}}{1 \text{ ft}}$$

$$\frac{x}{12} = \frac{\frac{5}{2}}{1}$$

$$1 \cdot x = 12 \cdot \frac{5}{2}$$

$$x = 30$$

$$2\frac{1}{2} \text{ ft} = 30 \text{ in}$$

$2\frac{1}{2}$ ft

14 in

The rectangle is 14 in $\times$ 30 in.

Part II: Find the area.

$$A = l \cdot w$$
$$A = 14 \cdot 30$$
$$A = 420$$

The area is 420 in².

APPLICATION   Pedro decides to fertilize his front lawn, which measures $\frac{1}{4}$ acre. If the product information on a bag of fertilizer states that one bag will cover 5000 square feet, how many bags will he need?

Practice Problems

Find the area of each rectangle.

7. l = 3 mm, w = 1.5 mm

8. l = 1 yd, w = 6 yd

9. l = 2 m, w = 51 cm

Find the missing dimension.

10. A = 15 cm², l = 6 cm

11. A = 110 yd², w = 12 yd

Part I: Convert $\frac{1}{4}$ acres to square feet.

Let x = number of square feet in $\frac{1}{4}$ acre.

$$\frac{x}{43{,}560} = \frac{\frac{1}{4}}{1}$$

$$1 \cdot x = \frac{1}{4} \cdot 43{,}560$$

$$x = 10{,}890$$

There are 10,890 square feet in $\frac{1}{4}$ acre.

Part II: Calculate how many bags are needed.

Let b = number of bags needed to cover 10,890 ft^2

$$b = \frac{10{,}890}{5000}$$

$$b = 2.178$$

Answer: Pedro needs to buy 3 bags of fertilizer. ◆

Area of a Square

Since a square is a rectangle with all sides equal, we will use the formula for the area of a rectangle.

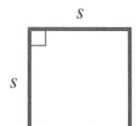

$A = l \cdot w$ Let s = length of one side of a square.

$A = s \cdot s$ Substitute s for l and w.

$A = s^2$

RULE **Area of a square**
The area of a square is side squared.

$$A = s^2 \quad \text{where } s \text{ is the length of one side}$$

EXAMPLE 6   Find the area of the square whose side is 21 feet.

$A = s^2$

$A = (21)^2$ 21 ft

$A = 441$

         21 ft

The area is 441 ft^2.

USING THE CALCULATOR

For the example just discussed,

Enter 21 x^2 **441** ← Display

Area of a Parallelogram

Before discussing the area of a parallelogram, we need to define a few terms.

DEFINITION   **Parallel lines:** Lines that are always the same distance apart.

DEFINITION   **Parallelogram:** A four-sided polygon with both pairs of opposite sides parallel. ABCD is a parallelogram. It can also be referred to as ▱ ABCD.

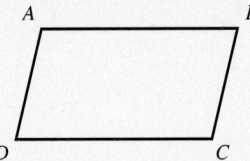

DEFINITION   **Altitude:** A line segment drawn from one side of a polygon which forms a 90° angle (also called a right angle) with the opposite side. $\overline{AE}$ is the altitude of ▱ ABCD. We refer to the length of the altitude as h (for height). We refer to the length of the opposite side as b (for base).

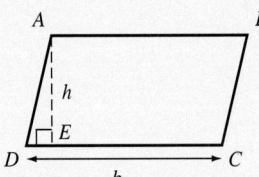

RULE **Area of a parallelogram**

The area of a parallelogram is base times altitude.

$$A = b \cdot h \quad \text{where } b \text{ is the base and } h \text{ is the altitude}$$

To gain a better understanding of why this formula works, consider the parallelograms below.

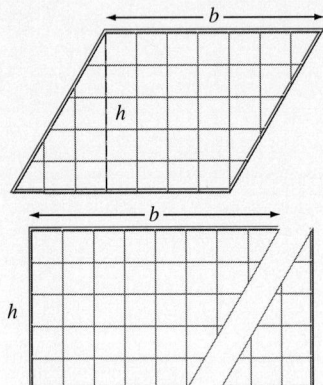

Separate the triangle from the figure.

Move it to the other side of the figure.

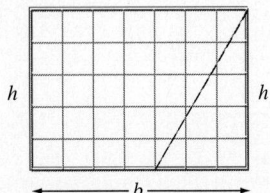

Reattach it.

Practice Problems

Find the area of the square.

12. $s = 16$ in

13. $s = 35.6$ dm

Find the length of the side of the square.

14. $A = 256$ ft^2

15. $A = 121$ mm^2

This new figure is a rectangle with length $= b$ and width $= h$.

$$A = l \cdot w \qquad \text{Apply the formula for area.}$$
$$A = b \cdot h \qquad \text{Substitute the values.}$$

The area of the new rectangle will be the same as the area of the original parallelogram. For this example, we have shown that the formula $A = b \cdot h$ works.

EXAMPLE 7 Find the area of a parallelogram with $b = 13$ yards and $h = 5$ yards.

$$A = b \cdot h$$
$$A = 13 \cdot 5$$
$$A = 65$$

The area is 65 yd².

Area of a Triangle

> **RULE** **Area of a triangle**
> The area of a triangle is one-half the base times the altitude.
>
> $$A = \frac{1}{2} b \cdot h \quad \text{where } b = \text{base and } h = \text{altitude}$$

To gain a better understanding of why this formula works, consider a parallelogram with base b and altitude h.

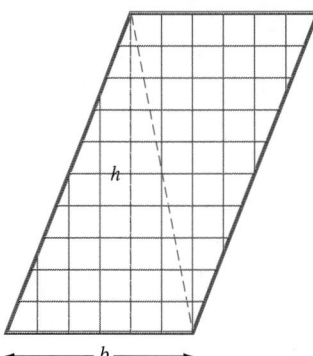

Draw a diagonal.

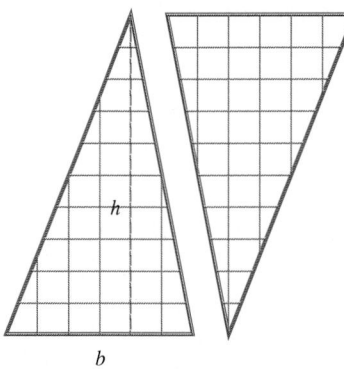

This diagonal will divide the parallelogram into two equal parts. Each part will form a triangle. The sum of the areas of the triangles will equal the area of the original parallelogram. Since the triangles are equal in area, each will have an area of one-half the area of the original parallelogram—written as a formula:

$$A = \frac{1}{2} b \cdot h.$$

Practice Problems

Find the area of the parallelogram.

16. $b = 5$ cm, $h = 39.2$ cm

17. $b = 0.7$ in, $h = 12.3$ in

EXAMPLE 8   Find the area of a triangle with a 12 centimeter base and a 5 centimeter altitude.

$$A = \frac{1}{2} b \cdot h$$

$$A = \frac{1}{2} \cdot 12 \cdot 5$$

$$A = 30$$

The area is 30 cm^2.

Area of a circle

> **RULE**   **Area of a circle**
> The area of a circle is π times the square of the radius.
> $$A = \pi \cdot r^2 \quad \text{where } r = \text{radius}$$

Note   Use $\pi \approx 3.14$ or use the $\boxed{\pi}$ key on the calculator.

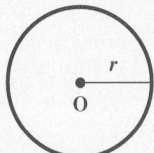

EXAMPLE 9.   Find the area of a circle with a radius of 13 inches.

$$A = \pi \cdot r^2$$
$$A \approx (3.14)(13)^2$$
$$A \approx (3.14)(169)$$
$$A \approx 530.66$$

The area is approximately 530.66 in^2.

 USING THE CALCULATOR

For the example above,

     Enter $\boxed{\pi}$ $\boxed{\times}$ $\boxed{13}$ $\boxed{x^2}$ $\boxed{=}$ **530.92916** $\leftarrow$ Display

Rounded to hundredths, the area is approximately 530.93 in^2.

Note   Using 3.14 for π and using the $\boxed{\pi}$ key on a calculator resulted in different answers. Only if we rounded both answers to units would we have gotten identical answers (531 in^2). This is to be expected since π is an irrational number and can only be approximated as a decimal.

Find the area of the triangle.

18. $b = 5$ in, $h = 2$ in

19. $v = 6.2$ dm, $h = 7$ dm

Find the base of the triangle.

20. $A = 224$ ft^2, $h = 14$ ft

EXAMPLE 10 The area of a circle is approximately 1256 mm². Find the radius.

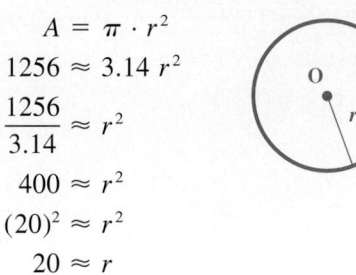

$$A = \pi \cdot r^2$$
$$1256 \approx 3.14 \, r^2$$
$$\frac{1256}{3.14} \approx r^2$$
$$400 \approx r^2$$
$$(20)^2 \approx r^2$$
$$20 \approx r$$

The radius is approximately 20 millimeters.

USING THE CALCULATOR

For the above example,

Enter 1256 ÷ $\boxed{\pi}$ $\boxed{=}$ $\boxed{\sqrt{x}}$ **19.99493** ← Display

Rounded to units, the radius is approximately 20 millimeters.

Areas of Other Plane Figures

To find the area of a plane figure formed by combining two or more polygons or a combination of polygons and circles or semicircles, separate the figure into regions. Find the area of each region, and then find the sum of the areas of all the regions.

EXAMPLE 11 Find the area of the accompanying figure.

1. The figure is formed by combining a semicircle with a diameter of 2 centimeters with a square of side 2 centimeters.
2. Separate the regions.
3. Find the area of the semicircle.

$$d = 2 \text{ cm, so } r = 1 \text{ cm}$$

Area of a circle $= \pi r^2$, so

$$\text{Area of a semicircle} = \frac{1}{2}\pi r^2$$
$$= \frac{1}{2} \cdot \pi \cdot 1^2$$
$$\approx 1.57 \text{ cm}^2$$

4. Find the area of the square.

$$s = 2 \text{ cm}$$
$$\text{Area of square} = s^2$$
$$= 2^2$$
$$= 4 \text{ cm}^2$$

5. Add the areas of the two regions.

$$\text{Area} \approx 1.57 + 4$$
$$\text{Area} \approx 5.57 \text{ cm}^2$$

Area of the figure is approximately 5.57 cm².

Practice Problems

Find the area of the circle.

21. $r = 35$ cm

22. $r = \frac{1}{3}$ yd

23. $r = 2.6$ m

ANSWERS TO PRACTICE PROBLEMS
1. 135 2. 432 3. 1472
4. 600 5. 1540 6. 270
7. 4.5 mm² 8. 6 yd²
9. 10,200 cm² or 1.02 m²
10. $w = 2.5$ cm 11. $L = 9\frac{1}{6}$ yd
12. 256 in² 13. 1267.36 dm²
14. 16 ft 15. 11 mm
16. 196 cm² 17. 8.61 in²
18. 5 in² 19. 21.7 dm²
20. 32 ft 21. 3848.451 cm²
22. ≈ 0.35 yd² 23. ≈ 21.24 m²

PROBLEM SET 12.2

Make the following conversions within the U.S. customary system.

1. 288 in² to ft²

2. 2 yd² to ft²

3. 1 mi² to acres

4. 22 yd² to ft²

5. 40 ft² to in²

6. 36 in² to ft²

7. 1 in² to ft²

8. 720 in² to yd²

9. 3 yd² to in²

10. 2 acres to ft²

Make the following conversions within the metric measure.

11. 23 cm² to mm²

12. 200 mm² to cm²

13. 10 m² to mm²

14. 1.5 m² to cm²

15. 1000 mm² to m²

16. 5 km² to m²

17. 9 m² to cm²

18. 898 mm² to m²

19. 8 m² to cm²

Find the area of the following figures. Let $\pi \approx 3.14$.

20.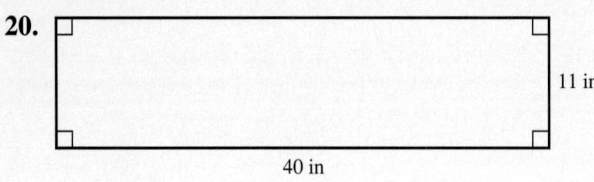
11 in
40 in

21.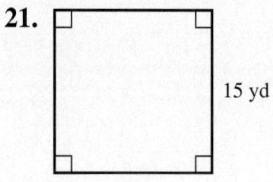
15 yd
15 yd

22.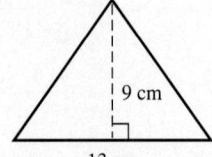
9 cm
13 cm

23.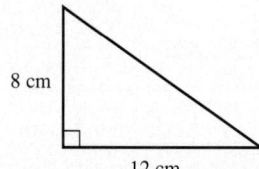
8 cm
12 cm

24.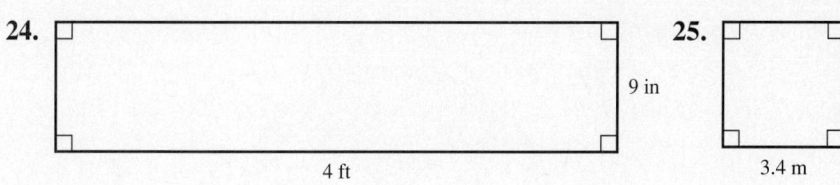

9 in

4 ft

25.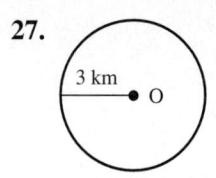

3.4 m

3.4 m

26.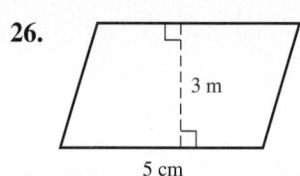

3 m

5 cm

27.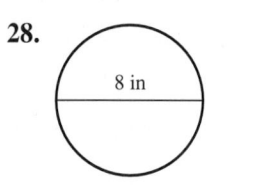

3 km

O

28.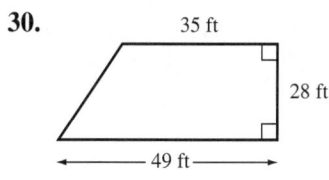

8 in

29.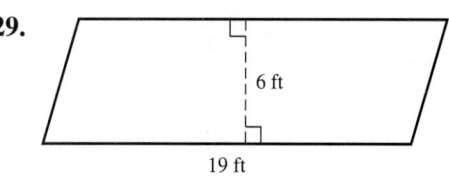

6 ft

19 ft

30.

35 ft

28 ft

49 ft

31.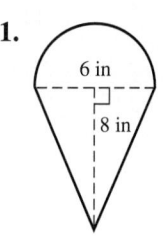

6 in

8 in

Applications:

32. How many square yards of carpet are needed to carpet a room 9 feet by 15 feet?

33. How many square centimeters of glass would be needed to make an 80-centimeter diameter round table top?

34. How many square feet of floor tile will be needed to re-cover a kitchen floor 17 feet by 9 feet?

35. A picture measures 12 centimeters by 16 centimeters. How many square centimeters of glass are needed to cover the picture?

36. How many square feet of fabric is needed to make a roller blind to cover a window 30 inches wide by 5 feet long?

37. How many square inches of paper in a notebook containing 80 pages, each $8\frac{1}{2}$ inches by 11 inches?

Find the area of the shaded region in each of the following figures.

38.

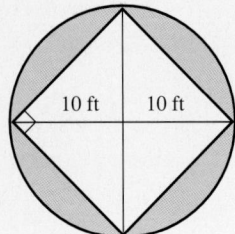

39.

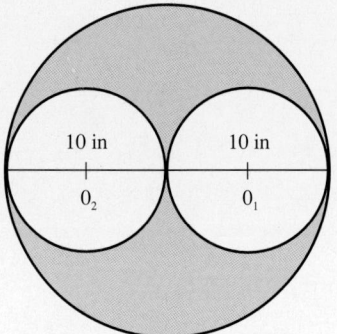

40.

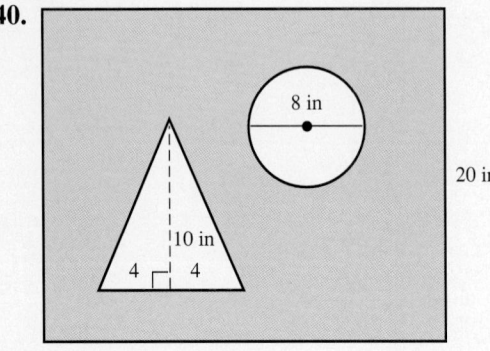

12.3 **Cubic Measure and Applications**

OBJECTIVES

- Perform conversions of cubic measures within both the U.S. and metric systems.

- Calculate the volumes of selected three-dimensional solids.

- Solve application problems involving volume.

We use the term volume to denote the amount of space a three-dimensional solid object occupies and to denote how much space a three-dimensional hollow object can contain. We use **cubic measure** to calculate volume. We measure by calculating how many cubes, each measuring 1 unit × 1 unit × 1 unit, could fit within the boundaries of a three-dimensional figure.

When we say that the volume of a box 2 ft × 4 ft × 3 ft is 24 cubic feet, we mean that 24 cubes, each measuring 1 ft × 1 ft × 1 ft, could fit within the boundaries of the box.

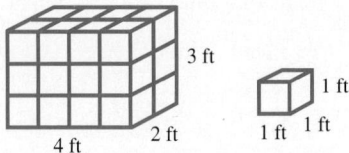

DEFINITION **Volume:** The number of cubic units enclosed by a three-dimensional figure.

U.S. Customary and Metric Systems of Cubic Measure

As we noted, cubic measure is used to calculate volume. We will use both the U.S. system and the metric system of cubic measure. By using the following charts, we can convert from one unit to another within each system.

U.S. Cubic Measures

$$1728 \text{ in}^3 = 1 \text{ ft}^3$$
$$27 \text{ ft}^3 = 1 \text{ yd}^3$$

   Abbreviations
   in^3 = cubic inches
   ft^3 = cubic feet
   yd^3 = cubic yards

EXAMPLE 1   Convert 3 cubic yards to cubic feet.

Method 1

$$1 \text{ yd}^3 = 27 \text{ ft}^3$$
$$3 \cdot 1 \text{ yd}^3 = 3 \cdot 27 \text{ ft}^3$$
$$3 \text{ yd}^3 = 81 \text{ ft}^3$$

Method 2
Use the procedure outlined in Section 12.1.

Let x = number of cubic feet in 3 cubic yards.

$$\frac{x \text{ ft}^3}{27 \text{ ft}^3} = \frac{3 \text{ yd}^3}{1 \text{ yd}^3}$$

$$\frac{x}{27} = \frac{3}{1}$$

$$1 \cdot x = 3 \cdot 27$$

$$x = 81$$

There are 81 ft^3 in 3 yd^3.

Metric Cubic Measures

$$1000 \text{ mm}^3 = 1 \text{ cm}^3$$
$$1000 \text{ cm}^3 = 1 \text{ dm}^3$$
$$1000 \text{ dm}^3 = 1 \text{ m}^3$$

   Abbreviations
   mm^3 = cubic millimeters
   cm^3 = cubic centimeters
   dm^3 = cubic decimeters
   m^3 = cubic meters

EXAMPLE 2   Convert 5.2 m³ to cubic decimeters.

$$1 \text{ m}^3 = 1000 \text{ dm}^3$$
$$(5.2) \cdot 1 \text{ m}^3 = (5.2) \cdot 1000 \text{ dm}^3$$
$$5.2 \text{ m}^3 = 5200 \text{ dm}^3$$

Volume of a Right Rectangular Solid

A right rectangular solid is what we commonly call a box.

DEFINITION   **Right rectangular solid:** A three-dimensional closed figure that has rectangles for its **bases** and for each of its **faces,** i.e., sides.

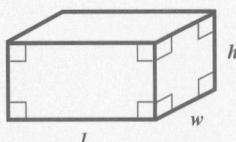

> **RULE** **Volume of a right rectangular solid**
> The volume of a right rectangular solid is length × width × height.
>
> $$V = l \cdot w \cdot h \quad \text{where } l = \text{length}, \ w = \text{width}, \ h = \text{height}$$

EXAMPLE 3   Find the volume of a 12 in × 9 in × 3 in box.

$$V = l \cdot w \cdot h$$
$$V = 12 \cdot 9 \cdot 3$$
$$V = 324$$

The volume is 324 in³.

Volume of a Cube

We mentioned the central role a cube plays in cubic measure at the beginning of this section. Now we will look at it more formally.

DEFINITION   **Cube:** A right rectangular solid with square bases (top and bottom) and square faces.

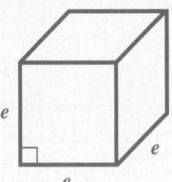

DEFINITION   **Edge:** Each of the line segments that form the boundaries of a three-dimensional figure.

> **RULE** **Volume of a cube**
> The volume of a cube is edge cubed.
>
> $$V = e^3 \quad \text{where } e = \text{length of an edge}$$

EXAMPLE 4   Find the volume of a cube with an edge of 6 centimeters.

$$V = e^3$$
$$V = 6^3$$
$$V = 216$$

The volume is 216 cm³.

USING THE CALCULATOR

For the preceding example,

Enter 6 y^x 3 = **216** ← Display

Volume of a Right Circular Cylinder

The easiest way to describe a right circular cylinder is to picture a can—as in, a can of tomatoes or a can of fruit juice. This three-dimensional figure has two congruent circles for its bases and is of uniform height.

> **RULE** Volume of a right circular cylinder
> The volume of a right circular cylinder is the area of the base times the height.
>
> $$V = A \cdot h \quad \text{where } A = \text{area of the base and } h = \text{height}$$
> $$\text{or} \quad V = \pi r^2 h \quad \text{where } r = \text{radius of the base and } h = \text{height}$$

EXAMPLE 5   Find the volume of a right cylinder that is 6 inches high and has a 5 inch radius.

$$V = \pi r^2 h$$
$$V = \pi \cdot 5^2 \cdot 6 \text{ or } V \approx 3.14 \cdot 5^2 \cdot 6$$
$$V \approx 150 \, \pi \text{ in}^3 \text{ or } V \approx 471 \text{ in}^3$$

Note   It is acceptable to leave π in the answer, unless the problem is an application type. In that case, approximate π.

APPLICATION   How much potting soil will a cylindrical pail hold? Its radius is 5 inches and the height is 9 inches. (Solve by using a calculator.)

USING THE CALCULATOR

For this application,

$$V = \pi \cdot r^2 \cdot h \quad \text{where } r = 5 \text{ and } h = 9$$

Enter π × 5 x^2 × 9 = **706.85835** ← Display

Answer: Volume of the pail is approximately 706.86 in³. ◆

Practice Problems

Find the volume of the box.

5. *l* = 5 cm, *w* = 12 cm,
   h = 3.1 cm

6. *w* = 3.9 ft, *h* = 2.7 ft,
   l = 11.4 ft

Find the volume of the cube.

7. *s* = 5 yd

Find the volume of the cylinder.

8. *r* = 1 cm, *h* = 3 cm

9. *r* = 2 ft, *h* = 11 ft

10. *r* = 30.6 mm, *h* = 43.8 mm

Volume of a Sphere

We think of a **sphere** as a ball-shaped object. Mathematically, we can think of it as a set of points in space a fixed distance from a given point called the center. The length of the radius determines the volume of the sphere.

> **RULE** Volume of a sphere
> The volume of a sphere is four-thirds pi times the radius cubed.
>
> $$V = \frac{4}{3}\pi r^3 \quad \text{where } r = \text{radius of the sphere}$$

EXAMPLE 6 Find the volume of a sphere whose radius is 6 centimeters.

$$V = \frac{4}{3} \cdot \pi \cdot 6^3$$

$$V = \frac{4}{\cancel{3}} \cdot \frac{\pi}{1} \cdot \frac{\cancel{6}^2 \cdot 36}{1}$$

$$V = 288\pi \text{ cm}^3$$

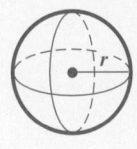

🖩 **USING THE CALCULATOR**

For this example,

Enter 4 $a^{b/c}$ 3 × π × 6 y^x 3 = **904.77868** ← Display

$V \approx 904.78$ cm^3

Practice Problems

Find the volume of the sphere.

11. $r = 12$ in

12. $r = 1$ m

13. $r = \frac{3}{4}$ ft

14. $r = 5.28$ cm

ANSWERS TO PRACTICE PROBLEMS

1. $2\frac{2}{3}$ 2. 35,000,000
3. 116,640 4. 2 5. 186 cm^3
6. 120.042 ft^3 7. 125 yd^3
8. ≈ 9.42 cm^3 9. ≈ 138.23 ft^3
10. 128,844.78 mm^3
11. ≈ 7238.23 in^3
12. ≈ 4.19 m^3 13. ≈ 1.77 ft^3
14. 616.58 cm^3

PROBLEM SET 12.3

Make the following conversions within the U.S. customary system.

1. 162 ft^3 to cubic yards

2. 8.5 yds^3 to cubic feet

3. 15,552 in^3 to cubic feet

4. 3 ft^3 to cubic inches

5. 93,312 in^3 to cubic yards

Make the following conversions within the metric system.

6. 2.2 cm^3 to cubic millimeters

7. 93 m^3 to cubic centimeters

8. 670 cm^3 to cubic meters

9. 1300 mm^3 to cubic meters

10. 5200 cm^3 to cubic meters

Find the volume. Let π ≈ 3.14.

11.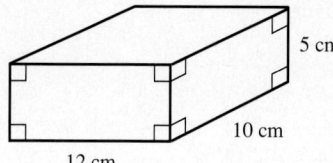

5 cm

10 cm

12 cm

12.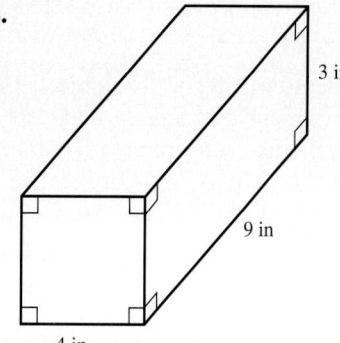

3 in

9 in

4 in

13.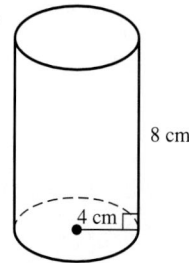

8 cm

4 cm

14.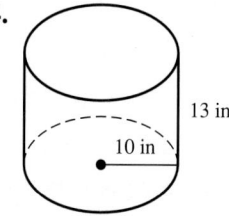

13 in

10 in

15. Find the volume of a sphere with a radius of 4 feet.

16. Find the volume of a sphere with a radius of 5 centimeters

17. Find the volume of a cube which measures 6 centimeters on each edge.

18. Find the volume of a cube which measures 7 inches on each edge.

19. Find the volume in cubic inches of a cylinder that has a radius of 8 inches and a height of 3 feet.

20. Find the volume in cubic centimeters of a rectangular solid with a length of 9 centimeters, a width of 7 centimeters, and a height of 2 meters.

21. A water tank is a cylinder. If a tank has a radius of 20 inches and a height of 4 feet, what is the volume of the tank in cubic inches?

22. Mulch is spread 8 inches thick on a garden 10 yards wide by 12 yards long. How many cubic yards of mulch will be needed?

23. How many cubic yards of cement will be needed to pour a patio 12 feet wide by 25 feet long and 4 inches deep?

24. How many cubic yards of gravel will be needed to make a path 15 yards long, 4 feet wide, and 6 inches deep?

25. How many cubic inches of air are there in a ball with a radius of 5 inches?

SMALL GROUP ACTIVITY *How large is a ____? It's about the size of a ____.*

Materials: Several measuring devices—yard sticks, meter sticks, rulers, tape measures.

Activities:

1. Each group of students will try to find objects or surfaces that have a volume or area approximately equal to a cubic or square measure.

2. At the end of the session, each group will submit a one-page report that answers the questions in the title of this activity and outlines the procedure they followed and the calculations they made to arrive at their conclusions.

Sample: "How large is an *acre*?" It's almost as big as a *football field*.

Rationale: A football field is approximately 100×50 yards (not counting the end zones).

$$A = l \cdot w$$
$$A = 100 \cdot 50$$
$$A = 5000$$

A football field is 5000 yd².

$$\text{Convert 5000 yd}^2 \text{ to ft}^2$$
$$1 \text{ yd}^2 = 9 \text{ ft}^2$$
$$5000 \cdot 1 \text{ yd}^2 = 5000 \cdot 9 \text{ ft}^2$$
$$5000 \text{ yd}^2 = 45{,}000 \text{ ft}^2$$

but

$$1 \text{ acre} = 43{,}560 \text{ ft}^2$$

Conclusion: An acre is almost as big as a football field.

CHAPTER 12 REVIEW

After completing Chapter 12, you should be familiar with the following concepts and be able to apply them.

- *Perform conversions of linear measure within both U.S. and metric systems (12.1).*
  Fill in the following conversions.

1. 3 mi = _____ ft

2. 13 yd = _____ in

3. 84 in = _____ ft

4. 10 yd = _____ ft

5. 48 ft = _____ yd

6. 54 in = _____ yd

- *Apply the Pythagorean Theorem (12.1).*

Use the Pythagorean Theorem to find the length of the missing side in the following right triangles.

7.

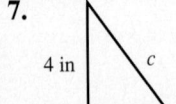

4 in c
3 in

8.

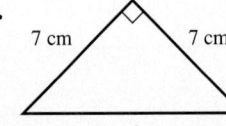

7 cm   7 cm
c

9.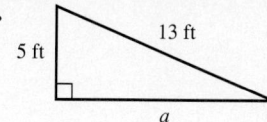

10.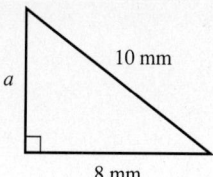

11. $a =$ ____ , $b = 13$ cm, $c = 18$ cm

12. $a = 27$ in, $b =$ ____ , $c = 45$ in

13. $a = 2$ ft, $b = 3$ ft, $c =$ ____

14. $a = 2.5$ mm, $b = 3.6$ mm, $c =$ ____

• *Calculate the perimeters of polygons, including triangle, rectangles, and squares (12.1).*

Calculate the following perimeters

15.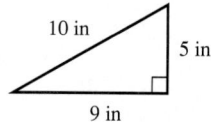

16.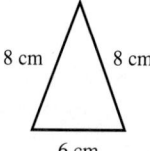

17.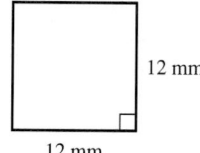

18.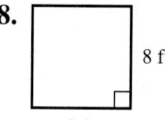

19.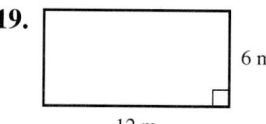

20.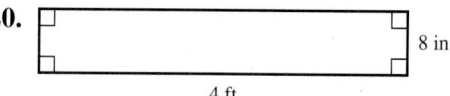

• *Calculate the circumferences of circles (12.1).*

Find the circumferences of circles with the following diameters or radii.

21. Diameter of 8 in

22. Radius of 2.5 cm

23. Diameter of $3\frac{1}{2}$ ft

24. Radius of 19 mm

25. Radius of 22 yd

• *Find the perimeters of certain irregular plane figures (12.1).*

Find the perimeters of the following figures.

26.
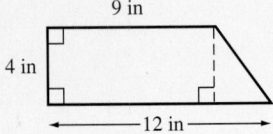
9 in
4 in
12 in

27.
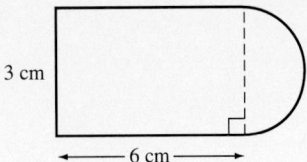
3 cm
6 cm

28.
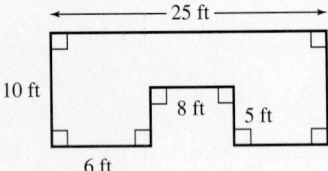
25 ft
10 ft
8 ft
5 ft
6 ft

29.
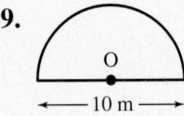
O
10 m

• *Solve application problems involving the concepts of perimeter and circumference (12.1).*

Solve the following application problems.

30. The outside measure of a picture is 18 inches long by 16 inches wide. How many feet of picture framing will be needed to make the frame?

31. How many yards of Halloween decoration will be needed to edge a tablecloth 8 feet long by 3 feet wide?

32. How many feet of fence will be needed to fence a vegetable garden 25 feet square?

33. How many feet of decorative border is needed to decorate a room 30 feet long by 17 feet wide?

• *Perform conversions of square measure within both the U.S. and metric systems (12.2).*

Fill in the following conversions.

34. $216 \text{ in}^2 = $ _____ ft^2

35. $3 \text{ yd}^2 = $ _____ ft^2

36. $3.5 \text{ ft}^2 = $ _____ in^2

37. $2.5 \text{ ft}^2 = $ _____ in^2

38. $\frac{1}{2} \text{ yd}^2 = $ _____ in^2

39. $400 \text{ mm}^2 = $ _____ cm^2

40. $4 \text{ m}^2 = $ _____ mm^2

41. $10 \text{ km}^2 = $ _____ m^2

• *Calculate areas of selected plane figures (12.2).*

Find the area of each of the following figures.

42.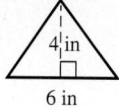

4 in
6 in

43.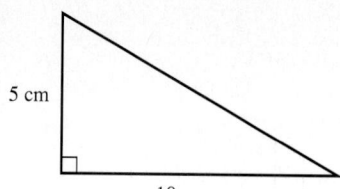

5 cm
10 cm

44.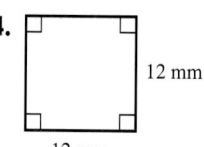

12 mm
12 mm

45.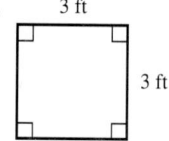

3 ft
3 ft

46.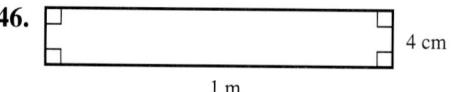

4 cm
1 m

47.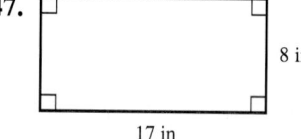

8 in
17 in

48.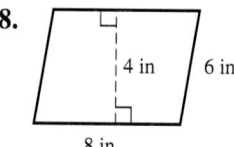

4 in   6 in
8 in

49.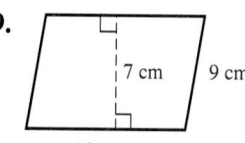

7 cm   9 cm
13 cm

50.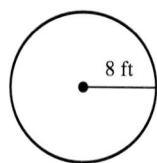

8 ft

51.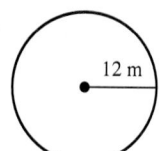

12 m

• *Calculate the areas of certain irregular plane figures (12.2).*

Calculate the area of each of the following figures.

52.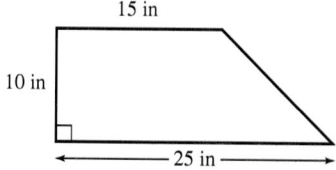

15 in
10 in
25 in

53.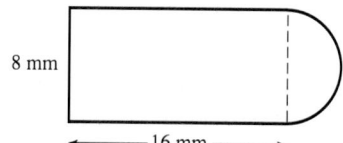

8 mm
16 mm

54.

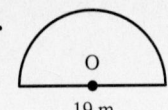

19 m

55.

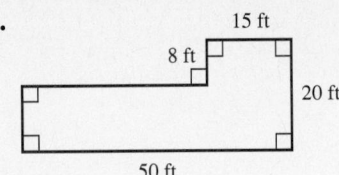

- *Solve application problems involving area (12.2).*

Solve the following application problems.

56. How many square yards of material are needed to carpet a room 25 feet long by 7.5 feet wide?

57. How many square centimeters of glass are needed to cover a picture 9.5 centimeters by 12 centimeters?

58. How many square feet of wood paneling would be needed to cover a family room wall 8.5 feet high by 12.5 feet long?

59. A room 8 meters by $7\frac{1}{2}$ meters contains a carpet 6 meters by $5\frac{1}{2}$ meters. What is the area of the uncarpeted floor?

- *Perform conversions of cubic measure within both the U.S. and metric systems (12.3).*

Make the following conversions.

60. 243 ft^3 = _____ yd^3

61. 10m^3 = _____ cm^3

62. 127 mm^3 = _____ cm^3

63. 4320 in^3 = _____ ft^3

64. 108 ft^3 = _____ yd^3

- *Calculate the volumes of selected three-dimensional solids (12.3).*

Find the volume of the following figures.

65. Find the volume of a cube which measures 16 inches on each edge.

66. Find the volume of a sphere with a radius of 3 centimeters.

67. Find the volume of a cylinder with a height of 27 inches and a radius of 10 inches.

68. Find the volume of a box 8 feet long by 3 feet wide by 7 inches deep.

- *Solve application problems involving volume (12.3).*

Solve the following application problems.

69. How many cubic yards of cement will be needed to pour a path 90 feet long, 3 feet wide, and 4 inches deep?

70. What is the volume of a cylindrical water tank with a radius of 30 inches and a height of 5 feet?

71. A cylindrical can has a radius of 3 centimeters and a height of 10 centimeters. Find the volume.

72. A box measures 10 by 6 by 4 centimeters. Find its volume. How many cubes, each having 2 centimeter edges, will fit in the box?

● CHAPTER 12 PRACTICE TEST

Perform the following conversions.

1. Convert 50 inches to feet

2. Convert 3 miles to yards

3. Convert 382 centimeters to meters

4. Convert 9.5 kilometers to meters

5. The ceiling was 8.5 feet, how many inches is this?

6. The door was 53 inches wide, how wide is the door in yards?

7. A carpet is 390 centimeters square, how many meters square is the carpet?

Find the missing side in the following right triangles.

8.

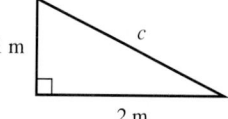

9.

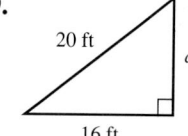

10.

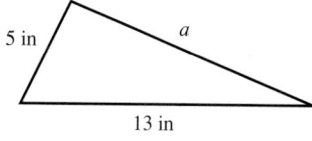

Applications:

11. How long must a wire be to reach from the top of a 12-foot flagpole to a point on the ground 8 feet from the base of the pole?

12. Find the perimeter of a square with a side of length 19 feet.

13. Find the length of a side of a square with a perimeter of 28 mm.

Find the perimeter of each of these figures.

14.

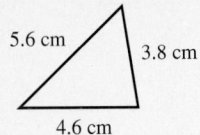

5.6 cm   3.8 cm

4.6 cm

15.

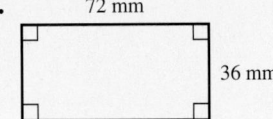

72 mm

36 mm

16.

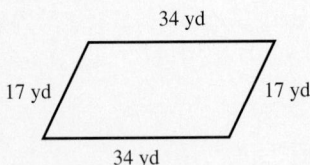

34 yd

17 yd   17 yd

34 yd

17.

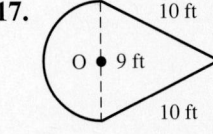

10 ft

O   9 ft

10 ft

Solve the following problems.

18. Find the circumference of a circle with a radius of 2.5 inches.

19. How many feet of decorative fence will be needed to edge a circular flower garden 7 feet in diameter?

20. Convert 720 square inches to square feet.

21. Convert 200 millimeters to centimeters.

22. Find the area of a square with a side 19 inches in length.

23. Find the area of this figure.

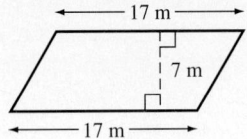

17 m

7 m

17 m

24. Find the area of this figure.

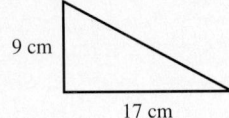

9 cm

17 cm

25. Find the area of a rectangle 3 yards long by 15 inches wide.

26. Find the area of a circle with a diameter of 5 inches.

27. How many square yards of carpet are needed to carpet a room 13 feet long by 9 feet wide?

28. Find the volume of a sphere with a radius of 7 centimeters

29. Find the volume of a cube that measures 3.5 inches on each edge.

30. How many cubic yards of cement will be needed to pour a rectangular path 75 yards long, 3 feet wide, and 5 inches deep?

THE MATHEMATICS JOURNAL

Future Experiences in Mathematics

Task: Consider the role that mathematics will play in your future. Write a one-page (150-word) entry on your thoughts.

Starters: Have you chosen a major? What mathematics courses do you need to take for it? How do you envision your performance in them? What plans can you make to maximize your success in these courses? What careers are you interested in? What role will mathematics play in your career?

CUMULATIVE REVIEW: CHAPTERS 10, 11, and 12

Directions *Test your mastery of the topics covered in Chapters 10, 11, and 12 by taking this practice test. All items are worth 2 points, unless indicated otherwise. The total number of possible points is 100. Remember, this is a review, so check your answers and then refer to the appropriate chapter and section to get assistance in correcting any mistakes.*

Completion *Choose the appropriate word or phrase from the list below.*

U.S. customary system
Distributive Property
Property of Reciprocals
Ratio
Zero Product Property
The product of the means equals the product
   of the extremes.

Metric system
Pythagorean Theorem
Property of Additive Inverses
Proportion
Identity Property for Multiplication

1. A _____ is an indicated quotient.

2. A _____ is an equation whose members are ratios.

3. The _____ states that $a \cdot 0 = 0$, where $a \in$ Reals.

4. The _____ is the system of measurement used by most of the world.

5. The _____ states that, for all $a \in$ Reals, $a \neq 0$, $a \cdot \frac{1}{a} = 1$.

6. In a right triangle, the square of the length of the hypotenuse is equal to the sum of the squares of the

lengths of the legs. This is called the _____ .

7. The statement _____ means that, for all $a, b, c,$
   $d \in$ Reals, $b \neq 0$, $d \neq 0$, if $\frac{a}{b} = \frac{c}{d}$, then $bc = ad$.

8. The _____ states that, for all a, b, $c \in$ reals, $a(b + c) = ab + ac$.

9. The _____ states that, for all $a \in$ reals, $a + (-a) = 0$.

10. The _____ states that, for all $a \in$ reals, $a \cdot 1 = a$.

Conversions. Complete the table.

	Fraction	*Decimal*	*Percent*
11.	$\frac{1}{8}$	_____	_____
12.		$0.\overline{6}$	_____
13.	_____	_____	_____
14.	_____	2.3	_____
15.	_____	_____	$\frac{1}{2}\%$
16.	_____	0.003	_____

Express each expression as a ratio or a rate. Label each rate appropriately.

17. 18 to 45

18. $\frac{1}{3} : \frac{1}{2}$

19. $\frac{0.6}{60}$

20. 20 seconds to 7 minutes

21. 385 miles per 15 gallons of gasoline

22. 2600 TVs to 1000 households

Percents. Solve these problems.

23. What is 60% of 152?

24. 55% of what number is 123?

25. What percent of 82 is 63.96?

26. 15.5% of what number is 45?

Measurement. Complete these statements by performing the appropriate conversions.

27. 3 ft = _____ in

28. 2 yd^2 = _____ ft^2

29. 81 ft^3 _____ yd^3

30. 300 mm^2 = _____ cm^2

31. 2.5 m = _____ cm

32. 2500 m = _____ km

Applications: Solve these problems (3 points each).

33. A record store pays $8.00 for each CD it buys. It then sells the CD for $14. Find the ratio of selling price to cost.

34. A college bookstore makes a profit of $10 on each textbook it sells for $50. Find the ratio of profit to selling price.

35. In a survey, 100 families reported having 300 children. How many children were there per family?

36. The new monthly charge for basic cable TV service is $32, a 2% increase over the previous monthly charge. What was the previous charge?

37. Jennie invites Max out to dinner for his birthday and plans to put the charges on her recently issued credit card. At dinner, she orders the $8.95 pasta special, and he orders the $10.95 seafood platter. Calculate the total that will be charged to her credit card account if the tax rate is 6%, and she plans to include a 15% tip on the total—excluding tax.

Geometry. For each problem, refer to the appropriate figure, state the formula, and answer the question (3 points each).

38. Find the length of side c of the right triangle.

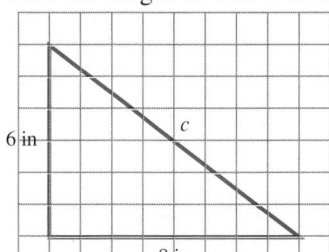

39. Find the length of side a of the right triangle.

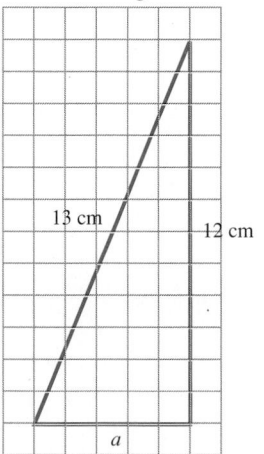

40. The perimeter of a rectangle is 20 feet. The width is 4 feet. Find the length.

41. Find the area of the triangle.

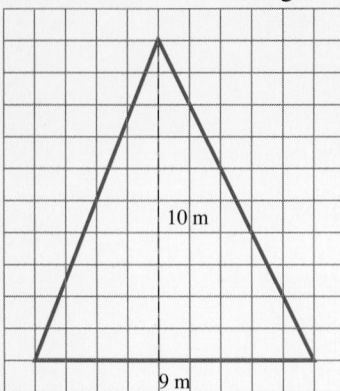

42. Find the volume of the cube.

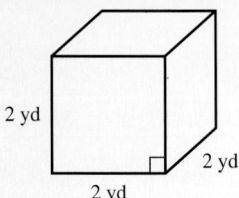

43. Find the circumference of the circle.

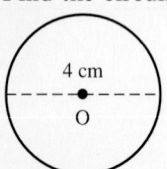

44. Find the area of this irregular figure.

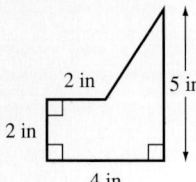

13 *Graphs and Charts*

13.1 The Rectangular Coordinate Plane
13.2 Graphing Linear Equations
13.3 Interpreting Statistical Graphs and Charts
13.4 Constructing Statistical Graphs and Charts

Introduction to Graphing in the Coordinate Plane

As we have seen in Chapters 3, 6, and 12, the early Babylonians developed algebra in the second millennium B.C., and the early Greeks organized geometry in the fifth century B.C. It was not until the seventeenth century A.D., however, that the Europeans combined these two branches of mathematics to form what we now call coordinate geometry. Though several mathematicians made significant contributions to this union, its origin is often attributed to René Descartes (1596–1650), a French philosopher and mathematician. He wrote a treatise entitled *La Geometrie,* which outlined his ideas for applying algebra to geometry and interpreting algebraic properties geometrically.

13.1 The Rectangular Coordinate Plane

OBJECTIVES

- Plot points on the rectangular coordinate plane.
- Identify the coordinates of specific points on the coordinate plane.
- Identify the quadrants that contain specific points.

In Chapter 4, we discussed in detail the uses of the horizontal and vertical number lines. In this chapter, we will merge them and discuss what we call the **rectangular coordinate plane.**

Recall that the horizontal number line looks like the following:

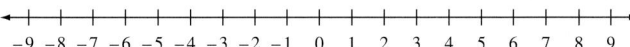

$$-9\ -8\ -7\ -6\ -5\ -4\ -3\ -2\ -1\ \ 0\ \ 1\ \ 2\ \ 3\ \ 4\ \ 5\ \ 6\ \ 7\ \ 8\ \ 9$$

Remember that the signs (+ or −) indicate direction. On a horizontal number line, positive numbers lie to the right of zero and negative numbers to the left of zero. On a vertical number line, the signs (+ or −) also refer to direction: positive numbers lie above zero and negative numbers lie below zero.

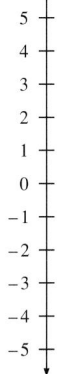

If we join these number lines by forming a right angle at their zero points, we form what is called a rectangular coordinate plane. It looks like the following:

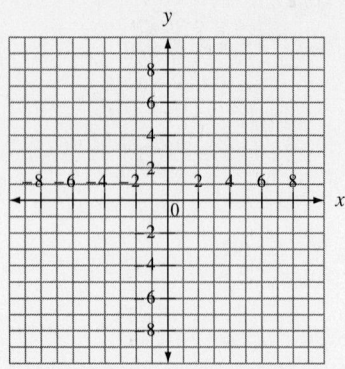

This plane is the subject of this chapter. In this plane, each point is named by two numbers, one on the horizontal line and one on the vertical line. The coordinate plane has its own terminology and notation.

DEFINITION **x-axis:** The name for the horizontal number line.

DEFINITION **y-axis:** The name for the vertical number line.

DEFINITION **Origin:** The name for the point where the x- and y-axes meet— the point paired with the number zero on both axes. It is usually designated by the letter O.

DEFINITION **Ordered pair:** The name given to the numbers on the x- and y-axes that uniquely identify a point. An ordered pair is written (x, y).

DEFINITION **x-coordinate:** The name given to the first value in the ordered pair (x, y).

DEFINITION **y-coordinate:** The name given to the second value in the ordered pair (x, y).

Note The coordinates of the origin are $(0, 0)$.

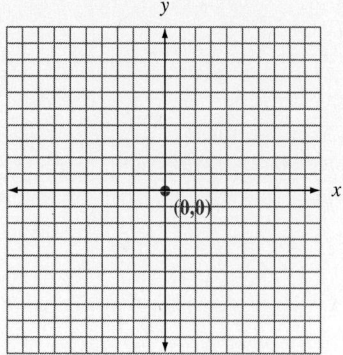

Note Locating a point on the coordinate plane given its coordinates is called **plotting** the point.

EXAMPLE 1   Locate the point P, which has coordinates (3, 2). Start at the origin. Move 3 units to the *right* (in a positive direction) on the x-axis, and then 2 units *up* (in a positive direction) on a vertical line parallel to the y-axis.

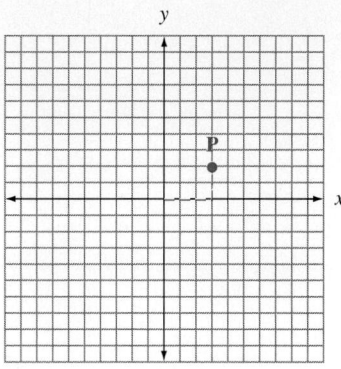

EXAMPLE 2   Plot the point A (4, −3).

     Start at the origin. Move 4 units to the *right* (in a positive direction) on the x-axis, then move 3 units *down* (in a negative direction) on a line parallel to the y-axis.

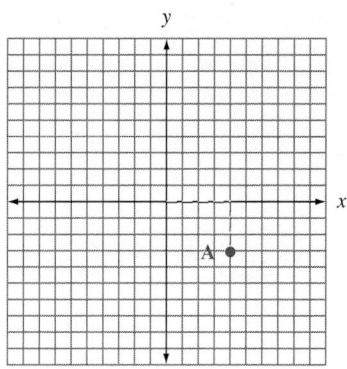

EXAMPLE 3   Plot C (−3, −4)

     Start at the origin. Move 3 units to the *left* (in a negative direction) on the x-axis, and then move 4 units *down* (in a negative direction) on a line parallel to the y-axis.

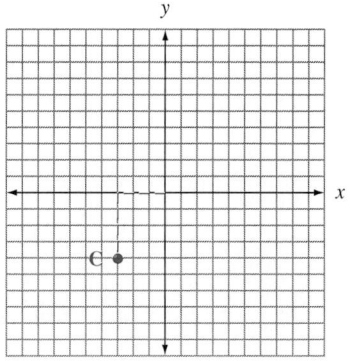

EXAMPLE 4 Plot D (−2, 1)

Start at the origin. Move 2 units to the *left* (in a negative direction) on the x-axis, and then move 1 unit *up* (in a positive direction) on a line parallel to the y-axis.

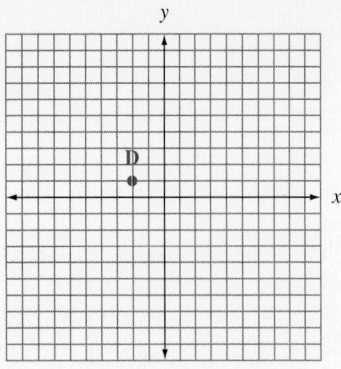

EXAMPLE 5 Plot E (0, 5)

Start at the origin. Move 0 units on the x-axis and then move 5 units *up* on a line parallel to the y-axis.

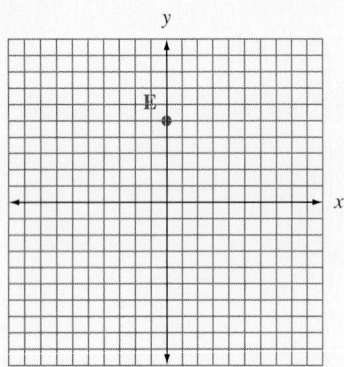

Note If the x-coordinate of a point is 0, the point lies on the y-axis. If the y-coordinate of a point is 0, the point lies on the x-axis.

EXAMPLE 6 Plot F $\left(\frac{1}{2}, \frac{3}{4}\right)$

Start at the origin. Move $\frac{1}{2}$ unit to the *right* on the x-axis and then move $\frac{3}{4}$ unit *up* on a line parallel to the y-axis.

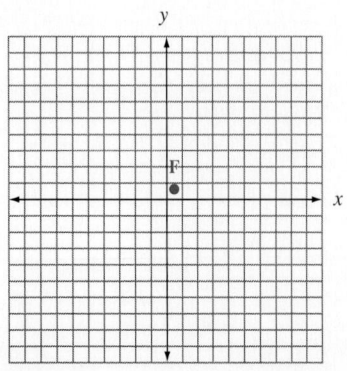

Practice Problems

Plot these points.

1. Q (3, 1)

2. R (−2, 5)

3. S (−1, −3)

4. T (4, −1)

5. U (0, 0)

6. V (2, 0)

7. W (−1, 0)

8. X (0, −3)

ANSWERS TO PRACTICE PROBLEMS
1.–8.

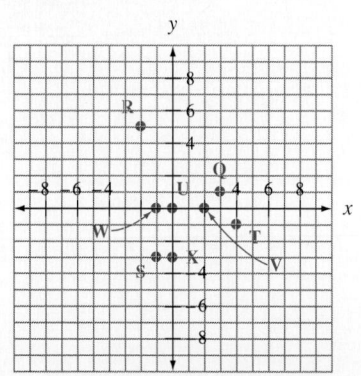

Note   When plotting fractions or decimals, use reasonable estimates.

EXAMPLE 7
          Plot G(−0.6, 0.8)
     Start at the origin. Move 0.6 units to the *left* on the x-axis and then move 0.8 units *up* on a line parallel to the y-axis.

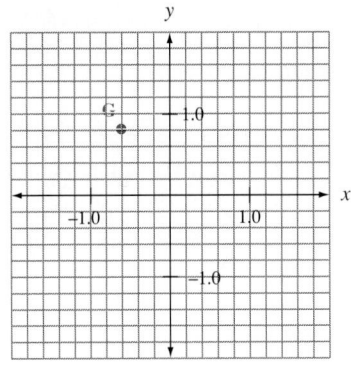

Naming the Coordinates of Points

To name the coordinates of a point

1. Count how many units on the x-axis the point is from the origin. *This number is the x-coordinate.* Recall from Section 4.1 that the sign of a number indicates direction. On the horizontal or x-axis, − indicates to the left of the origin, and + indicates to the right of the origin.
2. Count how many units on the y-axis the point is from the origin. *This number is the y-coordinate.* On the vertical or y-axis, − indicates below the origin, and + indicates above the origin.

EXAMPLE 8   Write the coordinates of the points A, B, C, H, J, and K plotted on the coordinate plane.

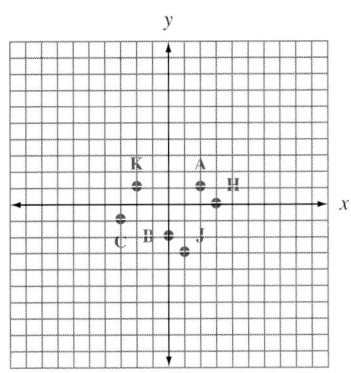

Answers: A (2,1), B (0, −2), C (−3, −1), H (3, 0), J (1, −3), K (−2, 1)

Practice Problems

Plot these points.

9. L $\left(-\dfrac{1}{2}, \dfrac{1}{4}\right)$

10. M $\left(\dfrac{2}{3}, -\dfrac{1}{3}\right)$

11. N (0.2, −0.5)

12. Z (−1.7, −2.6)

ANSWERS TO PRACTICE PROBLEMS
9.–12.

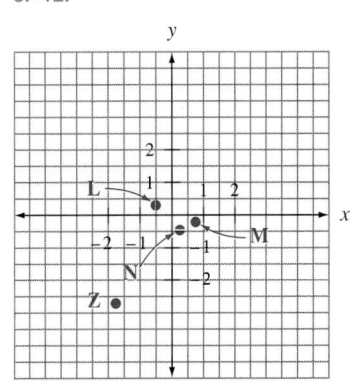

The Quadrants

Notice that the x- and y-axes divide the plane into four sections. We call the sections **quadrants** and name them Quadrant I, II, III, and IV in a counterclockwise fashion.

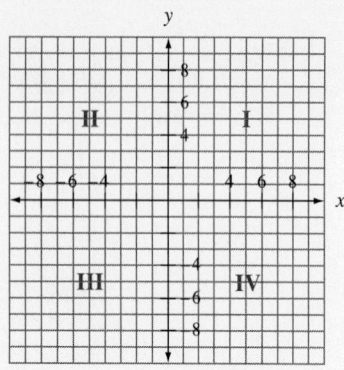

Note The signs of the coordinates in each quadrant follow:

$$I(+, +)$$
$$II(-, +)$$
$$III(-, -)$$
$$IV(+, -)$$

Note The x- and y-axes form the **boundaries** of the quadrants. They are not in any of the quadrants.

EXAMPLE 9 Name the quadrants that contain the points (5, 6), (−1, −7), (0, 9), (2, −11), (6, 0), (−1, 1), and (0, 0).

(5, 6) is in Quadrant I.

(−1, −7) is in Quadrant III.

(0, 9) is not in a quadrant. It is on the y-axis.

(2, −11) is in Quadrant IV.

(6, 0) is not in a quadrant. It is on the x-axis.

(−1, 1) is in Quadrant II.

(0, 0) is not in a quadrant. It is the origin.

Plot each point and name the quadrant it is in.

13. A (3, 1)

14. B (−2, 5)

15. C (−1, −3)

16. D (4, −1)

17. E (0, 0)

18. F (2, 0)

19. G (−1, 0)

20. H (0, 3)

ANSWERS TO PRACTICE PROBLEMS
13.–20.

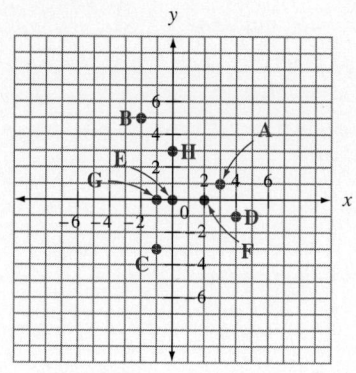

13. Quadrant I   14. Quadrant II
15. Quadrant III   16. Quadrant IV
17. Origin, on the boundaries of all the quadrants   18. On the boundary of Quadrants I and IV
19. On the boundary of Quadrants II and III   20. On the boundary of Quadrants I and II

PROBLEM SET 13.1

Plot and label these points on the rectangular coordinate grid provided.

1. Q $(-1, 4)$

2. R $(-3, 2)$

3. S $(2, 5)$

4. T $(5, 2)$

5. U $(-6, 0)$

6. V $(-1, 0)$

7. W $(0, -1)$

8. X $(0, -6)$

Plot and label these points on the grid provided.

9. Y $\left(\dfrac{1}{2}, -\dfrac{1}{2}\right)$

10. Z $\left(\dfrac{3}{5}, -\dfrac{1}{5}\right)$

11. A $(0.2, 0.4)$

12. B $(0.3, 0.6)$

13. C $\left(-\dfrac{1}{2}, \dfrac{1}{5}\right)$

14. D $\left(-\dfrac{1}{5}, \dfrac{1}{2}\right)$

Name the quadrant of each point. Refer to the graph.

15. J

16. K

17. H

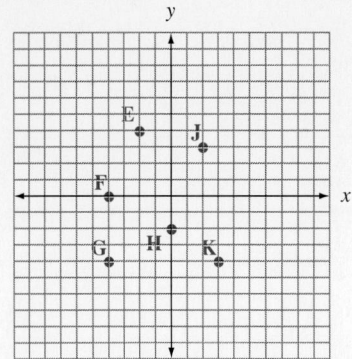

18. G

19. F

20. E

Name the coordinates of each point. Refer to the graph.

21. Z

22. Y

23. X

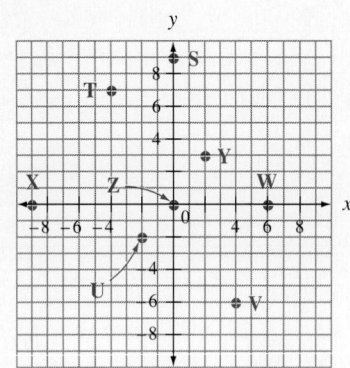

24. W

25. V

26. U

27. T

28. S

13.2 **Graphing Linear Equations**

OBJECTIVES
- Graph linear equations by plotting points.
- Use the graph of a line to estimate the coordinates of a point on it.

In Chapter 8, we discussed equations in one variable. In this section, we will discuss equations in two variables. In Chapter 8, we limited our discussion to first-degree equations—those in which the variable has an exponent of 1 (for example,

$3x + 2 = 5$). In this chapter, we will limit our discussion to first-degree equations in two variables—those in which each of the two variables has an exponent of 1 (for example, $2x + 4y = 5$).

A first-degree equation in two variables is also called a **linear equation** because its graph on the coordinate plane is a straight line. Several methods can be used to graph linear equations. In this section, we will consider only one method—plotting points. According to Euclidean geometry, "two points determine a line." Put another way, it is possible to draw one line and only one line through any two specific points. To graph linear equations, we'll use this assumption, the addition and multiplication properties of equalities, and the procedure we used in Chapter 6 to evaluate algebraic expressions.

> **RULE** To graph a linear equation in x and y by plotting points
>
> 1. Solve the equation for y in terms of x.
> 2. Choose any three real numbers as replacement values for x. (See the suggestions that follow the examples.)
> 3. Find the corresponding replacement values for y.
> 4. Construct a table of these x and y values.
> 5. Graph these ordered pairs (x, y) on a coordinate plane.
> 6. Draw a line through two of the points.
> 7. Check to see whether the third point lies on the line. If it does, the line drawn is the graph of the linear equation. If it does not, the line drawn is not the graph of the linear equation. Try again.

EXAMPLE 1   Graph $x + y = 3$

$$x + y = 3$$
$$\underline{-x \qquad\quad -x}$$
$$y = -x + 3$$

Solve for y in terms of x.

x	$-x + 3$	y
1		
0		
−1		

Choose three values for x.

x	$-x + 3$	y
1	$-1 + 3$	2
0	$-0 + 3$	3
−1	$-(-1) + 3$	4

Find the corresponding y values.

x	y
1	2
0	3
−1	4

Make a table of these x and y values.

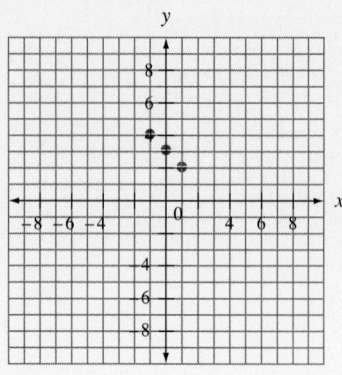

Graph these values.

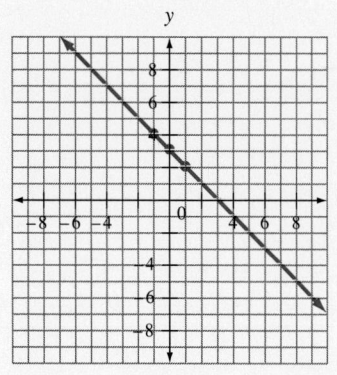

Draw a line through two of the points.

Check. Does the third point lie on the line?

EXAMPLE 2   Graph $-x + y = 2$

$$-x + y = 2$$
$$\underline{x \qquad = x}$$
$$y = x + 2$$

x	$x + 2$	y
3	$3 + 2$	5
0	$0 + 2$	2
-3	$-3 + 2$	-1

x	y
3	5
0	2
-3	-1

EXAMPLE 3   Graph $x - y = 1$

$$x - y = 1$$
$$\underline{-x \qquad = -x}$$
$$-y = -x + 1$$
$$(-1)(-y) = (-1)(-x + 1)$$
$$y = x - 1$$

x	y
4	3
0	−1
−4	−5

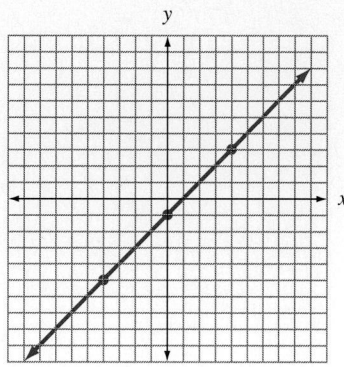

EXAMPLE 4　　Graph $2x + y = 3$

$$2x + y = 3$$
$$\underline{-2x \qquad = -2x}$$
$$y = -2x + 3$$

x	y
1	1
0	3
−1	5

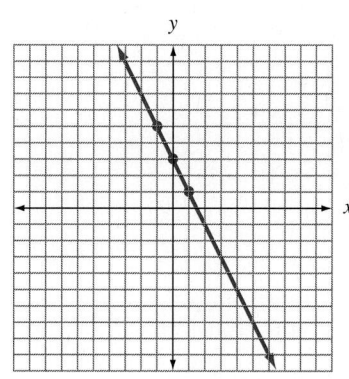

EXAMPLE 5　　Graph $\frac{1}{2}x + y = 4$

$$\frac{1}{2}x + y = 4$$
$$\underline{-\frac{1}{2}x \qquad = -\frac{1}{2}x}$$
$$y = -\frac{1}{2}x + 4$$

x	y
2	3
0	4
−2	5

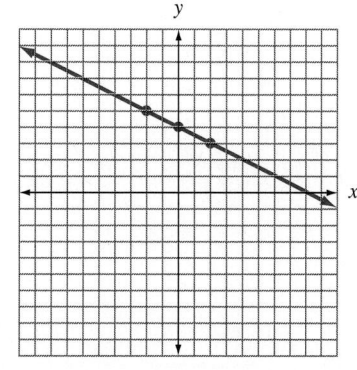

EXAMPLE 6 Graph $-x + \frac{1}{4}y = 1$

$$-x + \frac{1}{4}y = 1$$

$$\underline{x \qquad = x}$$

$$\frac{1}{4}y = x + 1$$

$$4\left(\frac{1}{4}y\right) = 4(x + 1)$$

$$y = 4x + 4$$

x	y
-1	0
0	4
1	8

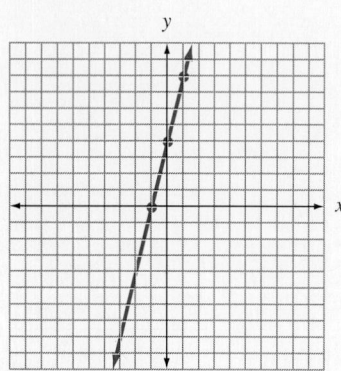

Suggestions for choosing replacement values for x.

1. To see how the graph behaves in more than one quadrant, choose zero, a positive number, and a negative number; $\{1, 0, -1\}$ will always work.
2. To achieve a wider "spread," choose zero, a positive number greater than one, and a negative number less than negative one.
3. Choose convenient numbers—not too large and not too small.
4. If, after solving the equation for y in terms of x, the coefficient of x is a fraction or the coefficient of x and the constant term are fractions, choose numbers that are multiples of the denominators. Doing so will eliminate the need to plot fractional values.

Horizontal and Vertical Lines

An equation that contains only a y variable has a horizontal line as its graph on the coordinate plane.

EXAMPLE 7 Graph $y = 3$

x	y
-1	3
0	3
1	3

Make a table.

Graph these equations.

1. $x + y = 1$

2. $-x + y = 3$

3. $4x + y = -2$

4. $x + 2y = 6$

5. $6x + 3y = 9$

6. $x - y = 5$

7. $x = 1$

8. $y = -4$

ANSWERS TO PRACTICE PROBLEMS

1.

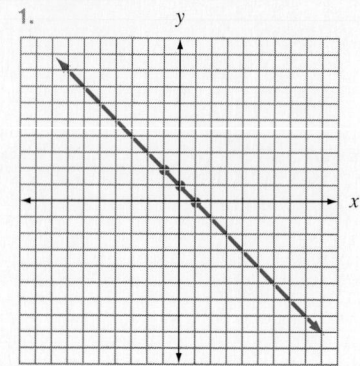

2.

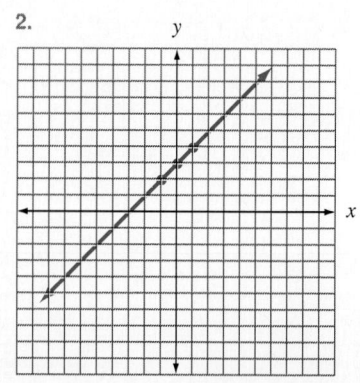

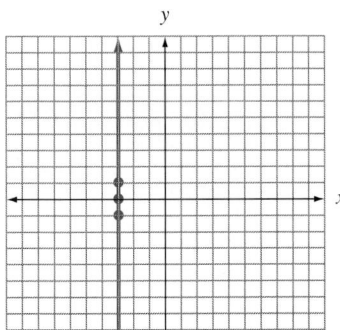

Notice that no matter what values of x you choose, the y values remain unchanged.

Draw the graph.

The graph of $y = 3$ is a *horizontal line*.

Note In general, the graph of $y = c$, where c is a constant (any real number), is a horizontal line.

An equation that contains only an x variable has a vertical line as its graph.

EXAMPLE 8 Graph $x = -2$

Since there is no y in the equation, we cannot solve for y in terms of x.

x	y
−2	1
−2	0
−2	−1

In this case, choose three values for y.

Notice that no matter what values of y you choose, the x values remain unchanged.

Draw the graph.

The graph of $x = -2$ is a *vertical line*.

Notes In general, the graph of $x = c$, where c is a constant (it can be any real number), is a vertical line.

ANSWERS TO PRACTICE PROBLEMS

3.
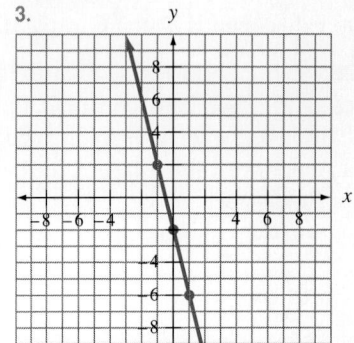

4.
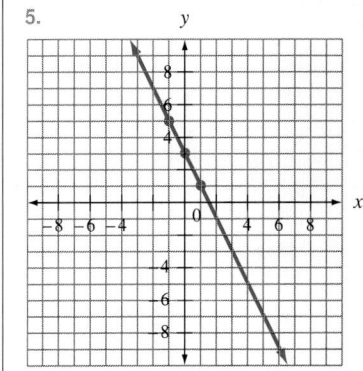

5.

6.
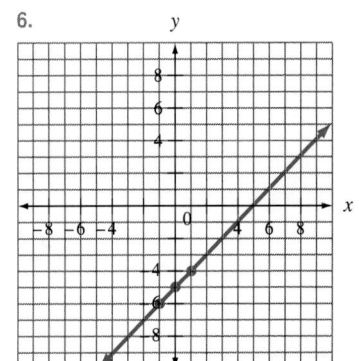

APPLICATION   The formula for converting the temperature from Celsius to Fahrenheit is $F = \frac{9}{5}C + 32$. By changing the names of the x- and y-axes of the rectangular coordinate plane to the C- and F-axes respectively, we can graph this equation. We can then read the graph to estimate the temperature in degrees Fahrenheit when it is given in degrees Celsius. When graphing large numbers, use a scale. For this example, let 1 unit = 5 degrees.

C	F
0	32
5	41
−5	23

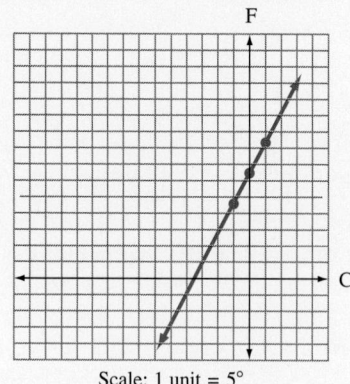

Scale: 1 unit = 5°

7.

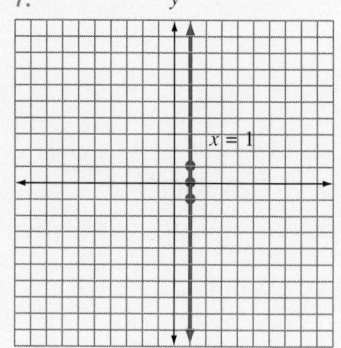

Now try estimating the temperature by reading the graph.

1. If the temperature is 10° C, what is the Fahrenheit reading?
   Answer: Approximately 50° F. (Check by using a calculator.) ◆
2. If the temperature is −4° F, what is the Celsius reading?
   Answer: Approximately −20° C. ◆

8.

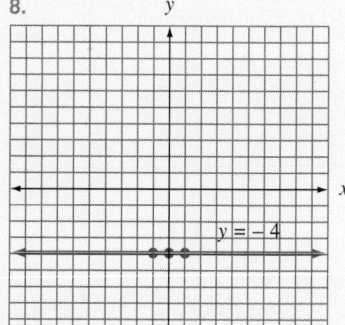

  In the previous application, we indicated that each unit on the grid represented 5° F rather than 1° F as we would have expected. Doing this is called "setting the scale" and is usually done when working with very large or very small numbers or in application problems. (Unless noted, the scale is assumed to be set at 1 unit on the grid = 1 unit of the variables being graphed.) We may set the scale of each variable independently, but doing so will distort the graph.

PROBLEM SET 13.2

Use graph paper to graph each of these lines on separate coordinate planes.

1. $x + y = 7$ **2.** $x + y = 10$

3. $-x + y = 2$

4. $-x + y = 5$

5. $4x + y = -1$

6. $2x + y = 3$

7. $x = -2$

8. $x = -12$

9. $y = \dfrac{1}{2}$

10. $y = \dfrac{3}{4}$

11. $30x + 5y = 18$

12. $40x + 8y = 24$

13. $2x - y = 11$

14. $3x - y = 9$

15. $2x + y = \dfrac{4}{5}$

16. $4x + y = \dfrac{1}{4}$

17. $-3x = 12$

18. $-2x = 20$

19. $-4y = -16$

20. $-5y = -25$

Use the accompanying graph to estimate the values of the variables.

21. $x = 2, y = ?$

22. $x = 5, y = ?$

23. $x = -1, y = ?$

24. $x = -3, y = ?$

25. $y = 0, x = ?$

26. $x = 0, y = ?$

27. $y = 5, x = ?$

28. $y = 3, x = ?$

29. $y = -1, x = ?$

30. $y = -2, x = ?$

31. $x = \dfrac{1}{2}, y = ?$

32. $x = \dfrac{1}{3}, y = ?$

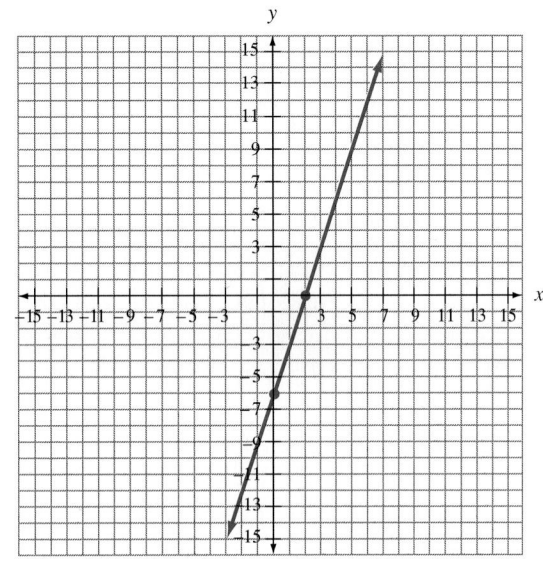

Introduction to Statistical Graphs and Charts

Historically, the term **statistics** referred to numerical information gathered by the government; it has the same root word as the term *state*. This branch of applied mathematics was well established by the time of the Roman Empire and continued to develop and expand through the centuries.

In our times, the "information explosion," coupled with the advent of calculators and computers, has played a big role in this expansion. George Gallup (1901–1984) popularized the study of statistics through his opinion polls. Today, we can't glance at a newspaper without seeing charts and graphs depicting the results of statistical studies. We can't watch the evening news on TV without hearing the results of the latest poll on the president's popularity or seeing a graph of current stock market trends. No longer the exclusive domain of the state, the science of statistics today permeates the sciences. It even follows us to the baseball park and football stadium.

13.3 Interpreting Charts and Graphs

OBJECTIVES

- Read and interpret bar graphs.
- Read and interpret line graphs.
- Read and interpret circle graphs.

In Sections 13.1 and 13.2 of this chapter, we discussed the rectangular coordinate plane. In this section, we will focus mainly, but not exclusively, on Quadrant I of that plane as we examine statistical graphs. We will begin our study by defining a few terms.

DEFINITION **Data:** Information that can be expressed in numerical form.

DEFINITION **Statistics:** The science of collecting, classifying, and interpreting data.

DEFINITION **Frequency:** The number of times a specific piece of information occurs.

Once the data are collected and sorted or classified, they are frequently represented in a graph. Besides helping us see the total picture at a glance, graphs help us see what the data are telling us. They help us analyze and draw conclusions from the data. "A picture is worth a thousand words," as the poet says.

DEFINITION **Bar graph:** A graph in which frequencies are represented by vertical or horizontal bars.

EXAMPLE 1 The bar graph at the top of page 447 represents the enrollment in a prealgebra course at State University from 1990 to 1994.

Notice that this bar graph is placed in Quadrant I of the coordinate plane. The x- and y-axes have been renamed. The units on the horizontal axis represent the academic years, and the units on the vertical axis represent the number of students

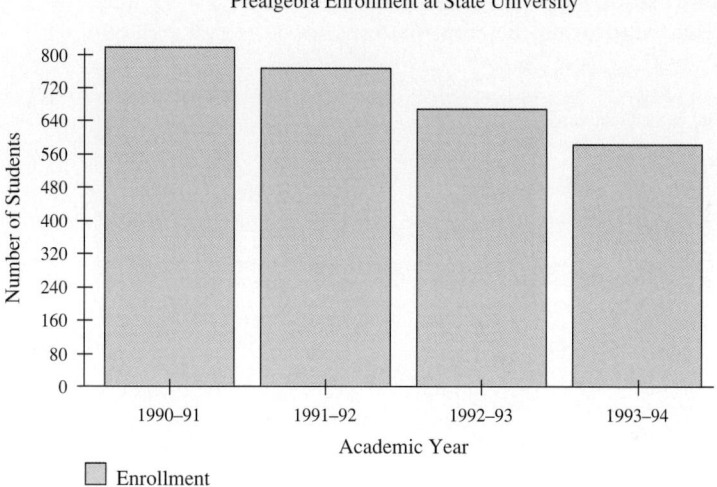

enrolled in the course. *By looking at this graph, we can immediately see that enrollment in the course has been steadily declining since the 1990–1991 academic year.*

EXAMPLE 2   Let's take another look at the statistics on the enrollment in this course over the same period. This time, the units on the horizontal axis represent semesters rather than years.

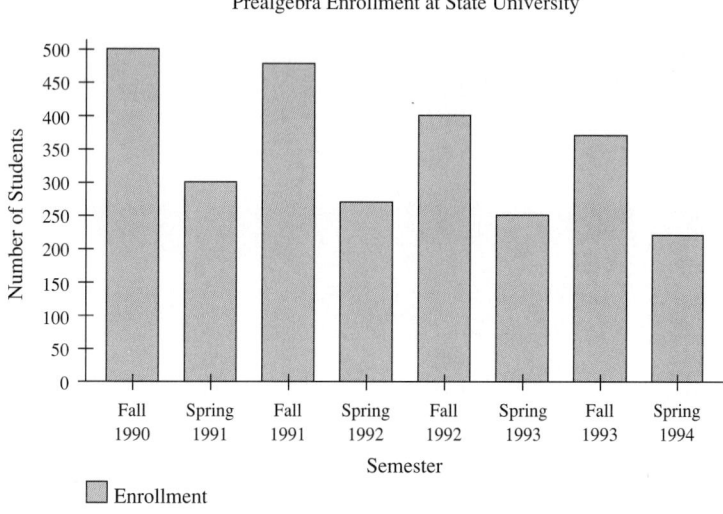

From this graph we can see that enrollment has been consistently lower during the spring semesters and that it has been steadily declining over both fall and spring semesters. Notice the differences between the graphs of Example 1 and Example 2. While the second one gives us more information, the first one gives us a more dramatic picture of the decline in enrollment.

EXAMPLE 3 We can, however, make the information stand out better by differentiating between the fall and spring semesters and using different patterns to shade them.

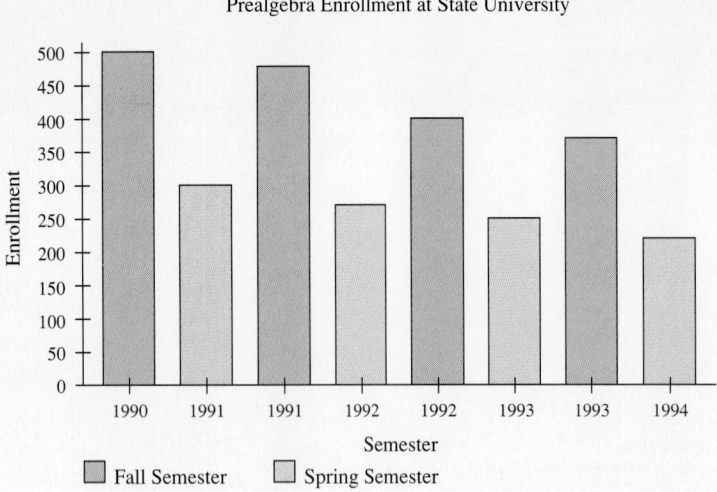

DEFINITION **Line Graph:** A graph in which the data are represented by points on a plane. The points are joined by line segments in a follow-the-dots manner. Statisticians formally call this type of graph a **frequency polygon.**

EXAMPLE 4 Let's refer back to the data on enrollment in the prealgebra course by year and draw a line graph of it.

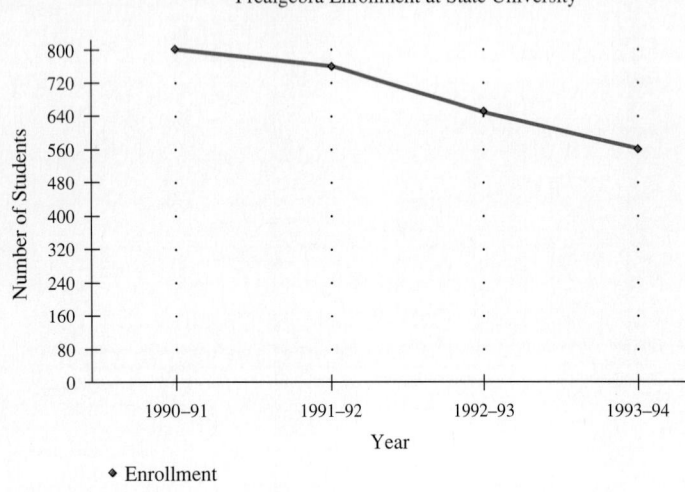

In your opinion, which graph—the line or the bar—shows the decline in enrollment better? Why?

EXAMPLE 5 This time let's look at the enrollment data by semester and draw two line graphs, one for the fall semesters and one for the spring semesters, using only one set of axes.

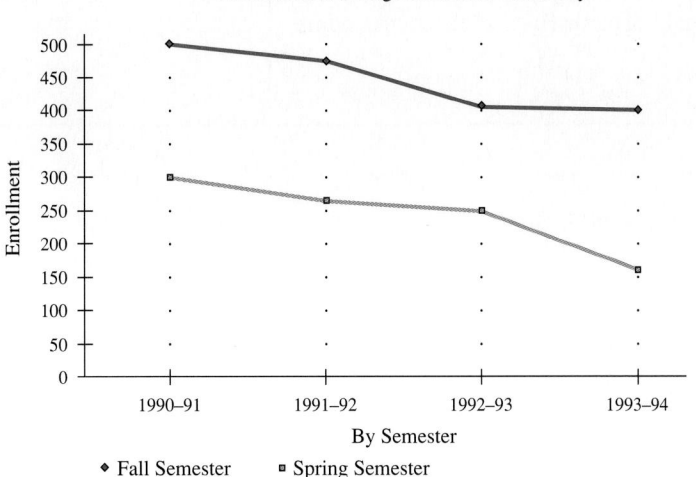

Enrollment in Prealgebra at State University

By Semester

◆ Fall Semester   ▫ Spring Semester

In your opinion, which semester graph—the line or the bar—does a better job showing the decline in enrollment? Why?

DEFINITION   **Circle (or Pie) Chart:** A graph in the form of a circle divided into parts called **sectors.** Each sector is drawn in proportion to the amount (expressed as a percent) of the quantity it represents. The entire circle represents 100%. (Refer to Chapter 10 for a discussion of proportions and to Chapter 11 for a discussion of percents.)

EXAMPLE 6   This pie chart shows the grade distribution for a prealgebra course.

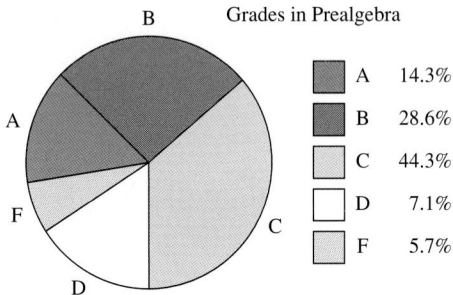

Grades in Prealgebra

▦	A	14.3%
▩	B	28.6%
▢	C	44.3%
☐	D	7.1%
▨	F	5.7%

Notice all the information we can get from this chart and how easy it is to compare percentages. More Cs were awarded than any other grade. There were more Bs than As. Notice also *what this graph does not state,* namely, how many students were enrolled in the course and how many received each grade.

EXAMPLE 7   The pie chart used in the previous example could be improved by including the total number of students in the class.

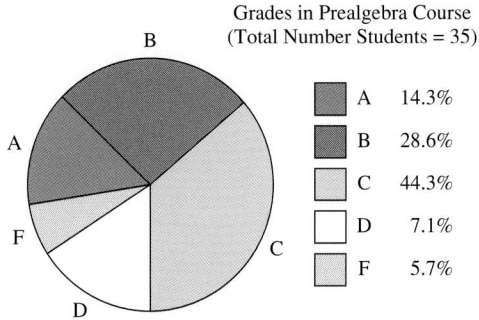

Grades in Prealgebra Course
(Total Number Students = 35)

▦	A	14.3%
▩	B	28.6%
▢	C	44.3%
☐	D	7.1%
▨	F	5.7%

With this new information, we could calculate the number of As, Bs, Cs, Ds, and Fs awarded if we needed to. (Refer to Chapter 11 for a discussion of the procedure to be used.)

 PROBLEM SET 13.3

Directions: Consider each of the following graphs and charts. Identify each as a bar graph, line graph, or pie chart. Write a paragraph on each, analyzing the information presented. Include your observations on the items that had the greatest and least frequencies. Note any patterns you see. Use the information given to form a conclusion of your own. Comment on the appropriateness of the type of graph for the given data.

1.

Median* Price of a House in April, 1992

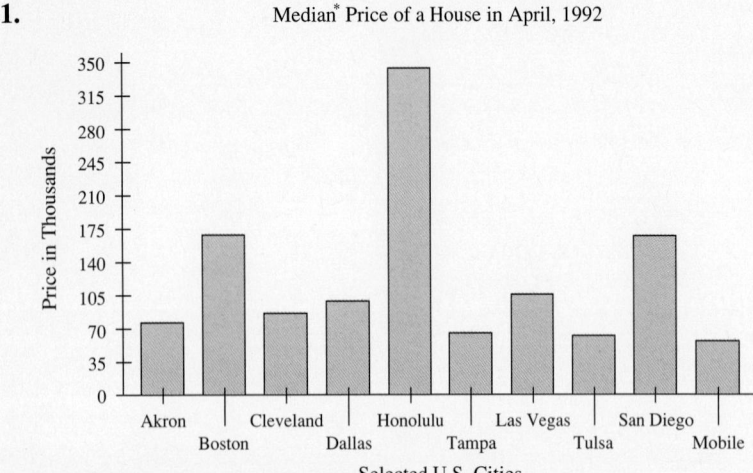

Price in Thousands of Dollars

Source: *The World Almanac and Book of Facts,* 1993.

*Median: The midpoint of a set of numerical data. It is obtained by arranging the data in numerical order, counting them, and then, if the number of items in the set is odd, selecting the middle one or, if the number of items is even, selecting the average of the two middle items.

2.

Racial and Ethnic Distribution in the United States

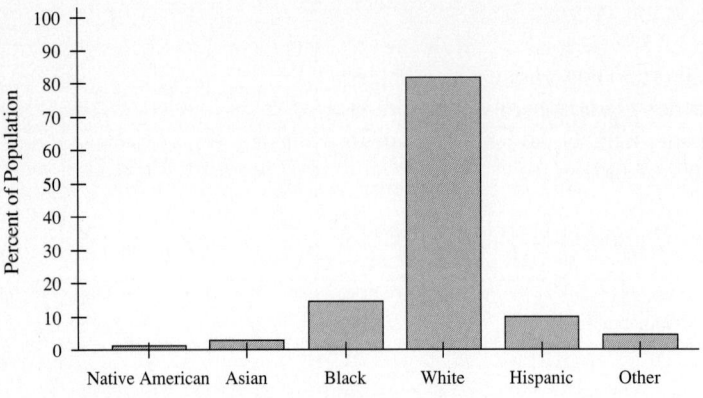

 Percent of U.S. Population in 1990

Source: U.S. Census Bureau, 1990.

3.

Age Distribution of U.S. Population
(Population 225,082,000)

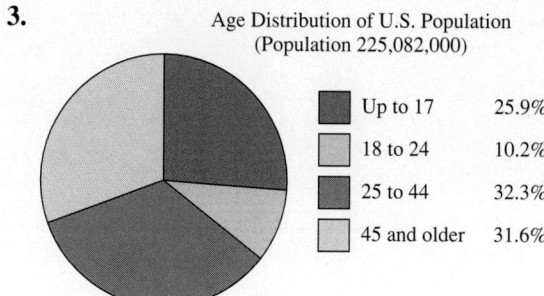

■	Up to 17	25.9%
■	18 to 24	10.2%
■	25 to 44	32.3%
□	45 and older	31.6%

Source: U.S. Census Bureau, 1990.

4. Weekly Schedule for a College Freshman
(Total Hours per Week = 168)

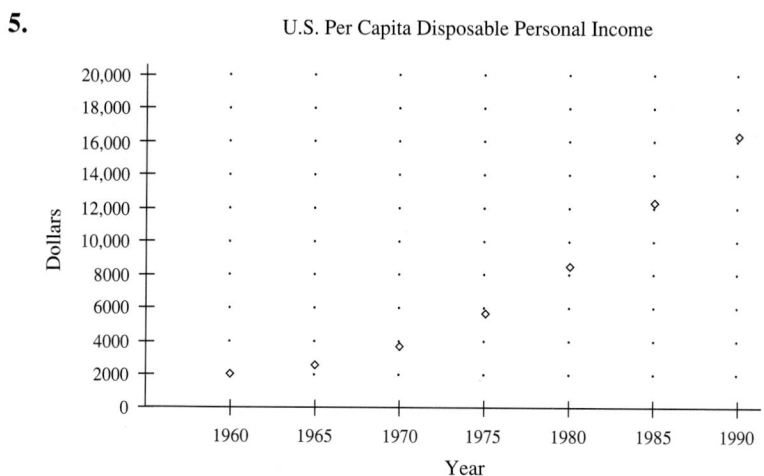

■	Classes	9.5%
■	Study	19.0%
■	Meals	12.5%
□	Sleep	33.3%
■	Personal hygene	9.5%
■	Fitness	2.4%
□	Recreation	6.0%
■	Chores	3.0%
■	Misc.	4.8%

5.

U.S. Per Capita Disposable Personal Income

◇ Disposal Personal Income

Source (for Graphs 3 and 5): U.S. Census Bureau, 1992.

6.

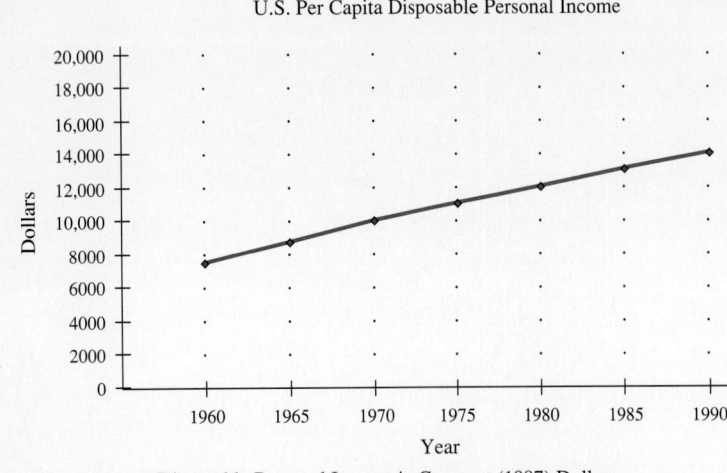

U.S. Per Capita Disposable Personal Income

◆ Disposable Personal Income in Constant (1987) Dollars

Work as a group to analyze this graph as fully as possible.

7.

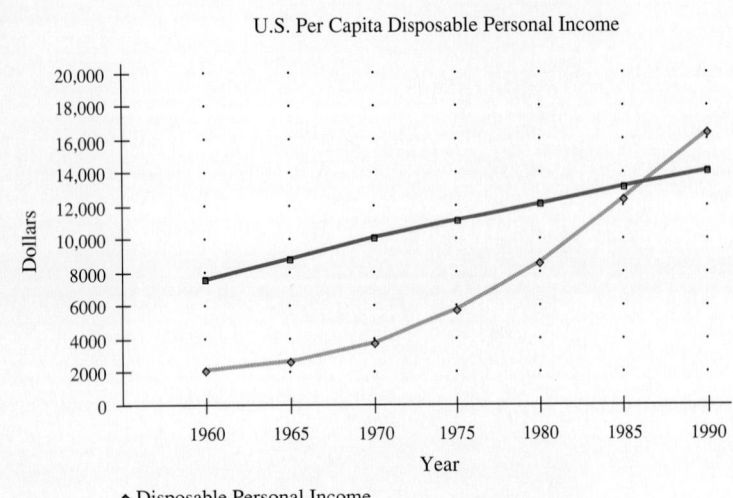

U.S. Per Capita Disposable Personal Income

◆ Disposable Personal Income

▫ Disposable Personal Income In Constant (1987) Dollars

Source (for Graphs 6 and 7): U.S. Census Bureau, 1992.

13.4   Constructing Graphs and Charts

OBJECTIVES

⬤ Construct bar graphs.

⬤ Construct line graphs.

⬤ Construct circle graphs or pie charts.

In this section, we will develop procedures to construct graphs and charts. After using these procedures a few times to get comfortable with them, we suggest you try using a computer software package to draw graphs and charts. Many spreadsheet programs will produce graphs of data.

Bar Graphs

> **RULE** **To construct a bar graph**
>
> **1.** Choose the data you wish to graph.
> **2.** Choose the two variables you wish to graph.
> **3.** Set the scale for each variable. Suggestion: Divide the range of each variable by an appropriate number—often ten—so the graph will be neither too large nor too small.
> **4.** Construct a two-column table with these variables as headings.
> **5.** Label the axes.
> **6.** Plot the points.
> **7.** Draw rectangles with the x-coordinates of the points $(x, 0)$ as the midpoints of the widths and the y-coordinates of the points $(0, y)$ as the lengths.

EXAMPLE 1   Given the following data, draw a bar graph of the total points scored in the Orange Bowl during the 1980s.*

1980 Oklahoma 24, Florida St. 7
1981 Oklahoma 18, Florida St. 17
1982 Clemson 22, Nebraska 15
1983 Nebraska 21, Louisiana St. 20
1984 Miami 31, Nebraska 30
1985 Washington 28, Oklahoma 17
1986 Oklahoma 25, Penn St 10
1987 Oklahoma 42, Arkansas 8
1988 Miami 20, Oklahoma 14
1989 Miami 23, Nebraska 3

Step 1. The data are the years and the sums of the scores.
Step 2. The variables: Horizontal axis: year
         Vertical axis: total points
Step 3. Set the scales: Horizontal axis: 1 unit = 1 year
         Vertical axis: 1 unit = 7 points

Step 4. Construct a table.

Year	Total points
1980	31
1981	35
1982	37
1983	41
1984	61
1985	45
1986	35
1987	50
1988	34
1989	26

*Source: The World Almanac and Book of Facts, 1993.

Steps 5, 6, and *7.* Label the axes, plot the points, draw the rectangles.

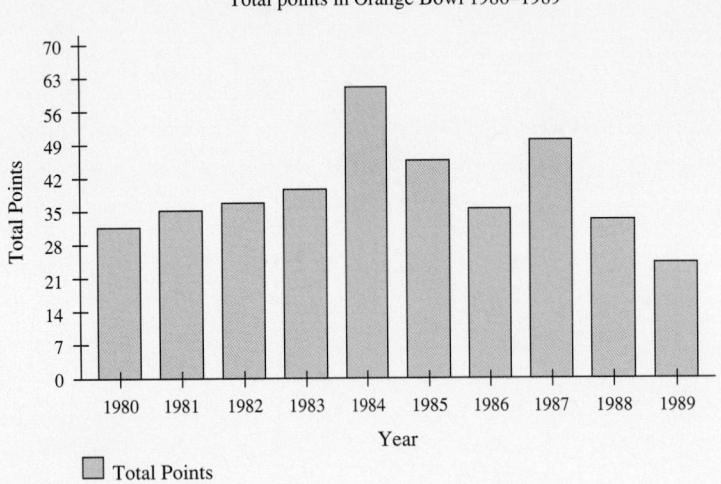

Total points in Orange Bowl 1980–1989

Note The vertical scale was set by increasing the highest number of points (61) to 70 and then dividing by 10. Scale: 1 unit = 7 points.

Line Graphs

RULE **To construct a line graph**

1. Choose the data you wish to graph.
2. Choose the two variables you wish to graph.
3. Set the scale for each variable.
4. Construct a two-column table with these variables as headings.
5. Label the axes.
6. Plot the points.
7. Connect the points by drawing line segments between them in a follow-the-dots manner.

Note Steps 1 through 6 in the procedures for drawing bar graphs and line graphs are the same. They differ only in Step 7.

EXAMPLE 2 Given the following data, draw a line graph of the resident population (rounded to thousands) of the United States from 1950–1990.*

Census year	Population
1950	151,326,000
1960	178,464,000
1970	203,302,000
1980	226,546,000
1990	248,710,000

*Source: U.S. Census Bureau, 1992.

Step 1. The data represent the years—and the population, rounded to thousands. (Refer to Chapter 2 for a discussion on rounding.)

Step 2. The variables: Horizontal axis: years
Vertical axis: population

Step 3. Set the scales: Horizontal axis: 1 unit = 10 years
Vertical axis: 1 unit = 25,000,000 persons

Step 4. Construct a table.

In this example, the data was given in table form. We just need to round the numbers so that they conform to the scale we've chosen.

Years	*Population (in millions)*
1941–1950	151
1951–1960	178
1961–1970	203
1971–1980	227
1981–1990	249

Step 5. Label the axes.

Step 6. Plot the points.

Step 7. Draw line segments between the points.

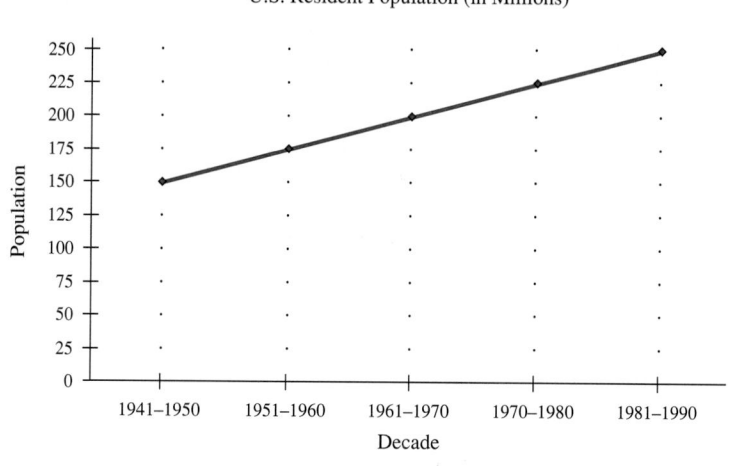

U.S. Resident Population (in Millions)[*]

Circle Graphs or Pie Charts (Optional)

Because of the geometry involved in constructing these graphs, you will want to study Chapter 12 before considering this material.

---

[*]Source: U.S. Census Bureau, 1992.

> **RULE** **To construct a circle graph**
>
> 1. Choose the data you wish to graph.
> 2. Choose the numerical data you wish to represent as percents.
> 3. Convert this data to percents. (Divide each value by the sum of the values. Refer to Chapter 11 as needed.)
> 4. Using a compass, construct a circle with a radius of convenient length.
> 5. Draw a radius of the circle.
> 6. Compute the degrees in the central angle of each sector (wedge) by using the following formula:
>
>   Degrees in Central Angle = Data Expressed as a Percent × 360°
>
> 7. Using a protractor and the degrees found in Step 6, divide the circle into sectors.
> 8. Shade and label the sectors.

EXAMPLE 3   On the first day of the semester, the instructor of the prealgebra course at City Community College announces that the grading system is based on the following distribution of points:

	Points
Unit tests	400
Quizzes	200
Projects	100
Final exam	300

Draw a circle graph representing these data.

Step 1. The data are the four components and the points awarded in each of them.

Step 2. We will represent the points in each component as a percent.

Step 3. Convert the data to percents:

	Points	*Percent*
Unit tests	400	40%
Quizzes	200	20%
Projects	100	10%
Final exam	300	30%
Total	1000	100%

Step 4. Construct a circle.

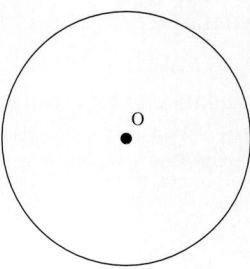

Step 5. Draw a radius of the circle.

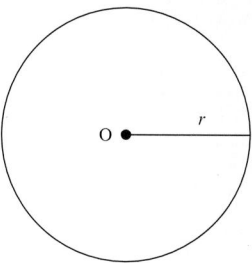

Step 6. Compute the degrees of the central angles by using the following formula:

Data as a Percent × 360° = Degrees in Central Angle

	Points	*Percent*	*Degrees*
Unit tests	400	40%	144°
Quizzes	200	20%	72°
Projects	100	10%	36°
Final exam	300	30%	108°
Total	1000	100%	360°

Step 7. Using a protractor, divide the circle into sectors whose central angles have the degrees computed. Each sector will represent one component of the grade.

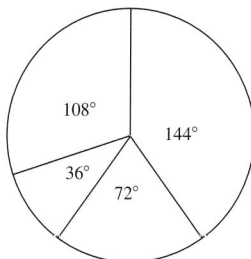

Step 8. Shade and label the sectors.

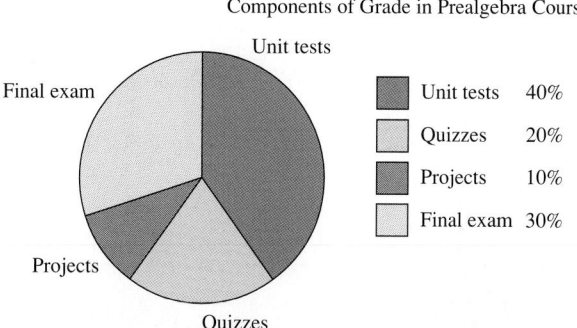

Components of Grade in Prealgebra Course

Choose the appropriate type of graph for the data you wish to represent.

1. Use a bar graph to compare data.
2. Use a line graph to show a trend and where the data can be arranged in chronological order.
3. Use a circle graph only if the data can be expressed as percents and only if the sum of the data expressed as a percent equals 100%.

In general, a bar graph is almost always appropriate. Most data can be represented by more than one type of graph. As a rule of thumb, use the type that best highlights the aspects of the data you want to stress.

 PROBLEM SET 13.4

Draw a bar graph and a line graph for each of the following sets of data.

1. U.S. Airline Traffic Fatalities from 1985–1990.*

Year	Fatalities
1985	197
1986	5
1987	231
1988	285
1989	278
1990	39

2. U.S. Airline Fatal Accidents from 1985–1990.*

Year	Fatal accidents
1985	4
1986	2
1987	4
1988	3
1989	11
1990	6

*Source: *The World Almanac and Book of Facts,* 1993.

3. U.S. Airline Safety from 1985–1990.*

Year	Fatal accidents	Fatalities
1985	4	197
1986	2	5
1987	4	231
1988	3	285
1989	11	278
1990	6	39

4. Motor Vehicle Accidental Deaths from 1985–1990.*

Year	Fatal accidents
1985	45,901
1986	53,172
1987	48,290
1988	48,900
1989	46,900
1990	46,300

*Source: *The World Almanac and Book of Facts, 1993.*

5. Persons Living Alone in the United States in 1990.*

Age	Males living alone (in thousands)	Females living alone (in thousands)
18–24	674	538
25–34	2395	1578
35–44	1836	1303
45–64	2203	3300
65–74	1042	3309
75 and older	901	3924

*Source: *The World Almanac and Book of Facts, 1993.*

6. Using the data from Problem 5, construct a chart and then bar and line graphs for all persons living alone in the United States in 1990 regardless of sex.

(Optional: Do these problems only if you studied the section entitled "Circle Graph or Pie Charts.")

Draw circle graphs for each of the following sets of data.

7. Types of Motor Vehicles Involved in Fatal Accidents in 1991.*

Type	*Fatal accidents*
Cars	70.3%
Trucks	24.2%
Buses	0.5%
Motorcycles	1.2%
Motor scooters, motorbikes	0.1%
Other	3.7%

SMALL GROUP ACTIVITIES   *Graphs and Charts*

1. Supply each group of students with a map of the campus, city, county, state, or region or have each student bring in a map. Instruct each student to select a location on the map, perhaps one that has a special meaning for him or her (hometown, birthplace, etc.). Have each student identify the selection to the others in the group by naming its coordinates. The other students must locate and name the place.

2. Design a follow-the-dots version of your school logo or a campus organization by listing the coordinates of a series of points. Have another group plot the points and connect them.

3. Using a section of the current issue of *USA Today,* analyze several of the graphs and charts included in the section. Choose one of them and report to the class on your discussion of it. Note: You may also use your college or local paper.

4. Use the data in Problem 5 of Problem Set 13.4 to construct two circle graphs: one for males in the United States living alone and one for females in the United States living alone.

5. Use the data from Problem 6 of Problem Set 13.4 to construct a circle graph for all persons (male and female) in the United States living alone.

CHAPTER 13 REVIEW

After completing Chapter 13, you should have mastered the objectives listed at the beginning of each section. Test yourself by working the following problems. Refer to the sections indicated as needed.

- *Plot points on the rectangular coordinate plane (13.1).*

Plot and label these points on the grid provided.

1. L(−2, 1)

2. M$\left(\dfrac{1}{2}, \dfrac{3}{4}\right)$

3. W(−1.2, −0.3)

4. A(0, 2)

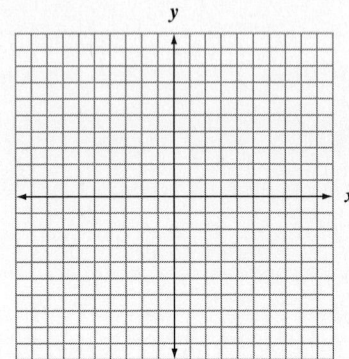

- *Identify the coordinates of specific points on the coordinate plane (13.1).*

Name the coordinates of each point. Refer to the graph.

5. B   6. F

7. G   8. D

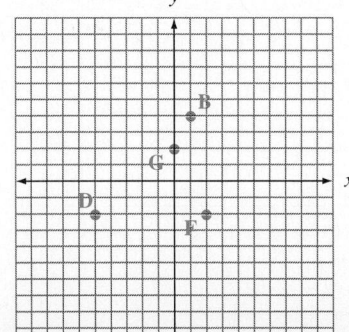

• *Identify the quadrants that contain specific points (13.1).*

  Name the quadrant of each point. Refer to the graph.

9. Z

10. Q

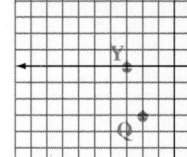

11. Y

12. U

• *Graph linear equations by plotting points (13.2).*

  Using graph paper, draw graphs of these equations.

13. $x + y = 3$

14. $-x + y = 0$

15. $4y = 12$

16. $-2x = -6$

17. $2x + y = 1$

18. $-2x + 4y = 8$

• *Use the graph of a line to estimate the coordinates of a point on it (13.2).*

Refer to the graph to estimate the values of the variables.

19. $x = 1, y = ?$

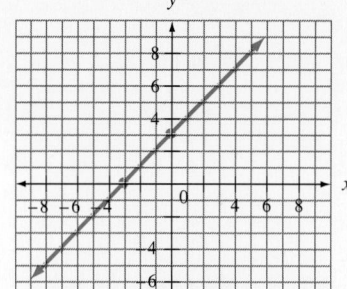

20. $x = ?, y = 5$

21. $x = 0, y = ?$

22. $x = ?, y = 0$

23. $x = -2, y = ?$

24. $x = ?, y = -1$

• *Read and interpret bar graphs (13.3).*

25. Answer these questions by examining the accompanying bar graph on the enrollment in fall 1991.

a. Which type of institution had the highest enrollment? _____

b. Which type had the lowest? _____

c. Between which two types was there the greatest difference in enrollment?

_____

d. Between which two types was there the least difference in enrollment? _____

Enrollment in Ohio Higher Education*

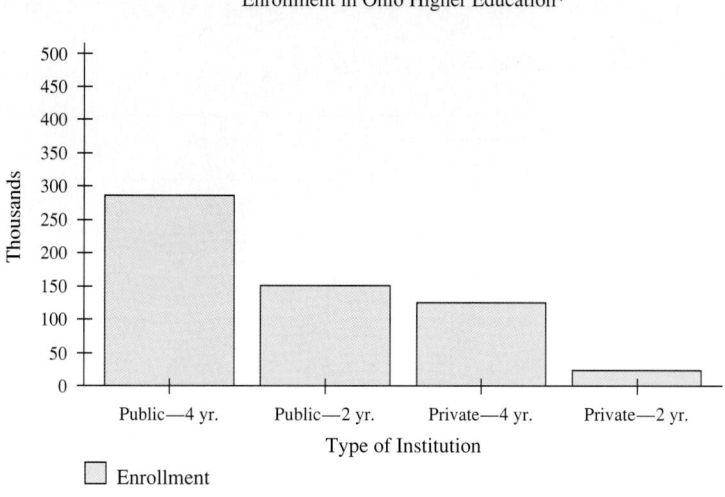

Enrollment

26. Answer these questions by examining the accompanying bar graph comparing enrollment in higher education institutions in Ohio and Pennsylvania in fall 1991.

a. In which types of institutions did Ohio have the higher enrollment? _____

b. In which types of institutions did Pennsylvania have the higher enrollment?

_____

c. In which type did the biggest difference in enrollment occur between the states? _____

d. In which type did the smallest difference in enrollment occur between the states? _____

Higher Education Enrollment in Ohio and Pennsylvania*

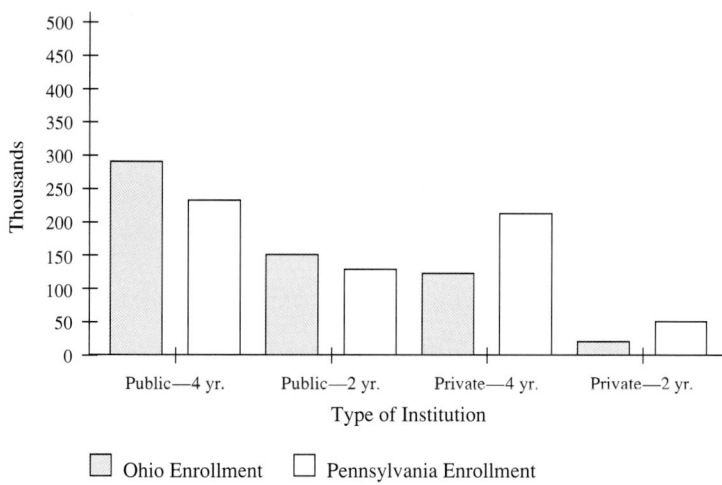

Ohio Enrollment   Pennsylvania Enrollment

*Source: *U.S. Department of Education, Chronicle of Higher Education*, Almanac Issue, August 25, 1993.

• *Read and interpret line graphs (13.3).*

27. Use the accompanying graph on U.S. marriages to answer the following questions:

    a. Which year had the most marriages? _____

    b. Which year had the least marriages? _____

    c. What is the trend? Is the number of marriages increasing or decreasing in the United States?

    _____

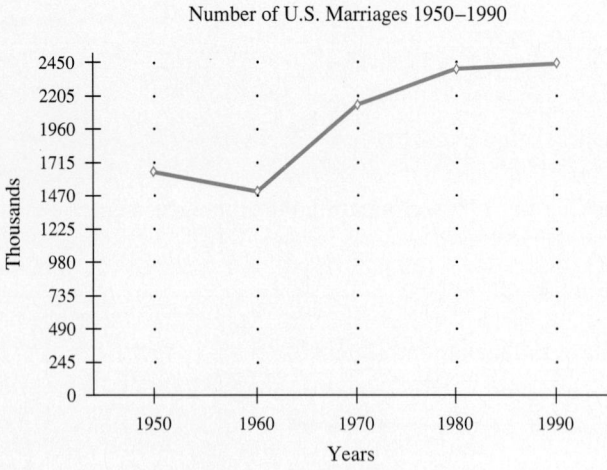

Source: *The World Almanac and Book of Facts*, 1993.

28. Use the accompanying line graphs on the growth of two mythical mutual funds from 1988 to 1992 to answer the following questions. Note: An initial investment of $10,000 was made to each fund in 1988.

    a. Which fund had the greater growth during this period? _____

    b. During what period did both funds lose money? _____

    c. Which fund appears to have lost the least money during that time? _____

    d. Approximate the 1992 value of each fund. _____

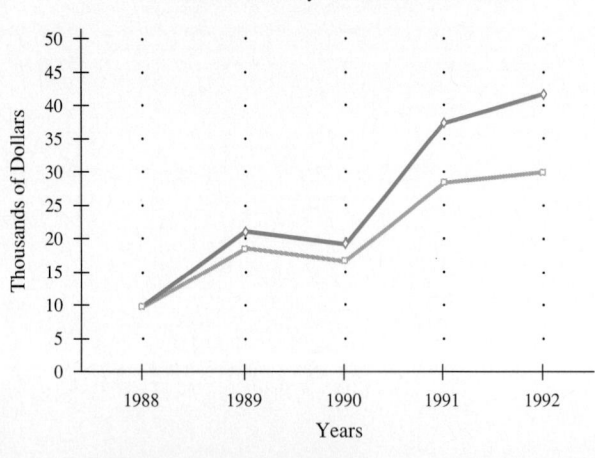

• *Read and interpret circle graphs (13.3).*

29. Suppose a survey was conducted with 300 freshmen asking them to name the sport they most like to participate in. Use the accompanying circle graph to answer the following questions about the students' responses:

a. Which sport did the most freshmen prefer? _____

b. Which sport did the least freshmen prefer? _____

c. How many freshmen chose golf as their favorite sport? _____

d. How many freshmen did not have a favorite sport? _____

Favorite Sport of Freshmen at State University

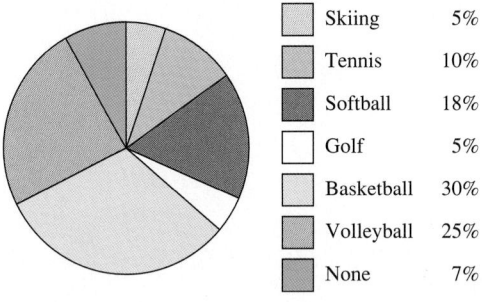

	Skiing	5%
	Tennis	10%
	Softball	18%
	Golf	5%
	Basketball	30%
	Volleyball	25%
	None	7%

• *Construct bar graphs (13.4).*

30. Use the data on number of bachelor's degrees conferred in selected fields in 1990–1991, as represented in the following chart, to construct a bar graph.

Field	*Bachelor degrees (in thousands)*
Architecture	10
Business	250
Computer science	25
Education	111
Health sciences	59

Source: *U.S. Department of Education, Chronicle of Higher Education,* Almanac Issue, August 25, 1993.

• *Construct line graphs (13.4).*

31. Construct a line graph on the number of divorces in the United States using the data in the following

table. _____

Year	Divorces (in thousands)
1970	708
1975	1036
1980	1182
1985	1187
1990	1175

Source: *The World Almanac and Book of Facts*, 1993.

• *Construct circle graphs (13.4). (Optional)*

32. Use the data in the accompanying chart on the results of a supposed survey of 100 seniors at Imaginary

University to construct a circle graph on their pizza preferences. _____

Preferred pizza	Percent of seniors
DEF Pizza	15
Mama Mio	30
Uncle Joe's	36
Roma Supreme	9
Roma with cheese	10

CHAPTER 13 PRACTICE TEST

Refer to the accompanying graph. State the quadrant and the coordinates of each point.

1. A

2. B

3. C

4. D

5. E

6. F

7. G

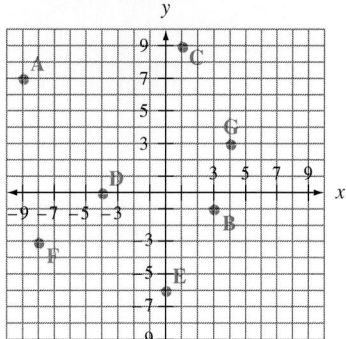

Plot these points on the grid provided.

8. H(0, −3)

9. J(−3, 0)

10. K(1, −4)

11. L(−2, −1)

12. M$\left(1\frac{1}{2}, 2\right)$

13. N(−4, 0.5)

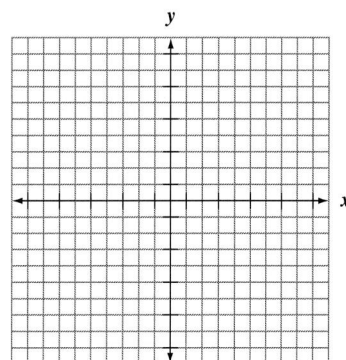

Use graph paper to graph these equations.

14. $x = 2$

15. $y = -1$

16. $x + y = -2$

17. $3x - y = -1$

18. $6x + 2y = 10$

Refer to the graph of the line. Estimate the coordinates of these points.

19. Q

20. R

21. S

22. T

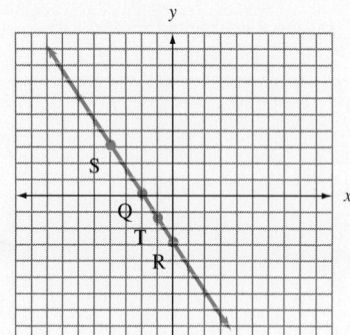

Refer to the graphs provided. Answer the questions.

23.

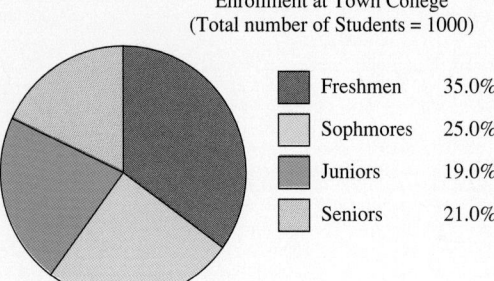

Enrollment at Town College
(Total number of Students = 1000)

Freshmen	35.0%
Sophmores	25.0%
Juniors	19.0%
Seniors	21.0%

a. What type of graph is this? _____

b. Which class has the largest enrollment? How many students? _____

c. Which class has the smallest enrollment? How many students? _____

24.

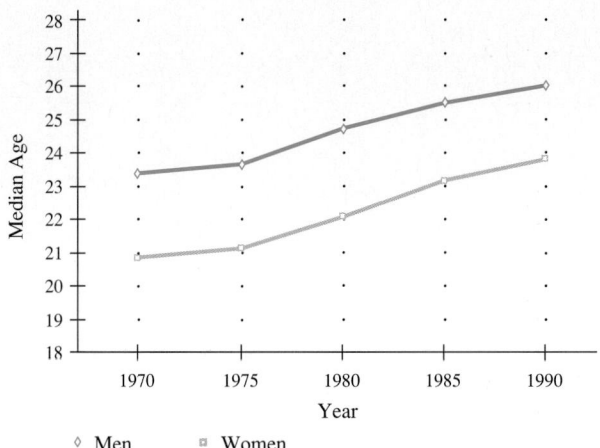

Median Age at First Marriage, 1970–1990*

◇ Men   □ Women

a. What type of graph is this? _____

b. What is the trend for men? For women? _____

_____

c. Is the difference between the median ages of men and women at first marriage widening, narrowing, or remaining constant?

_____

25.

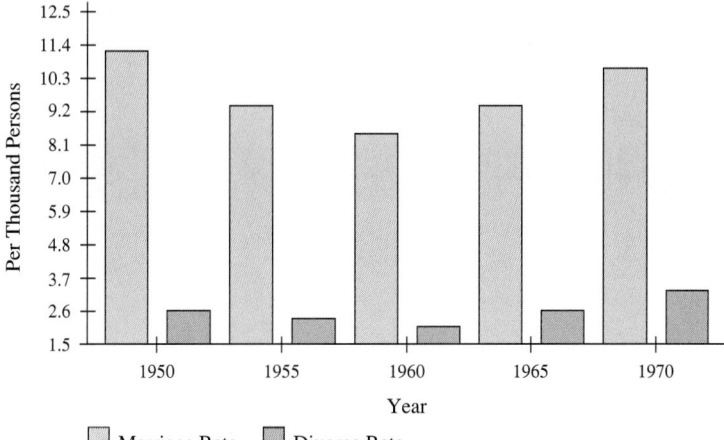

Marriage and Divorce Rates in the United States*

☐ Marriage Rate   ☐ Divorce Rate

a. What type of graph is this? _____

b. Did the marriage rate increase or decrease over this period? Explain.

_____

c. Did the divorce rate increase or decrease over this period? Explain.

_____

d. Did the divorce rate exceed the marriage rate for any year during this period? _____

26. Draw a bar graph representing the number of professional degrees awarded in selected fields by U.S. universities in 1990–1991.*

Field	Degrees (rounded to thousands)
Dentistry	4000
Law	38,000
Medicine	15,000
Pharmacy	1000
Theology	6000

Source: U.S. Department of Education, *Chronicle of Higher Education,* Almanac Issue, August 25, 1993.

27. Draw a line graph representing the National U.S. Health Expenditures from 1985 to 1991.*

Year	National expenditures (in billions of dollars)
1985	$422.6
1986	$454.8
1987	$494.1
1988	$546.0
1989	$602.8
1990	$666.2

28. (Optional) Draw a circle graph representing a college freshman's monthly budget: rent 25%, food 20%; phone 8%; transportation 20%; credit cards 12%; entertainment 10%; and miscellaneous 5%.

*Source: (for 24, 25, and 27): *The World Almanac and Book of Facts,* 1993.

STUDY SKILLS TIP

How to Take the Final Exam

The night before the exam, get a good night's rest. You will do your best only if you are well rested. *Do not* cram the night before the final; it will only confuse you. Arrive early at the room where the final is being given. Be sure you have several sharpened pencils and a working calculator. Find a seat where you feel comfortable, and make sure that the light is good and that you are not in the pathway of people arriving and leaving the room. It can break your concentration to have people walking past you during the exam.

Before you look at any questions on the exam, turn to the back and write out all the formulas you can remember and any other information that might be useful to you as you take the exam. Next, read the *entire exam.* As you do so, write on the back any other formulas and useful information that occur to you. When you are finished, check to see how many questions are on the exam, and divide the number by two. This will give you the halfway point in the exam. Mark the halfway question, and beside it write the time you should be halfway through the exam. Track your time accordingly.

Then start to work the problems at the beginning, skipping initially any problems you are unsure of. By working the problems you find easy, you will build your confidence and feel more relaxed. Work to the end of the exam, then take a good stretch in your seat. Turn to the back of the exam, and real all the information you wrote before you started. It is quite likely you will find something to help you with the remaining problems. Work back through all the problems you skipped.

After you have answered all the questions, return to the beginning again (if you have time) and check your exam. Have you entered all the answers in the answer column? Have you answered all the questions? Have you checked the answers thoroughly? Is your name and/or student number on the exam? If so, turn the exam face down and relax for a few minutes before you hand it in. In those few minutes, a new insight to a problem may occur to you. Ready? Now hand it in and enjoy your break!

THE MATHEMATICS JOURNAL

Opening-Day Advice to a Prealgebra Student

Task: Imagine that it is the opening day of next semester. You meet a friend who is going to a first class in prealgebra. Your friend asks you for advice on how to do well in the course. Write a one-page (150-word) entry on your reply.

PRACTICE FINAL EXAM

DIRECTIONS *Review for the final exam by taking this "practice test." Remember, this is a review, so check your answers and then refer to the appropriate chapter and section to get assistance in correcting any mistakes.*

Review all definitions, properties, and formulas.

COMPLETION *Choose the appropriate word or phrase from the list below.*

Integers Rational numbers
Composite number Prime number
Constant Variables
Real numbers Proportion
Absolute value Distributive Property
Associative Property of Addition Multiplication Property of Equality
Addition Property of Equality Ratio
Commutative Property of Multiplication

1. The _____ states that for all real numbers $a(b + c) = ab + ac$.

2. The _____ states that for all real numbers, $ab = ba$.

3. _____ are letters used to represent numbers.

4. A _____ is a whole number greater than one having only two factors, namely, itself and 1.

5. The _____ of a number represents the distance that number is from zero on the number line.

6. The _____ states that for all real numbers, if $a = b$, then $a + c = b + c$.

7. The _____ states that for all real numbers, if $a = b$, then $ac = bc$.

8. The _____ states that for all real numbers, $a + (b + c) = (a + b) + c$.

9. A _____ is an equation whose members are ratios.

10. In the expression $2y + 64$, the number 64 is a _____.

Perform the indicated operations. Follow the directions specified for each item.

11. Express 0.017 as a fraction in lowest terms.

12. Add: $21.68 + 0.563 + 1.4$

13. Subtract: $5.06 - 4.982$

14. Divide: $1.216 \div 0.04$

15. Multiply: $(0.004)(0.05)$

16. Express 8.7% as a decimal.

17. Express $18\frac{2}{3}$ as an improper fraction.

18. $4\frac{5}{6} - 2\frac{7}{8}$

19. Which is larger: $\frac{3}{5}$ or $\frac{5}{9}$?

20. Find the average of -15, 8, -16, and 3.

21. Round 78,907 to the nearest ten.

22. Simplify: $\dfrac{1\frac{2}{3}}{1\frac{1}{2}}$

23. Express 315 as a product of prime factors.

24. Express 24% as a fraction in lowest terms.

25. Simplify: $12 - 6 \div 3$

26. Simplify the ratio: $2\frac{2}{3}$ to $4\frac{4}{5}$

27. Solve the proportion for x: $\dfrac{1.8}{6.4} = \dfrac{x}{0.16}$.

28. Simplify: $12 + 3(6)^2$

29. Find the reciprocal of $9\frac{2}{5}$.

30. Round 897.00984 to the nearest hundredth.

31. Solve for x: $\frac{1}{2}x - \frac{1}{4} = 2$

32. Solve for y: $5 - 3y = 8 - y$.

33. Twenty is 8% of what number?

34. If the tax rate is $5\frac{1}{4}\%$, find the tax on a TV that costs $860.

35. Multiply: $(-1)(3)(-2)(-5)(-2)$

36. Divide: $(-15) \div (-3)$

37. Add: $|-9| + |-3| + |0| + |9|$

38. Find the value of $2b^2 - 5b - 3$, for $b = -2$.

39. What percent of $4.00 is $2.50?

40. Find the perimeter (P) and area (A) of a rectangle whose sides measure 5.3 inches and 2.7 inches, respectively.

41. Find the perimeter (P) and area (A) of this right triangle.

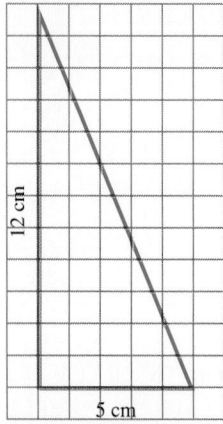

42. Find the area of a triangle with a base of 7 feet and an altitude of 3 feet.

43. Find the radius of a circle whose area is 50.24 square inches. Use $\pi \approx 3.14$.

44. Find the volume of a cube with an edge of 2 inches.

45. Solve for x: $\frac{0.05}{1.2} = \frac{x}{6}$

46. Solve for y: $\frac{y}{3} + \frac{2}{7} = -\frac{1}{2}$

47. Solve for x: $8\left(x - \frac{1}{2}\right) = 2\left(x - \frac{1}{4}\right)$

48. Express 1.2% as a fraction in lowest terms.

49. Refer to the figure. Find the length of b.

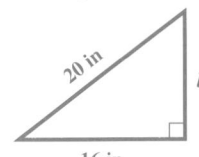

Complete the table.

	Fraction	Decimal	Percent
50.	$\frac{1}{250}$	____	____
51.	____	0.012	____
52.	____	____	66.6%
53.	$\frac{2}{9}$	____	____
54.	____	1.8	____
55.	____	____	0.14%

Perform the indicated operations:

56. $\left(2\dfrac{1}{5}\right)\left(5\dfrac{4}{9}\right)$

57. $3x + \dfrac{4x}{4}$

58. $x^4 \cdot x$

59. $(3xy^2)(-4x^2)$

60. Factor: $60xy^2 + 15x^2y$

61. Express in scientific notation: 0.000000837.

62. Express in standard notation: 9.378×10^8.

63. $(3x^2y)^3$

64. $\dfrac{4}{3x} + \dfrac{5}{4x}$

65. Simplify: $9(x - 6) + 7(8 - x)$

66. Multiply: $(x + 2)(x - 5)$

67. Find the sum: $(4a^3 + 2a^2 - 7) + (9a^2 + 4a + 9)$

68. Reduce to lowest terms: $\dfrac{19x^3y}{38x^3y^2}$

69. $3\sqrt{81} - 4\sqrt{36}$

70. Multiply: $\dfrac{2xy}{9} \cdot \dfrac{18x^3}{y^2}$

71. $(0.25)(4)\left(\dfrac{2}{9}\right)$

72. The point whose coordinates are $(-4, -8)$ is in which quadrant?

73. Solve and graph on the number line: $3x - 13 < 14$.

74. Graph $2x + y = 7$ on the accompanying grid.

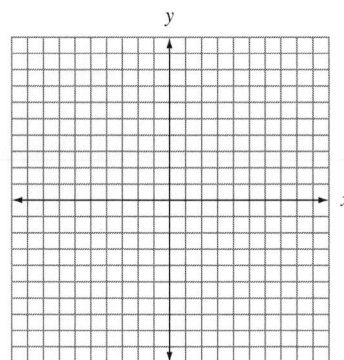

Write the proportion or equation, and solve each problem:

75. There are 450 students taking a math class. How many freshmen are taking the class if 60% of the class are freshmen?

76. If John rides his bike at 12.5 miles per hour for $2\frac{1}{2}$ hours, how far does he travel?

77. Jackie has $1.60 in dimes and nickels. If she has 7 more dimes than nickels, how many of each does she have?

78. Phillipe has $9.60 in dimes and quarters. How many of each type of coin does he have if he has twice as many quarters as dimes?

79. Lori and Peter are 4000 feet apart on a straight road when they start jogging toward each other. If Lori can jog 350 feet per minute and Peter can jog 400 feet per minute, how long will it take them to meet?

Graphs and charts.

80. Analyze the following graph. Interpret it for a college with an incoming freshmen class of 245. If the graph accurately portrays the incoming class, approximately how many of its freshmen students would be in each category?

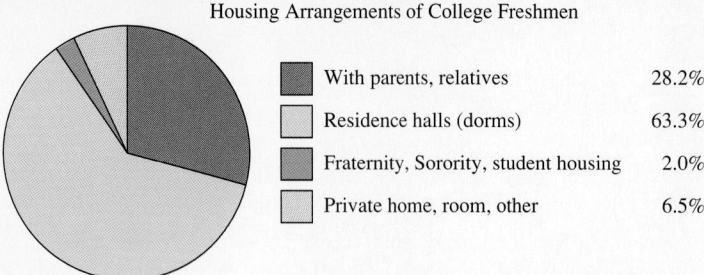

Housing Arrangements of College Freshmen

▨	With parents, relatives	28.2%
☐	Residence halls (dorms)	63.3%
▨	Fraternity, Sorority, student housing	2.0%
☐	Private home, room, other	6.5%

A The Arithmetic of Whole Numbers

A.1 Reading and Writing Numbers

In the English version of the Hindu-Arabic system, we group the places into sets of three, starting from the right, and often place a comma between each set of three. We also give each group of three places a new name.

> 1 is read one.
> 10 is read ten.
> 100 is read one hundred.
> 1000 is read one thousand.
> 10,000 is read ten thousand.
> 100,000 is read one hundred thousand.
> 1,000,000 is read one million.
> 10,000,000 is read ten million.
> 100,000,000 is read one hundred million.
> 1,000,000,000 is read one billion.

Notes
- The names units, tens, and hundreds, are used in each group of three.
- The new names for the groups are *thousand, million, billion*. The next new name is *trillion*.
- 1,000,000,000,000 is read one trillion.

EXAMPLE 1 Read the following numbers: 2002; 42,000,042; 7,010,403.

(**a**) 2002 is read two thousand, two.

(**b**) 42,000,042 is read forty-two million, forty-two.

(**c**) 7,010,403 is read seven million, ten thousand, four hundred three.

Note In writing numbers in words, a hyphen is placed between the tens digit and the units digit in each group. The exceptions to this rule are whole numbers less than 21 and numbers whose units digit is zero.

PROBLEM SET A.1

Write the following as numbers in the base 10 system.

1. Five thousand

2. Seven thousand

3. Sixty

4. Thirty

5. Thirteen

6. Seventeen

7. Thirty-five

8. Sixty-two

9. Eight hundred four

10. Six hundred two

11. One million, six hundred fifty-three thousand, nine hundred seventy.

12. Seven million, eight hundred forty-five thousand, seven hundred five.

Write these numbers in words.

13. 10

14. 20

15. 507

16. 408

17. 6000

18. 2000

19. 25,605

20. 32,804

21. 5,000,000,000

22. 6,000,000,000

23. 17,852,027

24. 24,739,054

A.2 Addition

In addition, the numbers being added are called addends. The answer is called the sum. Addition is a binary operation, which means you add only two numbers at a time.

To add two whole numbers:

1. Position one directly under the other, lining up the units digits, the tens digits, etc.
2. Add the units digits; add the tens digits; etc.

EXAMPLE 1

Tens	Units
4	2
+ 3	5
7	7

Note Addition makes good use of the base 10 place value system. To illustrate this, let's work the example given above by first writing the numbers in a modified expanded form and then performing the addition.

$$40 + 2 = 42$$
$$\underline{30 + 5 = 35}$$
$$70 + 7 = 77$$

EXAMPLE 2 Add: 67 + 41

$$60 + 7 = 67$$
$$\underline{40 + 1 = 41}$$
$$100 + 8 = 108$$

Notice what happened here; 60 + 40 = 100, so put a 0 in the tens column and a 1 in the hundreds column.

Let's look at the same problem and work it directly without using expanded notation. This problem requires a process called carrying.

Hundreds	Tens	Units
	6	7
+	4	1
1	0	8

$$\begin{array}{r} \overset{1}{} \\ 67 \\ + 41 \\ \hline 108 \end{array}$$

Practice Problems

Add.

1. 28 + 5   **2.** 63 + 17

3. 126 + 84   **4.** 918 + 97

5. 555 + 555

6. 217   **7.** 1560
    62   571
   301   89

8. 54   **9.** 78
    22   17
    88   27

10. 7890   **11.** 30,808
    1154   21,505
    5829   6 777
    6659   149,632

EXAMPLE 3   67 + 24

$$60 + 7 = 67$$
$$20 + 4 = 24$$
$$80 + 11 = 91$$ Study this example carefully!

$$\begin{array}{r} 1 \\ 67 \\ + 24 \\ \hline 91 \end{array}$$

Now try adding three or more numbers. Remember you only add two at a time. Keep track of the sum mentally, or write it down as you proceed.

EXAMPLE 4

$$\begin{array}{r} 1 \\ 251 \\ 172 \\ + 5 \\ \hline 428 \end{array}$$

USING THE CALCULATOR

Once you've mastered the operation of addition, use a scientific calculator to speed up the process. Consult your calculator manual. It contains detailed instructions and many examples. In this text, key strokes that indicate operations or functions are boxed.

25 + 32 = ?   Enter 25 + 32 = **57** ← Display

12555 + 7364 + 5385 = ?

Enter 12555 + 7364 + 5385 = **14004** ← Display

Note: Function keys may have different names on your calculator. Always consult the calculator manual.

ANSWERS TO PRACTICE PROBLEMS
1. 33 **2.** 80 **3.** 210 **4.** 1015 **5.** 1110 **6.** 580 **7.** 2220 **8.** 164 **9.** 122 **10.** 21,532 **11.** 208,722

PROBLEM SET A.2

Add.

1. 2 5
2. 5 4
3. 15 32
4. 23 71

5. 7 9
6. 8 7
7. 35 46
8. 47 24

9. 1 5 9
10. 2 6 8
11. 3 7 6 5
12. 4 8 7 6

13. 91
    19

14. 19
    91

15. 991
   119

16. 911
   199

Add. Use a calculator.

17. 2134 + 8976 + 11234 + 566 =

18. 2845 + 7832 + 54113 + 744 =

19.   1111
    234
   2987
 10,555
  8000

20.   2222
    567
  6754
 30,555
  5000

A.3 Subtraction

To subtract one whole number from another.

1. Position the subtrahend directly under the minuend, lining up the units digits, the tens digits, etc.
2. Subtract the units digit of the subtrahend from the units digit of the minuend; subtract the tens digit of the subtrahend from the tens digit of the minuend, etc.

EXAMPLE 1

Tens	Units
5	4
− 2	3
3	1

Check: 31 + 23 = 54

EXAMPLE 2   123 − 91 = ?

Hundreds	Tens	Units
1	2	3
−	9	1

Notice, in the set of whole numbers, you cannot subtract 9 tens from 2 tens, so you must convert the 100 to 10 tens and add it to the 2 tens. The number 123 can then be written as 12 tens + 3 units. Now subtract 9 tens from 12 tens.

Hundreds	Tens	Units
0	12	3
−	9	1
	3	2

Practice Problems

Subtract.

1. 9 − 5   **2.** 17 − 3

3.   69
 − 14

4.   72
 − 31

5.   156
 − 83

6.   291
 − 126

This procedure is often called borrowing. The usual way to indicate it follows.

$$\begin{array}{r} \overset{1}{\cancel{1}}23 \\ -\ 91 \\ \hline 32 \end{array}$$ Check: 32 + 91 = 123

EXAMPLE 3   This problem involves borrowing twice.

$$\begin{array}{r} 735 \\ -\ 168 \end{array}$$

Hundreds	Tens	Units	
7	3	5	
7	2	15	Convert 1 ten to 10 units, rewrite as
6	12	15	Then convert 1 hundred to 10 tens.
− 1	6	8	Now perform the subtraction.
5	6	7	

$$\begin{array}{r} \overset{6}{}\ \overset{12}{}\ \overset{1}{} \\ 7\ \cancel{3}\ 5 \\ -\ 1\ 6\ 8 \\ \hline 5\ 6\ 7 \end{array}$$ Check: 567 + 168 = 735

EXAMPLE 4   2000 − 1456 = ?

$$\begin{array}{r} 1\ 9\ 9\ 1 \\ 2\ \cancel{0}\ \cancel{0}\ 0 \\ -\ 1\ 4\ 5\ 6 \\ \hline 5\ 4\ 4 \end{array}$$ This is the usual way to write this procedure.

To analyze this example, consider rewriting the minuend as

Thousands	Hundreds	Tens	Units
2	0	0	0
1	10	0	0
1	9	10	0
1	9	9	10

Check: 544 + 1456 = 2000

USING THE CALCULATOR

Use a scientific calculator to rework the previous problem. (Remember, key strokes that indicate operations or functions will be boxed.)

2000 − 1456

2000 [−] 1456 [=] **544** ← Display

Practice Problems

Subtract:

7. $\begin{array}{r} 100 \\ -\ 9 \end{array}$

8. $\begin{array}{r} 100 \\ -\ 99 \end{array}$

9. $\begin{array}{r} 3000 \\ -\ 2345 \end{array}$

ANSWERS TO PRACTICE PROBLEMS
1. 4   2. 14   3. 55   4. 41   5. 73
6. 165   7. 91   8. 1   9. 655

PROBLEM SET A.3

Subtract the following problems mentally. Check answers by adding.

1. $\begin{array}{r} 5 \\ - \ 4 \\ \hline \end{array}$

2. $\begin{array}{r} 19 \\ - \ 8 \\ \hline \end{array}$

3. $\begin{array}{r} 38 \\ - \ 14 \\ \hline \end{array}$

4. $\begin{array}{r} 177 \\ - \ 64 \\ \hline \end{array}$

5. $\begin{array}{r} 32 \\ - \ 17 \\ \hline \end{array}$

6. $\begin{array}{r} 27 \\ - \ 19 \\ \hline \end{array}$

7. $\begin{array}{r} 123 \\ - \ 109 \\ \hline \end{array}$

8. $\begin{array}{r} 546 \\ - \ 517 \\ \hline \end{array}$

Subtract. Use a calculator.

9. $51{,}234 - 21{,}600$

10. $72{,}500 - 13{,}982$

11. $100{,}000 - 9199$

12. $1{,}500{,}000 - 259{,}768$

A.4 Multiplication

To multiply two whole numbers composed of only units digits, multiply them mentally.

To multiply two whole numbers when at least one of them is composed of two or more digits

1. Position one of the numbers directly under the other, lining up the units digits, then the tens digits, etc. For convenience, if the numbers do not have the same number of digits, place the number with fewer digits under the number with the greater number of digits.
2. Multiply the digits of the first number (starting with the units digit) by the units digit of the second number.
3. Write the result on the next line. Be careful to line up the units digit with the units digit of the second number.
4. Next, multiply the digits of the first number (starting with the units digit), by the tens digit of the second number.
5. Write the result on the next line. Be careful to start lining up the digits under the tens digit of the second number.
6. Repeat this procedure until all digits of the second number have been used as factors.
7. Add the resulting numbers.

EXAMPLE 1

Tens	Units

$$\begin{array}{r} 4 \quad 3 \\ \times \quad \ \ 2 \\ \hline 8 \quad 6 \end{array}$$

$2 \cdot 3 = 6$ and $2 \cdot 4 = 8$

Multiplication makes good use of the base 10 place value system. To illustrate what's happening in the previous example, we will write the first number in modified expanded form and use the Distributive Property to multiply it by the second number.

$$43 = 40 + 3$$
$$2(43) = 2(40 + 3)$$
$$= 2 \cdot 40 + 2 \cdot 3$$
$$= 80 + 6$$
$$= 86$$

Multiply.

1. 25   **2.** 68
   $\times$ 3 $\times$ 5

EXAMPLE **2**

$$\begin{array}{r} 432 \\ \times\ \ 21 \\ \hline 432 \\ 864 \\ \hline 9072 \end{array}$$

432   Steps 2 and 3 (432 × 1 = 432)
864   Steps 4 and 5 (432 × 2 = 864)
9072 Step 7

3. 165   **4.** 99
   $\times$ 0 $\times$ 1

To illustrate what's happening here, we will write the first number in modified expanded form and use the Distributive Property and the Associative Property of Multiplication to multiply it by the second number.

$$432 = 400 + 30 + 2$$
$$21(432) = 21(400 + 30 + 2)$$
$$= 21 \cdot 400 + 21 \cdot 30 + 21 \cdot 2$$
$$= 8400 + 630 + 42$$
$$= 9072$$

Use the calculator.

5. 123   **6.** 865
   $\times$ 15 $\times$ 10

7. 495   **8.** 1000
   $\times$ 68 $\times$ 52

EXAMPLE **3**   548 × 67   Notice all the carrying that occurs.

$$\begin{array}{r} 548 \\ \times\ \ 67 \\ \hline 3836 \\ 3288 \\ \hline 36{,}716 \end{array}$$

3836   Steps 2 and 3
3288   Steps 4 and 5
36,716 Step 7

9. 51,237
   $\times$ 123

USING THE CALCULATOR

To multiply, use the $\times$ key and the = key.

$$432 \times 21$$

Enter 432 $\times$ 21 = **9072** ← Display

$$21(65)$$

Enter 21 $\times$ 65 = **1365** ← Display

Note: On most calculators, the $\times$ is the only key used for multiplication.

---

1. 75 **2.** 340 **3.** 0 **4.** 99 **5.** 1845
6. 8650 **7.** 33,660 **8.** 52,000
9. 6,302,151

PROBLEM SET A.4

Multiply. Use a calculator to check each problem.

1. 21
$\times$ 4

2. 32
$\times$ 3

3. 48
$\times$ 6

4. 59
$\times$ 7

5. 74
$\times$ 21

6. 81
$\times$ 31

7. 93
$\times$ 48

8. 67
$\times$ 56

9. 81
$\times$ 0

10. 72
$\times$ 0

11. 153
$\times$ 32

12. 261
$\times$ 23

13. 8×15

14. 9×27

15. 7×64

16. 6×98

Multiply. (Look for shortcuts.)

17. 100
$\times$ 5

18. 200
$\times$ 7

19. 53
$\times$ 20

20. 72
$\times$ 10

Multiply. Use a calculator.

21. 5281×1234

22. 8964×2541

23. $100{,}005 \times 785$

24. $346{,}200 \times 958$

A.5 Division

The Division Process

Method 1

> Perform the division mentally.
> Check by using the definition of division.

Note   This method is suitable for dividing a one- or two-digit number by a one-digit number that is an *exact divisor* (no remainder).

EXAMPLE 1 $24 \div 6 = 4$. Check: $4 \cdot 6 = 24$

Method 2

Use long division. Check by using the division algorithm. Long division can be summarized as follows: divide, multiply, subtract, bring down the next digit or group of digits, divide, multiply, subtract, etc. This process is repeated until the final digit or group of digits has been brought down and the division, multiplication, and subtraction performed on this final digit or group of digits has resulted in a remainder smaller than the divisor.

Note   This method is suitable when the numbers are too large to divide mentally, and/or the dividend is not evenly divisible by the divisor.

EXAMPLE 2

$$\begin{array}{r} 1\,7 \leftarrow \text{Quotient} \\ 3\overline{)5\,2} \\ -3 \\ \hline 2\,2 \\ -2\,1 \\ \hline 1 \end{array}$$

↑ Note: The remainder 1 is smaller than the divisor 3.

So $52 \div 3 = 17$ remainder 1. Check using the division algorithm:

$$\text{Quotient} \cdot \text{Divisor} + \text{Remainder} = \text{Dividend}$$
$$17 \cdot 3 + 1 = 52$$

EXAMPLE 3 $1000 \div 15$

$$\begin{array}{r} 66 \leftarrow \text{Quotient} \\ 15\overline{)1000} \\ -90 \\ \hline 100 \\ -90 \\ \hline 10 \end{array}$$

Try dividing 15 into 1. One is too small. Try dividing 15 into 10. Ten is too small. Divide 15 into 100. $15 \times 7 = 105$, which is too large because it is greater than 100. $15 \times 6 = 90$ is less than 100, so write 6 in the tens position on the quotient line. Write 90 under the first two zeroes and subtract. Now bring down the next 0. Again try to divide 15 into 100. Choose 6, since $15 \times 6 = 90$. Write 6 in the units position on the quotient line. Write 90 under the two zeroes and subtract. The result is 10, which is the remainder.

$$1000 \div 15 = 66 \text{ r } 10$$

Check: $66 \times 15 + 10 = 990 + 10 = 1000$

USING THE CALCULATOR

To divide, use the ÷ key and the = key. $100 \div 5$

Enter 100 ÷ 5 = **20** ← Display

Note: If the division results in a remainder, using the ÷ key will not display the remainder; rather the quotient will be expressed as a decimal. (See Chapter 7.)

Practice Problems

Divide. Check answers.

1. $0 \div 5$

2. $5 \div 0$

3. $10 \div 10$

4. $0 \div 10$

5. $10 \div 0$

6. $92 \div 8$

7. $75 \div 3$

8. $112 \div 15$

9. $121 \div 14$

ANSWERS TO PRACTICE PROBLEMS
1. 0 2. Undefined 3. 1 4. 0
5. Undefined 6. 11 r 4 7. 25
8. 7 r 7 9. 8 r 9

PROBLEM SET A.5

Divide mentally. Check by multiplying.

1. $10 \div 5$ **2.** $12 \div 6$ **3.** $15 \div 5$ **4.** $18 \div 9$

5. $0 \div 10$ **6.** $0 \div 8$ **7.** $22 \div 0$ **8.** $36 \div 0$

Use long division. Check by using the division algorithm.

9. $13\overline{)78}$ **10.** $15\overline{)92}$ **11.** $21\overline{)153}$ **12.** $23\overline{)174}$

13. $20\overline{)600}$ **14.** $30\overline{)700}$ **15.** $15\overline{)605}$ **16.** $15\overline{)825}$

Divide. Use a calculator.

17. 1785 ÷ 17

18. 2821 ÷ 13

19. 16,422 ÷ 51

20. 19,920 ÷ 48

21. 322,091 ÷ 6853

22. 545,702 ÷ 1942

B

For Your Reference

B.1 Nonlinear Measures

Linear, square, and cubic measures are just a few of the measures we deal with almost every day. The following reference table includes the most common ones. Any good dictionary will have a complete listing. When working with these measures, remember that the procedure for conversions *within each system* is the same one we used throughout Chapter 12.

Reference Table

U. S. System	Metric System
Liquid Capacity	**Liquid Volume**
8 fl oz = 1 cup	1000 ml = 1 l
2 cups = 1 pt	1000 l = 1 kl
2 pt = 1 qt	
Weight	**Mass**
16 oz = 1 lb	1000 mg = 1 g
2000 lb = 1 T	1000 g = 1 kg
	1000 kg = 1 t
Abbreviations	*Abbreviations*
fl oz = fluid ounces	ml = milliliter
pt = pint	kl = kiloliter
qt = quart	mg = milligram
lb = pound	g = gram
T = ton	kg = kilogram
sec = second	t = metric ton
min = minute	

B.2 Conversions Between the U.S. Customary and Metric Systems

Unless we have lived in a country other than the United States or are professionals who use the metric system daily, we tend to think of measurement acccording to the U.S. customary system. Most of us can answer the following questions: "How long is an inch? How big is an 8-ounce cup? How heavy is a pound?" and even, "How far is a mile?" Many of us, however, are at a loss to answer these questions: "How long is a meter or centimeter? How far is a kilometer? How heavy is a kilogram? How big is a container that holds a liter?"

Learning a measurement system is similar to learning a second language. We don't really know the new language until we can think in it. While we're learning it, we must continually translate back and forth between the old language and the new one. For decades, the United States has been announcing that it is converting to the metric system, but Congress keeps extending the deadlines. Some of us keep putting off mastering the new system. As a result, we are forced to keep translating between the two systems—to keep converting measurements from one system to the other.

Meanwhile, here are a few handy comparisons to keep in mind: A meter is about 3 inches longer than a yard; an inch is approximately $2\frac{1}{2}$ centimeters. A kilometer is about 6 tenths of a mile (a little longer than $\frac{1}{2}$ mile) or stated the other way, a mile is just over $1\frac{1}{2}$ kilometers (5 miles equals 8 kilometers). A kilogram is just under a half pound. A liter is a little more than a quart.

The following reference table for conversions contains some of the more common equivalents. Consult a dictionary for a complete listing. When converting from one system to another, use the same procedure we've used for converting within a system.

Reference Table
Conversions Between U.S. Customary and Metric Systems

Linear

1 in $\approx$ 2.54 cm
1 mi $\approx$ 1.61 km
1 m $\approx$ 39.37 in
1 km $\approx$ 0.62 mi

Square Measure

1 in^2 $\approx$ 6.45 cm^2
1 ft^2 $\approx$ 929.03 cm^2
1 yd^2 $\approx$ 0.84 m^2
1 cm^2 $\approx$ 0.15 in^2
1 m^2 $\approx$ 1.20 yd^2

Cubic Measure

1 in^3 $\approx$ 16.39 cm^3
1 ft^3 $\approx$ 0.03 m^3
1 cm^3 $\approx$ 0.06 in^3
1 m^3 $\approx$ 35.31 ft^3

Weight

1 lb $\approx$ 454 g
1 kg $\approx$ 2.20 lb

Liquid Capacity

1 l $\approx$ 1.06 qt
1 qt $\approx$ 946 ml

B.3 Properties of the Real Numbers

Properties of Real Numbers for Addition

Name	*Mathematical Statement*	*English Version*	*Example*
Closure	$a + b \in R$	The sum of two real numbers is a real number.	$2 \in R, 3 \in R$, so $2 + 3 \in R$
Identity Property	$a + 0 = a$ $0 + a = a$	The sum of any real number and zero is the original real number. The sum of zero and any real number is the real number.	$5 + 0 = 5$ and $0 + 5 = 5$
Associative Property	$a + (b + c) = (a + b) + c$	One real number plus the sum of a second and third is equal to the sum of the first and second added to the third.	$2 + (3 + 4) = (2 + 3) + 4$ $2 + 7 = 5 + 4$
Property of Inverses	$a + (-a) = 0$ and $-a + a = 0$	The sum of any real number and its additive inverse (opposite) is zero.	$6 + (-6) = 0$ $-6 + 6 = 0$
Commutative Property	$a + b = b + a$	In adding two real numbers, either of them may be stated first.	$8 + 9 = 9 + 8$

Note: Let R represent the set of Real Numbers. Let a, b, c represent any elements of R.

Properties of Real Numbers for Multiplication

Name	Mathematical Statement	English Version	Example
Closure	$a \cdot b \in R$	The product of two real numbers is a real number.	$2 \in R, 3 \in R,$ so $2 \cdot 3 \in R$
Identity Property	$a \cdot 1 = a$ $1 \cdot a = a$	The product of any real number and 1 is the original real number. The product of 1 and any real number is the real number.	$5 \cdot 1 = 5$ and $1 \cdot 5 = 5$
Associative Property	$a \cdot (b \cdot c) = (a \cdot b) \cdot c$	One real number multiplied by the product of a second and third is equal to the product of the first and second multiplied by the third.	$2 \cdot (3 \cdot 4) = (2 \cdot 3) \cdot 4$
Property of Inverses	$a \neq 0,$ $a \cdot \dfrac{1}{a} = 1$ and $\dfrac{1}{a} \cdot a = 1$	The product of any non-zero real number and its reciprocal is 1.	$6 \cdot \dfrac{1}{6} = 1$ and $\dfrac{1}{6} \cdot 6 = 1$
Commutative Property	$a \cdot b = b \cdot a$	In multiplying two real numbers, either of them may be stated first.	$8 \cdot 9 = 9 \cdot 8$

Note: Let R represent the set of Real Numbers. Let a, b, c represent any elements of R.

Other Properties of Real Numbers

Name	Mathematical Statement	English Version	Example
Distributive	$a \cdot (b + c) = a \cdot b + a \cdot c$	The product of real number and the sum of two real numbers is equal to the sum of the products formed by multiplying the real number by each of the other two real numbers.	$2 \cdot (3 + 4) = 2 \cdot 3 + 2 \cdot 4$
Multiplication Property of Zero	$a \cdot 0 = 0$ and $0 \cdot a = 0$	The product of any real number and zero is zero.	$5 \cdot 0$ and $0 \cdot 5 = 0$
Other Special Properties:			
I	$(-1) \cdot a = -a$ and $a \cdot (-1) = -a$	The product of the additive inverse of one and any real number is the additive inverse of that real number.	$(-1) \cdot 6 = -6$ and $6 \cdot (-1) = -6$
II	$-(-a) = a$	The additive inverse of the additive inverse of any real number is that real number.	$-(-7) = 7$
III	$-(a + b) = (-a) + (-b)$	The additive inverse of the sum of two real numbers is the sum of the additive inverse of the numbers.	$-(3 + 5) = (-3) + (-5)$

Note: Let R represent the set of Real Numbers. Let a, b, c represent any elements of R.

Answers to Problem Sets

CHAPTER 1

Section 1.1–1.4 (page 9)

1. ∅, {0}, {1}, {2}, {0, 1}, {0, 2}, {1, 2}, {0, 1, 2};
3. ∅
5. ∅, {1}, {2}, {3}, {4}, {1, 4}, {2, 4}, {1, 2}, {2, 3}, {1, 3}, {3, 4}, {1, 2, 3}, {1, 2, 4}, {1, 3, 4}, {2, 3, 4}, {1, 2, 3, 4};
7. {0, 1, 2, 3, 4, 5};

9. 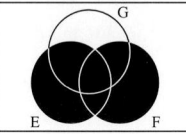   **11.**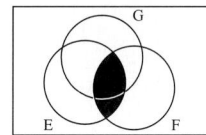

13. Infinite;   **15.** {1, 2, 3, . . .};   **17.** True;
19. True;   **21.** True;   **23.** False, M ⊆ S*;
25. True;   **27.** True.

Note: An * indicates answers may vary.

CHAPTER 1 REVIEW

1–22. Answers may vary.

CHAPTER 1 PRACTICE TEST

1. {1, 2, 3, . . .};   **2.** {0, 1, 2, 3, . . .};
3. {a, b, c, . . . , z};   **4.** {3};   **5.** {1, 2, 3, 4};
6. {1, 2, 3, 4};

7. 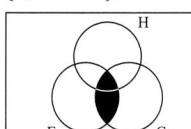   **8.**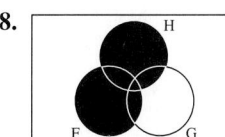

9. 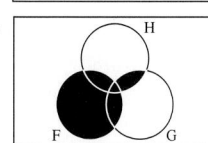   **10.** True;   **11.** True;

12. True;   **13.** True;   **14.** False, 2 ∉ H;
15. False;   **16.** ∅, {a}, {b}, {c}, {d}, {a, b}, {a, c}, {b, c}, {a, d}, {b, d}, {c, d}, {a, b, c}, {a, b, d}, {a, c, d}, {b, c, d}, {a, b, c, d};
17. Set of letters of the alphabet is finite; Set of whole numbers is infinite.   **18.** Answers may vary.

CHAPTER 2

Section 2.1 (page 18)

1. 2 × 10 + 3 × 1;   **3.** 7 × 100 + 8 × 10 + 9 × 1;
5. 1 × 1000 + 2 × 100 + 3 × 10 + 4 × 1;
7. 1 × 10,000 + 4 × 1;
9. 4 × 100,000 + 7 × 10,000 + 8 × 1000 + 2 × 100 + 9 × 10 + 1 × 1;   **11.** 21;   **13.** 567;
15. 10,101;   **17.** 100 + 1;   **19.** 400 + 50 + 7;
21. 4000 + 20:   **23.** 10,000;

Section 2.2 (page 21)

1. 10;   **3.** 40;   **5.** 60;   **7.** 100;   **9.** 100;
11. 600;   **13.** 2000;

	Number	Ten thousands	Thousands	Hundreds	Tens
15.	903		1000	900	900
17.	9084	10,000	9000	9100	9080
19.	479			500	480
21.	1005		1000	1000	1010
23.	10,654	10,000	11,000	10,700	10,650

25. 1200;   **27.** 1000;   **29.** 4 bags.

Section 2.3 (page 24)

1. Commutative Property of Addition;
3. Associative Property of Addition;
5. Identity Property of Addition;   **7.** 262 yards;
9. $272;   **11.** 747 students;   **13.** 26 points.

Section 2.4 (page 32)

1. 4(7) + 4(2) = 28 + 8 = 36;
3. 20(3) + 6(3) = 60 + 18 = 78;   **5.** 2;   **7.** 3;
9. 0;   **11.** Undefined;
13. Commutative Property of Multiplication;

15. Associative Property of Multiplication;
17. Identity Property of Multiplication; **19.** Prime;
21. $13 \cdot 7$; **23.** Prime; **25.** Prime; **27.** $5 \cdot 13$;
29. $2 \cdot 3 \cdot 5$; **31.** $3 \cdot 17$; **33.** $2 \cdot 5 \cdot 5$;
35. $2 \cdot 2 \cdot 2 \cdot 5 \cdot 5$; **37.** $2 \cdot 2 \cdot 3$;
39. $2 \cdot 2 \cdot 2 \cdot 2 \cdot 3 \cdot 3$; **41.** $2 \cdot 5 \cdot 31$
43. $2 \cdot 2 \cdot 2 \cdot 53$; **45.** $2 \cdot 61$; **47.** 1; **49.** 12;
51. 9; **53.** 26; **55.** 14; **57.** \$1800; **59.** \$60;
61. 50 mph.

Section 2.5 (page 36)

1. 1; **3.** 5; **5.** 16; **7.** 1000; **9.** 8;
11. 3^4; **13.** 5^4; **15.** 3375; **17.** 1024;
19. 11; **21.** 7; **23.** 15; **25.** $700^2 = 490,000$.

Section 2.6 (page 40)

1. 27; **3.** 14; **5.** 19; **7.** 22; **9.** 90;
11. 10; **13.** 1; **15.** 11; **17.** 53; **19.** 15;
21. 66; **23.** 152; **25.** 130; **27.** 6; **29.** 500.

Section 2.7 (page 43)

1. 43; **3.** 95; **5.** 240; **7.** 3333; **9.** 212;
11. 240; **13.** 400; **15.** 7000; **17.** 60,000;
19. 1000; **21.** 4; **23. a.** \$1894, **b.** Estimate \$1900.

CHAPTER 2 REVIEW

1.–26. answers may vary; **27.** 318; **28.** 5791;
29. 900; **30.** 8224; **31.** 6027;
32. $5 \times 10 + 7 \times 1$;
33. $2 \times 100 + 0 \times 10 + 3 \times 1$;
34. $8 \times 1000 + 5 \times 100 + 0 \times 10 + 2 \times 1$;
35. $1 \times 10,000 + 8 \times 1000 + 5 \times 100 + 1 \times 10 + 1 \times 1$;
36. $4 \times 100,000 + 0 \times 10,000 + 0 \times 1000 + 7 \times 100 + 4 \times 10 + 6 \times 1$; **37.** 70; **38.** 500;
39. 2370; **40.** 900; **41.** 1,000,010; **42.** 0;
43. Undefined; **44.** 1; **45.** 1; **46.** 100,000;
47. 19,683; **48.** Prime; **49.** Prime; **50.** 5, 57;
51. 2, 3, 5; **52.** Prime; **53.** 2, 3; **54.** 3;
55. Prime; **56.** Prime; **57.** 2, 5; **58.** 2, 7;
59. 3, 7, 11; **60.** 2; **61.** 10; **62.** 1;
63. 4; **64.** 8; **65.** 4; **66.** 45; **67.** 11;
68. 24; **69.** 25; **70.** 27; **71.** 10; **72.** 61;
73. 65; **74.** 69; **75.** 6^4; **76.** 10; **77.** 28;
78. 18; **79.** 229; **80.** 74; **81.** 1080;
82. 800; **83.** 320; **84.** 500;
85. Her bills are \$423. She has \$436, leaving her \$13. Even with the \$70 in savings, she cannot afford the bike.

CHAPTER 2 PRACTICE TEST

1. H; **2.** C; **3.** E; **4.** G; **5.** A; **6.** B;
7. F; **8.** D; **9.** 7152; **10.** 305;
11. $1 \times 10,000 + 7 \times 100 + 4 \times 1$; **12.** 7500;
13. 7000; **14.** 7480; **15.** 10,000; **16.** 6000;
17. 15; **18.** 216; **19.** 655,360,000; **20.** 23;
21. 13; **22.** 32; **23.** 146; **24.** 54; **25.** 5^3;
26. $2 \cdot 53$; **27.** 61, 67; **28.** 35; **29.** 12;
30. 6; **31.** 49;
32. Apt. 1 gives her \$25 extra per month while apt. 2 gives her \$159 extra per month. Apt. 2 seems a better arrangement financially.

CHAPTER 3

Section 3.1 (page 57)

1. $6x, 7y$; **3.** $4a, 5b, 6c, 7d$; **5.** $2a^2b, 3a^2, 4b$;
7. $4xy, 7y, 8x$; **9.** $9x^2, 6x, 7$; **11.** 1; **13.** 19;
15. 53; **17.** 8; **19.** 4; **21.** $70a$; **23.** $42y$;
25. $27ab$; **27.** $4x^2$; **29.** $81xy$; **31.** $9a^2b$;
33. $7x + 28$; **35.** $7x + 28$; **37.** $3x + xy$;
39. $32 - 8x$; **41.** $12x + 16y$; **43.** $9a + 18b$;
45. $3a^2 + 2ab$; **47.** $3bc + 6bd$; **49.** $67xz + 9yz$;
51. $135z + 105x$; **53.** $12x$; **55.** $2y$;
57. $3a^2 + 5a$; **59.** $8x^2 + 8x$; **61.** $42t$;
63. $7a^2 + 12a$; **65.** $28x^2 + 23x$; **67.** $15c + 13cd$;
69. $9j + 14k + 20jk$; **71.** $48x$; **73.** $4a + 30$;
75. $3x + 25$; **77.** $3x + 15$; **79.** $10x + 6y + 12$;
81. $29x + 7$; **83.** $4a^2 + 7ab + 19b$;
85. $19a + 57b + 51c$; **87.** $72c^2 + 45cd + 12d + 15$;
89. $39a + 40b$; **91.** $9x^2 + 29x$.

Section 3.2 (page 62)

1. 3^2; **3.** 9^4; **5.** 7^4; **7.** $(4y)^4$; **9.** $(ab)^3$;
11. 3^2; **13.** 9^2 or 3^4; **15.** 5^3; **17.** 1; **19.** 27;
21. 1; **23.** 729; **25.** $125y^3$; **27.** 10,000;
29. 343; **31.** a^8; **33.** x^6; **35.** 12^4; **37.** b^2;
39. z^3; **41.** 5^6; **43.** a^{40}; **45.** x^{49}; **47.** z^{20};
49. b^{15}; **51.** x^5y^5; **53.** $6561a^8$; **55.** $49a^2b^2$;
57. $x^4y^4z^4$; **59.** $729x^3$; **61.** a^{21}; **63.** y^{56};
65. s^{15}; **67.** x^8y^{16}; **69.** $a^{36}b^{54}$.

Section 3.3 (page 65)

1. 12; **3.** 29; **5.** 23; **7.** 6; **9.** 49; **11.** 2;
13. 20; **15.** 12; **17.** 23; **19.** 37; **21.** 84;
23. 44; **25.** 9; **27.** 34; **29.** 513; **31.** 373;
33. 3822; **35.** 0; **37.** 182; **39.** 287.

Section 3.4 (page 68)

1. $14 \div x = \frac{14}{x}$;   **3.** $x + 20$;   **5.** $4 - x$;

7. $8 + x$;   **9.** $\frac{x}{2} = x \div 2$;   **11.** $13 - x$;

13. $1 - x$;   **15.** $\frac{17}{x} = 17 \div x$;   **17.** $2x$;

19. $10x$;   **21.** $2x$;   **23.** $\frac{x}{25} = x \div 25$;   **25.** $6x$;

27. $x + 67$;   **29.** $x + 43$;   **31.** $19 - 9$;
33. $x + 10$;   **35.** $x + 10$;   **37.** $x - 19$;   **39.** $8x$;

41. $j(f)$;   **43.** $10x$;   **45.** $\frac{x}{7}$, $6x$, $52x$;   **47.** $\frac{150}{x}$;

49. $30 - x - y$;   **51.** $10x$;   **53.** $x + 6$;
55. $2x + x = 3x$;   **57.** $2x$.

CHAPTER 3 REVIEW

1.–8. See Section 3.1;   **9.–10.** See Section 3.2;
11. $3x$, $2y$, 92;   **12.** $4x^2$, $5x$, 7;   **13.** $6a$, $4ab$, $7b$;
14. $4x$, $6y$;   **15.** $9a$, 7;   **16.** $14b$, $3a$;
17. $2c$, $3d$, $4cd$;   **18.** $9x$, $7y$, 46;   **19.** $3jk$, $2j$, $7k$, 6;
20. x^3, x^2, x, xy, xy, y^2, y;   **21.** a^2, a, ab, b^2, b;
22. c^2, c, cd, d;   **23.** $6, 4, 5, 6, 1, 3, 1, 8$;
24. $3, 4, 6, 7, 1, 9$;   **25.** $4, 1, 1, 5, 7$;   **26.** $9, 46, 86$;
27. $19, 2$;   **28.** $15, 0$;   **29.** $16x$;   **30.** $20y + x$;
31. $6a + 4ab + 5b$;   **32.** $6y$;   **33.** $5x^2 + 11x$;
34. $12x$;   **35.** $14x^2 + 9x$;   **36.** $ab + 3a + 7b$;
37. $4y^2 + 13y$;   **38.** $9m^2 + 14m$;
39. $7c + 9d + 3cd$;   **40.** $13x^2 + 6$;
41. $5x + 2xy$;   **42.** $9ab + 96a$;
43. $16x + 12y + 8xy$;   **44.** $7x + 21$;
45. $8a + 40$;   **46.** $9y^2 - 72y$;   **47.** $xy - x$;
48. $7a + 7b$;   **49.** $6x + 12$;   **50.** $8y + 36$;
51. $12x + 4xy$;   **52.** $5x + 20$;   **53.** $6y - 18$;
54. $6x + 22$;   **55.** $4x + 27$;   **56.** $5y + 23$;
57. $5a + 14$;   **58.** $3x + 25$;   **59.** $5y + 12$;
60. $10a + 26$;   **61.** $6c + 5d$;   **62.** $11a + 11b$;
63. $36x + 51y$;   **64.** 49;   **65.** 125;   **66.** x^9;
67. 1;   **68.** 1;   **69.** a^{12};   **70.** x^{54};   **71.** g^6h^6;
72. y^{30};   **73.** $y^{16}z^8$;   **74.** $64x^{15}$;   **75.** 18;
76. 45;   **77.** 33;   **78.** 4;   **79.** 12;   **80.** 29;
81. 6;   **82.** 10;   **83.** 72;   **84.** 32;   **85.** 88;
86. 74;   **87.** 62;   **88.** 208;   **89.** 304;
90. $x + 7$;   **91.** $x - y$;   **92.** $2x$;

93. $9 \div x = \frac{9}{x}$;   **94.** $x + y$;   **95.** $72x$;

96. $\frac{15}{x} = 15 \div x$;   **97.** $19 - x$ **98.** $6 + x$;

99. $14 - 7$;   **100.** $x - 9$;   **101.** $x - y$;
102. $3 - 7$;   **103.** $x - 7$;   **104.** $9 + 10$;

105. $x - 21$;   **106.** $3x$;   **107.** $\frac{x}{y}$;   **108.** $x + 19$;

109. $\frac{36}{x}$;   **110.** $x - 9$;   **111.** $2x$;   **112.** $50 + x$;

113. $43 - x + y$.

CHAPTER 3 PRACTICE TEST

1. $6x$, $9xy$, $7y$;   **2.** 1;   **3.** 7;   **4.** 81;   **5.** 27;
6. $6a$;   **7.** $4y$;   **8.** $9ab$;   **9.** $16x + 6y$;
10. $4x^3 + 5x^2 + 6x + 7$;   **11.** $4x + 28$;
12. $27x + 9x^2$;   **13.** $5x - 40$;   **14.** $6y^2 - 12y$;
15. $25x + 8$;   **16.** $8a + 30$;   **17.** $14x + 63$;
18. 49;   **19.** 298;   **20.** 9;   **21.** $x + 10$;

22. $17 - x$;   **23.** xy;   **24.** $\frac{x}{y}$ or $x \div y$;

25. $\frac{x}{4}$ or $x \div 4$;   **26.** $6 - x$;   **27.** $8x$;

28. $x - 6$;   **29.** $\frac{x}{7}$ or $x \div 7$;   **30.** $x + 23$;

31. $343x^6$;   **32.** a^{21}.

CUMULATIVE REVIEW CHAPTERS 1, 2, & 3

1. Commutative Property of Multiplication;
2. Subset;   **3.** Associative Property of Multiplication;
4. Commutative Property of Addition;
5. Distributive Property;   **6.** Natural numbers;
7. Whole numbers;   **8.** Variable;
9. Prime number;   **10.** Constant;   **11.** $15x + 30y$;
12. $4a + 80$;   **13.** $26j$;   **14.** $13p + r$;   **15.** 5^3;
16. $76,080$;   **17.** $76,100$;   **18.** $76,000$;
19. $80,000$;   **20.** 60;   **21.** 9;   **22.** 216;
23. 0;   **24.** 1;   **25.** 34;   **26.** 37;   **27.** 4;
28. Undefined;   **29.** x^{30};   **30.** $27y^{12}$;   **31.** 4;
32. $16ab$;   **33.** $16x + 10$;   **34.** 35;   **35.** 148;
36. 5;   **37.** 2;   **38.** $12,700$;   **39.** 36;   **40.** 0;
41. $2l + 2w$;   **42.** $3r - 40$;
43. $\{x, y, z\}$ $\{x, y\}$ $\{x, z\}$ $\{y, z\}$ $\{x\}$ $\{y\}$ $\{z\}$, $\{\ \}$
44.
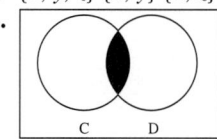
45.
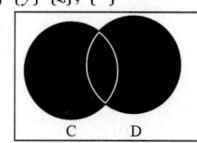

CHAPTER 4

Section 4.1 (page 85)

1. 4 is greater than 2;   **3.** 4 is less than 8;
5. 0 is greater than -4;   **7.** -43 is greater than -44;
9. 14 is greater than 0;   **11.** $9 > 5$;   **13.** $15 < 17$;
15. $0 > -4$;   **17.** $-64 < -63$;   **19.** $-1 > -10$;
21. $7 < 16$;   **23.** $-7 < 16$;   **25.** $7 > -16$;
27. $-7 > -16$;   **29.** $0 > -3$;   **31.** $30 = 30$;
33. $24 < 25$;   **35.** $-24 > -25$.

Section 4.2 (page 88)

1. 4; **3.** 0; **5.** 17; **7.** 27; **9.** 734;
11. 27; **13.** 9; **15.** 3; **17.** 1369; **19.** 6;
21. −8; **23.** 1; **25.** −4683; **27.** −15;
29. 5; **31.** −3; **33.** −1; **35.** 9; **37.** 17;
39. −19; **41.** 14; **43.** −23; **45.** −304;
47. −93; **49.** 23; **51.** −10; **53.** 7; **55.** 5;
57. −13; **59.** 4; **61.** 1; **63.** 99; **65.** −2;
67. −83; **69.** 55.

Section 4.3 (page 93)

1. 7

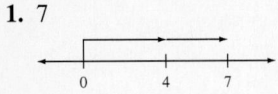

3. 3

5. −4

7. −10

9. −8

11. 14; **13.** −4; **15.** −14; **17.** −7;
19. −4; **21.** −30; **23.** −9; **25.** 169;
27. 49; **29.** −100; **31.** 40; **33.** 60;
35. 535; **37.** −8; **39.** −4; **41.** −1;
43. −7; **45.** −10; **47.** −17; **49.** 0;
51. −14; **53.** −2; **55.** 11; **57.** 10;
59. −14° F; **61.** $79; **63.** 92° F.

Section 4.4 (page 99)

1. 2; **3.** −2; **5.** 14; **7.** −4; **9.** 5;
11. 7; **13.** −23; **15.** −13; **17.** 5;
19. 5; **21.** 0; **23.** −18; **25.** −13; **27.** −3;
29. 10; **31.** 4; **33.** −20; **35.** −9; **37.** 0;
39. −230; **41.** 11; **43.** 19; **45.** −15;
47. 13; **49.** −1; **51.** 28,611 ft.

Section 4.5 (page 102)

1. 24; **3.** 24; **5.** 21; **7.** 63; **9.** −36;
11. −20; **13.** 24; **15.** −54; **17.** −98;
19. −96; **21.** −125; **23.** −84; **25.** 288;
27. 275; **29.** 0; **31.** −225; **33.** 198;
35. −168; **37.** 0; **39.** 1520; **41.** 96;
43. 0; **45.** 72; **47.** −36; **49.** −14;
51. $2079 lost.

Section 4.6 (page 105)

1. 4; **3.** −4; **5.** −9; **7.** 7; **9.** 9;
11. −7; **13.** 11; **15.** 8; **17.** Undefined;
19. 12; **21.** −11; **23.** −18; **25.** 49;
27. −1; **29.** −50; **31.** 6; **33.** 8; **35.** 5;
37. 9; **39.** 12.

Section 4.7 (page 109)

1. 80; **3.** −8; **5.** 4; **7.** 0; **9.** 126;
11. 110; **13.** −37; **15.** −68; **17.** −12;
19. −18; **21.** 8; **23.** 6; **25.** 4; **27.** −1;
29. −343; **31.** 81; **33.** 36; **35.** 32;
37. 29; **39.** 6; **41.** 60; **43.** 5; **45.** 45;
47. 6; **49.** 10; **51.** −22; **53.** −7;
55. −120; **57.** −84; **59.** −55; **61.** −30;
63. 6; **65.** 0; **67.** 0; **69.** −41; **71.** −35;
73. 57; **75.** $1 average gain; **77.** −4° F;
79. Average overbooked 2.

Section 4.8 (page 115)

1. Identity Property of Equality;
3. Transitive Property of Equality; **5.** 5; **7.** 168;
9. −4; **11.** −10; **13.** −4; **15.** −8;
17. −43; **19.** 23; **21.** No solution;
23. No solution; **25.** Any real number;
27. Any real number; **29.** −10; **31.** 44;
33. 34; **35.** 16; **37.** $15.

CHAPTER 4 REVIEW

1.–2. Section 4.1; **3.–4.** Section 4.2;
5.–10. Section 4.8; **11.–12.** Section 4.2;
13. Section 4.3; **14.** Section 4.4; **15.** Section 4.5;
16. Section 4.6; **17.** −3 > −23; **18.** −3 = −3;
19. 0 > −3; **20.** 19 = 19; **21.** −92 < 27;
22. −4 < 4; **23.** −10 < 6; **24.** −5 > −10;
25. 0 < 15; **26.** 0 > −15; **27.** 3; **28.** 54;
29. 6; **30.** 86; **31.** 342; **32.** 732; **33.** 693;
34. 67; **35.** 93; **36.** 0; **37.** −7; **38.** 63;
39. 14; **40.** −84; **41.** 0; **42.** −9 **43.** 76;
44. −23; **45.** −53; **46.** −6; **47.** 6;
48. −17; **49.** −27; **50.** 19; **51.** 3; **52.** 14;
53. 7; **54.** −2; **55.** −16; **56.** 2; **57.** 10;
58. −14; **59.** −32; **60.** −9; **61.** 0;
62. 2; **63.** 25; **64.** −14; **65.** −6; **66.** 14;
67. 12; **68.** −2; **69.** 0; **70.** 12; **71.** 0;
72. −19; **73.** 12; **74.** −4; **75.** −2;
76. −21; **77.** 72; **78.** 0; **79.** 72;
80. 18; **81.** −72; **82.** −162; **83.** −72;
84. 126; **85.** −180; **86.** 154; **87.** 0;
88. 96; **89.** −80; **90.** −90; **91.** −108;
92. 5; **93.** −8; **94.** 5; **95.** −9; **96.** −5;
97. 19; **98.** −5; **99.** −10; **100.** Undefined;
101. −34; **102.** −62; **103.** −23; **104.** 11;

105. -14; **106.** -49; **107.** 49; **108.** 4;
109. -28; **110.** 110; **111.** 12; **112.** -6;
113. -4; **114.** 0; **115.** 66; **116.** 0;
117. -44; **118.** -8; **119.** 27; **120.** -9;
121. -19; **122.** -4; **123.** 1; **24.** 2
125. 4; **126.** 4; **127.** 8; **128.** 3;
129. No Solution; **130.** Any real number.

CHAPTER 4 PRACTICE TEST

1. 0; **2.** $|-6|$; **3.** $-|-7|$; **4.** $-(-47)$;
5. 56; **6.** -10; **7.** 0; **8.** -30; **9.** -8;
10. -120; **11.** Undefined; **12.** -6;
13. -18; **14.** 252; **15.** 44; **16.** 1;
17. -36; **18.** 36; **19.** 0; **20.** 32;
21. 32; **22.** 11; **23.** -7; **24.** -15;
25. 8; **26.** 18; **27.** No Solution; **28.** 9;
29. 10; **30.** -8.

CHAPTER 5

Section 5.1 (page 129)

1. 3; **3.** 1; **5.** 8; **7.** 3; **9.** 13; **11.** 3;
13. 2; **15.** 0; **17.** 20; **19.** 5; **21.** Proper;
23. 6; **25.** Proper; **27.** Proper; **29.** $8\frac{1}{7}$;
31. Negative; **33.** Positive; **35.** Negative;
37. Positive; **39.** Negative; **41.** $3\frac{2}{5}$; **43.** $3\frac{2}{3}$;
45. $10\frac{6}{7}$; **47.** $-3\frac{1}{3}$; **49.** $-10\frac{2}{3}$; **51.** $35\frac{1}{3}$;
53. $\frac{31}{4}$; **55.** $\frac{65}{9}$; **57.** $\frac{14}{3}$; **59.** $\frac{19}{5}$.

Section 5.2 (page 134)

1. $\frac{2}{6}$; **3.** $\frac{1}{6}$; **5.** $\frac{2}{10}$; **7.** $\frac{2}{10}$; **9.** $\frac{3x}{12x}$;
11. $\frac{4}{12x}$; **13.** $3y$; **15.** $2a$; **17.** $\frac{1}{2}$; **19.** $\frac{2}{3}$;
21. $\frac{4}{5}$; **23.** $\frac{1}{9}$; **25.** $\frac{1}{2}$; **27.** $\frac{1}{4}$; **29.** $\frac{36}{25}$;
31. $\frac{x}{y}$.

Section 5.3 (page 140)

1. $\frac{8}{15}$; **3.** $\frac{6}{35}$; **5.** -1; **7.** $-\frac{5}{2}$ or $-2\frac{1}{2}$; **9.** $-\frac{2}{3}$;
11. $-\frac{9}{5}$ or $-1\frac{4}{5}$; **13.** $-\frac{2}{15}$; **15.** 4; **17.** $\frac{3}{2}$ or $1\frac{1}{2}$;
19. $-\frac{1}{2}$; **21.** 1; **23.** $\frac{3}{4}$; **25.** $\frac{1}{4}$; **27.** $\frac{1}{16}$;
29. $-\frac{1}{18}$; **31.** $\frac{ac^2}{b}$; **33.** $11\frac{2}{3}$; **35.** 33;

37. $-\frac{121}{225}$; **39.** 1; **41.** $\frac{3ac}{5b}$; **43.** $\frac{7}{16}$; **45.** $\frac{3}{5}$;
47. $1\frac{3}{4}$ cups; **49.** $\frac{1}{4}$ of the class gets an A;
51. $\frac{3}{8}$ of a ton; **53.** $7\frac{1}{2}$ cups; **55.** $\frac{7}{32}$ got an A;
57. $\frac{63}{80}$ of a ton; **59.** 2079 fish.

Section 5.4 (page 146)

1. $\frac{5}{6}$; **3.** $-\frac{2}{3}$; **5.** $-\frac{1}{18}$; **7.** -1; **9.** $-\frac{1}{4}$;
11. $-\frac{3}{7}$; **13.** $-\frac{2}{27}$; **15.** $-\frac{2}{3}$; **17.** 1;
19. $\frac{1}{4}$; **21.** $\frac{x^3y^2}{z^3}$; **23.** $\frac{a^2b^4}{c^3d^3}$; **25.** $\frac{4}{3}$ or $1\frac{1}{3}$;
27. $-\frac{10}{7}$ or $-1\frac{3}{7}$; **29.** 2; **31.** $\frac{5}{4}$ or $1\frac{1}{4}$;
33. 35 shirts; **35.** 10 measures; **37.** 32 shirts;
39. 22 measures.

Section 5.5 (page 157)

1. 1; **3.** $\frac{3}{7}$; **5.** $-\frac{1}{3}$; **7.** $\frac{1}{2}$; **9.** $-\frac{1}{35}$;
11. $\frac{7}{9}$; **13.** $\frac{3}{4}$; **15.** 1; **17.** $\frac{23}{6}$ or $3\frac{5}{6}$;
19. $-\frac{1}{30}$; **21.** $-\frac{1}{16}$; **23.** $\frac{13}{6}$ or $2\frac{1}{6}$;
25. $-\frac{83}{12}$ or $-6\frac{11}{12}$; **27.** $-\frac{8}{15}$; **29.** $\frac{17}{42}$;
31. $\frac{13}{3}$ or $4\frac{1}{3}$; **33.** $\frac{11}{5}$ or $2\frac{1}{5}$; **35.** $\frac{23}{30}$;
37. $1\frac{1}{8}$ cups; **39.** $\frac{3}{10}$ of its income; **41.** $3\frac{7}{12}$ mi;
43. $3\frac{5}{8}$ points; **45.** $3\frac{2}{3}$ in.

Section 5.6 (page 163)

1. $\frac{7}{6}$ or $1\frac{1}{6}$; **3.** $\frac{2}{5}$; **5.** $8\frac{2}{3}$; **7.** $\frac{4}{45}$; **9.** $\frac{3}{4}$;
11. $\frac{5}{16}$; **13.** $-\frac{1}{15}$; **15.** $\frac{241}{120} = 2\frac{1}{120}$;
17. $-\frac{68}{15}$ or $-18\frac{2}{5}$; **19.** $\frac{83}{82}$ or $1\frac{1}{82}$; **21.** $-\frac{13}{3}$ or $-4\frac{1}{3}$;
23. $\frac{51}{20}$ or $2\frac{11}{20}$; **25.** $-\frac{13}{5}$ or $-2\frac{3}{5}$; **27.** $\frac{39}{28}$ or $1\frac{11}{28}$.

Section 5.7 (page 168)

1. $x = 5$; **3.** $q = 5$; **5.** $z = 10$; **7.** $x = -7$;
9. $y = -8$; **11.** $m = -3$; **13.** $s = -4$;
15. $z = 13$; **17.** $x = 12$; **19.** $y = -8$;
21. $x = 28$; **23.** $y = -44$; **25.** $x = 36$;
27. $x = \frac{29}{98}$; **29.** $y = -\frac{155}{8}$ or $-19\frac{3}{8}$;
31. 229 calories.

CHAPTER 5 REVIEW

1.–6. See Section 5.1; **7.** See Section 5.2;
8.–9. See Section 5.4; **10.** See Section 5.5;
11. See Section 5.6; **12.** See Section 5.2
13. See Section 5.7;
14. Numerator 1, denominator 6;
15. Numerator 23, denominator 4;
16. Numerator 7, denominator 9;
17. Numerator 9, denominator 8;
18. Numerator 12, denominator 7;
19. Numerator 0, denominator 5;
20. Numerator 7, denominator 1;
21. Numerator 1, denominator 12;
22. Numerator 5, denominator 2;
23. Numerator 37, denominator 86; **24.** 9;
25. 9; **26.** 12; **27.** $6x$; **28.** 5; **29.** 1;
30. 7; **31.** 1; **32.** 5; **33.** 1; **34.** $\frac{1}{2}$; **35.** $\frac{1}{3}$;
36. $\frac{3a}{4b}$; **37.** $\frac{15x}{16y}$; **38.** $8x$; **39.** $\frac{x}{2}$; **40.** $\frac{1}{2}$;
41. $\frac{5}{9}$; **42.** $\frac{8}{3}$ or $2\frac{2}{3}$; **43.** $\frac{55}{3}$ or $18\frac{1}{3}$; **44.** $\frac{85}{7}$ or $12\frac{1}{7}$;
45. $\frac{5xy}{44}$; **46.** -1; **47.** $\frac{1}{2}$; **48.** xy; **49.** $\frac{1}{3}$;
50. $-\frac{1}{49}$ **51.** $\frac{1}{4}$; **52.** 1518 students; **53.** $\frac{2025}{77}$ or
$26\frac{23}{77}$; **54.** \$2; **55.** $\frac{8}{75}$; **56.** $\frac{3}{2}$ or $1\frac{1}{2}$; **57.** $-\frac{49}{96}$;
58. $\frac{2}{3}$; **59.** 4; **60.** 4; **61.** $-\frac{1}{22}$; **62.** $-\frac{25}{144}$;
63. $-\frac{7}{2}$ or $-3\frac{1}{2}$; **64.** $11\frac{1}{4}$ cans; **65.** $\frac{1}{3}$; **66.** 14
omelettes; **67.** $\frac{5}{4}$ or $1\frac{1}{4}$; **68.** $\frac{2}{3}$; **69.** $\frac{7}{4}$ or $1\frac{3}{4}$;
70. $\frac{17}{16}$ or $1\frac{1}{16}$; **71.** $-\frac{3}{35}$; **72.** $\frac{7}{24}$; **73.** $-\frac{8}{3}$ or $-2\frac{2}{3}$;
74. $-\frac{1}{16}$; **75.** $\frac{17}{12}$ or $1\frac{5}{12}$; **76.** $\frac{51}{40}$ or $1\frac{11}{40}$;
77. $\frac{19}{6}$ or $3\frac{1}{6}$; **78.** $\frac{7}{12}$; **79.** $\frac{1}{3}$; **80.** $\frac{7}{6}$ or $1\frac{1}{6}$;
81. $\frac{7}{5}$ or $1\frac{2}{5}$; **82.** $\frac{20}{27}$; **83.** 2; **84.** $\frac{14}{5}$ or $2\frac{4}{5}$;
85. 6; **86.** $1\frac{2}{9}$; **87.** $\frac{37}{20}$ or $1\frac{17}{20}$; **88.** $\frac{17}{11}$ or $1\frac{6}{11}$;
89. $\frac{13}{10}$ or $1\frac{3}{10}$; **90.** 5; **91.** $y = 18$;
92. $x = -17$; **93.** $t = -4$; **94.** $z = 160$;
95. $w = -30$; **96.** $x = -\frac{3}{4}$.

CHAPTER 5 PRACTICE TEST

1. a. 3, **b.** 9, **c.** 17, **d.** 1, **e.** 4;
2. a. 0, **b.** 19, **c.** 3, **d.** 5, **e.** 7;
3. a. $-$, **b.** $+$, **c.** $+$, **d.** $-$; **4.** 56; **5.** $60x$;
6. $\frac{8}{9}$; **7.** $\frac{1}{2}$; **8.** $\frac{x}{8y}$; **9.** $\frac{5}{3}$ or $1\frac{2}{3}$; **10.** $-\frac{6}{17}$;
11. $\frac{25}{24}$ or $1\frac{1}{24}$; **12.** $\frac{2}{3}$; **13.** $-\frac{13}{35}$; **14.** $\frac{7}{20}$;
15. $-\frac{50}{9}$ or $-5\frac{5}{9}$; **16.** $-\frac{5}{3}$ or $-1\frac{1}{3}$; **17.** $\frac{2}{3}$;

18. 3; **19.** $\frac{19}{12}$ or $1\frac{7}{12}$; **20.** $-\frac{5}{3}$ or $-1\frac{2}{3}$; **21.** $\frac{31}{12}$
or $2\frac{7}{12}$; **22.** $\frac{67}{24}$ or $2\frac{19}{24}$; **23.** $-\frac{38}{15}$ or $-2\frac{8}{15}$;
24. $\frac{1}{3}$; **25.** $\frac{19}{28}$ of the students will earn an A, B, or C;
26. 87 will earn an A; **27.** 84 bottles; **28.** $\frac{5}{12}$ left;
29. $x = 12$; **30.** $3 = x$; **31.** $x = \frac{16}{9}$ or $1\frac{7}{9}$;
32. $x = \frac{9}{2}$ or $4\frac{1}{2}$.

CHAPTER 6

Section 6.1 (page 161)

1. 3^2; **3.** 9^2 or 3^4; **5.** $\left(\frac{2}{3}\right)^2$; **7.** $\left(\frac{4}{9}\right)^2$; **9.** $\frac{1}{x^4}$;
11. $(a^2b)^2$; **13.** $\left(\frac{3pq}{z^3}\right)^2$; **15.** 1; **17.** 1;
19. 625; **21.** $-2{,}985{,}984$; **23.** $-537{,}824$;
25. -512; **27.** $-c^2$; **29.** a^4; **31.** $-f^3$;
33. $\frac{8}{27}$; **35.** $-\frac{25}{36}$; **37.** $\frac{1}{4}$; **39.** $\frac{1}{216}$; **41.** $\frac{1}{x^2}$;
43. $\left(\frac{1}{8}\right)q^3$; **45.** $324x^4y^2z^6$; **47.** $\frac{125}{2}q^4$ or $62\frac{1}{2}q^4$;
49. $\frac{282{,}475{,}249}{1{,}073{,}741{,}824}$; **51.** $-\frac{32}{243}$; **53.** $\frac{1}{3{,}486{,}784{,}401}$;
55. $\frac{1}{531{,}441}$.

Section 6.2 (page 188)

1. -1, 2nd; **3.** $\frac{2}{3}$, 3rd; **5.** 27, 0; **7.** $\frac{15}{31}$, 0;
9. 3, 1st; **11.** Binomial, 1st; **13.** Trinomial; 2nd;
15. Binomial; 3rd; **17.** $5a + 31$; **19.** $-b + 14$;
21. $-4x - y$; **23.** $4t^2 + 8t + 5$;
25. $7a^4 + a^3 + 2a^2 - 3a + 8$; **27.** $11t^2 + 4t + 4$;
29. $7x^4 - 9x^2 + 4$; **31.** $a^3 - 6a^2 - 7a + 4$;
33. $4x^2 + x - 7$; **35.** $15x^5$; **37.** $-48m^4$;
39. $27z^4$; **41.** $q^2 + 7q$; **43.** $-3z^3 - 3z^2$;
45. $3x^3 + 7x^2 + 5x$; **47.** $-6y^3 - 21y^2 - 3y$;
49. $-a^2b^2 - ab + 3a$; **51.** $-44x^3 - 34x^2 - 10x$;
53. $x^2 + 11x + 30$; **55.** $2y^2 + 9y + 7$;
57. $18r^2 + 36r + 10$; **59.** $14 + 5y - y^2$;
61. $3a^2 - 7ab + 2b^2$; **63.** $6t^2 - 7t - 20$;
65. $q^2 - 4$; **67.** $x^2 + \left(\frac{3}{4}\right)x + \frac{1}{8}$; **69.** $(x + 4)$;
71. $2(3z^2 + 9z + 2)$; **73.** $2b(6a^2 + 1)$;
75. $5ab(b + 3a - 7)$; **77.** $3(-x^2 + 2x - 8)$;
79. $9(3 - h^2)$.

Section 6.3 (page 194)

1. $\frac{y}{x}$; **3.** $\frac{3b}{2a}$; **5.** $-\frac{10}{x}$; **7.** $\frac{1}{4}$; **9.** $\frac{6}{ab}$;
11. $\frac{xy^4}{4}$; **13.** $\frac{1}{xy}$; **15.** $\frac{t^3}{3s^2}$; **17.** $3xy + 6y$;
19. $\frac{8}{a}$.

Section 6.4 (page 197)

1. $\frac{5c}{3}$;   **3.** $\frac{2f}{7}$;   **5.** $\frac{3x}{7}$;   **7.** $\frac{7}{a}$;   **9.** $\frac{x}{5}$;

11. $\frac{2x+5}{4}$;   **13.** $\frac{t-1}{5}$;   **15.** $\frac{x+6}{8}$;   **17.** $\frac{6-x}{3x}$;

19. $\frac{5}{24x}$;   **21.** $\frac{5x+17}{6}$;   **23.** $\frac{3c-2b}{bc}$;

25. $\frac{2x^2+3y^2}{xy}$;   **27.** $\frac{xy+6y^3+10x^3}{2x^2y^2}$;   **29.** $\frac{10s^2-st+33t^2}{11st}$.

Section 6.5 (page 200)

1. 5;   **3.** 1;   **5.** 64;   **7.** -59;   **9.** 15;

11. 41;   **13.** 4;   **15.** 13;   **17.** 55;   **19.** $-\frac{15}{8}$;

21. $\frac{59}{54}$ or $1\frac{5}{54}$;   **23.** $10\frac{85}{108}$;   **25.** $3\frac{11}{60}$;   **27.** $\frac{2}{3}$;

29. $-\frac{124}{81}$ or $-1\frac{43}{81}$;   **31.** 48;   **33.** 28.26;   **35.** 2;

37. $12.2°$ C.

CHAPTER 6 REVIEW

1.–4. See Section 6.1;   **5.** See Section 6.2;
6.–9. See Section 6.1;   **10.–12.** See Section 6.2;
13.–14. See Section 6.4;   **15.** See Section 6.5;
16. $\frac{5^3}{2^5}$;   **17.** $\frac{x^3}{3^3}$;   **18.** $-\frac{125}{216}$;
19. Trinomial, 2nd degree;   **20.** Monomial, 0 degree;
21. Binomial, 1st degree;   **22.** Monomial, 3rd degree;
23. $19y-13$;   **24.** $-2x+18$;
25. $7x^2-15x+4$;   **26.** $5y^2-7y-15$;
27. $-15y^3+21y^2-39y$;   **28.** $15z^3+5z^2+10z$;
29. $-2x^3-x^2+x$;   **30.** $z^2+11z+30$;
31. $27x^2+3x-2$;   **32.** q^2-4;
33. $6x^2-13xy+6y^2$;   **34.** $2(3w+6x-2y)$;
35. $3y(4+y)$;   **36.** $5xy(-3x+5y-1)$;
37. $-\frac{4}{5xy}$;   **38.** $-\frac{104h}{g}$;   **39.** $\frac{2y}{x^2}$;   **40.** $\frac{3hj}{2}$;

41. $\frac{6r}{s}$;   **42.** $\frac{1}{2r}$;   **43.** $10d^2$;   **44.** $\frac{5}{a}$;   **45.** $\frac{2}{xy}$;

46. $\frac{8x}{c}$;   **47.** $\frac{2y}{x}$;   **48.** $\frac{a+2}{b}$;   **49.** $\frac{5a}{6}$;   **50.** $\frac{b}{10}$;

51. $\frac{3x+2x^2}{12}$;   **52.** $\frac{37r}{6s}$;   **53.** $\frac{3-y}{y}$;   **54.** $\frac{6}{w}$;

55. $\frac{17y+8x^2+10xy}{x^2y}$;   **56.** 7, -1;   **57.** 41, 16;

58. 19, 1639;   **59.** $14,094\frac{3}{8}$;   **60.** 4.

CHAPTER 6 PRACTICE TEST

1. $17x-y$;   **2.** $2a^2+9ab-5b^2$;
3. $6x^2+13xy+6y^2$;   **4.** c^2-16d^2;
5. $5x^4+21x^3-11x^2$;   **6.** $-2y^2+20y-10$;
7. $a^3b+3a^2b^2+ab^3$;   **8.** $-6v-7w$;

9. $-28a+96b-25c$;   **10.** $4x^2+14x-12$;
11. $\frac{15y+12x}{xy}$;   **12.** $\frac{3x-2}{2xy}$;   **13.** $\frac{-a+7b}{10}$;

14. $-50x^2+12x-66$;   **15.** $\frac{x^2}{y^2}$;   **16.** $\frac{24a^3}{b^3}$;

17. $\frac{3}{m^2}$;   **18.** $-\frac{16}{9rs^4}$;   **19.** $\frac{4a}{b}$;   **20.** $16q^6$;

21. $\frac{1}{r^4}$;   **22.** $\frac{8g^3}{27h^3}$;   **23.** x^{17};   **24.** $-25x^2$;

25. 1;   **26.** $\frac{1}{16a^2b^6}$;   **27.** $9q^2$;   **28.** $\frac{5}{x^7}$;

29. -9;   **30.** 2;   **31.** 14;   **32.** 131;   **33.** 6;

34. $\frac{1}{10}$;   **35.** $\frac{27}{64}$;   **36.** $\frac{1}{108}$;

37. Monomial, 0 degree;   **38.** Binomial, 1st degree;
39. Trinomial, 4th degree;   **40.** Binomial, 2nd degree.

CUMULATIVE REVIEW CHAPTERS 4, 5, & 6

1. Variable;   **2.** Identity Property of Multiplication;
3. Trinomial;   **4.** Addition Property of Equality;
5. Polynomial;   **6.** Property of Additive Inverses;
7. Binomials;   **8.** Distributive Property;
9. Multiplication Property of Equality;
10. Property of Reciprocals;   **11.** 24;   **12.** $65\frac{7}{8}$;

13. $3\frac{4}{7}$;   **14.** 7;   **15.** 0;   **16.** Undefined;

17. 24;   **18.** -21;   **19.** $-\frac{7}{44}$;   **20.** 11;

21. -33;   **22.** $\frac{1}{2}$;   **23.** 9^5;   **24.** $\frac{23}{28}$;

25. $3\frac{1}{20}$;   **26.** $\frac{7}{11}$;   **27.** 37;   **28.** 2;

29. $3m^5-9m^3-5$;   **30.** $32x^4y^2$;   **31.** $9x^6y^2$;
32. x^2+3x+2;   **33.** $4a^2+2a-2$;
34. $x(2x^2+3)$;   **35.** 1;   **36.** x;   **37.** k^2m^{14};

38. $\frac{1}{x}$;   **39.** $\frac{x}{12}$;   **40.** $\frac{n}{60}$;   **41.** $\frac{208y}{11}$;   **42.** $\frac{26a}{35}$;

43. $\frac{(4x+6)}{x^3}$;   **44.** $\frac{(b+a)}{ab}$;   **45.** $a=-17$;

46. Any real number;   **47.** $y=-15$;   **48.** $x=8$.

CHAPTER 7

Section 7.1 (page 219)

1. Hundred;   **3.** Tenths;   **5.** Ten thousandths;
7. Thousand;   **9.** Thousandths;   **11.** Five hundredths;
13. Negative fifty and twenty-three hundredths;
15. Twenty-three ten thousandths;
17. Four and six tenths;
19. Seven and eight hundred seventy-five thousandths;
21. Negative five thousand, three hundred and thirty-five ten thousandths;
23. Four hundred sixty-three and four hundred sixty-three ten thousandths;

25. 0.04; **27.** 0.0004; **29.** −0.067;
31. −8004.02; **33.** −400.42; **35.** 16,000.016;
37. 0.0099; **39.** 1.4 > 1.04; **41.** 4.03 > 4;
43. 0.58 = 0.580; **45.** 0.96 > 0.149;
47. 0.33 > 0.3; **49.** 0.3 = 0.3;
51. −0.38 > −0.381; **53.** −0.495 > −0.4995;

	Number	Hundreds	Tens	Units	Tenths	Hundredths
55.	6.529	0	10	7	6.5	6.53
57.	0.04	0	0	0	0.0	0.04
59.	56.192	100	60	56	56.2	56.19
61.	76.001	100	80	76	76.0	76.00

63. 78,000; **65.** 0.0000846; **67.** 0.00021;
69. 39,100,000; **71.** 8.63×10^{-5}; **73.** 2×10^3;
75. 6.8×10^3; **77.** 3.86×10^{-3};
79. 2.07×10^8 tons of trash; **81.** 136,000,000 mi;
83. 1.0×10^{-8} of an in; **85.** 2.24×10^6 dollars;
87. 256,000,000 people; **89.** 610,000,000 mi;
91. 4.0×10^{-7} of an in;
93. 0.00000000000000000000000000665
− 0.00000000000000000000000000000919
= 0.00000000000000000000000006649081;
95. 1.3×10^{-5} erg; **97.** 3.5×10^{-12} g.

Section 7.2 (page 225)

1. 5.69; **3.** 12.16; **5.** 132.472; **7.** 8.742;
9. 93.1855; **11.** 10.04; **13.** 0.48; **15.** 9.23;
17. 1780.9001; **19.** −13.2041; **21.** 77.3076;
23. 77.659; **25.** −52.54; **27.** −529.704;
29. 14.25; **31.** 27.91; **33.** −1.71;
35. 0.843; **37.** 29.218; **39.** −80.74;
41. 67.655; **43.** 143.89; **45.** 0; **47.** 105.009;
49. −735.94; **51.** −97.26; **53.** 99.137;
55. −89.64; **57.** 8.6; **59.** −2.4; **61.** 2.32;
63. −37.4; **65.** −2.15; **67.** 14.57; **69.** −26.976;
71. 14.85; **73.** 3.068; **75.** 42.56; **77.** −38.37;
79. 306.65; **81.** 3.624; **83.** $161.67;
85. 1.09849 min; **87.** 54.86 yd², 4.86 yd²; **89.** 25.2 in.

Section 7.3 (page 231)

1. 0.63; **3.** 0.99; **5.** −0.2; **7.** 0.024;
9. 0.024; **11.** 0.0024; **13.** 0.00024;
15. 0.42; **17.** −0.000000009; **19.** 347.028;
21. −0.6489; **23.** 4.746; **25.** 0.04746;
27. −7543; **29.** 98999; **31.** −54.2;
33. 0.005; **35.** 0.5; **37.** 21.9234; **39.** 6.93;
41. 26.27; **43.** 15; **45.** 9.797;
47. −32.5241; **49.** 3.738; **51.** 0.0602;
53. 33.27; **55.** 0.0468; **57.** 14.68;
59. 642.06 meters of fabric would cost $3203.88;
61. 90.18 min;
63. 252.45 yards of fabric would cost $2259.43;
65. $20.35; **67.** $5; **69.** 31.28 in.

Section 7.4 (page 238)

1. 1.78; **3.** 58.29; **5.** 300.008; **7.** 40;
9. 4; **11.** 0.04; **13.** 0.0004; **15.** 400;
17. 40,000; **19.** −409.01; **21.** 45.9;
23. 405; **25.** −909; **27.** 4.683; **29.** 85.36;
31. 53.8646; **33.** 83.48; **35.** −0.08;
37. −23.78; **39.** 38.26; **41.** 25.83;
43. 1.11; **45.** 1.09; **47.** 0.595; **49.** 6.94;
51. 34 cents each; **53.** $60.47 each;
55. $8.43 each ticket; **57.** 13.805; **59.** 0.04;
61. −18.10; **63.** 2.42; **65.** −49.19;
67. 0.88; **69.** −2.85; **71.** 24.11.

Section 7.5 (page 240)

1. 0.5; **3.** 0.25; **5.** 1.5; **7.** $0.\overline{6}$; **9.** $3.8\overline{3}$;
11. $0.\overline{428571}$; **13.** $2.08\overline{3}$; **15.** 2.125;
17. 0.05; **19.** 0.15; **21.** 0.8; **23.** 0.75;
25. $0.1\overline{6}$; **27.** $0.8\overline{3}$; **29.** $0.\overline{1176470588235294}$;
31. $\frac{7}{10}$; **33.** $\frac{4}{5}$; **35.** $\frac{1}{20}$; **37.** $\frac{1}{2000}$; **39.** $2\frac{3}{4}$;
41. $\frac{3}{200}$; **43.** $\frac{7}{100}$; **45.** $\frac{3}{250}$; **47.** $\frac{17}{50}$;
49. $3\frac{1}{8}$; **51.** $3\frac{3}{25}$; **53.** 236; **55.** $7\frac{1}{250}$;
57. $\frac{1}{50}$; **59.** $\frac{1}{5000}$; **61.** 1.25; **63.** $\frac{7}{12}$ or $0.58\overline{3}$;
65. 0; **67.** 2.1; **69.** $\frac{1}{6}$ or $0.1\overline{6}$; **71.** $\frac{8}{15}$ or $0.5\overline{3}$;
73. $\frac{11}{15}$ or $0.7\overline{3}$; **75.** 1.476; **77.** −1; **79.** 0.72;
81. $1\frac{19}{24}$ or $1.791\overline{6}$; **83.** −0.625; **85.** $5.1\overline{3}$ or $5\frac{2}{15}$;
87. −10.1; **89.** 2.125 = $2\frac{1}{8}$ in.

CHAPTER 7 REVIEW

1.–4. See Section 7.1; **5.** Hundredths;
6. Tenths; **7.** Ten thousandths; **8.** Units;
9. Hundredths; **10.** Thousandths;
11. Ten millionths; **12.** Ten; **13.** 0.06 < 0.6;
14. 0.3 < $0.\overline{3}$; **15.** −0.6 > −0.65;
16. 1.7 > 1.17; **17.** 0.0004 < 0.004;
18. −0.72 > −0.73; **19.** 9.01 < 9.1;
20. 0.0006 > 0.00006; **21.** 27.027; **22.** 400.07;
23. 5,000.005; **24.** 302.02; **25.** 0.763;
26. Seventy-five and seventy-five hundredths;
27. Nine hundred thirty and three thousandths;
28. Four and five thousand, two ten thousandths;
29. Six hundred seventy-eight hundred thousandths;
30. Twelve and four thousand, four ten thousandths;
31. 6,730,000; **32.** 0.000024; **33.** 0.00608;
34. 7,000,000,000; **35.** 7.86×10^6;
36. 3×10^3; **37.** 9.8×10^{-7};
38. 9.684×10^{-3}; **39.** 36.3; **40.** 790;
41. 0.837; **42.** 100.00; **43.** 100; **44.** 700.0;
45. 382.23; **46.** 6.33; **47.** 2.5056;
48. 296.973; **49.** −676.7; **50.** 271.042;
51. −345.77; **52.** 972.99; **53.** 2.733;
54. 16.93; **55.** −91.54; **56.** −12.82;

57. 213.048;　**58.** −11.54;　**59.** 6.32;
60. −13.46;　**61.** 0.36;　**62.** 0.1683;
63. 0.0000002;　**64.** −10.881;　**65.** 1.0252;
66. 39,870;　**67.** 0.98;　**68.** 79009.87;
69. 26.2;　**70.** 75.8;　**71.** 0.24;　**72.** 18.8;
73. 5.2;　**74.** $4.0\overline{3}$;　**75.** 0.07;　**76.** 2.39;
77. 0.00;　**78.** 0.8;　**79.** $0.\overline{6}$;　**80.** $0.\overline{18}$;
81. 0.875;　**82.** 2.5;　**83.** 0.43;　**84.** 0.07;
85. 0.375;　**86.** $\frac{1}{4}$;　**87.** $2\frac{1}{8}$;　**88.** $\frac{1}{200}$;

89. $\frac{61}{100}$;　**90.** $\frac{3}{8}$;　**91.** $1\frac{3}{4}$;　**92.** $\frac{7}{8}$;　**93.** $3\frac{1}{2}$;
94. $1.91\overline{6}$;　**95.** 2.1875;　**96.** $-2.1\overline{6}$;
97. $2.91\overline{6}$;　**98.** 1;　**99.** 7.4;　**100.** −16.5;
101. 17.2.

CHAPTER 7 PRACTICE TEST

1. 0 is in the tenths place, 4 is in the hundredths place;

2. $0.\overline{6}$;　**3.** $\frac{1}{200}$;　**4.** −0.005;　**5.** 470.077;

6. $0.\overline{4}$;

7. Nine thousand, four and four hundred three ten thousandths;

8. 700;　**9.** $\frac{7}{8}$;　**10.** 939.9;　**11.** 0.429;

12. 654.723;　**13.** 26.81;　**14.** 277.104;
15. −409.8;　**16.** −4926.6;　**17.** 679.478;
18. 155.76;　**19.** −57.47;　**20.** 3.51;

21. 3.75;　**22.** −5.98;　**23.** $2\frac{2}{5}$;　**24.** 0.35;

25. 1;　**26.** 7.65×10^{-6};　**27.** 2,480,000;
28. $1.97 each;　**29.** $3.26;　**30.** $92.97.

CHAPTER 8

Section 8.1 (page 261)

1. $a = -13$;　**3.** $c = 5$;　**5.** $z = 39$;
7. $x = -22$;　**9.** $y = -12$;　**11.** $x = -3$;
13. $x = 6$;　**15.** $y = 8$;　**17.** $x = -1$;
19. $x = 3$;　**21.** $x = 3$;　**23.** $y = -4$;
25. $z = -\frac{11}{10}$;　**27.** $t = \frac{3}{11}$;　**29.** $y = 1$;
31. $x = -2$;　**33.** $y = 9$;　**35.** $x = -1$;
37. $x = 19$;　**39.** $q = \frac{15}{7}$;　**41.** $y = -1$;
43. $x = -11$;　**45.** $s = 12$.

Section 8.2 (page 265)

1. $x = -\frac{7}{3}$;　**3.** $y = \frac{11}{3}$;　**5.** $x = 13$;

7. $y = 14$;　**9.** $x = -\frac{1}{16}$;　**11.** $x = \frac{1}{6}$;

13. $y = -3$;　**15.** $x = \frac{10}{3}$;　**17.** $x = 8$;

19. $x = -\frac{14}{5}$;　**21.** $x = \frac{20}{53}$;　**23.** $y = -\frac{35}{11}$;

25. $y = \frac{15}{13}$;　**27.** $x = 12$;　**29.** $y = 11$;

31. $v = -\frac{21}{10}$;　**33.** $x = \frac{2}{3}$;　**35.** $y = \frac{56}{9}$;

37. $x = \frac{5}{7}$;　**39.** $y = \frac{8}{3}$;　**41.** $x = -\frac{86}{3}$;

43. $x = \frac{13}{3}$;　**45.** $x = -\frac{127}{24}$;　**47.** $q = -\frac{53}{40}$;

49. $x = \frac{7}{16}$.

Section 8.3 (page 270)

1. $5.624 = x$;　**3.** $2.723 = y$;　**5.** $z = 0.125$;
7. $q = -6$;　**9.** $-0.06 = s$;　**11.** $16.363 = x$;
13. $x = 12$;　**15.** $h = -31.5$;　**17.** $c = -10$;

19. $x = \frac{340}{19}$;　**21.** $y = \frac{70}{3}$;　**23.** $x = -8.5$;

25. $y = 0.5$;　**27.** $z = \frac{10}{7}$;　**29.** $q = \frac{11}{14}$;

31. $s = -14.4$;　**33.** $r = \frac{6}{5}$.

Section 8.4 (page 278)

1. $\{y \mid y = 56\}$;　**3.** $\{x \mid x = -17\}$;
5. $\{r \mid r \le 18\}$;　**7.** $\{c \mid c > -51\}$;　**9.** $\{x \mid x \ge 0\}$;
11. $x < 5$　　　　**13.** $x \ge 1$

15. $x \le 8$　　　　**17.** $x > 12$

19. $x \ge -4$　　　　**21.** $x > -8$

23. $x \le 2$　　　　**25.** $x \le 8$

27. $x < -1$　　　　**29.** $x \ge 3$

31. $x \ge 3$　　　　**33.** $x > -2$

35. $x \ge 1$　　　　**37.** $x > 0$

39. $x \le 0$　　　　**41.** $5 > x$

43. $-2 \ge x$

CHAPTER 8 REVIEW

1. See Section 8.1; **2.** See Section 8.1; **3.** See Section 8.4; **4. a.** Multiplication Property of Equality, **b.** Addition Property of Equality, **c.** Addition Property of Inequalities; **5. a, b, c.** See Section 8.1, **d. and e.** See Section 8.3, **f.** See Section 8.1, **g. and h.** See Section 8.4; **6.** $k = 24$ **7.** $x = 4$; **8.** $y = 8$; **9.** $a = 3$; **10.** No solution; **11.** All real numbers; **12.** $4 = r$; **13.** $y = 18$; **14.** $x = -17$; **15.** $t = -4$; **16.** $z = 160$; **17.** $w = -30$; **18.** $x = -\frac{3}{4}$; **19.** $y = \frac{7}{9}$; **20.** $y = -8$; **21.** $x = -5$; **22.** $y = \frac{5}{4}$; **23.** $q = -\frac{1}{3}$; **24.** $h = -1$; **25.** $w = -1$; **26.** $x = -5$; **27.** $y = -\frac{1}{12}$; **28.** $t = -\frac{17}{6}$; **29.** $w = -5$; **30.** $q = \frac{4}{5}$; **31.** $z = -\frac{3}{8}$; **32.** $x = 4$; **33.** $x = -4.7$; **34.** $x = \frac{1}{2}$; **35.** $x = 2$; **36.** $x = \frac{63}{2}$; **37.** $y = -\frac{29}{7}$; **38.** $x = \frac{15}{4}$; **39.** $x = -\frac{5}{2}$; **40.** $\frac{23}{6} = y$; **41.** $w = -\frac{11}{20}$; **42.** $n = \frac{5}{2}$; **43.** $y = 5$;

44. $x > -4$

45. $x < -4$

46. $x \geq -5$

47. $x \leq 0$

48. $x \leq 8$

49. $x > -5$

50. $x > -4$

51. $x \geq 4$

52. $x \geq 1$

53. $x < -3$

54. $x \geq -2$

55. $x > -4$

56. $9 < x$.

CHAPTER 8 PRACTICE TEST

1. B; **2.** A; **3.** A; **4.** D; **5.** C; **6.** C; **7.** $z = 5$; **8.** $y = 15$; **9.** $x = 2$; **10.** No solution; **11.** $x > -2$; **12.** $x \geq 4$; **13.** $y = -2$; **14.** $t = \frac{9}{32}$; **15.** $a = 2$; **16.** Any real number; **17.** $b = -1$; **18.** $c = 15$; **19.** $s = 9$; **20.** $14 = x$; **21.** $x = 4$; **22.** $a = -\frac{28}{3}$; **23.** $x = \frac{77}{12}$; **24.** $x = -30$; **25.** $x = -\frac{1}{5}$; **26.** $x = 14$; **27.** -10; **28.** $x = -\frac{15}{4}$; **29.** $x \geq -7$; **30.** $x > -3$; **31.** $x = -4.7$; **32.** $v = -2$; **33.** $c = 15$; **34.** $n = 12$; **35.** $q = -6$; **36.** $y = 0$; **37.** $4 = b$; **38.** $-\frac{1}{3} = y$; **39.** $z = 4.3$; **40.** $t = \frac{1}{2}$; **41.** $x = \frac{7}{2}$; **42.** $y = \frac{4}{5}$; **43.** $h = -\frac{93}{98}$; **44.** $k = -\frac{23}{47}$. **45.** $x \leq 4$; **46.** $x < -8$; **47.** $x < 1$; **48.** $x \geq -4$; **49.** $x < -1$; **50.** $x \geq 5$

CHAPTER 9

Section 9.1 (page 291)

1. & 3. Answers will vary.

Section 9.2 (page 296)

1. The yearly subscription is a better buy;
3. $6(8 - 3) + 2$. Answers may vary;
5. DCD;
7. Jaron has 2 males; Cliff has 10;
9. $947 + 256$;
11. 2802 people and animals;
13. Maria is 35, attends Red College, and takes Math 501; Evelyn is 21, attends Olive Community College, and takes Engl 700. LaTisha is 19, attends Green State University, and takes Psych 205.

Section 9.3 (page 304)

1. Information: Number of children expected to be born this year. Number of children who have been born this year. Question: How may more births are expected this year? **3.** Information: The results of series of arithmetic operations which are performed on a number. Question: What is the number? **5.** Information: Tuition cost for last 20 years. Question: How much has tuition increased? **7.** Let y = life expectancy for men in years; $y + 6.8$ = life expectancy for women in years; **9.** Let I = average household income of single parents in 1990; **11.** 2.5 million more births can be expected; **13.** 9 is the number; **15.** Christian earns approximately $17,711.67; **17.** $21,400 was the average household income of single parents in 1990; **19.** $20.63 is the approximate cost of 5 roses; **21.** 140.83 miles is the approximate distance you have traveled; **23.** $71.40 is the total cost of the pizzas; **25.** Parents live 180 miles from school; **27.** Each side is 16 feet; **29.** Area is approximately 50.27 in^2.

Section 9.4 (page 308)

1. Suda has 2 dimes and 6 quarters; **3.** They have 4 nickels and 8 dimes; **5.** 5 children are under 12 years old; **7.** Answers will vary.

CHAPTER 9 REVIEW

1.–3. Answers will vary. **4.** $\frac{2}{3}$ hr or 40 min;

5. Approximately 706 mph; **7.** 6000 ft^2; **9.** Ohio has 10 million, Indiana has 5 million, California has 30 million; **12.** 73 adult tickets, 142 student tickets.

CHAPTER 9 PRACTICE TEST

1. Ken has 100 CDs; **2. a.** 13.5 gal, **b.** 283.5 mi; **3.** Approximately 4 hr 52 min; **4.** 40 boxes; **5.** 247 ft^2; **6.** 2; **7.** C; **8.** "Give me liberty or give me death."; **9.** At least 618 flyers; **10.** Third prize: $50, second prize: $100, first prize: $200; **11. a.** Answers will vary, **b.** Answers will vary; **12.** 6 quarters and 4 nickels; **13.** Width: 100 ft.

CUMULATIVE REVIEW: CHAPTERS 7, 8 & 9

1. Greater than or equal to; **2.** Less than; **3.** Addition Property of Equality; **4.** Irrational number; **5.** Distributive Property; **6.** Multiplication Property of Equality; **7.** Multiplication Properties of Inequalities; **8.** Scientific notation; **9.** Repeating decimal;

10. Property of Reciprocals; **11.** 315,300; **12.** 8.32×10^{-5}; **13.** $\frac{3}{200}$; **14.** $\frac{1}{3}$; **15.** $\frac{2}{3}$; **16.** $0.\overline{2}$; **17.** 0.15; **18.** 5.25; **19.** 0.5; **20.** 1.01; **21.** 0.67; **22.** 3.6344; **23.** 165.44; **24.** -138.064; **25.** $-1.2\overline{6}$ or $-1\frac{4}{15}$; **26.** 2.058; **27.** $a = -17$; **28.** Any real number; **29.** $y = -15$; **30.** $x = 8$; **31.** $x = \frac{13}{5}$; **32.** $n = 19$; **33.** $b = -5$; **34.** $a = 4$; **35.** $y = -30$; **36.** $x = -\frac{1}{2}$; **37.** $y = -\frac{5}{2}$; **38.** $x = 38\frac{1}{3}$ or $38.\overline{3}$;

39. $y \le 3$

40. $x < 2$

41. $x \ge -6$

42. $y < -2$

43. $5 > y$

44. 227.5 miles; **45.** 1000 ft per min; **46.** 1.8$\overline{3}$ GPA; **47. a.** 31, 63, 127, **b.** 5 nickels and 6 dimes.

CHAPTER 10

Section 10.1 (page 321)

1. $\frac{1}{2}$; **3.** $\frac{3}{2}$; **5.** $\frac{6}{7}$; **7.** $\frac{46}{21}$; **9.** $\frac{8}{5}$; **11.** $\frac{2}{5}$; **13.** $\frac{6}{5}$; **15.** $\frac{4}{5}$; **17.** $\frac{2}{3}$; **19.** $\frac{1}{7}$; **21.** $\frac{2}{1}$; **23.** $\frac{1}{5}$; **25.** $\frac{1}{2}$; **27.** $\frac{6}{1}$; **29.** $\frac{8}{7}$; **31.** $\frac{1}{8}$; **33.** $\frac{1}{30}$; **35.** $\frac{5}{2}$; **37.** 1; **39.** $\frac{37}{7}$; **41.** $\frac{100}{1}$; **43.** $\frac{3}{10}$; **45.** $\frac{7}{100}$; **47.** $\frac{7}{3}$; **49.** $\frac{32}{7}$; **51.** $\frac{12}{1}$; **53.** $\frac{21}{2}$; **55.** $\frac{1}{12}$; **57.** $\frac{5}{8}$; **59.** $\frac{6}{1}$; **61.** $\frac{3}{2}$; **63.** $\frac{3}{5}$; **65.** $\frac{4}{1}$; **67.** $\frac{24}{7}$; **69.** $\frac{120}{19}$; **71.** $\frac{5}{18}$; **73.** $\frac{4}{9}$; **75.** $\frac{1}{9}$; **77.** $\frac{4}{3}$; **79.** $\frac{3}{16}$; **81.** $\frac{5}{14}$; **83.** $\frac{7}{2}$; **85.** $\frac{1}{6}$; **87.** $\frac{1}{4}$; **89.** $\frac{12}{1}$; **91.** $\frac{5}{36}$; **93.** 55 mph; **95.** $5.79 per hr; **97.** 23 students per section; **99.** 4 oz per cubic ft; **101.** The 28 oz jar; **103.** 13 students per professor; **105.** 74 mph; **107.** $3378 per mo; **109.** 2.5 gpm; **111.** 18-oz box costs 13 cents per oz, 13-oz box costs 17 cents per oz; **113.** 27 students per professor.

Section 10.2 (page 328)

1. True; **3.** True; **5.** False; **7.** $x = 2$;
9. $y = 24$; **11.** $y = 45$; **13.** $k = 50$;
15. $x = 7$; **17.** $a = 15$; **19.** $x = 60$;
21. $x = \frac{60}{7}$; **23.** $y = \frac{200}{9}$; **25.** $p = 31.5$;
27. $y = 14$; **29.** $x = \frac{92}{35}$; **31.** $x = 0.25$;
33. $y = 0.6$; **35.** $x = \frac{45}{8}$; **37.** $x = 0.42$;
39. $x = 0.045$; **41.** $x = \frac{81}{7}$; **43.** $x = \frac{21}{11}$;
45. $x = \frac{25}{6}$; **47.** $x = \frac{380}{13}$; **49.** $x = \frac{133}{24}$;
51. $x = 18$; **53.** $x = \frac{10}{3}$; **55.** $x = 2$;
57. $a = 4$; **59.** $x = 20$.

Section 10.3 (page 334)

1. 441 mi; **3.** 16 in; **5.** 720 mi; **7.** Living room is 26 ft, kitchen is 10 ft, garage is 24 ft, hallway is 15 ft; **9.** 5780 yes votes, the levy would fail; **11.** $350; **13.** 60 parts; **15.** 3.5 gal left in gas tank; **17.** 255 bricks, 3 lots with 45 left over; **19.** 28 yd; **21.** 91 games won, 63 games lost; **23.** 10 min; **25.** 12 in; **27.** $2\frac{1}{4}$ cups; **29.** $67\frac{1}{17}$ hits; **31.** 93 mi; **33.** 51,870 yes votes, the levy would pass; **35.** $1043; **37.** 42 parts; **39.** 19 gal, 2 gal left in tank; **41.** 309 bricks, 4 lots with 91 left over; **43.** 60 yd; **45.** 189 wins, 36 loses; **47.** $23\frac{1}{3}$ min; **49.** $e = 15, f = 9$.

CHAPTER 10 REVIEW

1.–2. See Section 10.1; **3.** See Section 10.2;
4. $\frac{1}{2}$; **5.** $\frac{3}{1}$; **6.** $\frac{9}{32}$; **7.** $\frac{4}{1}$; **8.** $\frac{1}{100}$;
9. $\frac{1}{10,000}$; **10.** $\frac{9}{8}$; **11.** $\frac{5}{3}$; **12.** $\frac{5}{6}$; **13.** $\frac{3}{1}$;
14. $\frac{1}{4}$; **15.** $\frac{1}{6}$; **16.** $\frac{6}{1}$; **17.** $\frac{5}{16}$; **18.** $\frac{1}{28}$;
19. $\frac{1}{6}$; **20.** 50 mph; **21.** $1250 per mo;
22. 23 mpg; **23.** 37 cents per lb; **24.** 19 students per section; **25.** 27.6 mpg; **26.** 65 mph;
27. $24.50 per cubic yd; **28.** $x = 6$;
29. $y = 60$; **30.** $x = 21$; **31.** $a = \frac{15}{16}$;
32. $x = 18$; **33.** $y = \frac{45}{2}$; **34.** $x = 8.4$;
35. $x = 16$; **36.** $1.32; **37.** 1500 mi;
38. 7.5 in; **39.** 19 gal; **40.** 54 in;
41. 50 fluid oz; **42.** 320 lb, 7 bags; **43.** $1095.

CHAPTER 10 PRACTICE TEST

1. $\frac{7}{12}$; **2.** $\frac{6}{7}$; **3.** $0.\overline{6}$; **4.** $\frac{11}{12}$; **5.** $0.\overline{6}$;
6. 1.4; **7.** $\frac{3}{1}$; **8.** 6.0; **9.** 0.25; **10.** $\frac{5}{8}$;
11. $2.\overline{6}$; **12.** $\frac{6}{1}$; **13.** 40 mph;
14. $1073 per mo; **15.** 20.5 mpg; **16.** The better buy is a package of 120 sheets; **17.** $x = 10$;
18. $y = 45$; **19.** $x = 14$; **20.** $x = 0.25$;
21. $x = \frac{45}{8}$; **22.** 75 hits; **23.** $2.09;
24. 36 lengths; **25.** 470.4 min.

CHAPTER 11

Section 11.1 (page 347)

1. $66\frac{2}{3}\%$; **3.** 125%; **5.** 37.5%; **7.** $183\frac{1}{3}\%$;
9. 37%; **11.** 14%; **13.** 7%; **15.** 106%;
17. 0.035; **19.** 0.115; **21.** 0.005; **23.** 0.31;
25. 0.625; **27.** 1.245; **29.** 0.0006; **31.** $\frac{2}{25}$;
33. $\frac{3}{20}$; **35.** $\frac{3}{10}$; **37.** $1\frac{3}{25}$; **39.** $\frac{3}{40}$;
41. $2\frac{13}{20}$; **43.** $\frac{33}{40}$; **45.** $\frac{1}{16}$; **47.** $0.25, \frac{1}{4}$;
49. $0.03, \frac{3}{100}$; **51.** $120\%, \frac{6}{5}$; **53.** $35\%, \frac{7}{20}$;
55. 75%, 0.75; **57.** $0.8, \frac{4}{5}$; **59.** $6.5\%, \frac{13}{200}$;
61. $0.05, \frac{1}{20}$; **63.** $1.45, 1\frac{9}{20}$; **65.** $25\%, \frac{1}{4}$;
67. $0.675, \frac{27}{40}$; **69.** $33\frac{1}{3}\%, 0.\overline{3}$; **71.** $0.035, \frac{7}{200}$;
73. 15%, 0.15; **75.** $0.5, \frac{1}{2}$; **77.** 40%, 0.4;
79. $0.037, \frac{37}{1000}$; **81.** $0.0012, \frac{3}{2500}$.

Section 11.2 (page 356)

	Base	Rate	Amount
1.	90	50%	45
3.	7	300%	21
5.	88	175%	154
7.	90	5%	4.5
9.	30	150%	45

11. 100; **13.** 1; **15.** 25%; **17.** 24;
19. 69; **21.** $26\frac{2}{3}\%$; **23.** 2.4; **25.** $133.\overline{3}\%$;
27. 6; **29.** 17.6; **31.** 800%; **33.** $133\frac{1}{3}\%$;
35. 2.34; **37.** 0.025; **39.** 15; **41.** $166\frac{2}{3}$;
43. $13\frac{1}{3}$; **45.** $66\frac{2}{3}\%$; **47.** 5000; **49.** $266\frac{2}{3}$.

Section 11.3 (page 361)

1. $2.70 tax; **3.** $300; **5.** 6% raise;
7. $6400; **9.** $5950; **11.** 8%; **13.** $12,500;
15. $120; **17.** $3.41; **19.** 12%; **21.** $144;
23. $375; **25.** 75%; **27.** $5200; **29.** $725;
31. 3 hr; **33.** 20 min; **35.** 9 mo; **37.** 39 wk;
39. $591; **41.** 45; **43.** 30% rent, 20% food, 15%
clothing; 10% incidentals, 18% saved, 7% entertainment.

Section 11.4 (page 367)

1. $47.70; **3.** $47.40; **5.** $15.96, $16.92;
7. 25%; **9.** $18.90; **11.** $3.38; **13.** 20%;
15. 25%; **17.** $42.20; **19.** 20%; **21.** $8\frac{1}{6}$%;
23. $27,000; **25.** $351; **27.** 25%;
29. $9.38; **31.** $5.85.

Section 11.5 (page 372)

1. $108; **3.** $93,000; **5.** $1200;
7. 6%; **9.** 3 mo; **11.** $159.38; **13.** $5\frac{1}{2}$%;
15. $2300; **17.** 9 mo; **19.** $9500;
21. $1827; **23.** $112.50; **25.** $74.25;
27. $1233.75; **29.** $6000; **31.** $28,890;
33. Stacie will have the larger balance.

CHAPTER 11 REVIEW

1.–4. See Section 11.1; **5.–6.** See Section 11.5;
7. $\frac{2}{5}$; **8.** $\frac{1}{25}$; **9.** $\frac{1}{3}$; **10.** $1\frac{1}{2}$; **11.** $\frac{1}{125}$;
12. 50%; **13.** $87\frac{1}{2}$%; **14.** $66\frac{2}{3}$%; **15.** 30%;
16. 22.2%; **17.** 0.27; **18.** 0.55; **19.** 2.5;
20. 0.004; **21.** 0.03; **22.** 80%; **23.** 4%;
24. 342%; **25.** 3.46%; **26.** 26.1%; **27.** 24%;
28. 12.5%; **29.** 975%; **30.** 0.4%; **31.** 10;
32. 50; **33.** 35; **34.** 25; **35.** 26.6; **36.** 1;
37. 525; **38.** $33\frac{1}{3}$%; **39.** 900; **40.** 9.5;
41. $1950; **42.** $82.50; **43.** 80%; **44.** $840;
45. 60%; **46.** 750; **47.** 80%; **48.** 20%;
49. $63; **50.** 70; **51.** 45; **52.** 20%;
53. $1950; **54.** $3428; **55.** $36,000;
56. 6 mo; **57.** 6%; **58.** $1200; **59.** 1.5 yr;
60. 4.5%; **61.** $15,156.25.

CHAPTER 11 PRACTICE TEST

1. 0.25, 25%; **2.** $\frac{5}{8}$, 0.625; **3.** $\frac{4}{5}$, 80%;
4. $0.\overline{3}$, $33\frac{1}{3}$%; **5.** $1\frac{3}{4}$, 1.75; **6.** $\frac{1}{250}$, 0.004;
7. $\frac{1}{200}$, 0.005; **8.** 0.625, 62.5%; **9.** $\frac{1}{200}$, $\frac{1}{2}$%;
10. 93; **11.** 128; **12.** 19%; **13.** 40;
14. 10.29; **15.** 0.1%; **16.** $180; **17.** 40%;
18. 90; **19.** $200; **20.** 5%; **21.** 26,051;
22. $7327; **23.** $5200; **24.** 5.5%; **25.** 9 mo.

CHAPTER 12

Section 12.1 (page 395)

1. 48 in; **3.** 4.16 ft; **5.** 15 ft; **7.** 25 yd;
9. 4400 yd; **11.** 9240 ft; **13.** 1200 cm;
15. 15,000 mm; **17.** 3000 mm; **19.** 7500 m;
21. 2 km; **23.** 830 mm; **25.** 90 in; **27.** 53 in;
29. 130 mm; **31.** 3 in; **33.** $2\frac{1}{4}$ in; **35.** 8 cm;
37. 5 cm; **39.** 5.5 cm; **41.** 5; **43.** 10;
45. 9; **47.** 11.62; **49.** 26 ft; **51.** 17 ft;
53. 27 in; **55.** 98 cm; **57.** 32 ft; **59.** 34 m;
61. 84 ft; **63.** 108 mm; **65.** 30 ft; **67.** 80 in;
69. 9.42 in; **71.** 7.85 ft; **73.** 21.98 cm;
75. 39.77 yd; **77.** 160 cm; **79.** 34 ft;
81. 21.42 ft; **83.** 8 cm; **85.** 21 ft; **87.** $6\frac{1}{3}$ ft;
89. $\approx$ 40.84 m; **91.** 26 ft; **93.** perimeter of pool
is 80 yd or 240 ft or 2880 in, 480 tiles

Section 12.2 (page 12.2)

1. 2 ft²; **3.** 640 acres; **5.** 5760 in²;
7. $\frac{1}{144}$ ft²; **9.** 3,888 in.²; **11.** 2300 mm²;
13. 10,000,000 mm²; **15.** 0.001 m²;
17. 90,000 cm²; **19.** 80,000 cm²; **21.** 225 yd²;
23. 48 cm²; **25.** 11.56 m²; **27.** 28.26 km²;
29. 114 ft²; **31.** 38.13 in²; **33.** 5024.5 cm²;
35. 192 cm²; **37.** 7480 in²; **39.** $\approx$ 157.08 in².

Section 12.3 (page 416)

1. 6 yd³; **3.** 9 ft³; **7.** 93,000,000 cm³;
9. 0.0000013 m³; **11.** 600 cm³; **13.** $\approx$ 401.92 cm³;
15. $\approx$ 267.95 ft³; **17.** 216 cm³; **19.** $\approx$ 7234.56 in³;
21. $\approx$ 60,288 in³; **23.** 3.703 yd³; **25.** $\approx$ 523.$\overline{3}$ in³.

CHAPTER 12 REVIEW

1. 15,840; **2.** 468; **3.** 7; **4.** 30; **5.** 16;
6. 1.5; **7.** 5 in; **8.** 9.90 cm; **9.** 12 ft;
10. 6 mm; **11.** 12.45 cm; **12.** 36 in;
13. 3.61 ft; **14.** 4.38 mm; **15.** 24 in;
16. 22 cm; **17.** 48 mm; **18.** 32 ft;
19. 36 m; **20.** 112 in; **21.** $\approx$ 25.12 in;
22. $\approx$ 15.7 cm; **23.** $\approx$ 10.99 ft; **24.** $\approx$ 119.32 mm;
25. $\approx$ 138.23 yd; **26.** 30 in; **27.** 19.71 cm;
28. 80 ft; **29.** $\approx$ 25.71 m; **30.** $5\frac{2}{3}$ ft;
31. $7\frac{1}{3}$ yd; **32.** 100 ft; **33.** 94 ft; **34.** 1.5;
35. 27; **36.** 504; **37.** 360; **38.** 648; **39.** 4;
40. 4,000,000; **41.** 10,000,000; **42.** 12 in²;
43. 25 cm²; **44.** 144 mm²; **45.** 9 ft²;
46. 400 cm²; **47.** 136 in²; **48.** 32 in²;
49. 91 cm²; **50.** 200.96 ft²; **51.** 452.16 m²;
52. 200 in²; **53.** 153.12 mm²; **54.** 141.6925 m²;
55. 720 ft²; **56.** 20.83 yd²; **57.** 114 cm²;
58. 106.25 ft²; **59.** 27 m²; **60.** 9;
61. 10,000,000; **62.** 0.127; **63.** 2.5; **64.** 4;

65. 4096 in³;   **66.** ≈ 113.04 cm³;   **67.** ≈ 8478 in³;

68. 14 ft³;   **69.** 3$\frac{1}{3}$ yd³;   **70.** ≈ 98.125 ft³;

71. ≈ 282.6 cm³;   **72.** 240 cm³, 30 cubes.

CHAPTER 12 PRACTICE TEST

1. 4.1$\overline{6}$ ft;   **2.** 5280 yd;   **3.** 3.82 m;

4. 9500 m;   **5.** 102 in.;   **6.** 1.47$\overline{2}$ yd;

7. 0.039 m²;   **8.** 2.23 m;   **9.** 12 ft;

10. 12 in;   **11.** 14.42 ft;   **12.** 76 ft;

13. 17 mm;   **14.** 14 cm;   **15.** 216 mm;

16. 102 yd;   **17.** ≈ 34.13 ft;   **18.** ≈ 15.7 in;

19. ≈ 21.98 ft;   **20.** 5 ft²;   **21.** 20 cm;

22. 361 in²;   **23.** 119 m²;   **24.** 76.5 cm²;

25. 1620 in²;   **26.** 19.625 in²;   **27.** 13 yd²;

28. 1,436.03 cm³;   **29.** 42.875 in³;   **30.** 10$\frac{5}{12}$ yd³.

CUMULATIVE REVIEW CHAPTERS 10, 11, & 12

1. Ratio;   **2.** Proportion;

3. Zero Product Property;   **4.** Metric system;

5. Property of Reciprocals;

6. Pythagorean Theorem;

7. The product of the means equals the product of the extremes;

8. Distributive Property;

9. Property of Additive Inverses;

10. Identity Property of Multiplication;

11. 0.125, 12.5%;   **12.** $\frac{2}{3}$, 66$\frac{2}{3}$%;   **13.** $\frac{13}{25}$, 0.52;

14. 2$\frac{3}{10}$, 230%;   **15.** $\frac{1}{200}$, 0.005;   **16.** $\frac{3}{1000}$, 0.3%;

17. 2 : 5;   **18.** 2 : 3;   **19.** 1 : 100;   **20.** 1 : 21;

21. 25$\frac{2}{3}$ mpg;   **22.** 2.6 TVs per household;

23. 91.2;   **24.** 223.63;   **25.** 78%;   **26.** 290.32;

27. 36;   **28.** 18;   **29.** 3;   **30.** 3;   **31.** 250;

32. 2.5;   **33.** 7 : 4;   **34.** 1 : 5;   **35.** 3 children per family;   **36.** $31.37;   **37.** $24.08;   **38.** 10 in;

39. 5 cm;   **40.** 6 ft;   **41.** 45 m²;   **42.** 8 yd³;

43. ≈ 12.56 cm;   **44.** 11 in².

Section 13.1 (page 435)

1.–8.

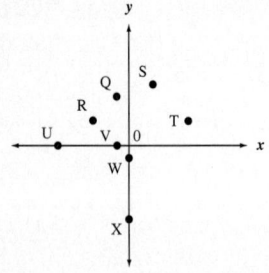

9.–14.

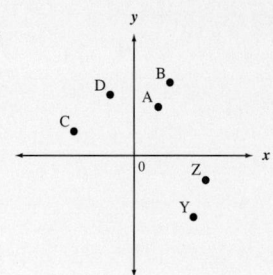

15. 1;   **17.** Between III and IV;

19. Between II and III;   **21.** (0, 0);   **23.** (−9, 0);

25. (4, −6);   **27.** (−4, 7);

Section 13.2 (page 442)

1.–2.

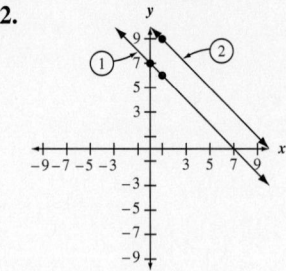

3.–4.

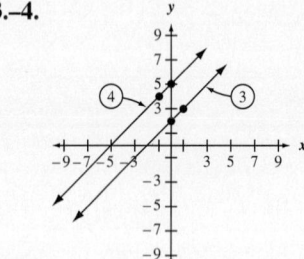

5.–6.

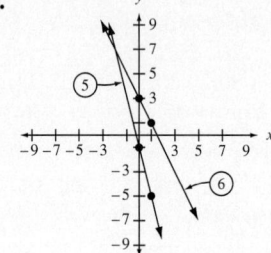

7.–8.

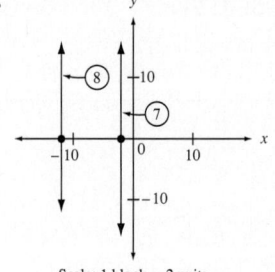

Scale: 1 block = 2 units

9.–10.

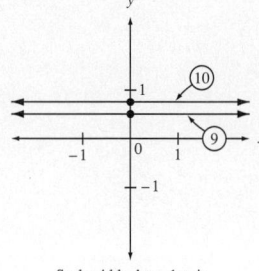

Scale: 4 blocks = 1 unit

11.–12.

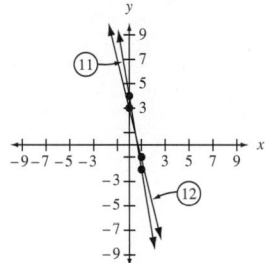

13.–14.

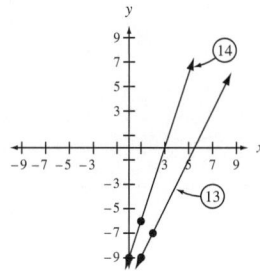

15.–16.

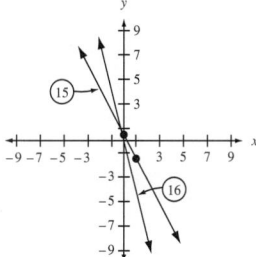

17.–18.

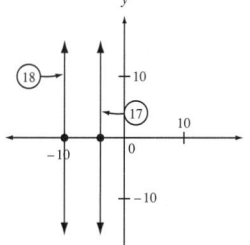

19.–20.

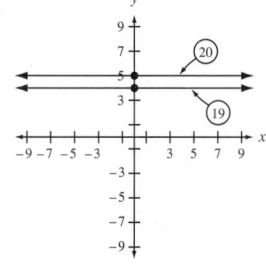

21. 0;　**23.** −9;　**25.** 2;　**27.** $3\frac{2}{3}$ or $\frac{11}{3}$;　**29.** $\frac{5}{3}$ or $1\frac{2}{3}$;　**31.** $-4\frac{1}{2}$ or $-\frac{9}{2}$.

Section 13.3 (page 450)

1. Answers may vary;　**3.** Answers may vary;
5. Answers may vary;　**7.** Answers may vary.

Section 13.4 (page 458)

1a.

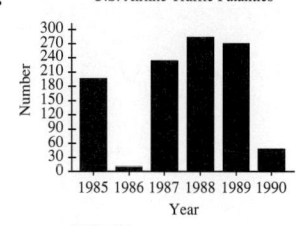

1b.

3a.

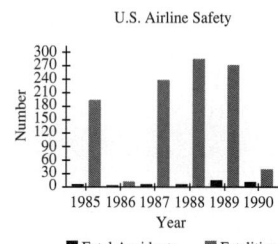

3b.

5a.

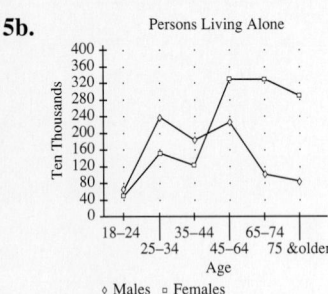

5b.

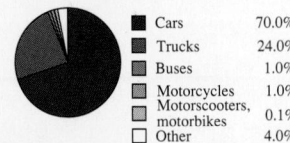

7.

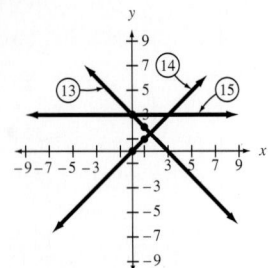

CHAPTER 13 REVIEW

1.–4.

Scale: 4 blocks = 1 unit

5. (1, 4);   **6.** (2, −2);   **7.** (0, 2);
8. (−5, −2);   **9.** IV;   **10.** III;
11. Between II and III;   **12.** I;

13.–15.

16.–18.

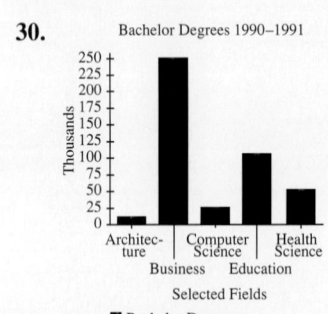

19. 4;   **20.** 2;   **21.** 3;   **22.** −3;
23. 1;   **24.** −4;
25. a. Public − 4yr,   **b.** Private − 2 yr,
c. Public − 4 yr and Private − 2 yr,
d. Public − 2 yr and Private − 4 yr;
26. a. Public − 4 yr and Public − 2 yr,
b. Private − 4 yr and Private − 2 yr,
c. Private − 4 yr,   **d.** Public − 2 yr;
27. a. 1990,   **b.** 1960,   **c.** Since 1960 the number
of marriages has been increasing;
28. a. ABC Fund,   **b.** From 1989 to 1990,   **c.** All As
Fund,   **d.** ABC Fund $41,000, All As Fund $30,000;
29. a. Basketball,   **b.** Tie: Skiing and Golf,   **c.** 15,
d. 21;

30.

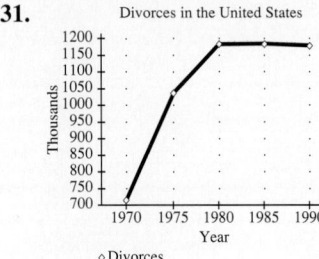

31.

32. Senior Pizza Preferences at IM University

■ DEF Pizza	15%
■ Mama Mio	30%
■ Uncle Joe's	36%
■ Roma Supreme	9%
□ Roma with cheese	10%

CHAPTER 13 PRACTICE TEST

1. II, (−9, 7);   **2.** IV, (3, −1);   **3.** I, (1, 9);
4. Between II and III, (−4, 0);
5. Between III and IV, (0, −6);   **6.** III, (−8, −3);
7. I, (4, 3);

8.–13.

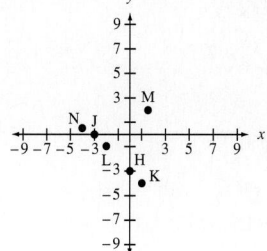

14.–16.

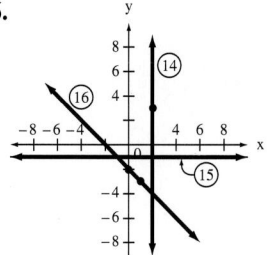

17.–18.

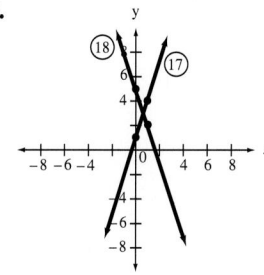

19. $(-2, 0)$;   **20.** $(0, -3)$;   **21.** $(-4, 3)$;
22. $(-1, -1.5)$;   **23. a.** Circle or pie, **b.** Freshmen,
350, **c.** Juniors, 190;   **24. a.** Line graph, **b.** Median
age at first marriage is increasing for men, median age at
first marriage is increasing for women, **c.** The difference
is narrowing;   **25. a.** Bar graph, **b.** The marriage rate
decreased from 1950 to 1960 and then increased from
1960 to 1970, **c.** The divorce rate decreased from 1950
to 1960 and then increased from 1960 to 1970, **d.** No.

26.

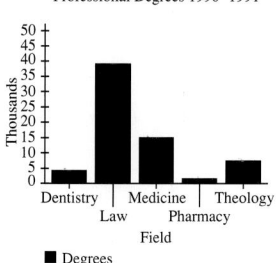

27.

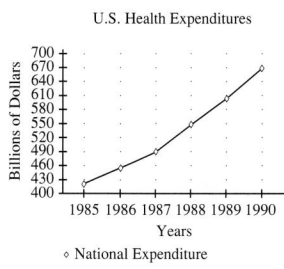

28.

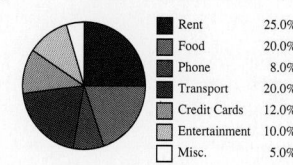

A Freshman's Monthly Budget

■	Rent	25.0%
▨	Food	20.0%
▨	Phone	8.0%
▨	Transport	20.0%
■	Credit Cards	12.0%
▨	Entertainment	10.0%
□	Misc.	5.0%

● PRACTICE FINAL EXAM

1. Distributive Property;
2. Commutative Property of Multiplication;
3. Variables;   **4.** Prime number;
5. Absolute value;   **6.** Addition Property of Equality;
7. Multiplication Property of Equality;
8. Associative Property of Addition;   **9.** Proportion;
10. Constant;   **11.** $\frac{17}{1000}$;   **12.** 23.643;
13. 0.078;   **14.** 30.4;   **15.** 0.0002;   **16.** 0.087;
17. $\frac{56}{3}$;   **18.** $1\frac{23}{24}$;   **19.** $\frac{3}{5} > \frac{5}{9}$;   **20.** -5;
21. 78,910;   **22.** $1\frac{1}{9}$;   **23.** $3 \cdot 3 \cdot 5 \cdot 7$;
24. $\frac{6}{25}$;   **25.** 10;   **26.** $\frac{5}{9}$;   **27.** 0.045;
28. 120;   **29.** $\frac{5}{47}$;   **30.** 897.01;   **31.** $4\frac{1}{2}$;
32. $-\frac{3}{2}$;   **33.** 250;   **34.** $45.15;   **35.** 60;
36. 5;   **37.** 21;   **38.** 15;   **39.** 62.5%;
40. $P = 16$ in, $A = 14.31$ in^2;
41. $P = 30$ cm, $A = 30$ cm^2;   **42.** 10.5 ft^2;   **43.** 4 in;
44. 8 in.^{3};   **45.** 0.25;   **46.** $-2\frac{5}{14}$;   **47.** $\frac{7}{12}$;
48. $\frac{3}{250}$;   **49.** 12 in;   **50.** 0.004, 0.4%;
51. $\frac{3}{250}$, 1.2%;   **52.** $\frac{333}{500}$, 0.666;   **53.** $0.\overline{2}$, $22.\overline{2}$%;
54. $1\frac{4}{5}$, 180%;   **55.** $\frac{7}{5000}$, 0.014   **56.** $11\frac{44}{45}$;
57. $4x$;   **58.** x^5;   **59.** $-12x^3y^2$;
60. $15xy(4y + x)$;   **61.** 8.37×10^{-7};
62. 937,800,000;   **63.** $27x^6y^3$;   **64.** $\frac{31}{12x}$;
65. $2x + 2$;   **66.** $x^2 - 3x - 10$;
67. $4a^3 + 11a^2 + 4a + 2$;   **68.** $\frac{1}{2y}$;   **69.** 3;
70. $\frac{4x^4}{y}$;   **71.** $\frac{2}{9}$;   **72.** Third;

73. $x < 9$

74.

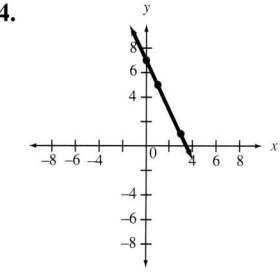

75. 270 are freshmen;   **76.** 31.25 mi;
77. 6 nickels, 13 dimes;   **78.** 16 dimes, 32 quarters;
79. $5\frac{1}{3}$ min;   **80.** Answers may vary.

APPENDIX A

Section A.1 (page 481)

1. 5000;   **3.** 60;   **5.** 13;   **7.** 35;   **9.** 804;
11. 1,653,970;   **13.** Ten;
15. Five hundred seven;   **17.** Six thousand;
19. Twenty-five thousand, six hundred five;
21. Five billion;
23. Seventeen million, eight hundred fifty-two thousand, twenty-seven.

Section A.2 (page 484)

1. 7;   **3.** 47;   **5.** 16;   **7.** 81;   **9.** 15;
11. 21;   **13.** 110;   **15.** 1110;   **17.** 22,910;
19. 22,887.

Section A.3 (page 487)

1. 1;   **3.** 24;   **5.** 15;   **7.** 14;   **9.** 29,634;
11. 90,801.

Section A.4 (page 489)

1. 84;   **3.** 288;   **5.** 1554;   **7.** 4464;   **9.** 0;
11. 4896;   **13.** 120;   **15.** 448;   **17.** 500;
19. 1060;   **21.** 6,516,754;   **23.** 78,503,925.

Section A.5 (page 490)

1. 2;   **3.** 3;   **5.** 0;   **7.** Undefined;   **9.** 6;
11. 7 r 6;   **13.** 30;   **15.** 40 r 5;   **17.** 105;
19. 322;   **21.** 47.

Index

A

Absolute value, 86–88
Addends, 23, 483
Addition, 483–484
  associative property of, 23,
    91–92
  commutative property of,
    23, 90–91
  distributive property of
    multiplication over, 26
  equality property of, 113,
    257–258
  identity property of, 23,
    92–93
  inequalities property of,
    275
  in order of operations, 38
  of algebraic fractions,
    195–196
  of decimals, 222–225
  of fractions, 148–153
  of integers, 89–93
  of mixed numbers, 153–154
  of polynomials, 184–185
  of whole numbers, 23
Additive inverse, 87–88
Algebra, 53
Algebraic expressions, 54,
  177–212
  evaluating, 64, 198–200
  simplifying, 53–57
  translating words into,
    67–68
Algebraic fractions, 191
  addition of, 195–196
  division of, 193
  multiplication of, 192–193
  simplifying, 191–192
  subtraction of, 195–196
Algorithm, division, 28
Al-jabr wa'l maqabalah, 53
Al-khwarizmi, 53
Altitude, 406
AMATYC standards, front
  and back inside covers
Amount, definition of, 350
Area, 401
  of circle, 408–409
  of other plane figures, 409
  of parallelogram, 406–407
  of rectangle, 403–405
  of square, 405
  of triangle, 407–408

Arithmetic mean, 42
Associative property
  of addition, 23, 91–92
  of multiplication, 26,
    54–55, 488
Averages, 41–42, 108–109

B

Babylonian number system,
  124, 177
Bar graphs, 446–448
  constructing, 453–454
Base, 34
  definition of, 350
Base 10 place value system,
  16, 487–488
Billion, 481
Binomials, 183
  multiplication of, 186–188
Boole, George, 1
Borrowing, 486
Boundaries of quadrants, 434
Braces in order of operations,
  38
Brackets in order of
  operations, 38

C

Canceling, 137
Cantor, Georg, 1
Capacity, units of, 318
Cardan, Gerome, 83
Center, of circle, 393
Centimeter, 383
Charts
  circle (pie), 449–450
  constructing, 452–458
  interpreting, 446–450
Chinese mathematicians, 83
Circle graphs, 449–450
  constructing, 455–457
Circles
  area of, 408–409
  center of, 393
  circumference of, 393–395
  closed, 273
  definition of, 393
  diameter of, 393
  filled-in, 273
  radius of, 393
  semicircle, 393

Circumference, of circle,
  393–395
Closed circle, 273
Coefficient of the variable, 54
Coin problems, solving,
  306–308
Common monomial factors,
  finding, 187
Commutative property
  of addition, 23, 90–91
  of multiplication, 26, 136
Complex fractions, 161–163
Composite number, 28
Constant, 54
Coordinate geometry, 429
Counting numbers, 7, 16
Cube
  of right rectangular solid,
    414
  perfect, 35
  volume of, 414–415
Cubic measure, 412–416
  Metric system of, 413–414
  U.S. customary system of,
    413–414

D

Data, 446
DaVinci, Leonardo, 316
Decimals, 213–256
  addition of, 222–225
  converting, to percents, 345
  converting, to reduced
    fractions, 346
  converting fractions to, 345
  converting percents to,
    344–345
  converting to fractions,
    240–245
  division of, 234–237
  multiplication of, 228–231
  place value and the
    comparison of, 213–218
  reading, 215
  rounding, 217–218
  solving equations
    containing, 267–268,
    269–270
  subtraction of, 222–225
  writing, 215
Deductive reasoning, 303
Degree of monomial, 54

Denominator, 125
Descartes, René, 53, 83, 429
Diameter, of circle, 393
Difference, 24
Digits, 16
Disjoint sets, 4
Distributive property of
  multiplication, 55–56,
    488
  over addition, 26
Dividend, 27
Divisibility, tests for, 28–29
Division, 489–490
  of algebraic fractions, 193
  algorithm for, 28
  of decimals, 234–237
  of fractions, 143–145
  of integers, 104
  of mixed numbers, 143–145
  in order of operations, 38
  of whole numbers, 27–28
  by zero, 27–28
Divisor, 27

E

Egyptian number system, 124
  fractions in, 169
Element, 1
Ellipses in sets, 8
Empty set, 3
Equality
  addition property of, 113,
    257–258
  identity property of, 112
  multiplication property of,
    258–261
  solving equations using
    multiplication property
    of, 165–168
  symmetric property of, 112
  transitive property of, 113
Equal sets, 2
Equal to (=) symbol, 84
Equation method in solving
  percent problems,
    350–353
Equations
  solving, containing both
    decimals and fractions,
    269–270
  solving, containing
    decimals, 267–268,
    269–270

515

solving, containing
   fractions, 262–265,
   269–270
solving, using addition
   property of equality,
   257–258
solving, using multiplication
   property of equality,
   165–168, 258–261
using, in one variable to
   solve coin and ticket
   problems, 306–308
using, in problem solving,
   298–304
Equivalent fractions, 130–134
Eratosthenes, 44
Eratosthenes's sieve, 44
Estimation, 42–43
Euclid, 381
Expanded form, 17
Expanded notation, 17
Exponents, 34–35, 60
   laws of, 60–61, 177–180
   in order of operations, 37,
   38
Expressions, rational, 191
Extremes, 327

F

Factors, 26
Factor tree, 30
Feet, 382
Fibonacci series, 124, 316
Filled-in circle, 273
Final exam
   preparing for, 379–380
   taking, 473
Finite set, 2
Flow chart, 289, 290
FOIL method for
   multiplication of
   binomials, 186–188
Ford, Gerald, 213
Formulas in problem solving,
   298–304
Fraction bar, 125
   in order of operations, 38
Fractions, 124
   addition of, 148–153
   algebraic, 191–193
   complex, 161–163
   converting decimals to
      reduced, 346
   converting percents to,
      343–344
   converting to decimals,
      240–245, 345
   converting to percents, 344
   division of, 143–145
   Egyptian, 169
   equivalent, 130–134
   expressing integers as,
      125–126
   expression of ratio as, 317
   improper, 126
   multiplication of, 136–138
   negative, 128–129
   positive, 128–129

proper, 126
   reducing, to lowest terms,
      133–134
   solving equations
      containing, 262–265,
      269–270
   subtraction of, 148–153
   types of, 126–128
   using the identity property
      of multiplication with,
      130–134
Frequency, 446
Frequency polygon, 448–449
Fundamental property of
   proportion, 326–328

G

Gallup, George, 446
Geometry
   coordinate, 429
   cubic measures in, 412–416
   history of, 381
   linear measures in,
      381–395
   square measures in,
      401–409
German mathematicians, 83
Golden Rectangle, 316
Graphs
   bar, 446–448
   circle, 449–450
   constructing, 452–458
   inequalities on the number
      line, 272–275
   interpreting, 446–450
   line, 448–449
   linear equations, 436–442
   order pair for, 430
   origin on, 430, 433
   plotting point on, 430–433
   quadrants in, 434
   rectangular coordinate
      plane, 429–434
   x-axis on, 430, 433
   x-coordinate on, 430, 433
   y-axis on, 430, 433
   y-coordinate on, 430, 433
Greater than ($>$), 84
Greater than or equal to ($\geq$),
   272
Greatest common factor
   (GCF), 31
Grouping symbols in order of
   operations, 38

H

Hindu mathematicians, 124
Homework, 50–51
Hypotenuse, 385

I

Identity property
   of addition, 23, 92–93
   of equality, 112
   of multiplication, 26,
      130–134
Improper fraction, 126

Inches, 382
Index, 35
Inequalities
   addition property of, 275
   graphing, on the number
      line, 272–275
   multiplication properties of,
      276–278
   solving, using addition and
      multiplication properties,
      277–278
Infinite set, 2
Integers, 8, 83–123
   addition of, 89–93
   averages of, 108–109
   comparison of, 84–85
   division of, 104
   expressing, as fractions,
      125–126
   multiplication of, 100–102
   order of operations using,
      106–108
   positive, 83
   set of, 84–85
   subtraction of, 96–98
Interest
   definition of, 370
   simple, 370–371
Intersection, 3
   of two sets, 4
Inverse, additive, 87–88
Irrational numbers, 9, 214
Islamic mathematicians, 213

J

Jefferson, Thomas, 213

K

Kilometer, 383

L

Legs, 385
Less than ($<$), 84
Like terms, combining, 56–57
Linear equations, graphing,
   436–442
Linear inequalities, solving, in
   one variable, 271–278
Linear measure, 381–395
   definition of, 382
   Metric system of, 383–384
   units of, 318
   U.S. customary system of,
      382–383
Line graphs, 448–449
   constructing, 454–455
Lines, parallel, 406
Line segment, 382
Logic problems, strategies for
   solving, 291–296
Lowest common denominator
   (LCD), 149–151

M

Mathematics problems,
   strategies for solving,
   291–296
Means, 42, 327

Member, 1
Meter, 383
Metric Conversion Act
   (1975), 213
Metric system, 213
   conversions between U.S.
      customary system and,
      492–493
   cubic measures in, 413–414
   linear measures in,
      383–384
   square measures in, 403
Miles, 382
Millimeter, 383
Million, 481
Mint Act (1792), 213
Minuend, 485
Mixed numbers, 126–128
   addition of, 153–154
   division of, 143–145
   multiplication of, 138–139
   subtraction of, 153–154
Modified expanded notation,
   17
Monomial, 54
   degree of, 54
Multiplication, 487–488
   associative property of, 26,
      54–55, 488
   by powers of 10, 230
   commutative property of,
      26, 136
   distributive property of,
      55–56, 488
      over addition, 26
   equality properties in,
      258–261
      solving equations using,
         165–168
   identity property of, 26,
      130–134
   inequalities properties in,
      276–278
   in order of operations, 38
   of algebraic fractions,
      192–193
   of binomials, 186–188
   of decimals, 228–231
   of fractions, 136–139
   of integers, 100–102
   of mixed numbers, 138–139
   of polynomials, 185–186
   of whole numbers, 25–27
   zero product property in, 27

N

Napier, John, 213
Natural numbers, 7, 16, 83
Negative fractions, 128–129
Negative integers, 83
Negative sign, 84, 87
Nonlinear measures, 492
Notation
   expanded, 17
   scientific, 216–217
   set builder, 272
   standard, 17, 216

Note taking, 15
Null set, 3
Number line, 7, 8, 85
  graphing inequalities on
    the, 272–275
Numbers
  as sets, 7–9
  composite, 28
  counting, 7, 16
  irrational, 9, 214
  mixed, 126–128
  natural, 7, 16, 83
  opposite of, 87–88
  prime, 28
  rational, 8–9, 124
  reading and writing, 481
  real, 7
  whole, 8
Numerators, 125, 148–149

O

One-to-one correspondence, 7
Operations
  order of, 37–39
  with sets, 3–7
Operators, order of, 37–39,
  106–108, 159–161
Ordered pair, 430
Order of operations, 37–39,
  106–108, 159–161
Origin, 430, 433

P

Parallel lines, 406
Parallelogram, 406
Parentheses, in order of
  operations, 37, 38
Parallelogram, area of,
  406–407
Pascal, Blaise, 202
Pascal's triangle, 202
Pedagogy, standards for, back
  cover
Percents
  application problems,
    358–361
  converting, to decimals,
    344–345
  converting, to fractions,
    343–344
  converting decimals to, 345
  converting fractions to, 344
  definition of, 343
  increase and decrease,
    365–367
  problem solving with, 350
    equation method,
      350–353
    proportion method,
      353–356
Perfect cube, 35
Perfect square, 35
Perimeters
  of other plane figures, 395
  of polygons, 387–390

of rectangles, 390–391
of squares, 391–392
of triangles, 387–389
Pie charts, 449–450
  constructing, 455–457
Place value, 16–17
  and the comparison of
    decimals, 213–218
Plane, 382
Plane figures, 384–385
Plotting, 430–433
Polygons, 385
  frequency, 448–449
  perimeters of, 387–390
Polynomials
  addition of, 184–185
  definition of, 183
  degree of, in one variable,
    183
  multiplication of, 185–186
  subtraction of, 184–185
Positive fractions, 128–129
Positive integers, 83
Powers of 10, multiplying by,
  230
Prime factorization, 29–30
Prime number, 28
Principal, in interest formulas,
  370
Problem solving, 288–312
  coin and ticket problems,
    306–308
  percent problems, 350,
    358–361
    equation method,
      350–353
    proportion method,
      353–356
  process of, 289–291
  strategies for logic and
    mathematics problems,
    291–296
  using formulas and
    equations in, 298–304
Product, 26
Proper fraction, 126
Proportions, 316
  applications of, 331–334
  definition of, 326
  fundamental property of,
    326–328
Proportion method in solving
  percent problems,
  353–356
Pythagorean theorem,
  385–387

Q

Quadrants, 434
Quadrilateral, 385
Quotient, 27

R

Radical sign, 35
Radicand, 35

Radius, of circle, 393
Rate, 319–321
  definition of, 319, 350
  in interest formulas, 370
Rational expressions, 191
Rational numbers, 8–9, 124
Ratios, 316–321
  and fitness, 338
  definition of, 316
  methods of writing, 317
Real numbers, 7
Reciprocals, 143
Rectangles, 385
  area of, 403–405
  perimeters of, 390–391
Rectangular coordinate plane,
  429–434
Remainders, 28
Right circular cylinder,
  volume of, 415
Roots, 35–36
Rounding
  decimal numbers, 217–218
  whole numbers, 19–21, 218

S

Scientific notation, 216–217
Sectors, 449
Semicircle, 393
Set builder notation, 272
Sets, 1–2
  disjoint, 4
  empty, 3
  equal, 5
  finite, 5
  infinite, 5
  intersection of two, 4
  notations and conventions
    used with, 2
  null, 3
  numbers as, 7–9
  of integers, 84–85
  operations with, 3–7
  relationships of, 2–3
  union of two, 4
  universal, 4
Set theory, 1–2
Seurat, Georges, 316
Sexagesimal system, 124
Similar terms, combining,
  56–57
Simple interest, 370–371
Simplication of algebraic
  fractions, 191–192
Sphere, volume of, 416
Square measures, 401–409
  Metric system, 403
  U.S. customary system,
    401–402
Squares, 385
  area of, 405
  perfect, 35
  perimeters of, 391–392
Standard form, 17
Standard notation, 17, 216

Statistics, 446
Stevin, Simon, 213
Study skills
  preparing for final exam,
    379–380
  taking the final exam, 473
Subset, 2–3
Subtraction, 485–486
  in order of operations, 38
  of algebraic fractions,
    195–196
  of decimals, 222–225
  of fractions, 148–153
  of integers, 96–98
  of mixed numbers, 153–154
  of polynomials, 184–185
  of whole numbers, 23–24
Subtrahend, 485
Sum, 23, 483
Symmetric property of
  equality, 112

T

Term, 53
Tests
  preparation for, 78–79
  strategies for taking, 79–80
Theorem
  definition of, 385
  Pythagorean, 385–387
Thousand, 481
Ticket problems, solving,
  306–308
Time
  in interest formulas, 370
  units of, 318
Time management, 50–52
Transitive property of
  equality, 113
Triangles, 385
  area of, 407–408
  perimeters of, 387–389
  Pythagorean Theorem for
    right, 385–387
Trichotomy, principle of,
  84–85, 215–216, 271
Trillion, 481
Trinomial, 183

U

Union, 3
  of two sets, 4
Universal set, 4
U.S. customary system
  conversions between Metric
    system and, 492–493
  of cubic measures, 413–414
  of linear measure, 382–383
  square measures in,
    401–402

V

Value, absolute, 86–88
Variable, 53
  coefficient of, 54
  solving linear inequalities in
    one, 271–278

Venn, John, 1
Venn diagrams, 1, 4–7, 8, 9
Volume
    definition of, 412
    of cube, 414–415
    of right circular cylinder,
      415
    of right rectangular solid,
      414
    of sphere, 416

W

Waist-to-hip ratio, 338
Weight, units of, 318

Whole numbers, 8
    addition of, 23
    averages in, 41–42
    division of, 27–28
    estimation in, 42–43
    exponents in, 34–35
    multiplication of, 25–27
    order of operations in,
      37–39
    place value and expanded
      notation, 16–17
    prime numbers and prime
      factorization, 28–31
    roots in, 35–36

    rounding, 19–21, 218
    subtraction of, 23–24

X

x-axis, 430, 433
x-coordinate, 430, 433

Y

Yards, 382
y-axis, 430, 433
y-coordinate, 430, 433

Z

Zero, division by, 27–28
Zero product property in, 27